Research, Quality, Competitiveness

Second edition

The Recommended Reading web site
http://stajano.deis.unibo.it/RQC.htm
is available on line

Attilio Stajano

Research, Quality, Competitiveness

European Union Technology Policy
for the Knowledge-based Society

Second edition

 Springer

Attilio Stajano
University of Bologna, Italy
attilio.stajano@unibo.it

ISBN: 978-0-387-79264-4   e-ISBN: 978-0-387-79265-1

Library of Congress Control Number: 2008934732

Printed on acid-free paper

springer.com

To Kathleen

Foreword

The European Union (EU) was launched as a response to the economic dominance of the United States and – to a lesser degree – the Soviet Union. The nations of Western Europe were too small to compete against large scale and diversified economies on their own. Six countries, eventually expanding to 27 (and counting), took a series of steps toward progressively deeper integration: the removal of internal tariffs, the construction of a common external tariff, the elimination of many (but not all) non-tariff barriers leading to a single market, and the adoption of a common currency by 15 of the member states. The EU today equals and even exceeds the U.S. on many key indicators of performance. In the process, two similar but nonetheless divergent models of social and economic life stand in contrast with each other. The U.S. is more committed to capitalism and does little to dilute its harsh edges while the nations of Europe support wider social safety nets and more active regulation of commercial activity to mute the crueller aspects of the free-market. Until recently, the economic dynamism of the U.S. called into question whether the so-called European social model was sustainable in an era of globalization. The EU was slipping in competitiveness and was being challenged by new global powerhouses like China and India. Although the U.S. economy has slowed, there is little indication that European countries are capable of leveraging the situation to their advantage.

This book by Attilio Stajano investigates the EU's competitiveness and the role played by research as its underlying engine. As such, it provides an important analysis on whether competition requires sacrifice of the traditional social safety net and its accompanying regulatory regime. The single market lies at the heart of the analysis.

The original proposal in 1987 was to eliminate 279 non-tariff barriers by 1992. About 95 percent of these were agreed upon by the target date. The number of barriers to be removed continued to escalate in the next decade as the member states of the EU agreed to ease the freedom of movement for goods, services, capital, and labour. The initial set of non-tariff barriers expanded from 279 to 1,475 by 2002. In its assessment on the operation of the single market after 10 years, the European Commission applauded what had been accomplished: "About 2.5 million jobs have been created in the EU thanks to the Internal Market, since the opening up of frontiers on 1 January 1993. The European Union's GDP in 2002 is 1.8 percentage points or € 164.5 billion higher than it would be without the Internal Market. Extra prosperity to the value of € 877 billion . . . has been created. That means € 5,700 per household on average."

This record of success masked unease about the pace at which the single market was being completed. Much had been achieved but the EU still languished in comparison to the U.S. In 2000 at the Lisbon Summit, the EU set for itself the goal of becoming the world's economic leader by 2010. The plan – called the Lisbon Agenda – outlined several areas for action: an 'information society' in which all citizens had access to the Internet, research, and innovation promoted at the EU level, and the removal of most regulations on the utility and transport sectors. The desire was to create more quality jobs, not just jobs. Yet it quickly became apparent that the aspirations of the Lisbon Agenda would not be fulfilled. A report prepared for the European Commission in 2004 warned that the failure to reach the Lisbon goals by 2010 risked "nothing less than the sustainability of the society Europe has built and to that extent, the viability of its civilization."

The European Commission soon acknowledged the futility of reaching the lofty objectives identified in Lisbon. It conceded in 2004 that "the Union cannot catch up on the United States" because member states were responsible for "inadequate implementation of the reforms." Indeed, less than 60 percent of the Lisbon Agenda's 40 directives had been enacted into national legislation. The EU's effort to build a single market, in short, was undermined by national regulations that too often ran contrary to the spirit of liberalization. There was no public pressure to change anything since the European social model filtered the impact of globalization. The adverse consequences of an anaemic European economy were not experienced by people in their daily lives.

In the face of such realities, the Lisbon Agenda was refocused on three more modest goals: (1) making knowledge and innovation the engines for sustainable growth; (2) transforming Europe into a more attractive location in which to invest and work; and (3) creating more jobs. Each member state was required to develop a 'national reform program' to achieve these goals but in a way that protects the European social model as well as the environment. It is beginning to look like even these more relaxed goals will be difficult to meet. Under the latest projections, the target of 70 percent employment by 2010 will probably be achieved 10 years after the deadline.

Rather than viewed as the catalyst for making Europe more competitive, the EU's effort to reduce economic regulation is considered a threat to a protected way of life. According to post-election analysis, the proposed constitution for the EU was defeated in France in 2005 to a large extent because voters regarded it as an attempt to erode the European social model. EU leaders absorbed the lesson and, in 2007, agreed to French President Nicholas Sarkozy's demand that the Reform Treaty – a much scaled down version of the constitution – should remove 'free and undistorted' competition as an objectives and banish the commitment to free competition to a separate protocol. The treaty is now before the member states for ratification.

Should this history be interpreted to mean that the EU is on a course to economic decline due to the lack of competition within its marketplace? The answer is a confident "no" as delivered by Attilio Stajano. The EU has advantages in high quality manufacturing and only needs to infuse more money into research and development to maintain an economic model that works. Quite clearly, Attilio Stajano has written a book worthy of serious consideration.

Professor Brian M. Murphy, January 2008
Dean, College of Liberal & Applied Arts
Professor of Political Science
Stephen F. Austin State University, Nacogdoches, TX

Foreword to the First Edition[1]

The European Union offers a profound challenge to all citizens interested in the future of Europe, while the ongoing processes of globalization and technological innovation disrupt traditional patterns of life and international commerce. The intersection between the policies of the European Union in the areas of research and industrial policy on the one hand and the increasing levels of competition faced by European industry on the other has often been ignored by traditional studies of the European Union. Yet the European Union has to be a critical actor if Europe is to meet the economic challenges presented by the United States as well as by increasingly important economies such as that of China. It is impossible to think of the European political economy without considering the role of the European Union.

In fact, Europe's political economy as we know it in 2004 has been fashioned by the European Union and her predecessor, the European Economic Community. The creation of a customs union, and granting that customs union a unitary voice in the General Agreement on Tariffs and Trade (GATT), gave the EEC a power in the world of international commercial diplomacy matched only by the United States. It is probably fair to say that the world outside of Europe was cognizant of a united Europe's economic power before Europeans themselves were. That power increased as the European Union enlarged, and the admission in 1973 of the important United Kingdom economy in particular made the integrating Europe an even more critical actor in the world of international economics. Both the United States and Europe had to agree to the multilateral rules that have shaped trade in the period since 1958; the Community essentially wielded a powerful veto at the level of multilateral commercial diplomacy.

Europe then coupled her power at the international level with the construction of a single market. The creation of the single market marked a historic turning point, for the economic forces that had been contained within national boundaries were now to be permitted to work across borders. The single market has been underestimated, for its effects were not immediately apparent in 1992, the date by which

---

[1]This foreword was written in June 2004 for the first edition of this book, published in Italian by Clueb.

most of the relevant legislation had been adopted. The single market is in many ways equivalent to a *time release capsule*, for its effect is seen over a long period of time. Yet if we compare the nature of the European economy now with its counterpart in the 1960s, the difference is absolutely startling. Sector after sector have been liberalized so that entire sectors are nearly unrecognizable. Some of that transformation would have occurred because of the pressures of globalization, but much of it is due directly to the work of the European Union's institutions and legislation.

Many of the readers of this volume will have recently taken a flight on a low-cost airline. That reduced cost is a consequence of the European Union's policies. Whether one studies the beer business or airlines or telecommunications or financial services, the impact of the single market is clear. A decade from now its impact will be even more so.

The decision to create a common currency, the first in Europe since the Roman Empire, reinforced the economic effects of the single market. Membership in the Euro-zone has created new pressures and tensions, but in general the existence of a single currency has given citizens in the member-states that have adopted it an economic instrument that facilitates economic exchange and heralds the creation of a true European economic space.

Yet the remarkable achievements of Europe's new political economy have not solved all of Europe's problems. In fact, Europe faces a set of challenges that the founders of the European experiment could not have imagined. Most dramatic of all, perhaps, is Europe's demographic profile. Other challenges, however, are directly related to features of Europe's political economy that are more amenable to policy interventions. The Union's activities in the area of research are particularly noteworthy, for it is through the EU's programs that Europe has in fact created what might be termed a European research community. Those programs have tried to produce the conditions that would lead to an increase in competitiveness for European goods and services. Yet there is still much to be done if Europe is to compete with the United States and Japan.

Italy's future in an integrating Europe will be shaped by how it responds to the challenges of Europe's political economy and the opportunities presented by the European Union's policies dealing with competitiveness. Scholars have long been fascinated by the flexibility and attention to design and quality that are hallmarks of Italy's small firms. Yet as the nature of the global economy shifts and the role of technological innovation become ever more important for the advanced industrial economies, the ability of Italian industry to compete will depend far more than in the past on its ability to make use of the resources and networks provided by the European Union. An enlarged Union now provides more opportunities for Italian firms but also increased competition for access to those EU programs that can help firms and governments provide the foundations for future economic growth and international competitiveness.

The European Union has become and will remain a critical actor for all those in any member state concerned with the competitiveness of firms and the creation of wealth within the framework of sustainable development in an increasingly competitive global economy. The European Union is of special importance, however, for Italy. As this volume explains so well, Italy's challenge is a particularly difficult one. Policymakers, academics, and businesspeople all will need to participate in EU programs in new ways in order to maximize the opportunities that the EU provides. Italy's future is inextricably tied to that of the European Union, and her leaders in all sectors of life must grasp the opportunities provided by the EU in order to meet their own home-grown challenges.

Professor Alberta M. Sbragia, June 2004
Director, European Union Center and Center for West European Studies
UCIS Research Professor of Political Science
University of Pittsburgh, Pittsburgh, PA

Preface

At the end of the Second World War, the creation of the European Economic Community was seen as the answer to the quest for peace, freedom, and prosperity by the citizens of the European countries wracked by the war. In the year 2007 the European Union celebrated the 50th anniversary of the signing of the founding Treaties of Rome. On that occasion, Angela Merkel, Chancellor of the Federal Republic of Germany and President of the European Council, said: "Half a century ago a number of Europe's political leaders set about building a European peace project the like of which had never been seen before. [...] For centuries Europe had been an idea, no more than a hope of peace and understanding. Today we, the citizens of Europe, know that hope has been fulfilled. It has been fulfilled because the founding fathers of Europe were thinking in terms well beyond their own generation."

The European Union is an ever-changing political reality in the making: the European Economic Community with six founding members has grown from 1957 to 2007, into the European Union of 27 member states, whose prime ministers committed, by signing the Lisbon Treaty in December 2007, to continue promoting peace, democracy, stability, and prosperity in a Union facing the 21st century challenges. To cope with the new challenges the European people need more than ever, political leaders that – like the founding fathers – think in terms well beyond their own time.

In this book we deal with the challenge of competitiveness. In the initial decades after the signing of the Treaty of Rome, European competitors were mainly in the U.S., and later within the Triad. Starting in the 1990s, it became apparent that economies of emerging Asian countries would change the name of the game, initially competing on products and services that could take advantage of cheap labour, but more recently also in businesses requiring advanced technologies and qualified workforce.

The sectors where European industry beats competition are mostly mature sectors where the challenges concern quality rather than price: the European Union is a region with high labour costs and can hardly compete on price. The social costs for the European welfare state and the high European salaries can be only partly compensated by efficiency in the public administration and the benefits of the internal market and the monetary union. The possibility of competing in world markets depends on the capacity to characterize European products and services as superior in their quality, design, innovativeness, and ability to satisfy the requirements of a diversified and ever-changing market. The superior quality of European products and services can make them competitive despite the high labour costs and standard of living within the EU.

The whole world is faced with the challenge of sustainability of development in today's globalized society. Energy and technology continue to be essential for economic growth but growth is now conditioned by two other factors: information and knowledge. While energy is limited and can be used only once, information is widely available and overabundant and can be used by several users at the same time. The new challenge is managing and exploiting information and structuring it into knowledge that can support a new approach to sustainable development and trigger an improvement in the quality of life.

This book shows that the future competitiveness of the European economy with respect to both traditional competitors and the new great economies of emerging Asian countries depends on the European capacity to seize the opportunities of the knowledge society and to ensure a competitive advantage in terms of quality. This could be achieved through a series of strategic actions, the most relevant of which are the increase in public and private investments in education, lifelong learning, research, and innovation.

These actions cannot be implemented at member state level, since their success depends upon the complementarity and synergy across the Union. A move in the right direction was made in the year 2000 by the European Council by formulating the Lisbon strategy, meant to build in Europe the most competitive economy of the world, based on the knowledge society. However, the move started with the wrong foot, as we discuss in Chapter 8 'Competitiveness in the Knowledge-based Society', and the pace of the member states towards the partial results achieved so far suggest that they are not moving cohesively towards the Lisbon strategy objectives.

In this book we focus on *research policy* and we prove that it does not only strengthen the scientific and technological base of EU industrial activities while qualifying the workforce, but it also contributes to the realization of other EU policies beyond industrial competitiveness: internal market, cohesion and integration of member states, sustainable growth, and enlargement, to name a few. We advocate the need for higher investments in education, lifelong learning, research and innovation. This need is ever so much pressing for the accession and candidate countries, where the ongoing changes in the societal structure are creating redundancies and skills mismatches in the labour market that demand for a strong focus on education at all levels, vocational training, life long learning, and research to help the young people as well as the adults and the elderly fit in the new fabric of the society as it converts to market economy. We acknowledge, however, that investments in education and research can lead to beneficial effects only in the very long term, that is, in no less than 20 or 30 years (see Fig 1). And this is precisely the reason why these actions should be carried out as soon as possible, before the positioning of the European economy is irremediably compromised by the aggressive presence

Fig. 1 A primary school class in the Montessori School in Waterloo, Belgium. The building of Europe's future starts from school. But education, as well as research, generates visible benefits in terms of a competitive presence in the international marketplace only in the long term. Education and research programs demand farsighted policymakers

of other actors. Political leaders must be farsighted and not conditioned only by the ephemeral pressures linked to the next elections. Each administration should be able to assure the children who today are attending their first classes in primary school that it is preparing for them a peaceful, multiethnic, multicultural, and competitive Europe with a high standard of living and a high employment rate.

This book is not a scholarly monograph on political science, but rather a pragmatic description of EU R&D policy and its implications on competitiveness. It aims at making readers aware that European citizens belong to a wider community than that of their own country, highlighting some aspects of the evolution of the European society transformed by technology, globalization, and networking.

The book is divided into three parts: the first part is an overview of the EU member states from the point of view of the competitiveness of their economies; the second part addresses EU research and innovation policy within the context of the knowledge society; the third part is written for readers looking for basic information on the institutional structure of the European Union: it introduces the reader to the origins of the EU, her ongoing enlargement to 30 members, her institutions, and her policies for sustainable development, mainly the internal market (including the economic and monetary union) and competition. Readers familiar with the Union's organization and its latest changes may want to only browse the third part and restrict their reading the Section 11.5 'The 2004 Enlargement Three Years on' addressing successes and challenges faced by the new members and the Section 11.8.3 'Turkey' covering the accession negotiations with Turkey, a process that unveils different visions on future and role of the Union as expressed by the various political leaders.

This book originates from the lecture notes for the courses on research policy in the European Union given by the author at the Georgia Institute of Technology in Atlanta, Georgia, in 1999; at the Faculties of Political Science and Engineering of the University of Bologna, Italy, from 1999 to 2007; and at the Faculty of Economics of the University of Ferrara, Italy, from 2001 to 2005. This book was initially published in Italian by Clueb in 2004. Springer published a first edition in 2006. The present edition is a cover-to-cover rewrite, with updates and extensions in particular on: the reform of the Treaty on European Union; enlargement; internal market and competition; the seventh Framework Program for Research and Development; the knowledge society and the Lisbon strategy.

The primarily audience of this book are teachers of courses on EU sustainable growth policy and on research and technology policy; they may use the book as textbook. Other categories of potential readers include economists and policy makers interested in competitiveness, and industrial and academic researchers who are planning to submit research proposals for Community funding under the Framework Program for research and technological development. By fully understanding the final objectives of the programs and the proposal selection criteria, they should be able to develop and draw up research proposals that have a better chance of being considered for funding. The book is also addressed to scholars of EU policies, particularly policies relating to research and competitiveness, who will find in the book not an abstract academic discourse but rather a pragmatic description of the current situation by a former EU officer with extensive industrial experience.

The courses held by the author at the Faculty of Engineering of the University of Bologna are part of a series of courses described at the URL http://www.elearning.unibo.it exploring the potentialities of e-learning. They are organized in such a way as to create a learning and training community in a situation where face-to-face lessons are integrated by online asynchronous activities. Students play an active role, empowered to the creation of contents and to the development of skills. The use of this book and of an e-learning platform for a university course on research policy is presented in Appendix B.

The first edition of this book included a cd-rom of recommended reading, which is substituted in this edition by a companion web site at the URL http://stajano.deis.unibo.it/RQC.htm containing reference papers, landmark papers, recommended reading, updates, an *ERRATA*, book reviews, tables and figures to build course materials, examples of slides for a course, and other materials as described in Appendix C.

The author welcomes any comments or notifications of errors to be sent to the following e-mail address: <attilio.stajano@unibo.it>.

Bologna, Italy   *Attilio Stajano*

Acknowledgments

The author thanks Professor Alberta Sbragia, Professor Patrizio Bianchi, and Professor Tibor Palankai, who read some early drafts of this book at different stages of its development, providing suggestions and criticisms; and the many colleagues and friends who read the first edition, offering valuable and appreciated comments. The author is deeply indebted to Leyla Tunç Yeltin, Secretary General of İktisadi Kalkınma Vakfı (Economic Development Foundation) in Istanbul, Turkey, who read a draft of the section on Turkey and offered comments and valuable input. Obviously, the author takes personal responsibility for any content of the book. The author also thanks his students, whose interest and involvement have greatly stimulated him.

The author expresses his deepest thanks to all those who, with patience, accuracy, and high professionalism, helped him with the language, style, layout, and production of the book: Ms. Leslie Haigh of Actionline; and Ms. Marilea Polk Fried, Ms. Barbara Fess, Ms. Marian Scott, Ms. Gillian Greenough, Ms. Deborah Doherty, and Mr. Gerry Geer from the Springer staff.

Finally, the author thanks the *Financial Times*, *Il Sole 24 Ore*, *La Repubblica*, and *The Economist*, which, as owners of the copyright on images and texts reproduced in the book and on the recommended reading web site, have allowed him to reproduce such materials.

The Author

At the moment, Attilio Stajano is holding a course on Research and Technology Policies of the European Union at the Faculty of Engineering at the University of Bologna. He has lectured on these topics also in other Italian universities and at the Georgia Institute of Technology of Atlanta. Previously he worked for 13 years as a civil servant of the European Commission. At first he was responsible for business applications in the information technology research and development program and then for the technological transfer of the whole program. He has also worked for over 20 years in the information technology industry, both in Italy and in several European countries as well as in America, dealing with software development, research, and training.

Contents

Part I
Competitiveness of the European Union

Chapter 1
Origins of the European Union

1.1 After the Second World War

The ideal of peace and coexistence in prosperity was formulated during the first half of the 20th century, when a large part of Europe was governed by dictatorial regimes that were preparing the bloody Second World War. In Italy, Altiero Spinelli (1907–1986), a future European Commissioner and later a member of the first European Parliament elected with direct suffrage, wrote the *Manifesto di Ventotene* [Spinelli 1941], which outlined the project of a Europe where citizens would peacefully cooperate in democratic growth. Spinelli had been jailed by the fascist regime for crimes of conscience at the age of 20, in 1927, and stayed in prison till 1937, when he was interned in the forced confinement of Ventotene for six more years. Ventotene is a small island in the Tyrrhenian Sea, facing the shore of Anzio, where the U.S. troops landed on 22 January 1944; here the fascist dictatorship had one of its confinement places. In the late 1930s the island hosted the élite of the opponents to the regime. The number of internees was 800, and they included members of all the political and intellectual movements that would later build the Republic. Among them – beyond Spinelli – were many patriots whose names are dear to Italian democrats: Ernesto Rossi (1897–1967), Eugenio Colorni (1909–1944), and Sandro Pertini (1896–1990). Ventotene became a *clandestine proletarian university* [Paolini 1996] where, while the European continent was plunging into the horrors of death and destruction, a new vision of Europe took shape, overcoming national divisions and aiming at creating for future generations the conditions for peace and democracy. Rossi, a journalist, and Colorni, a philosopher, contributed to the conception and the drawing up of the Manifesto that was initially endorsed also by Pertini, a future president of the Italian Republic. Pertini had been convicted by the Special Tribunal of clandestine antifascist activities and spent over 14 years in prison and later in confinement between the late 1920s and the fall of the fascist regime. Pertini later withdrew his support for the Manifesto because of pressures from his Socialist Party fellow-partners, but ultimately (in the early 1980s) expressed his regret for this step backwards [Paolini 1989]. The Manifesto was handwritten on cigarette paper by Rossi [Paolini 2005] and sneaked out of Ventotene to the clandestine community of opponents to the regime by Ursula Hirschmann, who had been authorized to visit her husband Colorni. She hid the document in her shoulderpads in order to pass the frisking at the confinement gates.

Spinelli's federalist vision suggested an ideal of union and solidarity, where *peace* did not mean the time interval between two wars during which the military would prepare and get equipped for the next conflict. Peace, for Spinelli, is rather a spiritual condition suggesting a new approach to international political relations that leads to negotiated conflict resolutions and makes war *impossible*.

After the Second World War (1939–1945), industrial rebuilding started with the support of the U.S. Marshall Plan (1947). It went along with the reconstruction of civil society, which had been torn apart by the bloody conflict. Fifteen years after the finalization of the Manifesto, the European Economic Community (EEC) was created on the basis of a plan drawn up by the French foreign minister Robert Schumann, who had been inspired by the visions of Spinelli and of the French economist and diplomat Jean Monnet. The European Union was on its way. The EEC was meant to be a space for democracy, freedom, and solidarity, where citizens would cooperate for prosperity in peace within a context of sustainable growth.

The ideals of peaceful coexistence and democratic growth were first attained through the agreements between the winners and the defeated. The objective of these agreements was the production of energy and steel. Agreements in the research field followed.

The development of the European Union, which has witnessed no conflicts in her territory over the past 50 years after hundreds of civil wars in past centuries, pursues the ideals of Schumann, Spinelli, and Monnet. However, what happened in 2002 and 2003 regarding the issue of war in Iraq, with the opposing positions of Germany and France on the one hand and of the United Kingdom and several pro-American countries on the other, indicates that, unfortunately, we are still far from unanimously achieving that goal of peace implied by the federalist theory. Nevertheless, we will see that great and irreversible strides towards it have been made. A major one is enshrined in the Lisbon Treaty (2007), instituting the "High Representative of the Union for Foreign Affairs and Security Policy," the unanimous and unique voice of the EU on foreign policy, a dream that had been waiting to happen for 50 years and will be in office by 2009, provided that the Lisbon Treaty – signed by the heads of state and government of EU27 in December 2007 – is ratified by the member states before the next European Parliament elections.

1.2 The Treaty of Rome

In the year 2007, the 50th anniversary of the signature of the Treaty of Rome was celebrated, at a point in time when the European Union was faced with the uncertainties about the outcome of the process of ratification of the Constitutional Treaty, that had been signed in Rome by the European Council in 2004, see Chapter 10, "From the Treaty of Rome to the Reform Treaty of Lisbon". The Treaty of Rome in

1957 was the beginning of a long process of development, moving from a customs agreement to the creation of an internal market and a political union.

This process, which has taken 50 years and to some extent is still ongoing, includes five main stages:

1. Customs agreement   Freedom of movement of goods
2. Customs union   Common external tariffs for trade with third countries
3. Common market   Free movement of labour, capital, and services
4. Economic union   Common policies and monetary union
5. Political union   Single currency, internal affairs, foreign policy, defence, and social policy

The first step, the *customs agreement*, approved the elimination of duties and taxes for goods exported between two countries that had signed the treaty. These taxes had a double role: to fund the national budget and to protect national production in all sectors, including agriculture, industry, and services. Consumers were induced to purchase goods and services produced in the national territory. The lack of competition before the customs agreement did not protect the consumers, and while it guaranteed a high level of profits, it did not guarantee quality production. However, the elimination of duties did not create a harmonious situation within the Community with regard to goods and services coming from third countries, as the tariffs of the customs duties for imports from third countries varied from country to country.

The *customs union* led to an agreement regarding common external tariffs on goods and services moving to and from third countries. This was an important step towards the unification of the market. However, the creation of a single internal market was still incomplete because noncustoms obstacles and barriers to the free movement of goods and services were generated by forces opposing the creation of such a market. Examples of these obstacles are the safety standards, which in some cases are still different from one country to another, or interface standards which at times have actually caused the separation of the markets. Let us think about electric plugs alone: in the past, French and German manufacturers of electrical appliances safeguarded their national oligopolistic market by opposing the standardization of electric plugs and by issuing guarantee contracts that bound customers by not allowing them to change the power supply cables. A recent example of forces opposing the full implementation of the internal market is presented in Chapter 13, "Internal Market and Competition", and concerns the difficulty experienced in reaching an agreement on the liberalization of services [EC 2004-8] offered in the 15 member states that constituted the European Union (EU) up to 2004 (EU15) by service providers from the 10 countries that accessed the EU in 2004 (AC10).

In 1992, the creation of a *common market* completed the free movement of goods by providing the free movement of labour, capital, and services. This was an important and decisive step. The four freedoms of movement are inseparable: for example, the free movement of labour and the possibility of residing in another country are

realistic only if a migrating citizen can sell his or her house and transfer the capital
to the new country of residence in order to buy a new house.

The *economic union* has been another step that enlarged the community beyond
trading agreements, ensuring the convergence of the economic policy of member
states and introducing a common currency. One of the founding elements of the economic union is the single currency established by the Treaty of Maastricht (1992),
introduced in 1998 and entered into circulation in the year 2002. This step is studied
in detail in Chapter 14, "Economic and Monetary Union".

Finally, the *political union* (currently *in fieri*) has led to the accomplishment of
the internal market and introduced a common policy for foreign affairs, defence,
and security. Some aspects of this development are described in the third part of
this book.

1.3 Enlargement of the European Economic Community

The institutional transformation took place concurrently with an enlargement of the
European Economic Community (see Figs. 1.1, 1.2 and 1.3).

The sequence of enlargements is as follows:

- 1957: Belgium, France, Germany, Italy, Luxembourg, The Netherlands
- 1973: Denmark, Ireland, United Kingdom
- 1981: Greece
- 1987: Portugal, Spain
- 1995: Austria, Finland, Sweden
- 2004: Cyprus, Malta, and eight countries in central and eastern Europe: the
  Czech Republic, Estonia, Hungary, Latvia, Lithuania, Poland, Slovakia, Slovenia
- 2007: Bulgaria and Romania.

Details follow in Chapter 11, "Enlargement of the European Union".

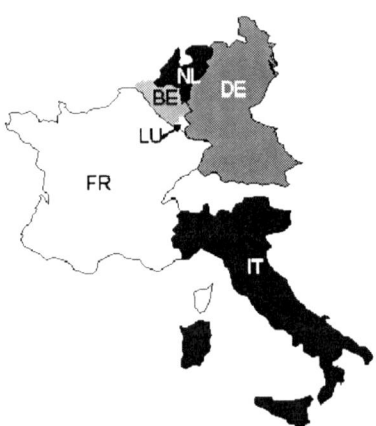

Fig. 1.1 The six *Founding Members* of the European Economic Community were Belgium (BE),
France (FR), Germany (DE), Italy (IT), Luxembourg (LU), and The Netherlands (NL)

Fig. 1.2 From 1995 to 2004 The European Union had 15 member states: Austria (AT), Belgium (BE), Denmark (DK), Finland (FI), France (FR), Germany (DE), Greece (EL), Ireland (IE), Italy (IT), Luxembourg (LU), The Netherlands (NL), Portugal (PT), Spain (ES), Sweden (SE), and the United Kingdom (UK)

1.4 Main Steps in the Construction of the European Union

The main steps in the construction of the European Union are schematized in Table A.1, in Appendix A, where only events that are significant from the point of view of the study of competitiveness and research policy are mentioned. The construction of the Union is masterfully summarized, covering all the policies, in [Fontaine 2003]. This text is available on the recommended reading web site that accompanies this book.

Part 3 of this book offers to newcomers to the study of the European Union the background information on her history, institutional structure, and policies that is needed for the study of EU competitiveness. Readers familiar with these topics might skip Part 3, although they might find interesting the reading of various sections in Chapter 11, "Enlargement of the European Union", in particular: the survey of new member states in Section 11.5, "An Overview of the 2004 Enlargement" and in Section 11.7, "Second Wave: an Overview of the 2007 Enlargement"; Section 11.6, "The 2004 Enlargement Three Years on", addressing successes and challenges faced by the new members; and of Section 11.8.3, "Turkey", covering the accession negotiations with Turkey, a process that unveils different visions on future and role of the Union as expressed by the various political leaders.

1.5 Recommended Reading

Recommended reading for this chapter are listed in Appendix C and are available on the web site companion to this book, at the URL http://stajano.deis.unibo.it/RQC.htm

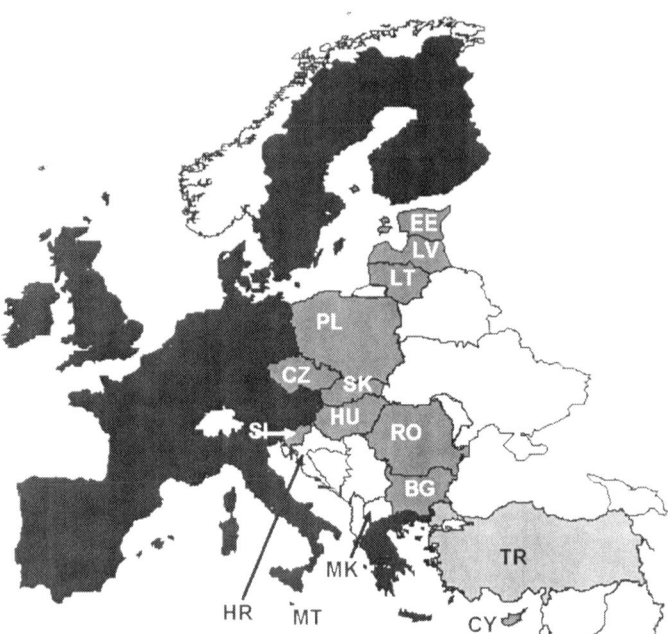

Fig. 1.3 From 1 May 2004 to 31 December 2006, the European Union has had 25 member states: Austria (AT), Belgium (BE), Cyprus (CY), the Czech Republic (CZ), Denmark (DK), Estonia (EE), Finland (FI), France (FR), Germany (DE), Greece (EL), Hungary (HU), Ireland (IE), Italy (IT), Latvia (LV), Lithuania (LT), Luxembourg (LU), Malta (MT), The Netherlands (NL), Poland (PL), Portugal (PT), Slovakia (SK), Slovenia (SI), Spain (ES), Sweden (SE), and the United Kingdom (UK). Bulgaria (BG), Romania (RO) accessed the EU on 1 January 2007, the current members being now 27. Croatia (HR), the Yugoslav Republic of Macedonia (MK), and Turkey (TR) are candidates for accession

Chapter 2
Overview of Member States

2.1 Geographic Data

The surface area of the European Union (EU27) is 4.3 million km^2. The surface area of the U.S. is well over twice that of the European Union. The surface area of Japan is less than 10 % of that of the Union (see Fig. 2.1). The enlargements of the Union in 2004 and 2007 included states that are smaller in size than Greece, with the exception of Poland and Romania, which have a surface area comparable respectively with that of Italy and of the UK (see Fig. 2.2). Should Turkey become part of the Union it would be the largest EU country.

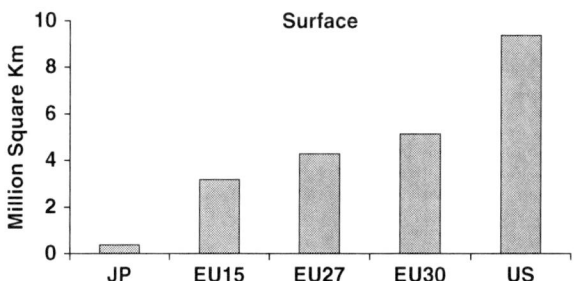

Fig. 2.1 Surface areas in the Triad (million square kilometres). EU30 stands for EU27 plus the three countries candidate to accession. (Eurostat 2001)

2.2 Demographic Data

In the year 2006, the European Union had 464 million inhabitants (see Table 2.1) and after the enlargement to 27 states (2007), the number of inhabitants rose to 493 million. If the enlargement goes on with the candidate countries and were also to include Turkey that number would then reach 572 million inhabitants. In Table 2.1 and in the text following, EU30 stands for the 30 countries of the EU after the further possible accession of Croatia, Macedonia, and Turkey.

A. Stajano, *Research, Quality, Competitiveness*,
© Springer Science+Business Media, LLC 2009

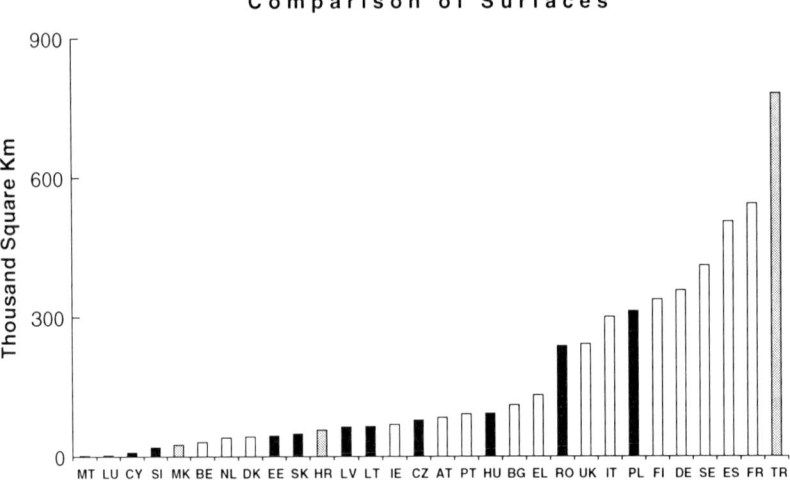

Fig. 2.2 Comparison of the surface area (thousands square kilometres) of EU member states and of candidate states. EU15 data are shown in *white*, the accession states in *black*, and the candidate countries in *grey*. (European Commission 2002)

Table 2.1 Populations in comparison (2006)

Population (million)	
EU15	390
EU25	464
EU27	493
EU30	572
U.S.	301
Japan	127
World	6,602

Source: Eurostat, 2007, WorldFactbook, 2007.

Figure 2.3 compares population sizes for the three regions in the Triad (Europe, U.S., and Japan). Figure 2.4 compares population sizes in the Triad with the global world population. It shows that only one-sixth of the worldwide population lives in the Triad. Later, we will see that the Triad produces about half of the wealth of the world.

Figure 2.5 shows the population sizes of the member states of EU15 (light bars), of the 12 accession states (black bars), and of the three candidate countries (light grey bars). Most of the enlargement states are less populated than the states of EU15. Only Romania and Poland have a population of over 10 million. If, on the other hand, Turkey should become a member of the Union, it would be one of the most populated states – probably the most populated, considering that Turkey has a higher birth rate than Germany (see Fig. 11.7).

The previous observation introduces the topic of demographic dynamics in Europe. The population in Europe is getting older because of the reduction in birth

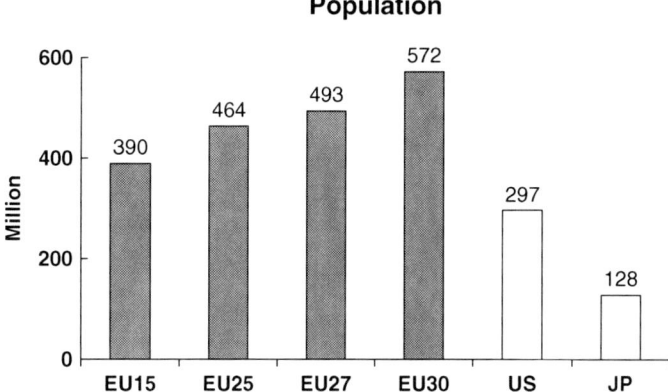

Fig. 2.3 Populations in comparison in the year 2006 (million). EU30 stands for EU27 plus the three countries for which accession negotiations are ongoing in 2007. (Eurostat, 2007. For JP, U.S., and the World, WorldFactbook, 2007)

Fig. 2.4 Populations in comparison. AC12 are the 10 accession countries of the 2004 and 2007 enlargements; the candidate countries are Croatia, Macedonia, and Turkey. (Eurostat, 2007. For JP, U.S., and the World, WorldFactbook, 2007)

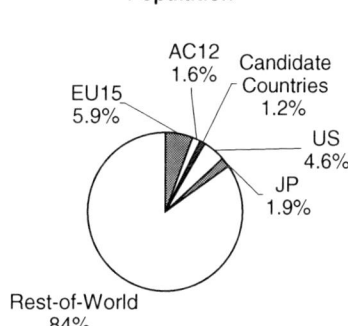

rate and the improvement in the conditions of life, a factor that leads to an increase in longevity. This is a worldwide phenomenon that has an impressive impact on age distribution in Europe.

Let us analyse in detail, as an example, the Italian case. Figure 2.6 shows that the number of children per woman of fertile age in Italy dropped from over 2.6 in 1960 to less than 1.2 in 2000. This trend has changed the age distribution in Italy in a dramatic way. Figure 2.7, based on data from the Italian national institutes of statistics (IsTAT), shows the age distribution of males and females in Italy in 1911 and in 2001. In 1911 the number of youngsters aged 0–9 made up 23 % of the population. In 2001, the number of youngsters aged 0–9 was only 9 %.

The situation at the beginning of the 20th century was one of economic growth and – thanks to the automation of agriculture – it was possible to feed an increasing number of citizens. Furthermore, health and sanitation conditions at that time did not guarantee the level of survival and ageing that was achieved later in the 20th century. The demographic situation at the beginning of the 21st century has radically

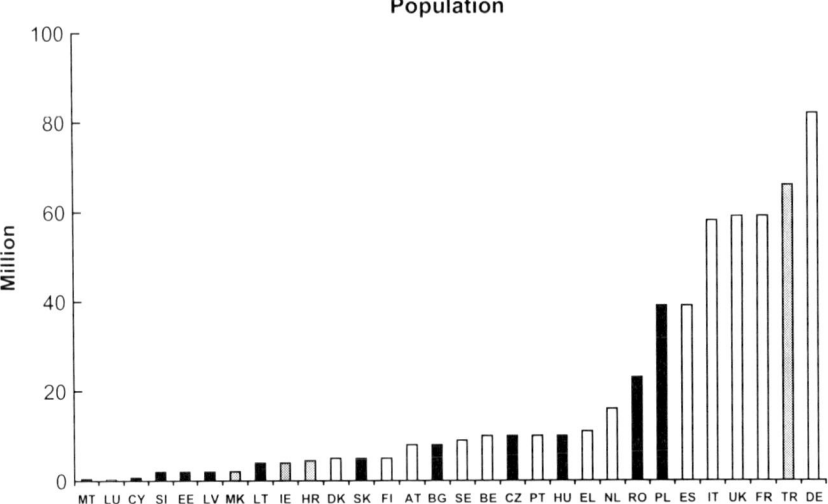

Fig. 2.5 Comparison of population sizes (million inhabitants) of EU15 (*white*), accession countries (*black*) and candidate countries(*grey*). (Eurostat, 2007)

Fig. 2.6 Number of children per fertile woman in Italy in the period 1960–2010. ([IsTAT 2002])

changed and this is very significant for the purpose of evaluating the age distribution of citizens of working age, that is to say in the 20–65 age group, highlighted by the two dark horizontal lines in Fig. 2.8.

The age distribution of the working class (20–65 years of age) has changed and today there is a lower percentage of young people, fresh from their studies, bearers of creativity, well disposed to occupational mobility, and capable of coping with a changing business environment. Figure 2.8 shows that this situation will worsen in the next 20 years, when the working class will be made up of those citizens who in 2001 were aged between 0 and 45 (box on the bottom right).

The problem is European in dimension: in 10 years time in Europe, there will be a deficit of 12 million workers aged less than 40 compared to the present situation, while those workers over 40 will increase by 13 million units (see Fig. 2.9). It is

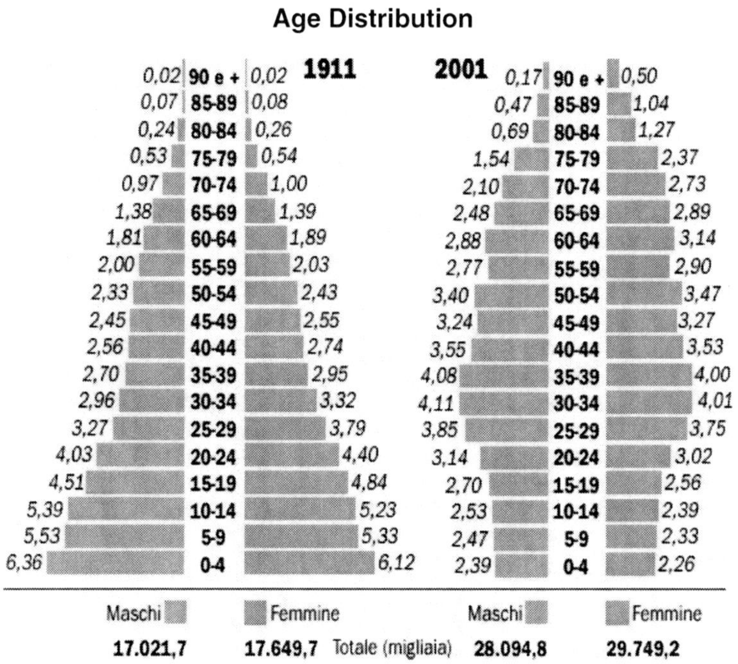

Age Distribution

	1911			2001	
0,02	90 e +	0,02	0,17	90 e +	0,50
0,07	85-89	0,08	0,47	85-89	1,04
0,24	80-84	0,26	0,69	80-84	1,27
0,53	75-79	0,54	1,54	75-79	2,37
0,97	70-74	1,00	2,10	70-74	2,73
1,38	65-69	1,39	2,48	65-69	2,89
1,81	60-64	1,89	2,88	60-64	3,14
2,00	55-59	2,03	2,77	55-59	2,90
2,33	50-54	2,43	3,40	50-54	3,47
2,45	45-49	2,55	3,24	45-49	3,27
2,56	40-44	2,74	3,55	40-44	3,53
2,70	35-39	2,95	4,08	35-39	4,00
2,96	30-34	3,32	4,11	30-34	4,01
3,27	25-29	3,79	3,85	25-29	3,75
4,03	20-24	4,40	3,14	20-24	3,02
4,51	15-19	4,84	2,70	15-19	2,56
5,39	10-14	5,23	2,53	10-14	2,39
5,53	5-9	5,33	2,47	5-9	2,33
6,36	0-4	6,12	2,39	0-4	2,26

Maschi	Femmine		Maschi	Femmine
17.021,7	17.649,7	Totale (migliaia)	28.094,8	29.749,2

Fig. 2.7 Age distribution (percentage) for males and females in Italy in 1911 and in 2001. The population is ageing. The 1960s baby boom is visible in the 2001 distribution as a bump of those aged 30–40 in 2001. ([ISTAT 2002])

estimated that in the period 2005–2030, there will be 21 million fewer EU citizens in working age and that the dependency ratio[1] will go up by 35–66 % [Muldur et al. 2006]. Population ageing creates two types of problems; first of all there is an increased need for lifelong learning to make up for the lack of influx of young people fresh out of school and to keep up with the fast changes in the working world. However, lifelong learning can only be effective in a society where a high level of basic education is guaranteed for everyone, as shown in Fig. 2.10 where the capacity to take advantage of lifelong learning appears to be dependent on the basic level of education achieved. Secondly, population ageing creates a serious problem of sustainability of health care and pension expenditures. It is foreseen that (see Fig. 2.11) between the year 1995 and the year 2035, pension expenditure in Italy will grow by 3 percentage points on the GDP[2], breaking through the upper limit of 16 % of the GDP.

---

[1] The ratio of the number of elderly persons of an age when they are generally economically inactive to the number of persons of working age.

[2] See Glossary

Age Distribution in the Working Population

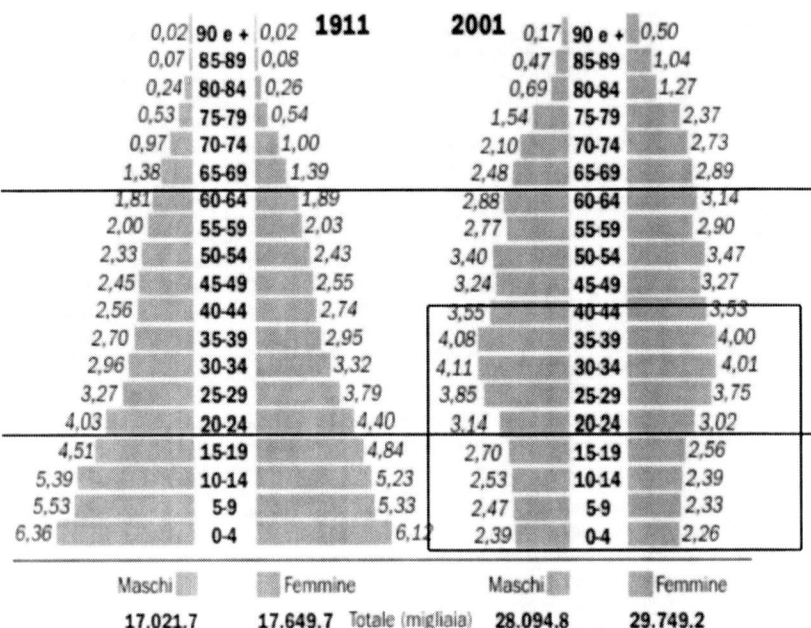

Fig. 2.8 Age distribution (percentage) of the working population (20–65) in Italy in 1911 (*left*), in 2001 (*right*), and in 2021 (box on the *bottom right*). The working population is ageing: in 2021 there will be twice as many workers aged 55–59 as workers in the 20–24 age bracket. This ratio is expected to be six times higher in 2021 than in 1911. ([Istat 2002])

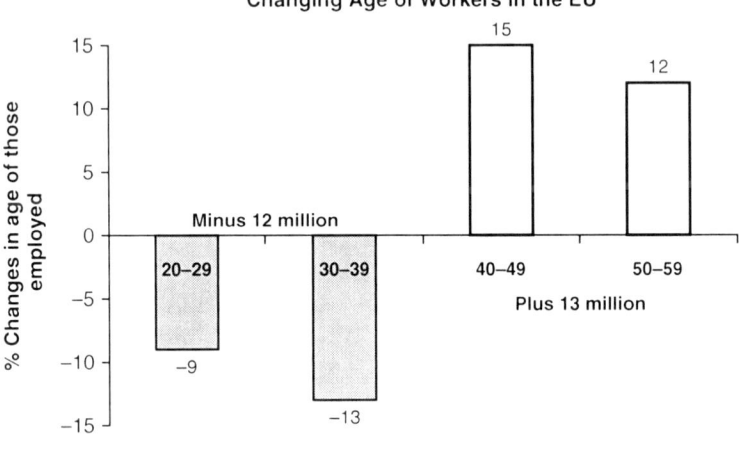

Fig. 2.9 Population of working age in the EU: changes between the year 2000 and 2010. The total working population will increase by 1 million, but there will be 12 million fewer workers in the age brackets 20–29 and 30–39. (European Commission 1999)

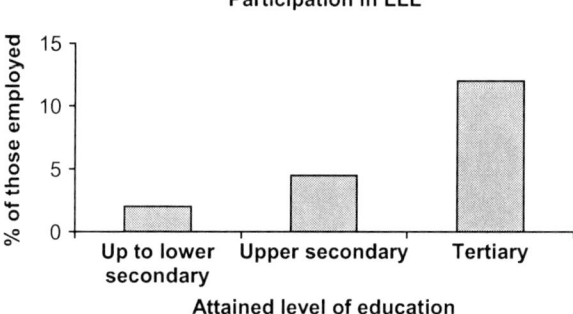

Fig. 2.10 Participation in lifelong learning (% of those employed) as a function of the level achieved in education. The fruition of lifelong learning depends on the level and quality of the educational platform one can build on. Data are based on a Eurostat survey measuring training attendance during the four weeks prior to the interviews (1997). (European Commission 1999)

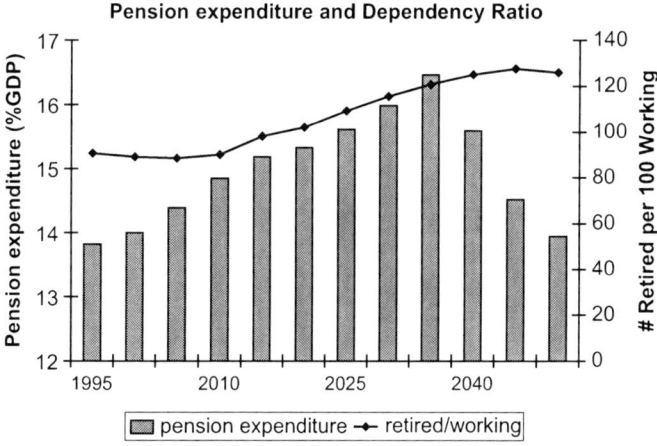

Fig. 2.11 Trends of the expenditure for pensions (bars, *left* scale) and of the ratio between retired and working citizens (line, *right* scale) in Italy in the period 1995–2050. (*La Repubblica* reporting data from Is TAT 2002)

In the year 2000 in some states, the pension expenditure was not as high as in Italy, but without drastic measures the situation in those countries is likely to worsen and become unsustainable over the long term (see Fig. 2.12).

In Italy, the demographic balance is positive only thanks to the contribution of immigrants (see Fig. 2.13). Otherwise, there would not have been as many children born as the number of Italian citizens who die.

The unsustainability of pension expenditure is highlighted in Figs. 2.14 and 2.15. Following population ageing and the extension of life expectancy the ratio between the number of elderly people of retirement age and the number of citizens of working age has increased by $2\frac{1}{2}$ times in a century. Pension expenditure has different

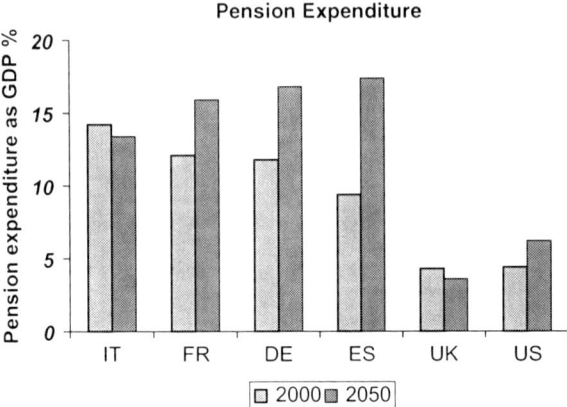

Fig. 2.12 Pension expenditure as GDP percentage for some EU states and the U.S. in the years 2000 and 2050. (Blanchard 2003, reported by *La Repubblica*, 18 January 2004)

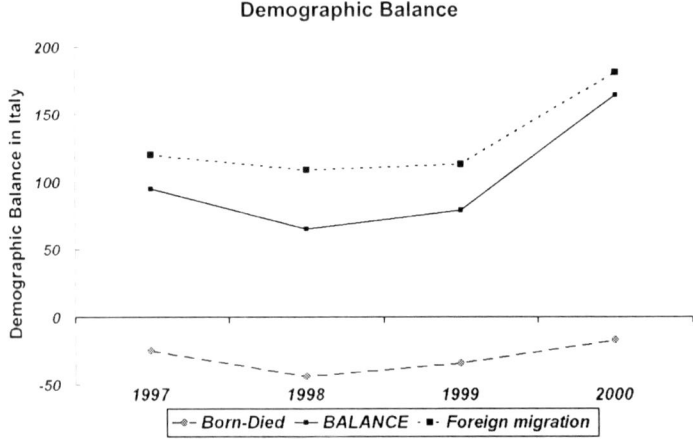

Fig. 2.13 Demographic balance in Italy (thousands of citizens). The birth–death balance (*bottom*) is negative, but the overall balance is positive (*middle*), thanks to the contribution of immigrant citizens (*top*). ([ISTAT 2002], edited by the author)

weights in the budget of different states: see the comparison of the ratio between pension expenditure and GDP for some EU states and for the U.S. in the years 2000 and 2050 in Fig. 2.12.

As mentioned previously, this is not only an Italian problem (see Table 2.2): it affects, in particular, medium-term economic relations between the European Union and the U.S.

The cartoon in *The Economist* [Economist 2002-2] shown in Fig. 2.16 illustrates a pregnant American as opposed to an old fat European. However, the American situation also has its problems, because fertility levels differ depending on social class.

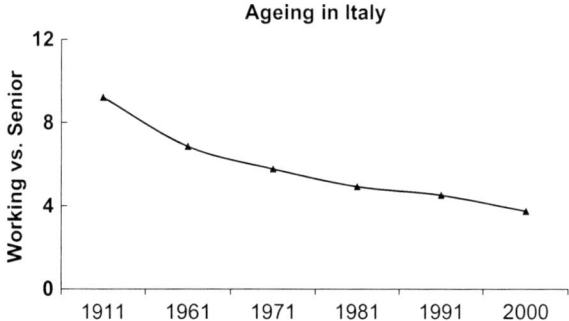

Fig. 2.14 Average number of citizens of working age per senior citizen (65+) in Italy. In a century, this ratio has dropped by a factor of $2\frac{1}{2}$, with a major impact on the sustainability of pension schemes. ([ISTAT 2002])

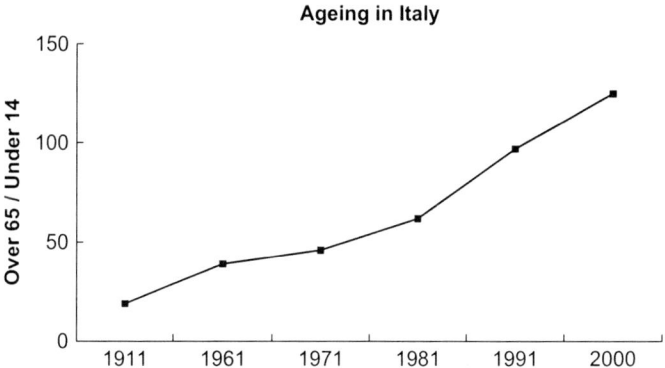

Fig. 2.15 Ageing in Italy: number of people over 65 per 100 people under 14. ([ISTAT 2002])

Table 2.2 Ratio of senior citizens (65+) to working-age citizens (15–64) from 1960 to 2030 in a number of developed countries and for OECD

Country	1960	1990	2000	2010	2020	2030
France	18.8	20.8	23.6	24.6	32.3	39.1
Germany	16.0	21.7	23.8	30.3	35.4	49.2
Italy	13.3	21.6	26.5	31.2	37.5	48.3
Japan	9.5	17.1	24.3	33.0	43.0	44.5
The Netherlands	14.7	19.1	20.8	24.2	33.9	45.1
UK	17.9	24.0	24.4	25.8	31.2	38.7
U.S.	15.4	19.1	19.0	20.4	27.6	36.8
OECD	14.9	19.3	20.9	23.5	29.8	37.7

Source: OECD Data reported by *La Repubblica*, 2003.

Fig. 2.16 U.S. versus Europe: fertility versus ageing. ([Economist 2002-2] ©*The Economist,* reproduced with kind permission)

We can expect to see that in both Europe and the U.S., the demographic contribution of immigrants will change the aspect of society within a few generations.

The age pyramid of white U.S. citizens in the year 2000 is similar to that of Italian citizens in the year 2001, while that of nonwhite U.S. citizens is similar to the Italian pyramid at the beginning of the 20th century (see Fig. 2.17). A projection

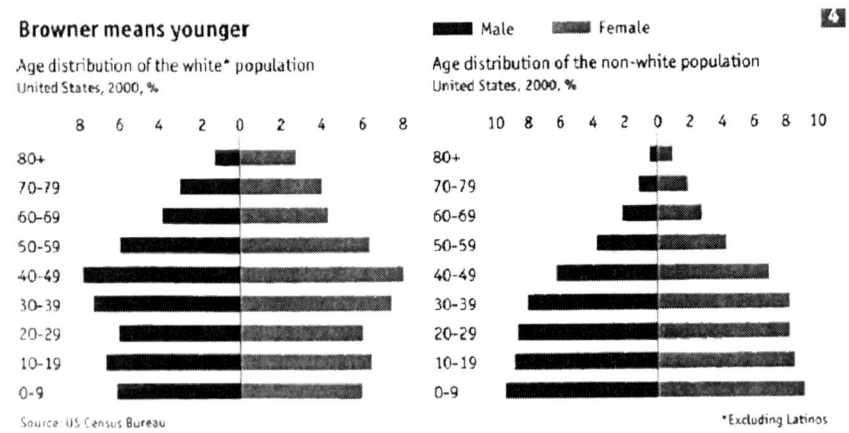

Fig. 2.17 Demography in the U.S. shows a different pattern for white and nonwhite citizens (Black, Asiatic, Latinos). ([Economist 2002-2] ©*The Economist,* reproduced with kind permission)

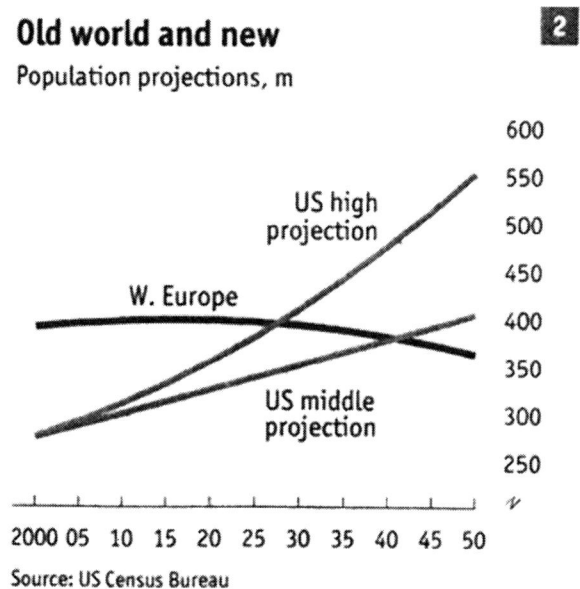

Old world and new

Population projections, m

Source: US Census Bureau

Fig. 2.18 Demographic projections for the U.S. and western Europe. ([Economist 2002-2] © *The Economist*, reproduced with kind permission)

of the U.S. Census Bureau foresees that the population of the U.S. will be higher than the population of western Europe between the years 2030 and 2040 (see Fig. 2.18).

2.3 Economic Data

The gross domestic product (GDP) is a quantification of the overall flow of goods and services produced by an economic system in a particular period of time (usually one year). It is obtained by estimating the sum of the market value of the final goods and services (that is to say without taking into account intermediate products). Capital goods are also included. Depreciation is not applied for replacement of capital goods – hence the adjective *gross*. Income generated from investments abroad is not included – hence the adjective *domestic*. The GNP (gross national product) is the GDP plus income produced by domestic residents with investments abroad, less income on the domestic market produced by foreign residents.

The calculation of the GDP for the EU countries is defined by a regulation of the European Union and carried out by national institutes of statistics (ISTAT in Italy), supervised by the Statistics Office of the EU, Eurostat. In 2005, one hundred indicators were used in 101 sectors. Activities that do not produce financial flows are not measured by the GDP, such as, for example, housework, voluntary work, child care, and growing vegetables for domestic consumption, while goods and services produced by the shadow economy are indirectly estimated. GDP accounts not only for added value generated by productive activities but also for social costs connected with criminal activities, pollution, natural disasters, and exhaustion of nonrenewable natural resources. Economists are debating on which other parameter could better measure a country's wealth but have not found so far a satisfactory GDP replacement. We will see that GDP by itself cannot be the only parameter on which to assess a country's standard of living.

The ratio between GDP and population is used to normalize GDP with respect to the number of inhabitants of different countries. This is the GDP per head or GDP pc (*per capita*) or pc GDP. The cost of living is not the same in different countries: GDP comparison may be normalized to take into account the different cost of living by converting it to purchasing power parity (ppp). Purchasing power parities are obtained by comparing the price levels for a basket of comparable goods and services that is selected to be representative of consumption patterns in the various countries. Table 2.3 and Figs. 2.19 and 2.20 show the GDP distribution in the world at current prices and at purchasing power parity. The study of the pc GDP shows a great disparity in the distribution of production means in the world. The pc GDP ppp in Japan is 4 % higher than the one in the EU15; the U.S. has a pc GDP equal to 140 % of the one in EU15 (see Fig. 2.21). The rest of the world taken together attains 19 % of the European value. It is important to consider that the rest of the world includes very rich economies such as Canada and Australia, emerging economies, and many poor countries, including the poorest ones, where the pc GDP is below 1 % of the European value.

In 2006, the GDP of the whole world was of about \$ 47 trillion at current prices (€ 36 trillion) and \$ 64 trillion at purchasing power parity (€ 49 trillion), according to the International Monetary Fund (IMF) data (see Table 2.3). Based on the ppp data, the Triad produced 51 % of the world's GDP, the European Union (EU25) 20 %,

Table 2.3 GDP in current prices and in purchasing power parity in the World (billion dollars) in the year 2006. The Triad is EU30 plus EFTA, U.S., and JP

Region	GDP current prices	%	GDP ppp	%
EU15	13,773	29	11,714	18
Euro-zone	10,700	23	9,328	15
AC10	754	1.6	1, 204	1.9
AC+2	116	0.3	273	0.4
EU27	14,643	31	13,191	21
CC3	410	0.9	688	1.1
EU30	15,054	32	13,879	22
EFTA	698	1.5	464	0.7
U.S.	13,153	28	13,049	21
JP	4,897	10	4, 168	6.6
Triad	33,802	72	31,561	49
Rest of the World	13, 132	28	32,643	51
World	46,933	100	63,516	100

Source: International Monetary Fund [IMF 2007].

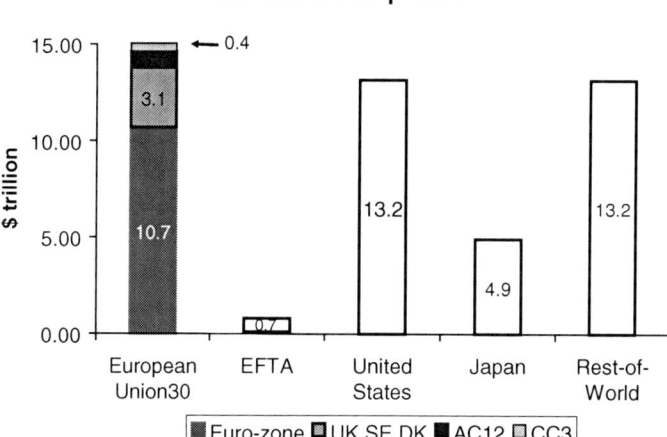

Fig. 2.19 Comparison of GDP at current prices ($ trillion) in the year 2006. The exchange rate $-€ may introduce differences from one year to the next in EU-U.S. comparisons. (International Monetary Fund [IMF 2007])

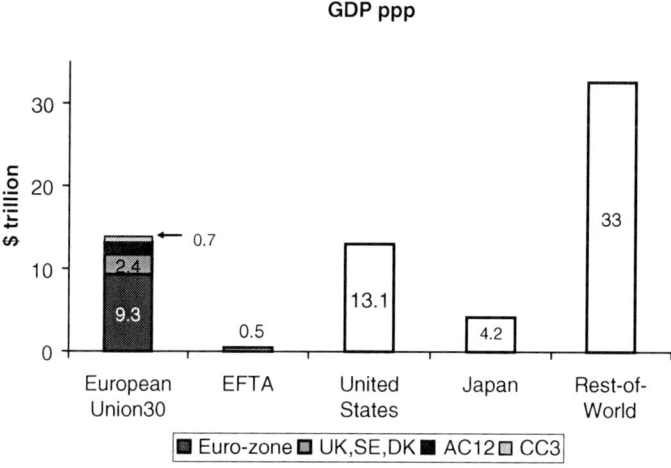

Fig. 2.20 Comparison of GDP at purchasing power parity ($ trillion) in the year 2006. Comparing to the Fig. 2.19, we notice that the relative weight of the less rich countries is increased and the weight of the richest countries is decreased. The total value of the world's GDP ppp is much higher than its value at current prices, as displayed in Table 2.3. (International Monetary Fund [IMF 2007])

and the U.S. 21 %. The time series calculated at current prices give higher weight to developed economies and present the Triad's GDP at 71 % of the world's GDP.

Figure 2.21 shows a comparison of GDP (per head, purchasing power parity) in the Triad, normalized with regard to the pc GDP ppp value in EU15. Table 2.4 shows GDP of the EU30 countries in current prices in euro (billion) in the years 2002–2006.

GDP Comparison

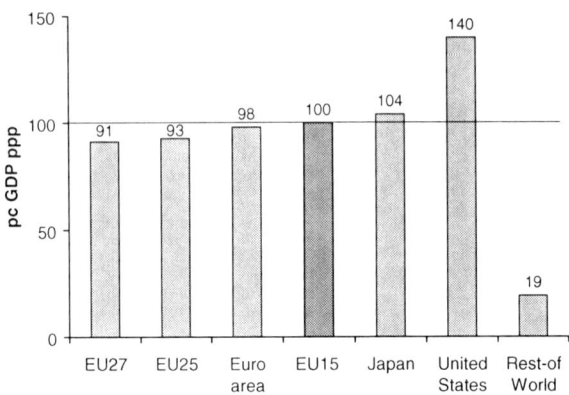

Fig. 2.21 pc GDP ppp (purchasing power parity) in the year 2007, relative to EU15=100. The absolute value for EU15 is € 26,900. (Eurostat 2007)

Table 2.4 GDP in current prices (€ billion) in EU30 during the years 2002–2006. Data for MK are in $ billion

Region	2002	2003	2004	2005	2006
EU27		10,072	10,567	11,000	11,567
EU25	9,843	10,002	10,486	10,899	11,445
EU15	9,385	9,550	9,997	10,337	10,825
Euro12	7,247	7,457	7,753	8,015	8,370
BE	268	275	290	299	313
BG	17	18	20	22	25
CZ	80	81	87	100	113
DK	185	189	196	208	219
DE	2,143	2,162	2,207	2,241	2,307
EE	8	8	9	11	13
IE	130	139	148	161	176
EL	143	156	168	181	195
ES	729	783	840	905	976
FR	1,549	1,595	1,660	1,718	1,786
IT	1,295	1,335	1,391	1,423	1,475
CY	11	12	13	14	15
LV	10	10	11	13	16
LT	15	16	18	21	24
LU	24	26	27	29	33
HU	71	75	82	89	89
MT	4	4	4	5	5
NL	465	477	490	506	528
AT	221	226	236	245	256
PL	209	192	204	244	271
PT	135	139	144	149	155
RO	0	53	61	80	97

Table 2.4 (continued)

Region	2002	2003	2004	2005	2006
SI	22	24	26	28	30
SK	26	29	34	38	44
FI	144	146	152	157	168
SE	259	270	281	288	307
UK	1,694	1,635	1,767	1,827	1,929
HR	24	26	28	31	
MK	4	5	5	5	5
TR	193	212	242	291	

Source: Eurostat 2007 for all countries except MK: IMF 2007.

However, even within the European Union there are significant differences in pc GDP ppp (see Fig. 2.22). Figure 2.23 compares the pc GDP (light columns) with the pc GDP at purchasing power parity (dark columns) in EU30. The pc GDP ppp varies in EU30 from a minimum of about $ 8,000 per year in Macedonia to nine times that amount in Luxembourg; at current prices the range of pc GDP values is from $ 3,000 to $ 81,000. Italy is one of the last, with € 18,000. Figure 2.23 shows that, as regards the pc GDP at purchasing power parity, the differences between the various member states are reduced with respect to values at current prices (the standard deviation in down to 13 from 20). For example, in terms of value at purchasing power parity, the pc GDP in Greece is about twice the value in Bulgaria, but in current prices it is five times as much. In the next chapter we will, however,

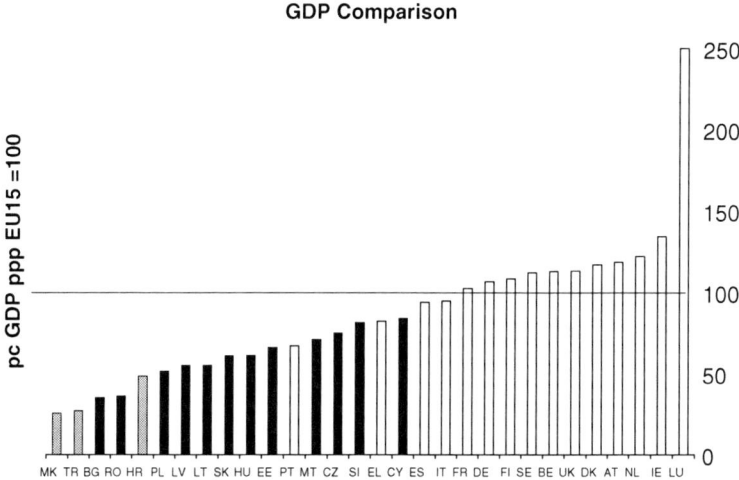

Fig. 2.22 pc GDP ppp (purchasing power parity) in the EU27 Countries normalized to EU15=100 in the year 2007. EU15 states are shown in *white*, AC12 in *black*, and candidate countries in *grey*. The absolute value for Macedonia is € 6,800 and for Luxembourg is € 67,200; the average for EU15 is € 26,900. (Eurostat 2007)

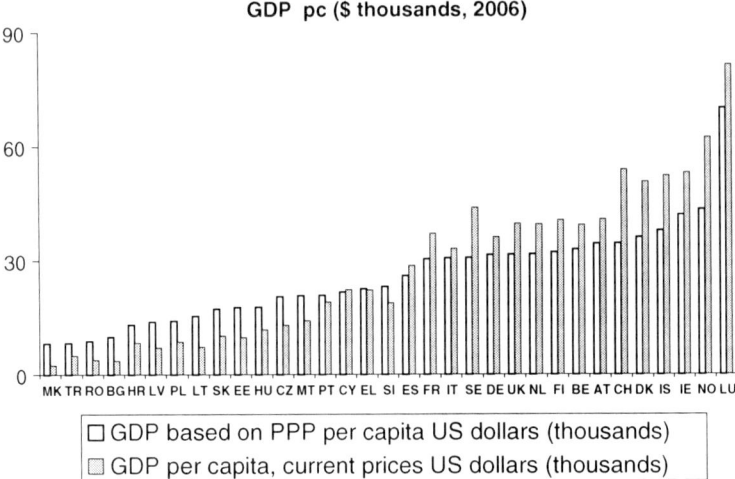

Fig. 2.23 pc GDP ppp (*white*) and pc GDP current prices (*grey*) for EU and EFTA in the year 2006 (\$ thousands). The range of values are {3,81} and {8,70} respectively for current prices and ppp; the averages are 28 and 26; the standard deviations are 20 and 13. (IMF 2007)

that the GDP parameter by itself alone cannot exhaustively measure a country's standard of living.

Public expenditure is the totality of the expenditures of central and regional governments, local authorities, and public companies. The expenditures items include: goods, services, benefits, subsidies, capital formation, and payment of interests on the national debt. The public expenditure of the Union (see Fig. 2.24) is equal to the 2.2 % of the overall public expenditure of the member states, which in

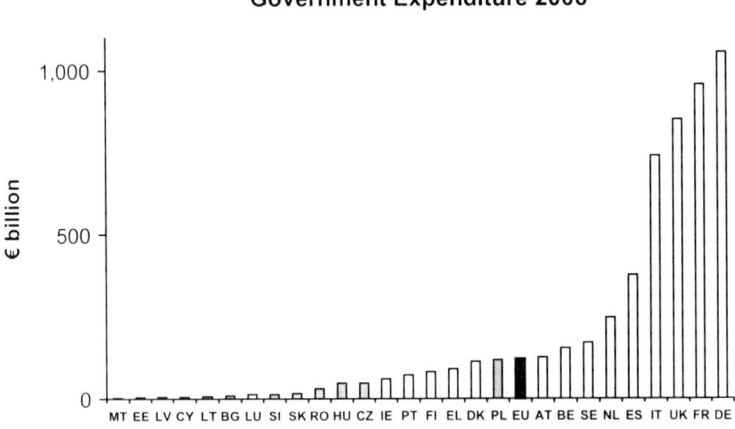

Fig. 2.24 Public expenditure of EU member states (€ billion) at market prices in the year 2006 compared with the EU budget. (Eurostat 2007)

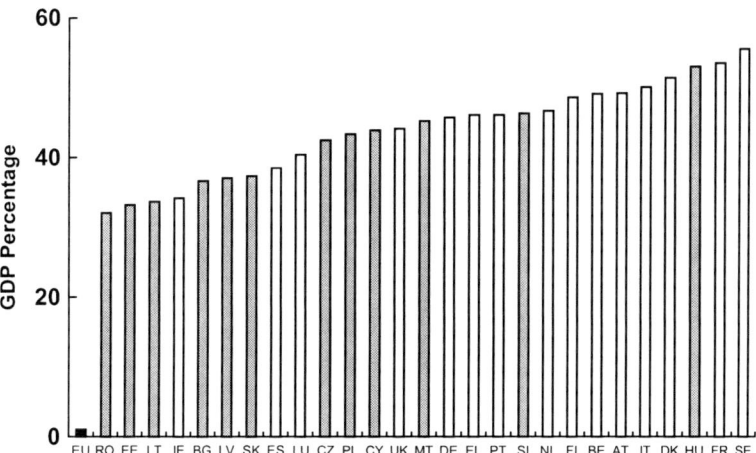

Fig. 2.25 Public expenditure as a percentage of GDP for the EU27 states in the year 2006. The value displayed for the EU is the ratio of the EU budget to the sum of GDPs of the member states. (Eurostat 2007)

total equals € 5.5 trillion. The budget of the Union (about € 120 billion) is about the same amount as the public expenditure of a country like Austria. For comparison purposes, the federal budget of the U.S. in the year 2002 was $ 1.9 trillion.

It is interesting to compare the ratio between public expenditure and GDP, shown in Fig. 2.25. This ratio varies from 32 % in Ireland to 58 % in Sweden. Italy is close to Germany with a value at 47 %. Among the most populated countries, Spain and the United Kingdom have low ratio values: 40 %. The reduced weight of pension expenditure on public expenditure (see Fig. 2.12) is not irrelevant in positioning Spain and the UK towards the low end of the ranking. Public sector activities carry a large weight in the overall output of an economy: public employment accounts for between 10 % (Germany) and 30 % (Sweden) of all jobs [EC 2004-7].

The enlargements in 2004 and 2007 entailed significant changes even in terms of the range of pc GDP across the EU member states. However, the wealth produced by the accession countries is much less than the wealth produced in the EU15 member states (see the structure of the first column in Fig. 2.19). The gross domestic product generated by the twelve accession countries is about equal to 6 % of the EU27 GDP. If the enlargement were to be extended to the three candidate countries, including Turkey, the GDP would increase by an additional 3 %, to about € 12 trillion in current prices.

The difference in wealth production is not necessarily alarming; on the contrary, it can be the origin of synergies and benefits for both new and existing member states. The situation was similar when the enlargements to Ireland, then Greece and later Spain and Portugal took place (see Fig. 2.26). Great advantages have resulted from these enlargements for both the joining states and the existing

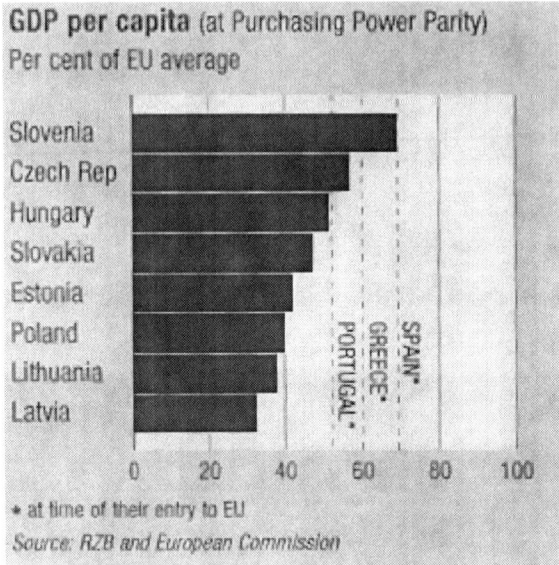

Fig. 2.26 pc GDP ppp as EU percentage for some of the accession countries: comparison with data relative to Greece, Portugal, and Spain at the time of their accession. (Data from the European Commission, reported by ©*Financial Times*, reproduced with kind permission)

member states. The former have taken advantage of Community programs for sustainable development and the synergy with EEC countries originating new companies, joint ventures, and mergers thereby creating more innovative and advanced structures; "old" member states have benefited from the enlargement of the internal market and new industrial synergies through the delocalization of production units in geographically close regions where qualified labour was available at lower costs.

By the end of the 21st century, social and economic transformations in Europe will fade out the present difference in pc GDP in the countries of EU30+ and the ranking of Fig. 2.23 may take a different shape. Countries that ensure long term prosperity and a bright future to their citizens are those that invest now in education, lifelong learning, research, and innovation; have a high percentage of young people; and run public finance with balanced budgets and low public debt.

Enlargement countries attract foreign capital because they are characterized by reduced cost of skilled labour (see Fig. 2.27) and favour the delocalization of industrial plants from western and Asian economies. Enlargement countries can become one of the driving forces of the development of EU27, provided that they are able to combine growth with curbing inflation.

The composition of the domestic product in Europe is analysed in Figs. 2.29, and 2.30 on the basis of the contribution of three production sectors: agriculture, industry, and services. In Italy, as in all advanced economies, there has been a progressive reduction in the percentage of production in the agricultural sector,

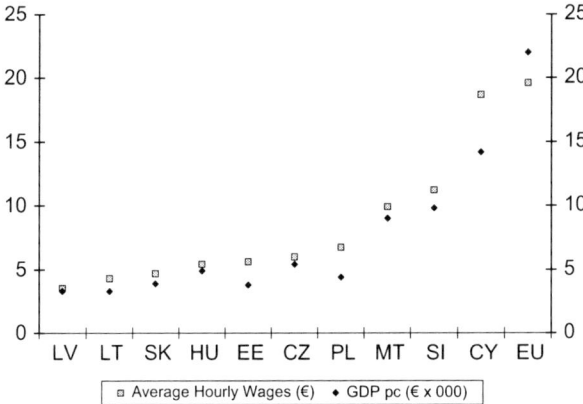

Fig. 2.27 Average hourly wages (euro, year 2001) and GDP pc (thousand euro, year 2000) in the accession countries. Comparison with EU15. (Data OECD and Institut der deutschen Wirtschaft, reported by Salzburg Agentur (2003))

which has resulted in a reduction of people working in agriculture from 20 % to 5 % in 30 years (see Fig. 2.28). The transformation of the economy has led to a reduction in the percentage of added value produced through agriculture, while the percentage of domestic product in industry and services has grown. Alongside

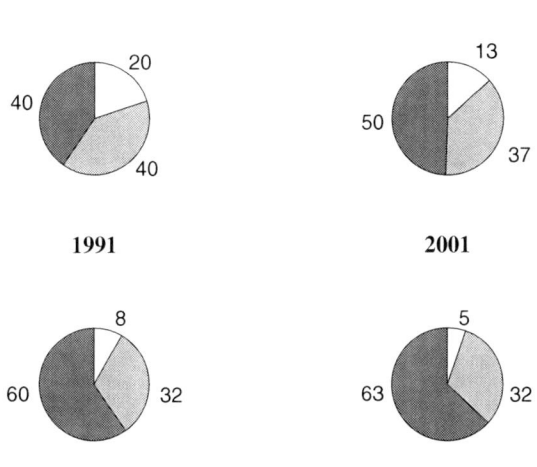

Fig. 2.28 Percentage of Italian workers in the three activity sectors from 1971 to 2001. ([Istat 2002])

GDP in Agriculture

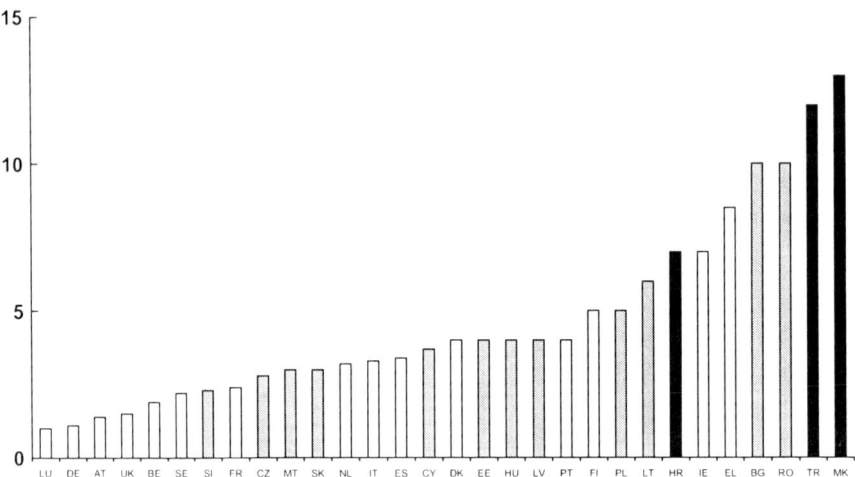

Fig. 2.29 Contribution of agriculture to GDP (%) in EU15 (*white bars*) in the year 1998 and in the rest of EU30 in 2005 or the latest year for which data are available. Accession countries are represented with *grey bars*, candidate countries with black bars. ([Istat 2002])

this phenomenon there has been a migration from the countryside towards urban centres. The agricultural policy of the Union has ensured a balance between the income of workers in the agricultural sector and that of workers in other sectors and has thereby limited the abandonment of agricultural activities by guaranteeing investments, modernization, infrastructures, and subsidies.

Figure 2.29 shows the percentage of GDP coming from agriculture in EU30. Figure 2.30 shows the apportionment of the GDP between agriculture, industry, and

GDP % produced in the three sectors

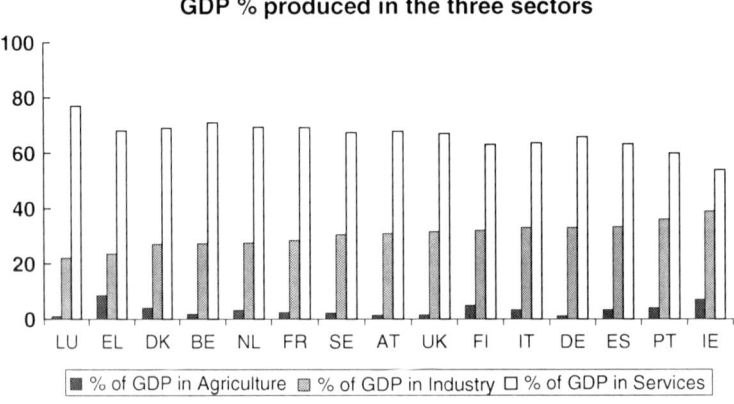

Fig. 2.30 Contribution of sectors to GDP (%) in EU15 in the year 1998. ([Istat 2002])

services for the countries of the Union in 1998. The countries are in growing order of value of GDP in industry. Ireland has the highest share of GDP in industry and the lowest in services. In Ireland, there has been a rapid development of industrial sites; thanks to various factors such as a tax incentive policy to call for foreign investments, which indeed attracted considerable capital, in particular from the United States. Later we will see that the growth in GDP in Ireland has been exceptionally high, beyond the limit of sustainable development.

A comparison of *productivity* between the various sectors can be made by calculating the ratio between the GDP in a sector and the number of workers in that same sector. In Fig. 2.31, member states are shown in growing order of productivity in industry (grey squares). The grey diamonds refer to agricultural productivity. The relatively low value of GDP per worker in agriculture does not indicate a lower income of agricultural workers compared with other workers but instead indicates the production of a different added value. The circles represent productivity in services. For half of the member states, productivity in agriculture (represented with diamonds) varies between 40 % and 60 % of the productivity in the other sectors. There is a considerable exception in the United Kingdom, where this value is 140 %.

Significant public investments in agriculture have been made in Great Britain that have resulted in extreme and unnatural conditions of exploitation, in particular in zootechnics. It is not by chance that bovine spongiform encephalopathy (BSE) and foot-and-mouth disease occurred in Great Britain in a zootechnics context characterized by extreme conditions contrary to ecological and organic farming principles.

An important element to measure the trend of the economy is given by the yearly variation in GDP (see Figs. 2.33, 2.34, and 2.35). A growth in GDP, insofar as the variation in GDP is higher than the increase of labour productivity, is loosely associated with an increase in income available to families and with the creation

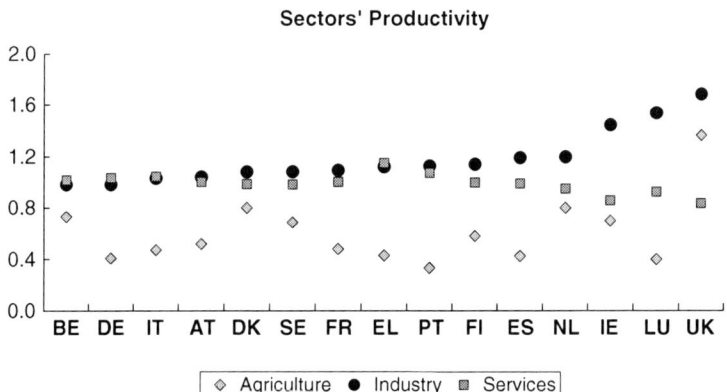

Fig. 2.31 Comparison of productivity in production sectors in the EU (1998). (From data Eurostat 2001)

GDP growth and Unemployment

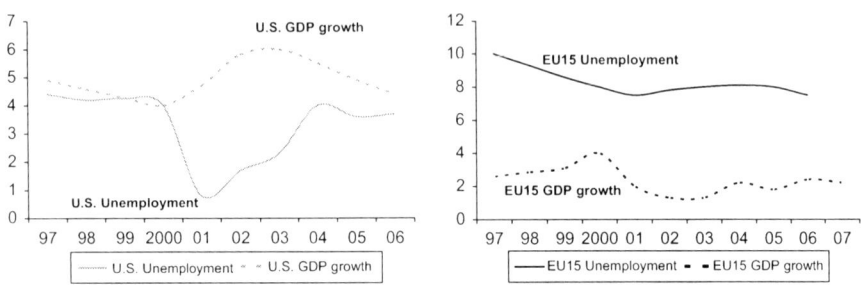

Fig. 2.32 Yearly change of GDP (*dotted line*) and Unemployment (*continuous line*) in the U.S. (*left*) and in the EU (*right*) in the period 1997–2007. The inverse correlation is visible in both the U.S. and the EU, but it is higher in the U.S. case. (The Economist [Economist 2007-25])

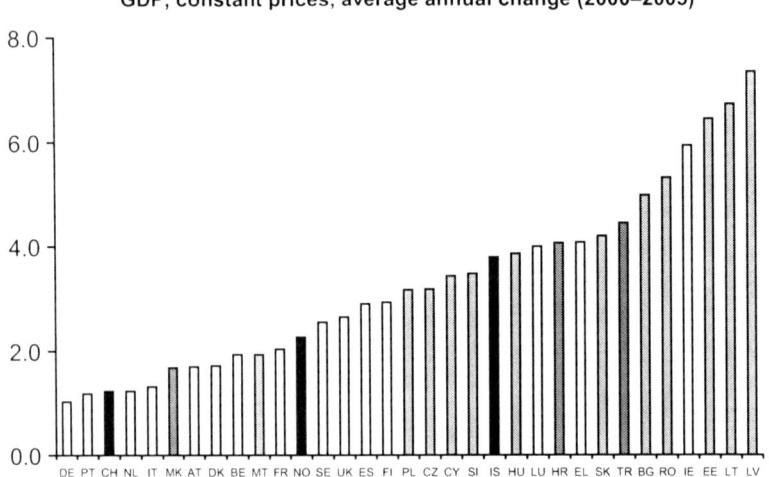

Fig. 2.33 Average yearly change of GDP in the EU30 and EFTA in the period 2000–2005. EU15 countries are represented with *white bars*, AC12 with *light grey bars*, CC3 with *dark grey bars*, EFTA with *black bars*. (Eurostat 2007)

of new jobs (see Figs. 2.32). An empirical verification of the correlation of GDP changes and levels of unemployment is shown in Fig. 2.32 for the U.S. and the EU economies. The inverse correlation is visible in both the U.S. and the EU, but it is higher in the U.S. case, which might be caused by a more homogeneous economy in America. An increase in GDP also determines an increase in tax collection with benefits for the state budget. In the 1990s and in the first years of the new century, the growth in GDP in Italy was one of the lowest in Europe. In contrast the growth in GDP in Ireland in the 1990s was exceptional and was constantly at a level of 2 % for about 10 years. Such a level of growth has never been seen before in any other

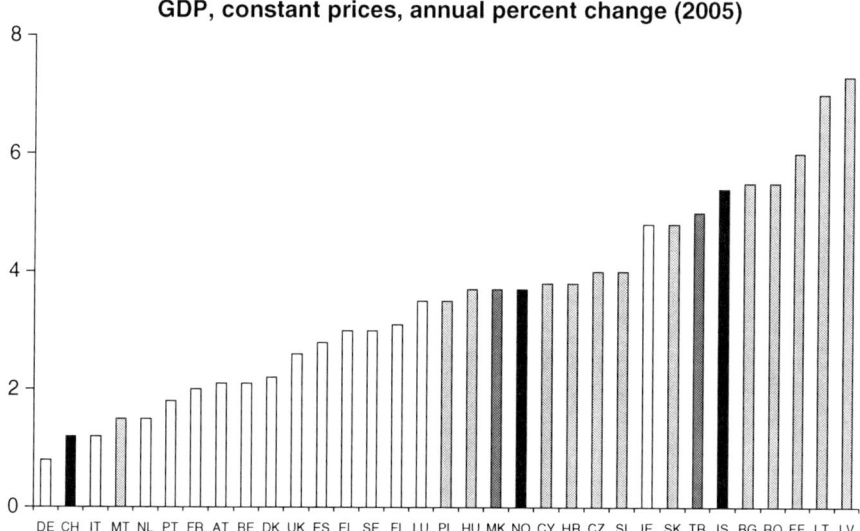

Fig. 2.34 GDP growth in EU30 and EFTA in the year 2005. EU15 countries are represented with *white bars*, AC12 with *light grey bars*, CC3 with *dark grey bars*, EFTA with *black bars*. (IMF 2007)

western economy for such a prolonged period of time. Such rapid growth resulted in unsustainability phenomena.

Before joining the Union, Ireland was a rather poor agricultural country. The national policy exploited the opportunities provided by taking part in Community programs. But such a rapid growth (see Fig. 2.35) has created an imbalance due to the lack of infrastructures, in particular as regards transport and training, which

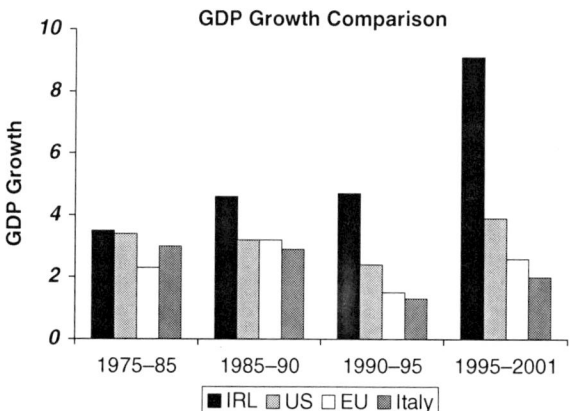

Fig. 2.35 GDP growth in Ireland, in the U.S., in the EU and in Italy from 1975 to 2001. (Eurostat 2001)

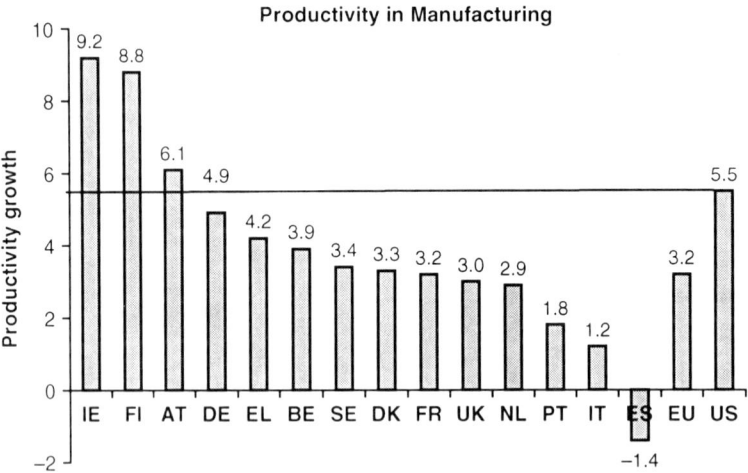

Employment growth

	low	average	high
high	Greece Portugal Austria	Sweden *United States*	Ireland Finland Luxembourg
average	Germany *Japan*	Belgium UK Denmark France	The Netherlands
low		Italy	Spain

(Productivity growth — vertical axis label)

Fig. 2.36 Employment and Productivity in the period 1995–2001. Positioning of the U.S., Japan and the EU15 member states. (European Commission 2001)

Productivity in Manufacturing

IE 9.2 FI 8.8 AT 6.1 DE 4.9 EL 4.2 BE 3.9 SE 3.4 DK 3.3 FR 3.2 UK 3.0 NL 2.9 PT 1.8 IT 1.2 ES −1.4 EU 3.2 US 5.5

Fig. 2.37 Growth of productivity in the manufacturing industry in the EU during 1996–2000. A comparison with the U.S. (European Commission 2001)

resulted in lack of qualified labour for the high-tech activities on which Irish development was based. One of the driving forces of the Irish growth was the policy of tax incentives for foreign investments, which in 2001 was declared by the European Commission as being incompatible with the Stability and Growth Pact of the Economic and Monetary Union.

In 2005 the seven EU30 countries with highest GDP growth (see Fig. 2.34) are among AC12 and CC3, while two large EU15 economies are at the bottom of the ranking. The new EU members can give a contribution to and pull the EU growth, provided that their development is achieved with control on inflation.

As we have seen, Italy's ranking in terms of GDP growth is at the tail end of the states of the EU. Even in relation to labour productivity growth (changing added value per worker) and to increase in employment, Italy is ranked as one of the last among the industrialized countries. In Fig. 2.36, Japan, the U.S., and EU countries are clustered into groups of low/medium/high growth rate with reference to the two criteria of productivity and employment growth. The most advanced countries are Ireland, Finland, and Luxembourg, which enjoyed a high growth rate under both aspects. Italy is in one of the bottom positions and, for different reasons, so are Germany and Japan.

The ranking of the countries in Fig. 2.36 is confirmed by a comparison of productivity growth in the manufacturing industry (see Fig. 2.37), where Italy and Spain are in the bottom positions.

2.4 Recommended Reading

Recommended reading for this chapter are listed in Appendix C and are available on the web site companion to this book, at the URL http://stajano.deis.unibo.it/RQC.htm

Chapter 3
Competitiveness of the European Economy

3.1 European Economy and Global Market

The European Union's (EU) prosperity is dependent on her capacity to compete in the global market. For this reason, we need to measure and study EU economy's position in terms of competitiveness. Competitiveness creates the necessary conditions for sustainable development, for the creation of new production activities and new jobs, and for a better quality of life.

When we talk about competitiveness, what do we really mean? If we refer to business competitiveness, then we mean market success and the acquisition of new market shares. However, in this book, we rather refer to competitiveness as related to countries and economic systems. The European quest for competitiveness is qualified: it should be compatible with the European dream [Rifkin 2004]. Europeans value quality of life higher than the accumulation of wealth and have a vision of the society they want to live in as a society that values solidarity, the well being, and the personal development of her citizens; that respects the environment, the less favoured individuals, and the less developed countries. Europeans want to export their values with solidarity, financial and technical aid, and international cooperation, as opposed to those countries that pretend to export democracy with the instruments of war and destruction. The EU approach to sustainable development affects our European attitude towards competitiveness: our struggle is up-hill because we mean not only the acquisition of new shares on the international market but also, and more importantly, the capacity to attain sustainable development while safeguarding the welfare of the citizens, protecting the natural environment, guaranteeing proper management of nonrenewable resources, and creating in the long term prosperity, employment, and improved living standards.

When we talk about the EU competing in the world, who are our competitors? EU competitors used to be the U.S. and Japan; however, in the last 15 years, new challenges have been coming from emerging economies, as we will see later in this chapter. Together with the United States and Japan, the European Union is part of a group of countries, called the Triad, accounting for 10 % of the world's population and having one of the world's highest per capita GDPs. Over three-fourths of European Union foreign trade is within the Triad. For the countries in the Triad, the main competitors in the market have been traditionally the most technologically advanced and industrialized countries. For example, in Italy, 71.6 % of exports (and 77.6 % of imports) are to (and from) economically advanced countries [ISTAT 2001]. However, the situation is definitely not stable, as will be explained in Section 3.6, "Emerging Economies". In the globalized society of the third millennium, we are faced with new challenges.

Globalization changed international economy by increasing the world's work-force by one-third and by introducing the demand of many hundred million new consumers now above the starving level of poverty in which they lived up to a few decades ago. Of these consumers, 2.4 million are Asian millionaires that add up to the 5.7 million millionaires of the western world. The new elite-consumers increased the demand for high price/high quality goods, while the bulk of the new consumers demands for good products at budget prices, creating a new phenomenon referred to as "good enough is better than best" that we will cover in the next Chapter 4 on "Competitiveness and Quality", Further to a Council resolution, every year the European Commission issues an annual report on competitiveness.[1] Another regular source of information on competitiveness comes from the annual report of the World Economic Forum (WEF).[2] The EC reports are part of the background knowledge base on which Community policies for sustainable development are conceived. These reports shed light on the role and effect of public initiatives that can stimulate competitiveness. Such initiatives include: education, lifelong learning, research and technological development, standardization, innovation, technology transfer, facilitating the access to financing, taxation, public spending, infrastructures, and the regulatory framework. The WEF reports will be analysed in Section 3.15, "Comparing Countries' Competitiveness". They offer a snapshot of the world economy with special focus on competitiveness and an in-depth analysis reflecting the views of both the public sector and the business community.

3.2 GDP in the World

As we have seen, the gross domestic product (GDP)—the estimate of the value of the overall flow of goods and services produced by an economic system in a year—is an indicator of an economy's prosperity. GPD can be measured at current prices or at purchasing power parity (ppp). We showed in Section 2.3, "Economic Data" that the two GDP measures can give quite a different picture of the relationship among economies when the respective countries are at different levels of development and of purchasing power. GDP at current prices in the world is indicated in Table 3.1 and in Fig 3.1 The Triad's GDP at current prices is 72 % of the world GDP, while the Triad's GDP ppp is about half of the world GDP. The annual GDP growth in the

---

[1] [EC 1997-3, EC 1998-3, EC 1999-2, EC 2000-8, EC 2001-1, EC 2002-2, EC 2003-5, EC 2004-7, EC 2006-10, EC 2007-10].

[2] [WEF 2003, WEF 2004, WEF 2005, WEF 2006].

Table 3.1 GDP at current prices in the World in trillions of dollars in the year 2006

GDP in the World		
EU15	13.7	29 %
EU27	14.6	31 %
EFTA	0.7	1 %
U.S.	13.2	28 %
JP	4.9	10 %
Triad	33.8	72 %
Rest of the World	13.2	28 %
World	46.9	100 %

Source: International Monetary Fund [IMF 2007].

industrialized countries was 2.4 % in the period 1999–2004, while it was 4.0 % in the world and 7.5 % in continental Asia (not including Hong Kong and Korea).

The European Union ranks behind the U.S. in terms of pc GDP by 30 % and behind Japan by 4 % (see Fig.7.21). Gross domestic product figures are not stable in time [Maddison 2001], nor are they perfect indicators of living standards. In talking about the stability of the World GDP geographical distribution, it is interesting to compare recent data (1998) with those dating back to the 19th century, expressed in 1998 U.S.$ (see Fig. 3.2). Europe and Africa accounted more or less for the same share in 1820 as in 1998 of a total GDP that was $ 0.7 trillion, 67 times lower than it is today. The U.S. and Japan had basically no economic power, whereas the large majority of goods were produced in mainland Asia.

A study of the evolution of GDP in Europe over the last two thousand years [Economist 1999-2] shows a stable pc GDP value around € 1,000 for 15 centuries.

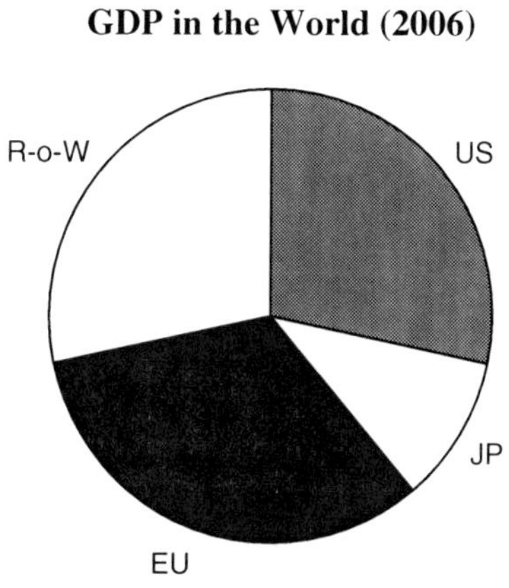

GDP in the World (2006)

Fig. 3.1 GDP at current prices in the World in the year 2006
Source: International Monetary Fund [IMF 2007].

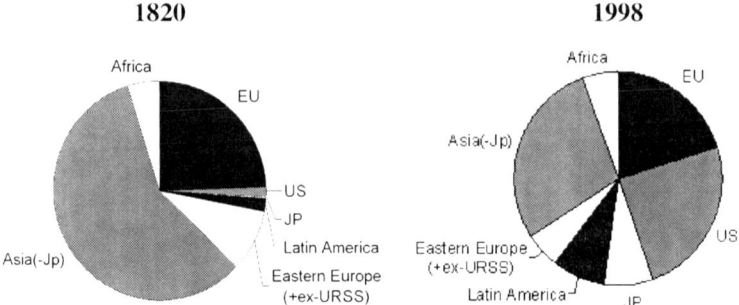

Fig. 3.2 Changes in the distribution of GDP in the world from 1820 to 1998. GDP is expressed in U.S.$ of 1998
Source: [Maddison 2001].

The absolute value of GDP grew slowly during that period, allowing more people to live, but the level of pc GDP stayed the same for 1,500 years. It was followed by two discontinuities, in the 16th and 18th centuries, the first resulting of the geographical discoveries and the introduction of printing and the second resulting of the exploitation of steam power. The second of the above-mentioned discontinuities led to a roaring growth of pc GDP supporting the wealth of a 20-fold population. Also in the third millennium, knowledge, energy, and technology continue to be the engines that drive economic growth. Growth is, however, not smooth even on a short period of time and within the relatively homogeneous set of European countries (Figs. 3.3 and 3.4).

A legend says that James Watt's steam engine (1763) was inspired by an observation in his kitchen, when the boiling water over an open fire threw the lid off a pot cooking potatoes. Potatoes had been boiling on cookers for a long time: why

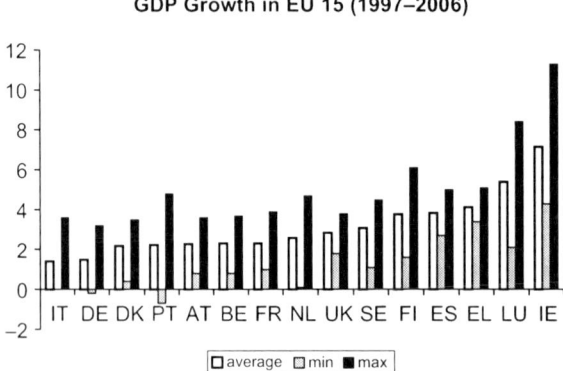

Fig. 3.3 GDP growth in EU15 in the period 1997–2006. The average growth in *white* bars, the minimum in *grey*, and the maximum in *black*
Source: Eurostat 2007.

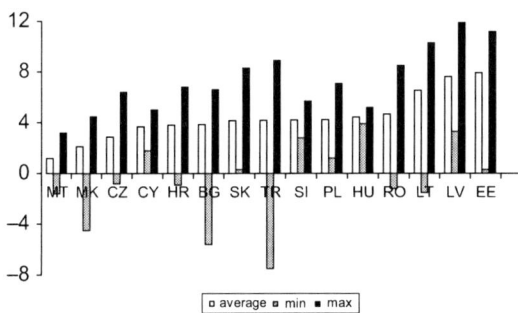

Fig. 3.4 GDP growth in the accession and in the candidate countries in the period 1997–2006. The average growth in white bars, the minimum in grey, and the maximum in black. The data for Malta cover only 2001–2006
Source: Eurostat 2007.

had steam power not been exploited earlier? The answer is that times were not ripe; the scientific culture to support the exploitation of that natural phenomenon had not been elaborate enough. Technological progress is driven by scientific knowledge, and Watt's invention was only possible because of previous scientific research by Galileo, Newton, and the many other founding fathers of modern science. This observation supports the thesis of this book: scientific knowledge and technology are, together with energy, the engines of economic growth. The conditions to achieve competitiveness through technological development are education and research, and—by the way—not only applied research, but also long-term fundamental research.

Coming back to comparisons within the Triad, the author would like to propose a contest in noneconomic terms in order to find a way to show, for once, the EU positioned well ahead of the other two regions. Sometimes Europeans are so proud of their national identities that they underestimate the advantage of presenting themselves united under the European flag. Let us look at the outcome of the Olympic Games. The U.S. is almost always at the top of the Olympic ranking by number of gold medals. In 2000 and 2004 the United States won the highest number of gold medals, followed by Russia, China, and Australia. But if we consider the Union as a whole, the situation is quite different (Table 3.2).

Table 3.2 Gold medals won by the Triad countries in the 2000 and 2004 Olympic Games

	Gold Medals	
	Sydney 2000	Athens 2004
EU	80	82
U.S.	39	35
Japan	5	16

Source: Olympic Committee 2003, 2005.

3.3 Industrial Structure

The European Commission paper, "EU industrial structure—Competitiveness and Economic Reforms" [EC 2007-10] reports that

> the industrial structure of the economy and the distribution of value added across sectors is the result of long-term trends, in particular productivity developments, the increase in the standard of living, changes in the structure of demand and in international trade. The sectoral trends in the EU are characterized by the dynamism of market services, which record growth rates higher than the economy as a whole. Services industries—market and non-market—account for 71 % of total value added in EU-25, while the share of manufacturing amounts to less than one fifth (18.3 %). The rest is divided among utilities and construction (all together 7.7 %), mining (0.8 %), and agriculture and fishing (2.2 %).

Over 80 % of value added in manufacturing comes from large enterprises, while SMEs dominate in the services sector. There is a correlation between country size and sectoral specialization: large countries are diversified and balanced in the sectoral distribution of economic activities, while smaller countries have specialized production structures. Countries can be classified by the different levels of labour skills required by the activities in which they are specialized, as displayed in Table 3.3, where some countries may appear twice when their specialization is in two distinct labour-skills groupings. Countries under the heading *Balanced* have an industrial structure which accommodates the four levels of labour skills. Accession countries are concentrated in the "low" and "low-intermediate" columns.

The following sectors are characterized by low labour skills and account for the highest share of EU exports (34 %): food, drink and tobacco, textiles, clothing, leather and footwear, rubber & plastics, non-metallic mineral products, basic metals, motor vehicles, furniture, miscellaneous manufacturing, recycling. Exports of high

Table 3.3 Labour-skills specialization in EU257. AC in *italic*

High	High-Intermediate	Low-Intermediate	Low	Balanced
BE	DK	AT	AT	DE
FR	FI	CZ	CZ	*HU*
IE	*MT*	DK	EE	NL
LU	SE	*EE*	EL	UK
		EL	ES	
		ES	IT	
		FI	*LT*	
		IT	*LV*	
		LT	*MT*	
		LV	PL	
		PL	PT	
		PT	*SI*	
		SE	*SK*	
		SI		
		SK		

Source: European Commission [EC 2007–10].

labour-skills products also account for a significant (27 %) part of EU sales abroad. More than half of EU trade with low income countries is in products with low levels of labour skills, while in trade with low-medium and upper-medium income countries, the share of products involving high-labour-skills is higher.

The globalization of trade, the increasing trade from emerging economies, and the opening of EU trade beyond the Union's borders have reduced the relative share of intra-EU trade that, however, still accounted in 2005 for as much as 59 % of EU manufacturing exports. In the sectors of textiles, clothing, leather, and footwear, EU trade was particularly affected. The integration of the EU market is visible via the increase of intra-EU15 Foreign Direct Investments (FDIs), whose stock rose in

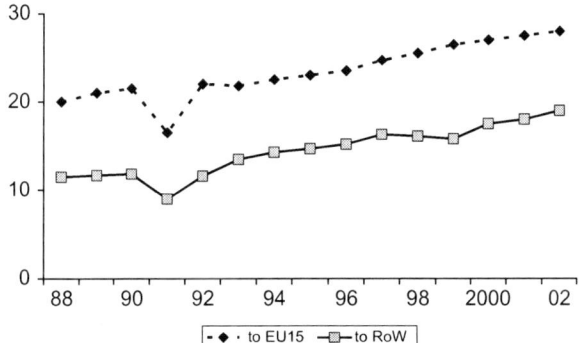

Fig. 3.5 Manufacturing export in the EU15: Intra-EU and to the rest of the World between 1988 and 2002
Source: European Commission [EC 2007-10].

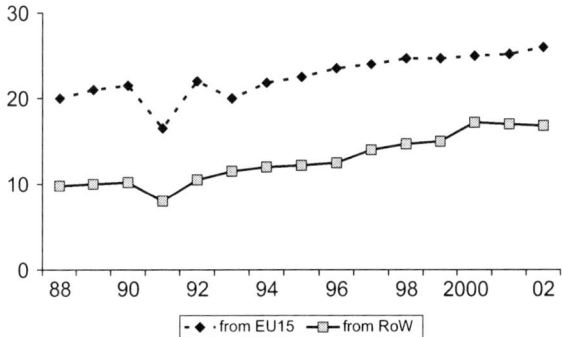

Fig. 3.6 Manufacturing import in the EU15: Intra-EU and from the rest of the World between 1988 and 2002
Source: European Commission [EC 2007-10].

2006, to 52 % of total foreign investments, up from 46 % in 1995 [EC 2007-10]. As a result of combined effects of the internal market legislation and of globalization, trade exchanges among EU countries have intensified between 1988 and 2002 and international trade grew as a result of increasing openness of European manufacturing, as shown in Figs. 3.5 and 3.6. The discontinuity in 1991 should be associated with the temporary slow down of German economy in the transition phase towards the unification of the Federal Republic and the Democratic Republic.

3.4 Shares in the High-Tech Market

To identify the sectors in which the European economy is most competitive, we shall look at the 15 sectors that have registered the highest market shares in each of the three regions of the Triad. In the U.S., the highest market shares belong to high-tech industries, whereas in the European Union the highest market shares are concentrated in other sectors. In Tables 3.4 and 3.5, we present a list of the 15 industries in which the set of all enterprises from in each region of the Triad have the largest world market share. High-tech industries are shown in Table 3.4 and low-tech industries in Table 3.5.

Each of the columns in the two Tables 3.4 and 3.5 has 15 stars, differently partitioned among the two tables: there are more U.S. or Japanese stars than EU stars in the high-tech table, as summarized in Table 3.6. The specialization of the EU in low-tech makes her more vulnerable to competition from emerging economies.

Table 3.4 High-tech industries among the 15 industries in which the set of all enterprises from in each region of the Triad have the largest world market share

High-tech	U.S.	JP	EU
Aerospace	*		
Electrical appliances		*	
Medical equipment	*		
Weapons	*		
Audiovisual		*	
Vehicles			*
Electronic components	*	*	
Steam generators			*
Media			*
Motorcycles/bicycles		*	
Vehicle engines	*		
Watches		*	
Vehicle parts		*	
Precision instruments	*		
Optical instruments	*		
Trains			*

Source: European Commission [EC 1998-3].

Table 3.5 Low-tech industries among the 15 industries in which the set of all enterprises from in each region of the Triad have the largest world market share

Low-tech	U.S.	JP	EU
Clothing	*		
Leather clothing			*
Agro-Chemical	*		
Domestic appliances		*	
Sport equipment	*		
Batteries		*	
Paper	*		
Cables			*
Cement			*
Ceramics			*
Knives and tools	*		
Textile fibres			*
Leather manufacture			*
Wood manufacture		*	
Construction materials			*
Minerals	*		
Fur			*
Fishery		*	
Stones		*	
Mill products	*		
Metal products		*	
Shoes			*
Musical instruments		*	
Tobacco	*		
Textile		*	*
Tissues		*	
Suitcases			*

Source: European Commission [EC 1998-3].

Table 3.6 Position of the 15 sectors with the largest world market share, with respect to their high-tech or low-tech content

Top market share sectors vs. high-/low-tech			
SECTORS	U.S.	JP	EU
High-tech	7	6	4
Low-tech	8	9	11

Source: European Commission [EC 1998-3].

3.5 Industrial Specialization

To study industrial specialization in the Union, we have identified the industries that have generated the highest percentages of added value in each EU15 state. Table 3.7 contains a list of the five major industries that have generated the highest percentages of added value. High-tech industries are in *italics*. It should be noted that Belgium,

Table 3.7 Specialization in the European Union: the first five industries that, in each country, have generated the highest percentages in added value. High-tech industries in *italics*

Specialization in the European Union	
Austria	Knitted crocheted fabrics; sport goods; *machine-tools;* made-up textile articles; *TV, radio recording apparatus*
Belgium and Luxembourg	Jewellery; other textiles; made-up textile articles; first processing of iron and steel; glass and glass product
Denmark	*Toys;* fish; meat; *ships and boats;* other transport equipment
Finland	Pulp paper; sawmilling; wood and panels of wood; leather clothes; *TV, radio, telephony*
France	*Steam generators; watches;* wooden containers; sport goods; detergents and perfumes
Germany	*Electricity distribution and control; industrial process control equipment; motor vehicles; machine-tools; electrical equipment*
Greece	Cement lime and plaster; textile fibres; tanning and dressing of leather; other wearing apparel; knitted crocheted fabrics
Ireland	*Recorded media;* jewellery; *medical equipment;* other chemical products; *office machinery and computers*
Italy	Ceramic tiles; tanning and dressing of leather; leather clothes; *motorcycles and cycles;* luggage bags and footwear
The Netherlands	*TV and radio recording apparatus; man-made fibres; Recorded media;* animal feed; vegetable and animal oils and fats
Portugal	Articles of wood and cork; footwear; knitted crocheted fabrics; other wearing apparel; tanning and dressing of leather
Spain	Fur; cutting and shaping stone; ceramic tiles; vegetable and animal oils and fats; cement, lime, and plaster
Sweden	Sawmilling; pulp and paper; *weapons and ammunitions; TV, radio, and telephony;* first processing of iron and steel
United Kingdom	Agro-chemical products; *aircraft and spacecraft;* grain mill products; processed fruits and vegetables; *ships and boats*

Source: European Commission [EC 1998-3].

Greece, Italy, Portugal, and Spain, unlike Germany, Ireland, or The Netherlands, are characterized by a specialization in low-tech industries.

3.6 Emerging Economies

In this section, we analyse five emerging economies among the countries with the largest GDP in the world. In Section 3.13, "Foreign Trade and Competition with Emerging Economies", we will analyse how their economies affects EU international trade. As Napoleon had anticipated, the waking up of China has shaken the whole world. Fifty years from now, the world's distribution of wealth could be very different from the pie chart on the right in Fig. 3.2, and Asia could have regained the leading position it used to have in the early 19th century [Wolf 2003].

Some industrial sectors in Europe are more vulnerable than others and may be severely affected by the competition from emerging economies. However, we claim

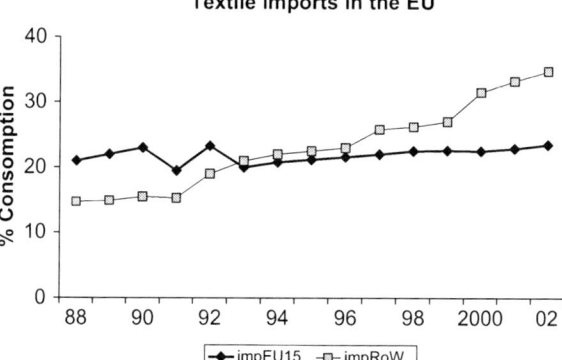

Fig. 3.7 Textile imports (as a percentage of consumption) in the EU from intra EU trade and from the rest of the world in 1988–2002
Source: European Commission [EC 2007-10].

that there is a way for Europe to face the challenge by investing on quality products and processes. Textile is for sure a case in point: there is no way to compete on price with cheap textile products imported from Asia; Asian import has been growing, and will continue to grow (see Fig. 3.7). There is, however an intra-EU trade for the top-of-range products that is not declining because it answers to a demand for fashion and design that the Asian products do not address. Exports within Europe of textile products of European manufacture grew during the period, as well as exports to the rest of the World, despite of the aggressive price competition from emerging economies (see Fig. 3.8). We will expand on this case in Section 3.12, "EU Foreign Trade in the Textile Industry".

Fig. 3.8 Textile exports (as a percentage of production) from the EU15 countries intra EU and to the rest of the world in 1988–2002
Source: European Commission [EC 2007-10].

A group of emerging economies is currently referred to as the BRIC, from the initials of Brazil, Russia, India, and China. Together, the latter two countries accounted for 38 % of the world population and 17 % of its GDP ppp in the year 2004 (while the Triad's population was 13 % and her GDP ppp 49 %). In this section, we shall cover the BRIC*K* economies—namely BRIC and (South-) Korea—and in particular the two major ones, i.e. China and India. In the Section 3.13, "Foreign Trade and Competition with Emerging Economies", we will expand on EU trade with the BRIC*K* economies.

3.6.1 Chindia

Chindia is a neologism to address the two emerging gigantic economies in Asia. They are perceived in the western world as one challenge, but they have more elements of differentiation than commonalities. Justin Yifu Lin, an economist from Beijing University, quoted by John Tornhill [Tornhill 2002], states that *China* has a GDP growth potential ranging from 7 to 8 %, which is going to last for the next 20 or 30 years. A study conducted by Goldman Sachs in 2003 [Goldman Sachs 2003] supports Yifu Lin's estimates and forecasts a 3.5 % growth rate in 2040. As for *India*, according to the paper by Goldman Sachs, and confirmed by *The Economist* [Economist 2004], this country has a 5 % growth potential that is expected to last for the following 50 years. Such GDP growth rates are far higher than population growth rates in the respective countries and are therefore responsible for an increase in the pc GDP.

On the basis of the assumptions presented in the above-mentioned paper, the Goldman Sachs economists have calculated the spectacular growth experienced by the BRICs and have made an estimate, in 2003, of the GDP in 2050, expressed in 2003 U.S. dollars of the largest economies in the world (see Fig. 3.9). In Fig. 3.10, this estimate is compared with the GDP estimate for the G6 countries (the U.S., Japan, Germany, France, the UK and Italy, namely, the G7[3] excluding Canada). The BRICs are expected to overtake the G6 in 2040, while China is expected to overtake the U.S. in 2041, after having overtaken all the other countries in 2016. India's economy is projected to overtake Japan's economy in 2032. If G7 meetings will still be called in 2050, presumably the countries invited will be, in descending order of GDP, China, U.S., India, Japan, Brazil, Russia, and the UK [Balls 2003]. The countries from the present Euro-zone, if considered separately, would not be present at the G7 meetings in the year 2050.

---

[3] The heads of state or government of the major industrial democracies have been meeting annually since 1975 to deal with the major economic and political issues facing their domestic societies and the international community as a whole. The six countries at the first summit, held at Rambouillet, France, in November 1975, were France, Germany, Italy, Japan, the United Kingdom, and the United States. They were joined by Canada at the 1976 Summit. Starting with the 1994 Summit, the G7 met with Russia. The 1998 Summit saw full Russian participation, giving birth to the G8.

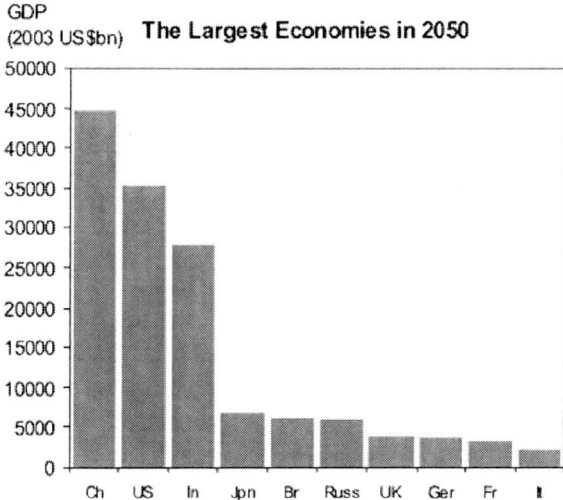

Fig. 3.9 The largest economies in the world: An estimate for 2050. In this figure, *Ch* stands for China
Source: [Goldman Sachs 2003].

The Goldman Sachs paper estimates are based on the assumption that the BRICs will maintain their economic growth in a politically stable environment. This assumption is more realistic for India than for China. India can claim to be the world's largest and most populated democratic country. Recent public

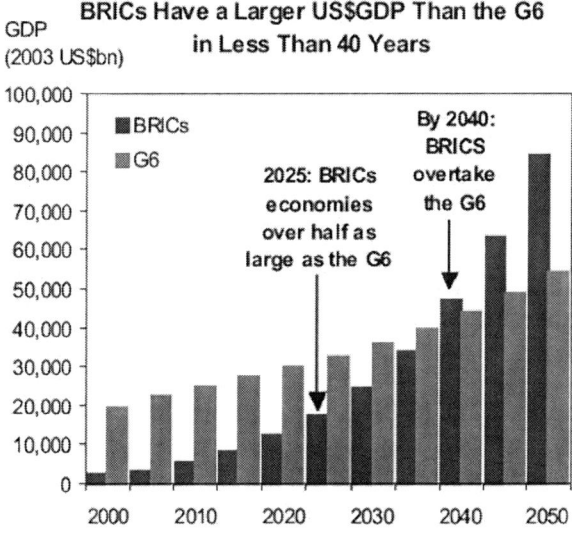

Fig. 3.10 Emerging BRIC economies could overtake G6 in 40 years. BRIC stands for Brazil, Russia, India, and China; G6 stands for the U.S., Japan, Germany, France, the UK and Italy
Source: [Goldman Sachs 2003].

administration reforms are virtually dismantling the caste system, which had already been outlawed after independence. Such reforms introduced a quota system [Kynge 2003] ensuring a 50% participation in public life of citizens from lower social classes. In China, the ruling communist regime uses authoritarian means to impose its social and economic views. This method is certainly quicker than the one enforced by the Indian democracy, but it leads to social malaise. If one day the awakening of China's economy is matched by an awakening of freedom and democracy, we could witness a period of transition characterized by unrest, political instability, and unsteadiness in the country's economic development.

In addition to 60 years of democratic ruling, there is another factor in favour of India in terms of a predictably high chance of enjoying a continued growth for several decades: the Indian society is founded on the deep roots of a millenary tradition, still nurtured by the Indian people. Both India and China were the cradle of civilizations that prospered from time immemorial, several millennia BC. While India developed with continuity from her past and maintained her traditions despite centuries of British ruling, China during the Mao Zedong revolution repudiated her glorious past and destroyed most of her cultural heritage depriving the country of an asset that could have contributed to the development of the Chinese society.

The forecasts of the Goldman Sachs paper are presented as dreams. However, they are less unrealistic than they might sound. For example, already in 2003, Volkswagen made a joint venture in China thanks to which it is selling more cars on the Chinese market than in Germany [Burgess 2003], and it has even started producing cars in China not only for the Chinese market but also for the Australian market [McGregor 2003]. China is sustaining its exports with regulations limiting the number of cars that can be registered in the national market. DaimlerChrysler is reported [Dyer 2005] to be considering plans to set up a production base in China, with a view to producing a subcompact to be exported to the U.S., while Chery, a Chinese manufacturer, is reported to be planning to build cars in Europe.

In 2007, Eurostat reported that China became the first European supplier: the ratio of Chinese to U.S. exports to the EU grew from 46% in 1996 to 109% in 2006. Chinese exports to the EU grew by 310% from 1996 to 2006, up to € 191 billion and the first sector in Chinese export is electro-mechanical products; in the same period the U.S. exports to the EU grew by 75%. [Rampini 2007].

We should not make the mistake of thinking that in China changes occur at the same pace as in EU countries. The speed of change in China is simply striking and western visitors are amazed when only six months or a year after their previous visit they are no longer able to recognize the urban environment and skyline. The Beijing Olympic Games of 2008 have triggered a rapid and profound transformation of the capital city's infrastructures: the seventh ring around the capital city was under construction in 2006. In addition, the high GDP growth in China generates a domestic demand that will soon be entirely satisfied by the Chinese production system, except for top quality or very advanced technology products and services. The huge Chinese investments may lead to a substantial change in the balance of trade with the West, even in the short term, that is, before 2009.

3.6.2 Brazil

Characterized by large and well-developed agricultural, mining, manufacturing, and service sectors, *Brazil*'s economy outweighs that of all other South American countries and is expanding its presence in world markets. Brazil in the 1960s and 1970s was the second-fastest-growing large economy. At the beginning of the new century, Brazil growth was slow compared to most developing countries; however since 2004, Brazil has enjoyed enough growth to yield increases in employment and real wages. In some ways, Brazil is the steadiest of the BRICs. Unlike China and Russia, it is a full-blooded democracy; unlike India it has no serious disputes with its neighbours. It is the only BRIC without a nuclear bomb. The Heritage Foundation's "Economic Freedom Index", which measures such factors as protection of property rights and free trade, ranks Brazil ("moderately free") above the other BRICs ("mostly unfree"). One of the main reasons why Brazil's growth has been slower than China's and India's is that Brazil is richer and more urbanized. Government tax revenues are at 33 % of GDP in 2006, about twice as much as Russia or China. Public spending is not very efficient; pensions are granted also to retired shadow-economy-workers who did not contribute to any pension scheme and there are redundant and inefficient civil servants protected from dismissal until a 2005, when a reform cracked down on nepotism. After hyperinflation in the last decades of the past century, curb on inflation since 2003 has changed lifestyles and was most beneficial to the national economy. The yield of Brazilian bonds is now only two percentage points higher than the U.S. one, down from 24 % in 2002 [WorldFactbook 2007, Economist 2007-21].

The climate in Brazil is supportive of profitable development of agricultural products, including—from the 1970s oil crisis—biofuel from sugar cane as a replacement for gasoline: Brazil can rely on ethanol for 40 % of the internal demand for fuel for vehicles, and exports 4 billion litres, half of world exports. Brazil is rich in raw materials and the growing international demand, in particular from China, is pushing Brazilian exports up. This compensate for increasing manufacturing import for a growing domestic demand. The long term development of the Brazilian economy depends on Brazil commitment to invest a GDP percentage in R&D higher than the present 1 % and on involving industry as a major research partner to universities.

3.6.3 Russia

Russia ended 2006 with its eighth straight year of growth, averaging 6.7 % annually since the financial crisis of 1998. Although high oil prices and a relatively cheap ruble initially drove this growth, since 2003 consumer demand and, more recently, investments have played a significant role. GDP growth will continue to be boosted by the high oil price. Until the beginning of the 1990s, Russian economy was centrally planned; today the transition to market economy is still incomplete, with visible market distortions. A change will result of Russia joining the WTO, which could be within 2008. Over the last five years, fixed capital investments have averaged real gains greater than 10 % per year and personal incomes have achieved real gains more than 12 % per year. During this time, poverty has declined steadily and the middle class has continued to expand, increasing the internal demand. The wealth of Russia comes from capital intensive industries with limited employment, contrary to the manufacturing economy of China that involves millions of new workers. For this reason, the Russian GDP growth has not generated a financial benefit to the majority of the population and the revenue ratio between the high income earners and the low income groups has increased. Another major difference with China is in demography: population is declining by 0.46 % in Russia while it is growing by 0.65 % in China after the end of the one-child-per-family policy of the 1980s. Despite Russia's recent success, serious problems persist. Oil, natural gas, metals, and timber account for more than 80 % of exports and 32 % of government revenues, leaving the country vulnerable to swings in world commodity prices. Russia's manufacturing base is dilapidated and must be replaced or modernized if the country is to achieve broad-based economic growth and be competitive in the global market. Next year, 2008, is the end of the second term of presidency for Mr. Putin, who cannot be reelected as president but might still be in the limelight as prime minister. However the transition might create instability, turbulence, and unrest affecting negatively the economic growth [WorldFactbook 2007, Economist 2007-23].

Europe has officially sought a "strategic partnership" with Russia since 1999, but this quest has failed, driven by Russia's recent behaviour. *The Economist* [Economist 2007-22] reports:

> Encouraged by its ability to divide Europe, Russia has expanded its bilateral disputes. The original list was dominated by ex-Soviet satellites (rows over Polish meat exports, Lithuanian oil imports, the siting of an Estonian war memorial, and Czech and Polish plans to host an American anti-missile defense system). But Russia is now growling at old Europe, too. Ask Sweden and Italy (trade squabbles), Spain (a Russian spying scandal), and Britain, locked in the fallout from the radioactive poisoning in London of Alexander Litvinenko.

3.6.4 Korea

Since the 1960s, *South Korea* has achieved an incredible record of growth and integration into the high-tech modern world economy. For this reason in this book we often speak of BRIC*K*s instead of BRICs. Four decades ago, GDP per capita was comparable with levels in the poorer countries of Africa and Asia. In 2004, South Korea joined the trillion dollar club of world economies, ranking 11th in 2006, followed by Canada, Mexico, and Spain. Today her budget is positive and her GDP per capita is equal to that of the lesser economies of the EU. This success was achieved by a system of close government/business ties, including directed credit, import restrictions, sponsorship of specific industries, a strong labour effort, and very high investments in R&D that have been growing at 16.5 % a year from 1977 to 2000, reaching 40 % in 2004 [Mahlich et al., eds. 2007]. The government promoted the import of raw materials and technology at the expense of consumer goods and encouraged savings and investment over consumption. GDP growth averaged 4.8 % in 2002–2006 and was 5.1 % in 2006; moderate inflation (2.2 % in 2006), low unemployment, an export surplus, and fairly equal distribution of income characterize this solid economy [WorldFactbook 2007, Economist 2007-24].

Korea raised her competitiveness in a few sectors to world level and penetrated western markets with electronics and automobile products. The national champion is Samsung. R&D in Korea rose from 0.31 % of GDP in 1972 to 2.85 % in 2005, a figure comparable to the percentage in the U.S., and higher than that of Germany, France, and the UK. Industry accounts for 76.7 % of the R&D expenditure of which 40 % is coming from the five largest firms. The Korean export in the OECD electronic industry market is 12.5 % and in the OECD computer industry is 8.6 % [Mahlich et al. eds. 2007].

3.6.5 The BRICKs: Five Quite Different Stories

We will present in this section some features of the five BRICK countries to show similarities and differences among them. The strong points of the Indian economy are as follows:

- a democratic tradition guaranteeing political stability also in the future;
- a well-developed private sector, offering good opportunities to foreign operators wanting to set up associations and partnerships;
- a legal system modeled on that of Great Britain, providing a context of legal certainties to foreign operators;
- a financial system and a stock exchange market that is more developed than in China;
- international competitiveness in different sectors, from information technology to media, entertainment, company management contracts, and manufacturing of car components;
- a qualified workforce speaking correct English;
- 30 years of accelerated pc GDP growth; and
- a demographic situation characterized by a continuous growth of the working-age population up until 2020.

The strong points of the Chinese economy are as follows:

- an almost unlimited resource of labour at a low cost;
- public administration capable of acting effectively and quickly;
- the capacity to accomplish structural and infrastructural reorganizations in shorter timeframes than in western industrial economies;
- a huge domestic market and an expanding demand;
- GDP growth rates higher than population growth rates;
- the capacity to attract offshore companies and rapidly acquire know-how;
- a fiscal policy that favours, supports, and stimulates business creation and development;

- public administration aware of the problems linked to urban development and environmental control;
- a large public works scheme resulting in new infrastructures and employment creation;
- sustained economic growth expected to continue for at least the next 10 years and exceeding 3 % in the long term; and
- huge cash reserves in the National Bank.

Within the BRIC*K*s, the situation is very different in terms of GDP growth (see Fig. 3.11), and pc GDP (see Figs. 3.14 and 3.15), as well as to the economic structure, the contribution of different sectors to GDP creation, and the rapid change (see Fig. 3.16) in GDP structure.

GDP growth rates in India are lower than in China, but according to the Goldman Sachs model, long-term growth sustainability is expected to improve in India and be around 5 % until 2050. India's GDP is expected to overtake Italy's GDP in 2016, and in 30 years, India could rank third, after China and the U.S. Figure 3.12 shows that in India the services sector has registered a much higher GDP growth rate than the other sectors. Service does not mean only ICT; first-class advanced health care services are also offered in India to wealthy Indians and to wealthy foreigners at budget prices in comparison with what the same treatment would cost in the Triad. The growth of the Indian software and services industry is shown in Fig. 3.13. According to a study conducted by Nasscom and McKinsey, reported by *The Economist* [Economist 2004-1], the Information Technology sector in India will grow by 34 % every year until 2008, when it will total $ 77 billion and account for 30 % of total exports. Pasquale Pistorio told *La Repubblica,* when he was chairman and managing director of ST-Microelectronics (STM)—the Italo-French giant and world-leading manufacturer of microprocessors—that STM is increasing its investment in Noida, India—near Delhi, where 1,400 microprocessor designers and programmers are presently working. Another centre will be opened in Bangalore with

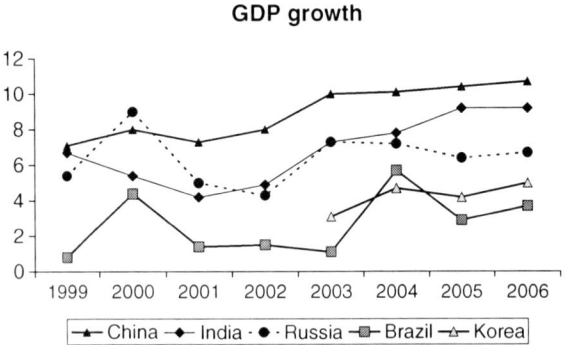

GDP growth

Fig. 3.11 GDP growth (percentage) in the BRIC*K*s and Korea from 1999 to 2006. BRIC*K* stands for Brazil, Russia, India, China, and Korea
Source: European Commission (2007).

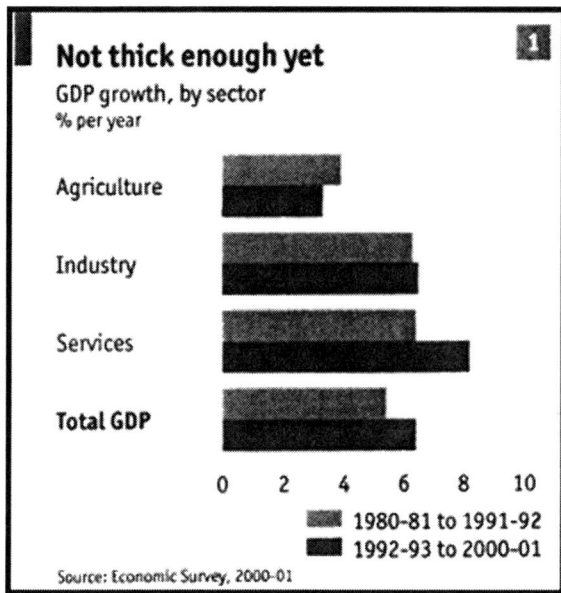

Fig. 3.12 GDP annual growth in India in the three sectors during 1980–1991 (*upper bars, dark grey*) and 1992–2000 (*lower bars, black*)
Source: [Economist 2001-6] "The Plot Thickens" © *The Economist vol.* 359, n. 8224 (2 June 2001), reproduced with kind permission.

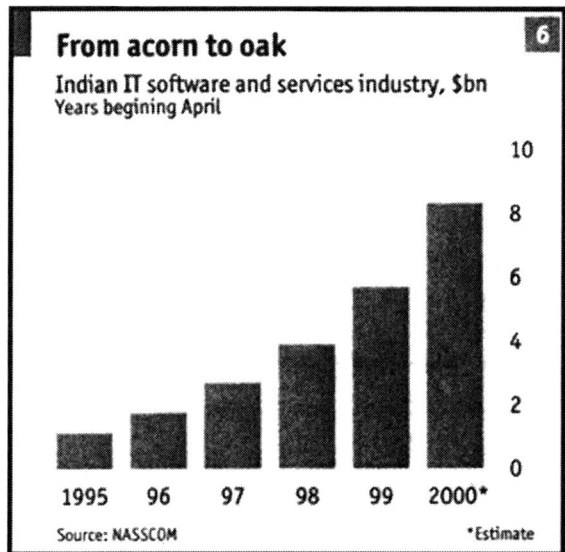

Fig. 3.13 Growth of the value of software and services in India from 1995 to 2000 (billion dollars)
Source: [Economist 2001-7] "Islands of Quality" ©*The Economist vol.* 359, n. 8224 (2 June 2001), reproduced with kind permission.

100 engineers to begin with. Then, by 2010, Pistorio expects to have 5,000 STM employees working in India, a country in which STM has been present since the 1980s [Tessa 2004].

The BRICs have a far lower pc GDP (see bars in Fig. 3.14, left scale) than the European Union or the Triad. In 2002, India's pc GDP equalled nearly 3 % of Italy's pc GDP, and China's pc GDP was almost double this value. Brazil and Russia had higher pc GDPs, namely, about 15 % of Italy's pc GDP. GDP growth rates are far higher than population growth rates and, consequently, according to the Goldman Sachs model, a significant increase in pc GDP is expected in the BRICs. For example, China is expected to soar from 2.4 % of U.S. pc GDP in 2000 to 37 % in 2050. India is expected to move—in the same period—from 1.3 % to 15 % of U.S. pc GDP.

The change in the ranking of the richest countries only applies to absolute GDP and not to pc GDP, which, even though it grows strongly in the BRICKs, will stay well below the G7 values in the foreseeable future, except perhaps for Korea. Supposedly, in 2050 the world economy will be very different from today's economy, at least from two perspectives. First, the group of the six countries with the highest GDP will be substantially different, as four new countries are expected to join the group of the six richest nations in the world. Second, while today the countries with the highest GDP are also among those with the world's highest pc GDP, presumably in 2050 this will only be the case for the U.S. and Japan. Third, the production of wealth in low pc GDP countries may change the demand in world market, privileging affordable products with good enough quality (see next Chapter 4: "Competitiveness and Quality").

The BRICKs, particularly India and China, are also very different from each other in terms of the ratio between foreign trade and GDP (see Figs. 3.14 and 3.15).

2002

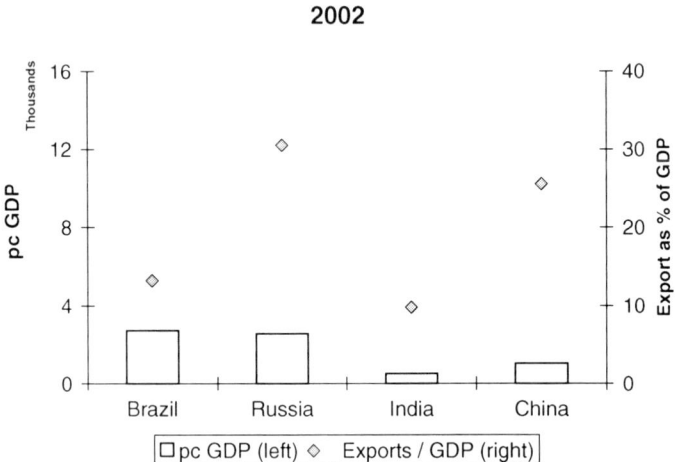

Fig. 3.14 pc GDP (*white bars*, scale on the left, thousand euro) and foreign trade as a percentage of GDP (*grey diamonds*, scale on the right) in the BRICs during 2002
Source: European Commission (2004).

China's and Korea's economies are more oriented towards the rest of the world (59 % of Chinese exports originate from foreign investments in China, mainly from the U.S., EU, and Japan [Rampini 2005-2]), and so the exports-to-GDP ratio is almost three times higher than that of the Indian economy. China's balance of trade has been favourable for 10 years. The national office for statistics reports exports for $ 507 billion in 2001 and a favourable balance of trade of $ 22 billion. At present, China's exports to western countries include mainly ironmongery, furniture, textiles, and industrial products (for example, electrical appliances and components), with an increasing presence of technological commodities, as witnessed by the sales of the IBM PC business for $ 1.7 billion to Lenovo, China's leading manufacturer of personal computers [Economist 2005-5]. A number of leading industries in China were originated by joint ventures with advanced technology firms in Taiwan. A case in point is Huawei, operating in the area of 3G mobile telephony, and 22 thousand employees strong, 46 % of whom are engineers. From its premises in Shenzen, Huawei is expanding throughout the world and is penetrating Europe with subsidiaries in Russia, The Netherlands, and Portugal. Another example is the electronic components company Foxconn that delivers mobile telephony components to Nokia and Motorola from Shenzen: Foxconn is present in Germany, Italy, The Netherlands, and the UK [Gerino 2005]. It is to be expected that, in the medium term, China will be on the world market also in other industrial segments. China's competitiveness is sustained by the low cost of labour and—increasingly—by *good enough* quality (we will expand on this point in Chapter 4, "Competitiveness and Quality"), to the point that a motto is spreading around Chinese technological commodities: *It's cheap, it works, it's Chinese.* Chinese economic growth is supported

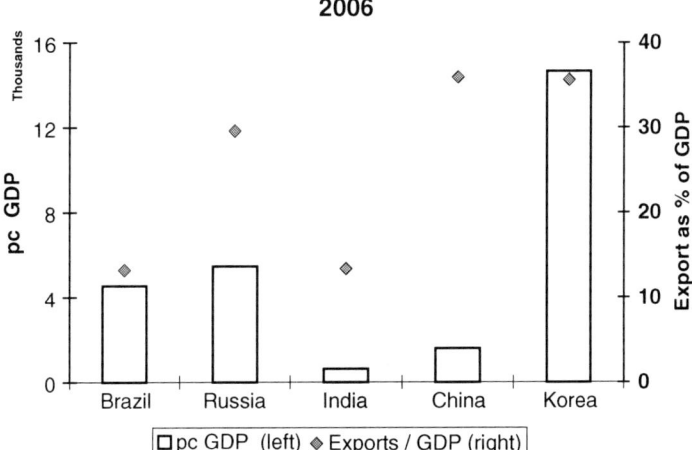

Fig. 3.15 pc GDP (*white bars*, scale on the left, thousand euro) and foreign trade as a percentage of GDP (*grey diamonds*, scale on the right) in the BRICKs during 2006
Source: European Commission (2007).

Changing Structure of GDP in the BRICKs

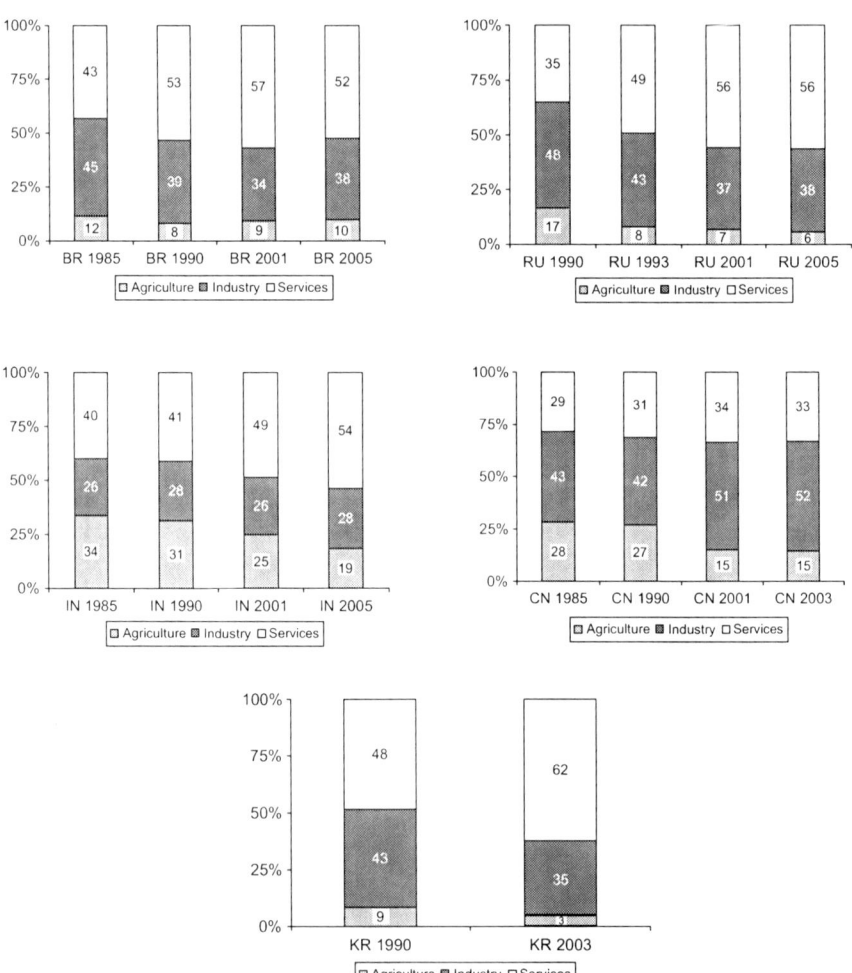

Fig. 3.16 The different and changing structure of GDP in the BRIC*K* countries from 1985 to 2005. In all countries except Brazil the percentage of GDP originated in agriculture is reduced over the years: in Brazil a non negligible percentage of agricultural activities are supporting the production of fuel. China has a high and increasing GDP share in industry, while in India the reduction of the agricultural share is compensated by an increase in services. South Korea's services percentage is approaching the EU value: the services sector in the European Union accounts for 69 % of GDP, followed by industry with 28 % of GDP, and agriculture with 2.3 % of GDP
Source: European Commission, 2004, 2007.

also by a huge public works scheme, business incentives, tolerance towards tax evasion, and growing domestic demand.

As regards competitiveness of European products on the Chinese market, investments on research in China are not expected to cause a crowding-out of the European high-quality products in the medium term (before 2009); however, in the same period, joint ventures and foreign investments from neighbouring R&D intensive economies may change the picture. Conversely, China is seeing a growing domestic demand for commodities (domestic appliances, clothing, and means of transport) and is developing the right industrial capacity to satisfy such a demand. In this way, China could become self-sufficient even before 2009 in the production of electric appliances, clothing, and budget cars.

In January 2004, according to WTO agreements, quotas on imports from China into western economies were removed, with dramatic effects in the U.S. and in Europe. Trade unions are asking the Administration to implement remedial actions. The problem is exacerbated by the significant amount of counterfeited western products manufactured in China and sold worldwide. Some of the Chinese dumping is generated by delocalized western industries, which take advantage of lower labour costs. An interesting (and somehow utopian) solution was proposed by Valeria Fedeli, an Italian textile trade unions officer, who suggested that western enterprises operating in China should apply western workers' rights standards to the Chinese workforce [Rampini 2005-1]. This voice is not isolated: western consumers' pressures have induced the multinational U.S.-based shoe manufacturer Nike to adopt a transparency policy and identify cases (and take remedial action) where its subcontractors around the world fail to respect basic workers' rights [Nike 2005].

Overall, the BRICK agriculture share of GDP has registered a reduction in the last decades (see Fig. 3.16). In India, Brazil, and Russia, the services percentage has increased, while China is the only country that has seen a substantial increase in industry's share. These changes are also triggered by the availability of top-quality, competitively priced services in India (call centres, software development, other ICT services, health care) and, in China, by the fact that many offshore companies are moving their industrial production to that part of the world because of the low cost of labour and the large dimensions of the internal market.

More will be said about the BRICKs in Section 3.13, "Foreign Trade and Competition with Emerging Economies", where we will also be drawing some conclusions on EU competition with the BRICKs today and in the long term.

3.7   Comparing Productivity in the Triad

In Chapter 2, "Overview of member states", we have compared data relating to population, surface area, and foreign trade in the European Union and in the Triad. Let us now make a comparison based on productivity.

Productivity in the Triad

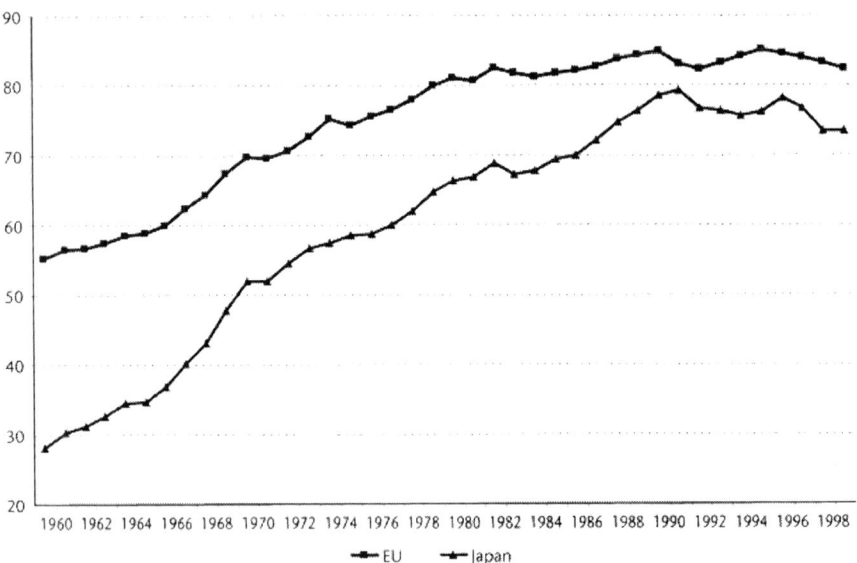

Fig. 3.17 Productivity in the EU and Japan versus U.S. (= 100). Productivity is calculated as GDP per person employed
Source: European Commission [EC 2000-8].

Labour productivity is defined as GDP per person in employment. In the period from 1975 to 1990, in the European Union and even more so in Japan, labour productivity (see Fig. 3.17) approached that of the United States, as it rose from 55 to 83 % (EU) and from 28 to 80 % (Japan) of the U.S. value. The relentless growth of the U.S. economy in the 1990s stopped and reversed such recovery, and in 2001 labour productivity was 78 % of the U.S. value in the European Union and 67 % in Japan [EC 2002-2]. The latter half of the 1990s showed a decline of labour productivity growth in Europe while growth in the U.S. accelerated. This may suggest Europe has not exploited the productivity effects from ICT to the same degree as the United States. In general, European countries are heavily investing in ICT equipment, but the ICT share in total investment in the EU remains considerably lower than in the United States [van Ark et al. 2003].

The decline of labour productivity in EU15 leads to her deindustrialization and drives entire industries to low-cost/high-tech countries in eastern Europe and in Asia. As examples, 70 % of Siemens factories are outside Germany; Volkswagen manufactures more cars in its Chinese joint venture than in Germany; 20,000 out of 122,000 Philips employees are in China, and not just for cheap labour, since 900 of them are R&D scientists and engineers [Muldur et al. 2006].

The pc GDP in the European Union and in Japan, as compared with that in the U.S. for the period 1960–1998, follows (see Fig. 3.18) the same trend as labour productivity (see Fig. 3.17).

GDP pc Comparison in the Triad

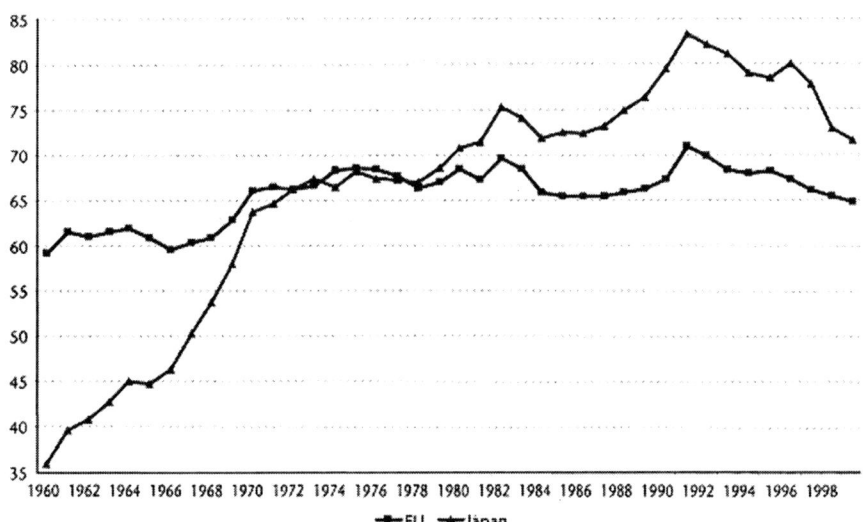

Fig. 3.18 pc GDP ppp in EU (*rectangles*) and Japan (*triangles*), versus U.S. (=100) from 1960 to 1999
Source: European Commission [EC 2000-8].

In the previous chapter, we looked at pc GDPs in the Triad in comparison with the rest of the world (see Fig. 2.20). In the European Union, pc GDP is lower than in the U.S. by one-third and its value relative to U.S. pc GDP is decreasing (see Fig. 3.19).

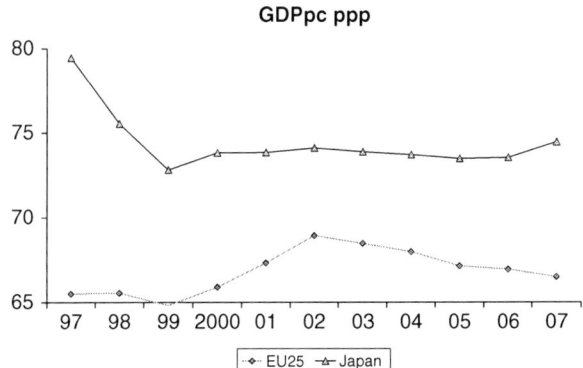

Fig. 3.19 pc GDP ppp in EU (*lower curve, diamonds*) and Japan (*upper curve, triangles*), versus U.S. (=100) from 1997 to 2007
Source: Eurostat 2007.

Participation in the Labor Market: Comparison in the Triad

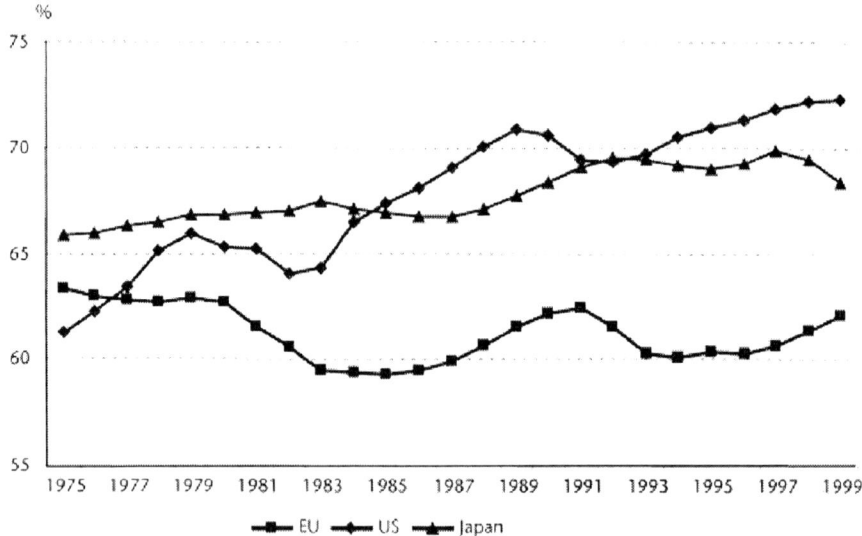

Fig. 3.20 Level of participation in the labour market (% of those employed in working age population) during the period 1975–1999 in the EU, Japan, and the U.S. The EU line is significantly below those of the other regions
Source: European Commission 2002.

Why does the Union have a lower per capita GDP? There are at least two factors to be considered. First, besides lower labour productivity, in Europe the fraction of working-age population in employment (see Fig. 3.20) was 5–10 percentage points lower than that in the U.S. in the period 1980–2000. As a matter of fact, in the U.S. there is no age limit after which one leaves the labour market, and in families there is seldom only one income receiver. Moreover (see Fig. 3.21), the shadow economy (illegally unreported employment) is a small percentage in the U.S., and there are many paid jobs with low professional content that help keep marginalization and unemployment under control. For example, U.S. supermarkets have personnel paid to stay next to the checkout counter to put the purchases in plastic bags, and some staff takes the trolleys outside to the parking lot. In the airports, one can find shoe polishers, a profession not carried out in Europe. The second factor that contributes to a lower pc GDP in Europe is related to the number of hours worked in a year. EU workers in manufacturing work 10% fewer hours than U.S. workers (see Table 3.9). In several EU member states, the number of working hours per week were reduced in the 1990s, under pressures from the unions that expected to reduce unemployment with this measure, which instead reduced competitiveness.

The lower employment level, as well as the lower number of annual hours worked per person in employment, is caused by structural differences and social preferences, which could be hard to overcome. A low level of participation (see Fig. 3.20 and

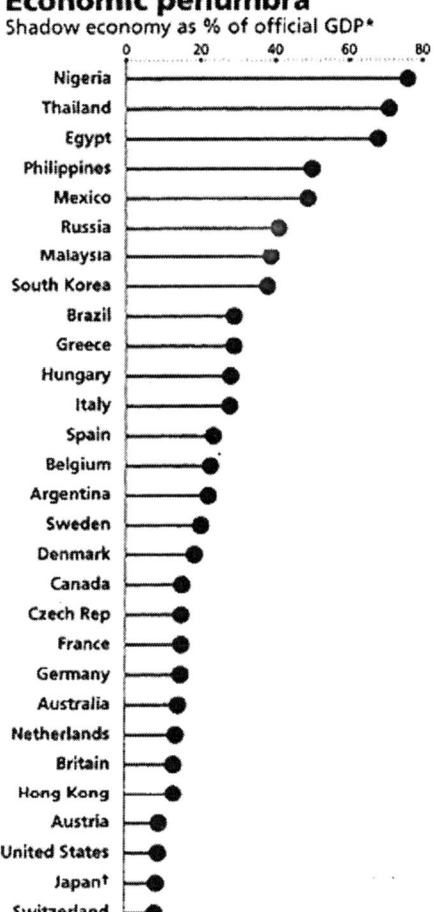

Economic penumbra
Shadow economy as % of official GDP*

Fig. 3.21 Shadow economy in the industrialized countries and in some emerging economies. Among EU25, Greece, Hungary, Italy, Spain, and Belgium are high in the ranking
Source: [Economist 1999-1] "Finance and economics: Black hole" © *The Economist,* vol. 352, n. 8134 (28 August 1999), reproduced with kind permission.

Fig. 3.92) may be an indicator of a higher percentage of shadow economy (see Fig. 3.21). A small rate (lower than 5 %) of shadow economy is physiological and difficult to eradicate. It may also play a positive role in integrating casual or part-time workers into the official job market and in providing a temporary safety margin during short periods of economic crisis. However, it should not be forgotten that illegally unreported employment also means child labour, no social security, labour accidents, deaths on the job, tax evasion, and no investment in research, innova-tion, and lifelong learning. Unfortunately, southern Italy is at the forefront of the shadow economy, and the trade union CGIL estimated a record illegally unreported employment of 40 % in Sicily [Repubblica 2003].

In Section 3.16, "Policy guidelines,", we shall provide more information on the subject of participation in production activities.

3.8 Standard of Living

Is pc GDP a good measure of the standard of living? What else affects the standard of living besides pc GDP? Factors affecting the standard of living include the following:

- purchasing power
- economic stability
- economic growth
- growth sustainability
- employment rate
- percentage of shadow economy
- extent of corruption in public administration
- education and lifelong learning
- quantity of free time
- social security
- social infrastructure
- control over organized crime
- degree of environmental protection

All the above, and more, affect the standard of living: pc GDP alone is not enough to assess a region's living conditions.

Apart from pc GDP, the Union registers some best performances in other fields, although the situation is quite different in each of the various member states. For example, see Table 3.8 concerning environmental pollution. In the European Union, the law favours a balanced relationship between economic growth and the protection of the environment. However, the situation is quite different in the member states in terms of the actual transposition of Community directives on environmental protection in the national legislation, the actual enforcement of the resulting national legislation, and the monitoring of compliance of industrial processes [Sbragia 2000].

Table 3.8 Level of per capita polluting emissions in the Triad

Polluting emissions			
Region	CO_2 tons	SO_x Kg	NO_x Kg
EU15	8.9	31.3	32.7
Japan	9.2	7.2	11.1
U.S.	19.9	63.1	75.1

Source: OECD, IFO cited by the European Commission [EC 1998-3].
Data from the year 1996.

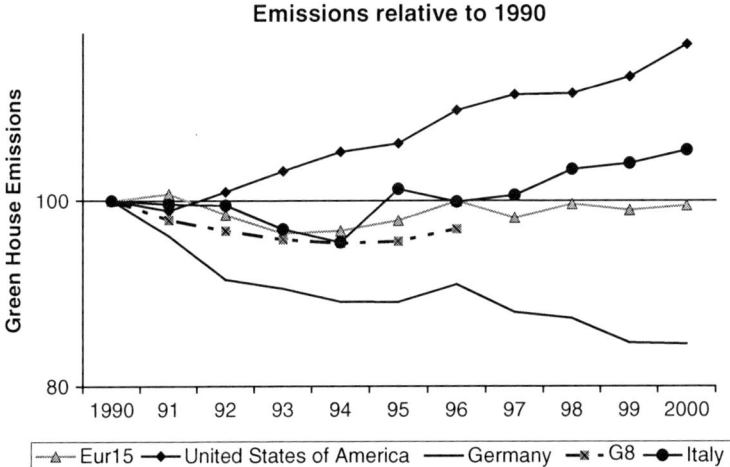

Fig. 3.22 CO_2 emissions relative to 1990 (=100). EU15 stays on the 1990 level thanks to Germany, which reduced emissions, while other countries, among them Italy, exceed—at the end of the century—the 1990 values. Notably, the U.S. has increased emissions from the 1990 values
Source: UN Convention on Climate Change, [UN 2003].

In practice, as regards the reduction of CO_2 emissions, the situation is summed up in Fig. 3.22. In short, the EU15 managed to cut its emissions so as not to exceed the 1990 emission level, whereas, in the same decade, the U.S. increased emissions by 15 %. But this good result of EU15 is not equally shared within Europe, as some countries (Germany, the UK) reduced emissions, while others (such as Greece, Italy, and Spain) increased them. At the United Nations Conference on Climatic Changes, which took place in Milan, Italy, in November 2003, participants took stock of the situation regarding the Kyoto Protocol ratification. It was to be enforced only after the ratification of the protocol by a number of signatory countries from the Convention on Climatic Changes accounting for at least 55 % of CO_2 1990 total emissions. On 26 November 2003, the protocol had been ratified by 87 countries responsible for 44.2 % of CO_2 emissions in 1990.

Among the countries that had not ratified the Kyoto Protocol by November 2003 were the U.S. (responsible for 36.6 % of total CO_2 emissions) and the Russian Confederation (17.4 %). If only one of these two countries had ratified the protocol, the 55 % threshold would have been overcome and the Kyoto protocol would have been enforced. This happened only at the end of September 2004, when the Russian Confederation ratified the protocol, which was then enforced. The current level of ratification can be viewed at the URL indicated in [UN 2003]. Countries not complying with the emission reduction objectives will face severe economic sanctions.

If we continue to compare the European Union with the other two regions of the Triad on the basis of factors other than pc GDP, we will notice a sensible difference in terms of free time (see Table 3.9), relative to workers in the manufacturing industry.

Table 3.9 Hours worked per year and free time for workers in manufacturing in the Triad

Free time

Region	hours/ year	Hours/ week	holidays (days/year)	Bank holidays (days/year)
EU15	1715	38.6	27.7	10.1
Japan	1832	40	18	13
U.S.	1896	40	12	11

Source: OECD, IFO, cited by European Commission, [EC 1998-3]. Data from 1995.

The EU – U.S. gap in terms of annual hours worked per person is equivalent to almost five weeks:

$$1896 - 1715 = 181 \text{ h} =\sim 5 \text{ weeks}$$

More recent data (see Fig 3.23) point to a widening of the gap between the U.S. and the other industrialized countries. The difference in terms of hours worked per person is the most significant factor determining the difference in pc GDP, which is further exacerbated by a lower productivity in the EU.

In conclusion, pc GDP alone cannot be taken as a reference point to establish a region's living conditions, nor should any generalizations be made by equating different elements belonging to a far-reaching and complex system. However, from

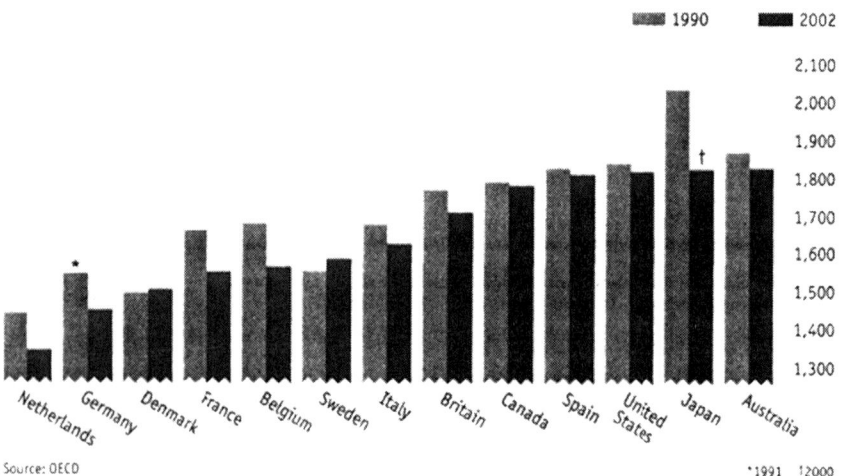

Fig. 3.23 Average hours worked per year per person in full-time employment in 1990 and 2002
Source: [Economist 2003-1] "Working Hours" © *The Economist* vol. 368, n. 8338 (23 August 2003), reproduced with kind permission.

these brief and sketchy comparisons, we can see how quality of life is made up of a complex series of elements, often difficult to quantify and to some extent subjective.

3.9 European Union's Foreign Trade

The European Union is the world's largest trade partner. Excluding trade within the Union, the EU accounts for one-fifth of world trade in goods and one-fourth of world trade in services (see Figs. 3.24 and 3.25). Trade with the rest of the world is increasing (see Fig. 3.26).

The Euro-zone registered a favourable balance of trade of $ 18.8 billion in September 2001 [Economist 2001-8] (Fig. 3.27). The balance of trade in the Triad over the period 1994 to 2005 is shown in Fig. 3.28 for goods and in Fig. 3.29 for services

EU trade balance is favourable both for goods and for services, while U.S. goods trade balance is increasingly unfavourable over the period at levels that are not compensated by the favourable service trade balance. Conversely, the Japanese goods trade balance is favourable throughout the period and largely compensates the unfavourable service trade balance. The figures relative to single member states are shown in Fig. 3.27 for 1998. The UK has a negative balance of trade; the same as the U.S. (see Fig. 3.28).

Trade of Goods

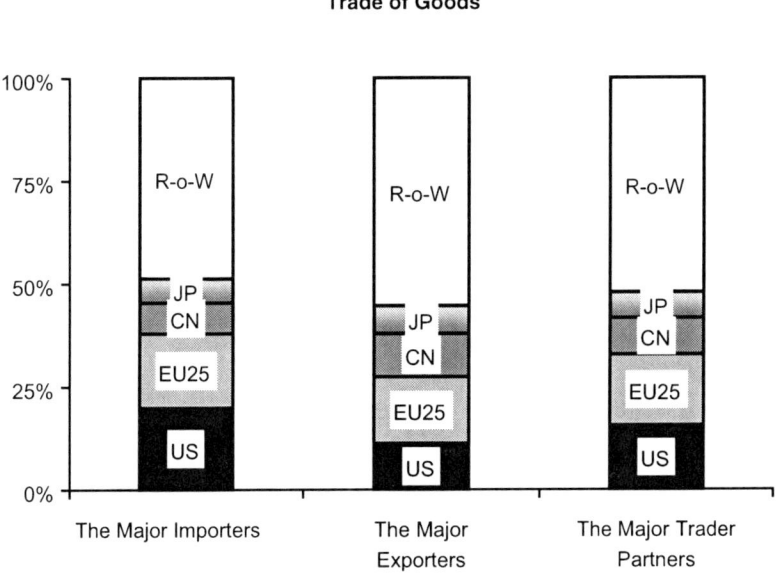

Fig. 3.24 Trade of goods: Major actors in the world in 2006. EU25 is the major exporter, followed by the U.S. China follows at a reducing distance
Source: European Commission 2007.

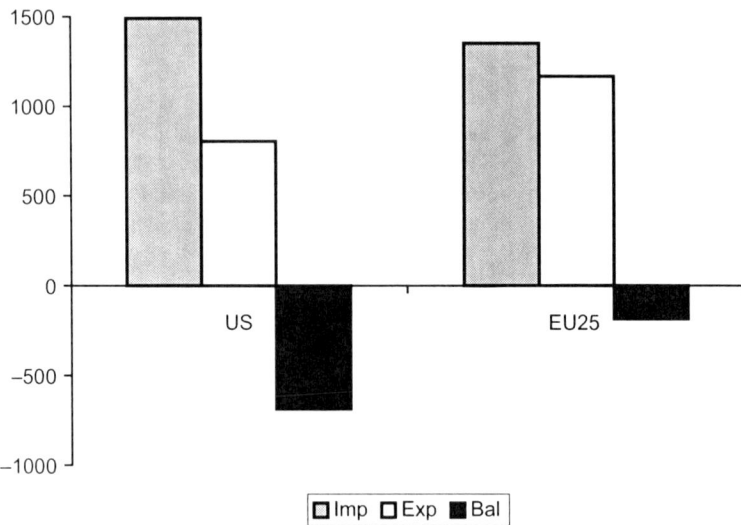

Fig. 3.25 Import, export, and trade balance in the U.S. and in EU25 in 2006 (€ billion)
Source: European Commission 2007.

Figure 3.30 presents total (viz. goods plus services) trade balance for the U.S., EU15, and the Euro-zone for the year 2000. The Euro-zone balance is favourable, while the EU15 balance is slightly unfavourable, mainly due to the contribution of the UK, whose performance reminds more the American one than that of the rest of the EU.

Fig. 3.26 EU25 trade of goods and services with the rest of the world in 1990–2006 (Hundreds of billion euro)
Source: European Commission 2007.

European Union Trade

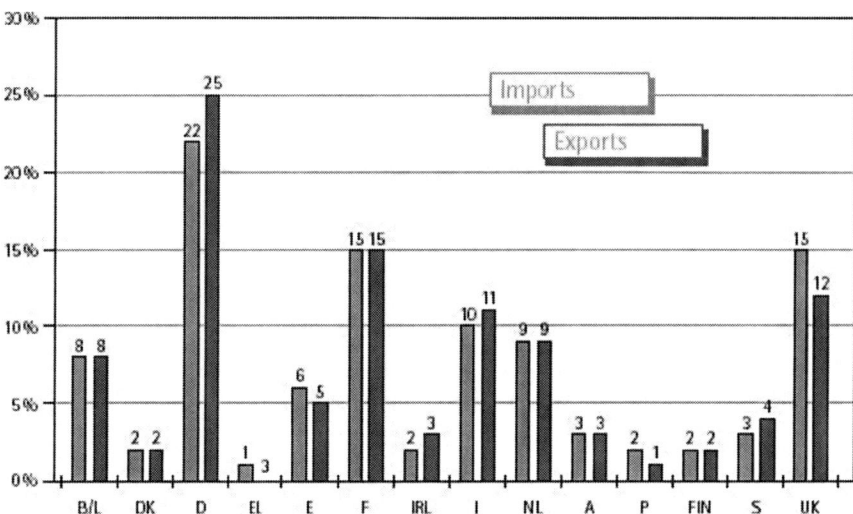

Fig. 3.27 Share of EU imports (*light grey*) and EU exports (*dark grey*) of goods and services, inclusive of internal trade among the member states (1998). The trade balance in 1998 was active for Germany, Ireland, Italy, and Sweden
Source: European Commission 1999.

Figure 3.31 shows that from January 1999 to July 2001 the Euro-zone foreign trade was almost always favourable. It is interesting to look at foreign trade in more detail to see which region in the Triad register a favourable balance of trade in the different sectors.

Figure 3.32 shows export market shares by sector relative to the three regions in the Triad. The European Union has the highest market share in

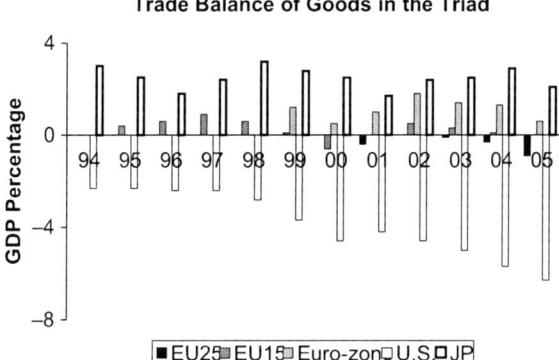

Fig. 3.28 Balance of trade in goods as a percentage of GDP for worldwide trade of goods (1994–2005) in the Triad. Data for Euro-zone are available from 1999. Data for EU25 are available from 2001
Source: Eurostat 2007.

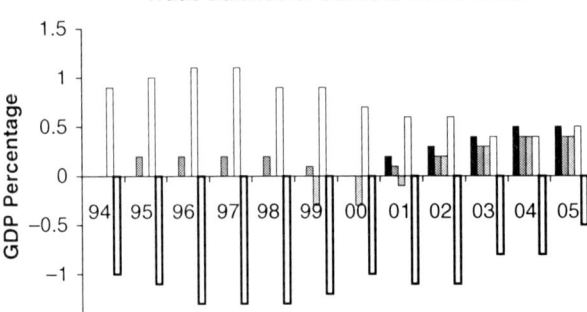

Fig. 3.29 Balance of trade in services as a percentage of GDP for worldwide trade of services (1994–2005) in the Triad. Data for Euro-zone are available from 1999. Data for EU25 are available from 2001
Source: Eurostat 2007.

nonelectrical/electronic machinery, chemicals, and pharmaceuticals. EU exports of aerospace and scientific instruments are also quite good, although these two sectors have a strong U.S. presence. The U.S. is also the leader in the computer and office equipment sectors, whereas Japan prevails in the electrical/electronic machinery sector. Finally, the three regions are on an equal footing in the electronics and telecommunications sectors.

The most thriving sectors in terms of European Union international trade are indicated in Fig. 3.33 for the years 1989 (dark) and 1996 (light). These are machinery, motor vehicles, chemicals, other means of transport, and metal products.

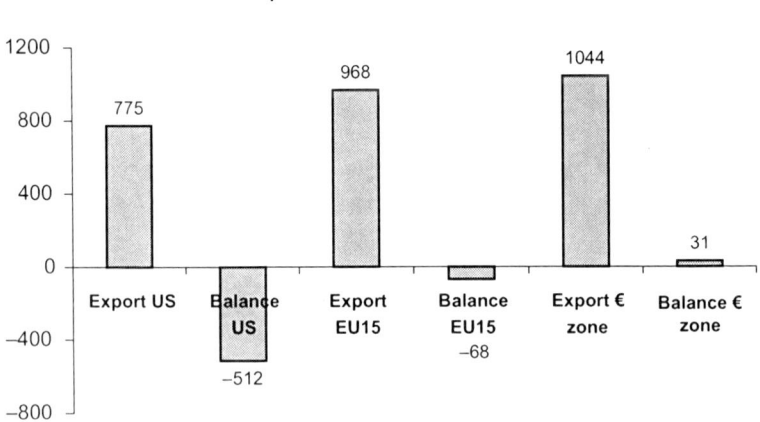

Fig. 3.30 Export and trade balance in the U.S., EU15, and the Euro-zone (billion euro, 2000)
Source: U.S. Census, Eurostat 2001.

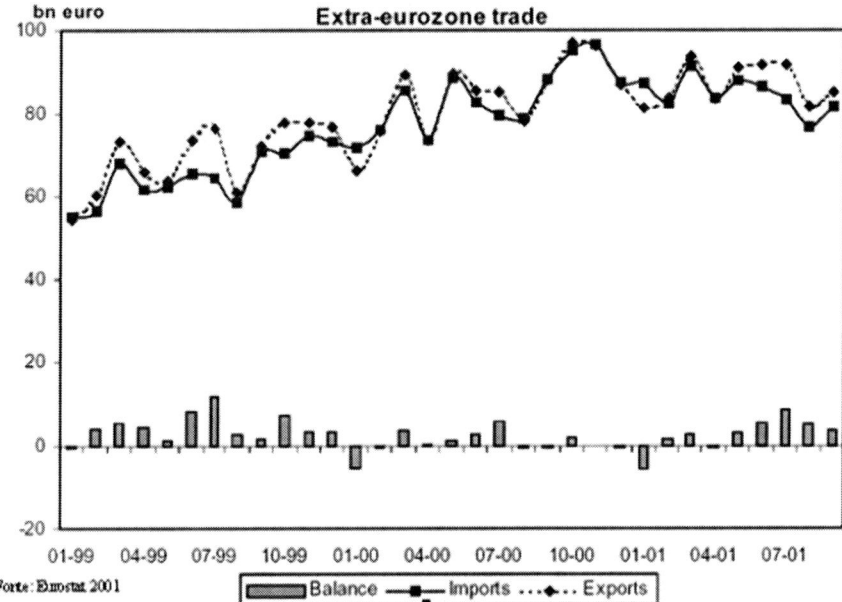

Fig. 3.31 Foreign trade (billion euro) in the Euro-zone (1999–2001)
Source: Eurostat 2001.

If we look at the balance of trade between the Union and the rest of the World
(see Fig. 3.34), we can see that the Union's economy beat its competitors, in the
year 1996, more on the basis of quality (by € 161 billion) than on the basis of
price (€ 8 billion). The highest European market shares are in sectors with modest

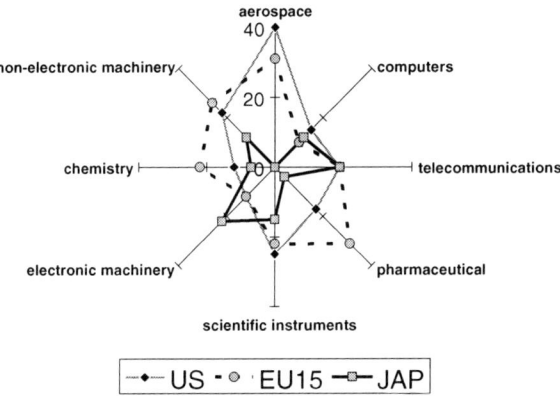

Fig. 3.32 Shares of foreign trade in the Triad, by product group (1966). The lines in the figure
have the sole purpose of facilitating the reading by joining points that belong to the same region
Source: Edited by author from data in [EC 1999-3].

Best Sectors for EU Exports

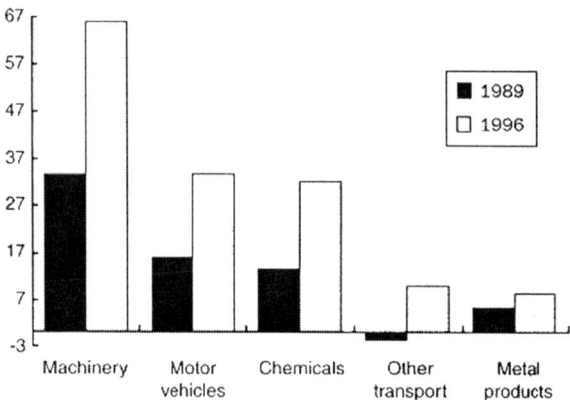

Fig. 3.33 Sectors with the most favourable trade balance (European Union)
Source: European Commission [EC 1998-3].

technological sophistication. European strength in the area of machinery is vulnerable in those countries where mechanical manufacturing has not been upgraded by the integration of information technology and artificial intelligence in the hardware. The globalization of markets requires products that self-repair or at least self-diagnose so that they can be serviced throughout the whole world, also outside the geographical reach of the specialized technical staff of the manufacturers. Notable exceptions have been those business groups (e.g. Daimler) that embedded information technology in their mechanical products and created a multimedia wideband connection of their customer sites with a unique service centre operating at the manufacturing plant offering a 7/24 consulting service that analyses the self-diagnosis

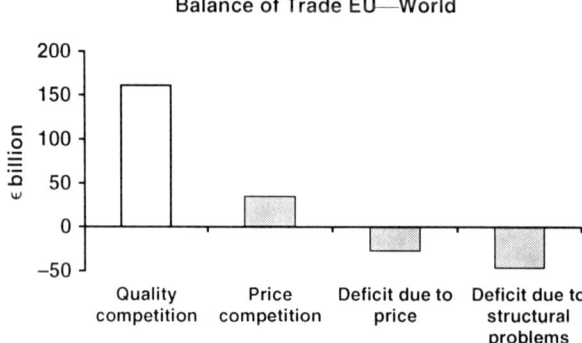

Fig. 3.34 Trade balance EU – world, billion euro, year 1996
Source: Edited by author from data in [Aiginger (ed.) 2001].

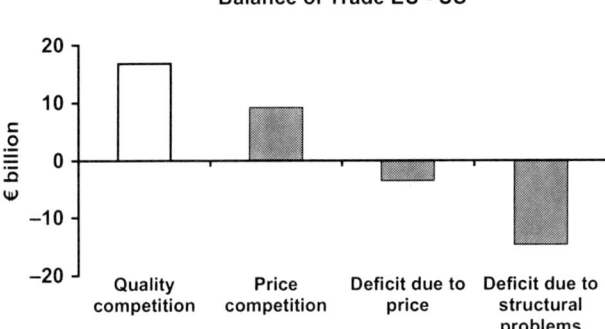

Fig. 3.35 Trade balance EU – U.S., billion euro, year 1996
Source: Edited by author from data in [Aiginger (ed.) 2001].

of malfunctions and enables local engineers at the customer premises to maintain complex equipment under remote supervision.

With respect to trade with the U.S. (see Fig. 3.35), we can observe the same supremacy based on quality, although it is worth noticing the highly negative balance of trade in sectors with structural weaknesses, where the cost of basic services and the lack of infrastructures make Europe's position weaker.

As regards trade with Japan (see Fig. 3.36), the situation changes and the balance is generally unfavourable because of the strong price competition both in the audiovisual sector and in mature sectors such as motor vehicles.

More recent data show a darker picture for Europe due to the world wide increase of trade in high-tech sectors. The European market for traditional products is nearly saturated and is mainly a replacement market, while technological and high-tech products are booming. The trade of high-tech manufactured products of European origin is increasing three times less than the World average (Fig. 3.37). Europe lags behind in the ranking of added value generated in high-tech sectors with respect to

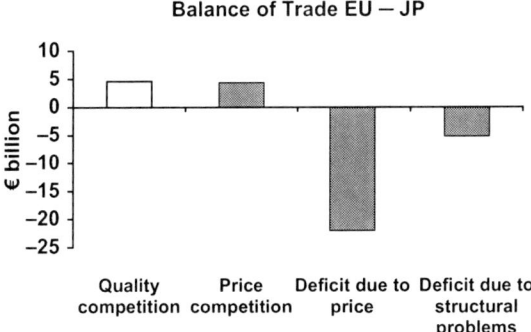

Fig. 3.36 Trade balance EU – Japan, billion euro, year 1996
Source: Edited by author from data in [Aiginger (ed.) 2001].

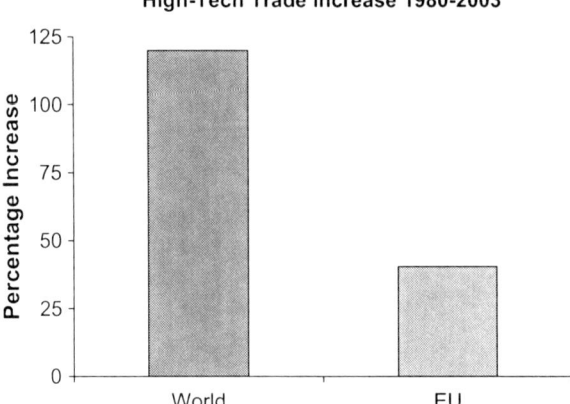

Fig. 3.37 Trade increase in high-tech products manufactured in the World and in the EU, in the period 1980–2003
Source: [Muldur et al 2006].

China, but also with respect to the other regions in the Triad (Fig. 3.38). Among the large EU economies, DE and FR had a positive high-tech trade balance in 2004, while UK, IT, and ES had a negative balance. In the period 1999–2004 the average growth rate of high tech exports was +4.9 for EU25 and +9.2 for DE, +7.5 for ES, and only +3.9 for IT; high-tech import growth was +3.4 for EU25 and +6.1 and +3.8 for DE and IT respectively. The world market share for high-tech exports was 9.7 for DE, 4.9 for FR, 4.7 for UK, and 1.4 for IT and 0.6 for ES [Eurostat 2006, Eurostat 2008].

In conclusion, Europe achieved, in the period 1994–2004, a favourable balance of trade in goods, compensating her relative weakness in high-tech manufacturing

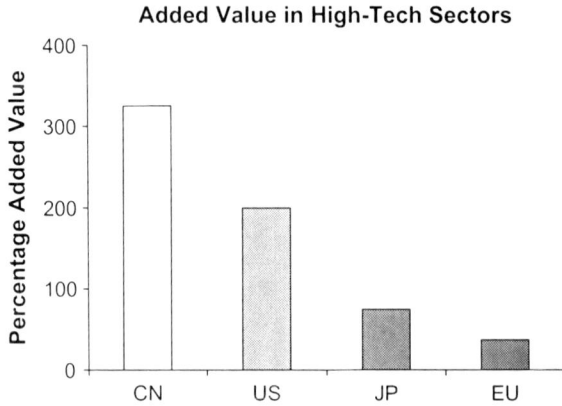

Fig. 3.38 Added value in manufacturing in high-tech sectors as a percentage of added value in total manufacturing in the Triad and in China in 2003
Source: [Muldur et al. 2006].

thanks to its quality products in mature sectors, such as BMW, Ferrari, and Porsche
cars, or thanks to luxury items such as clothing, cosmetics, shoes, wines, and
watches, with brand names such as Bulgari, Dior, Dom Pérignon, Gucci, Hermès,
Luis Vuitton, Magli, Trussardi, and Versace. Italy, thanks to the added value of
MADE IN ITALY, could from 1993 to 2003 (see Fig. 3.43), make up for its lack
of raw materials and manage to strike a favourable balance of payments.

3.10 Foreign Trade with Enlargement Countries

Enlargement countries started trading with Union member states well before 2004.
Figure 3.39 shows imports in the year 2000 from the accession countries divided by
sector. Member states have found a new market in enlargement countries, as shown
in Fig. 3.40.

Figures 3.41 and 3.42 provide some quantitative data on trade between the Union
and some of the largest accession and candidate countries in the years 1999 and
2000. In 2000, the EU–AC10 balance of trade was favourable to the EU by € 17
billion.

Imports to the EU from the Accession Countries

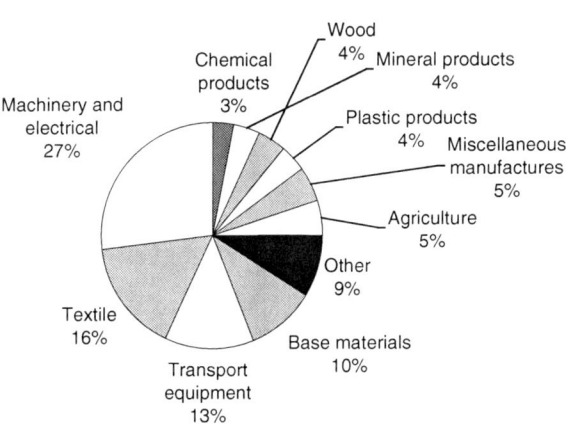

Fig. 3.39 Imports to the EU from the accession and candidate countries (2000). Data for Croatia
and Macedonia are not available
Source: European Commission 2001.

3.11 Italy's Foreign Trade

To complete this overview, let us look at Italian imports and exports in more detail.
In 2003, the Italian balance of trade was favourable by € 1.62 billion, and it had
been favourable for 11 years (see Fig. 3.43), with an average balance of € 16 billion.

Exports from the EU to the Acession Countries

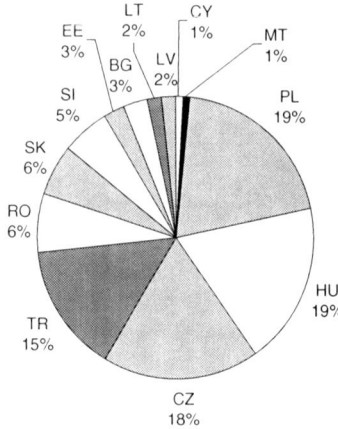

Fig. 3.40 Exports from the EU into accession and candidate countries (2000). Data for Croatia and Macedonia are not available
Source: European Commission 2001.

EU Trade with Accession Countries (1999)

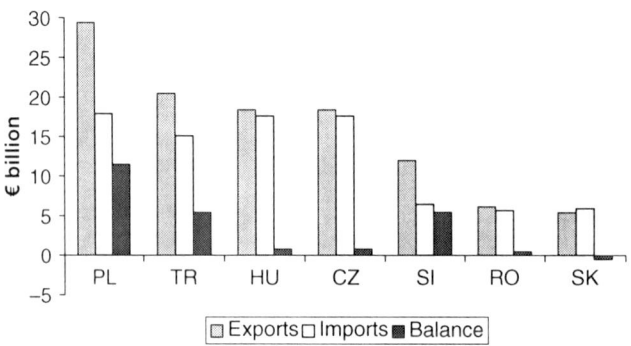

Fig. 3.41 Foreign trade between the EU and the major accession and candidate countries in 1999 (billion euro). Slovakia has a slightly positive balance with the EU
Source: European Commission 2001.

Italy struck a favourable trade balance; thanks to the measures implemented by the first Giuliano Amato government (1992) to reorganize public finances; until the mid-1990s, it also benefited from the Lira's weak exchange rate as well as from devaluations aimed at supporting competitiveness. Policies to bring the Lira in line with the stronger European Union currencies have reduced the competitive edge of Italian exports. The trade balance was negative in 2004 by € 7.7 billion (almost 3 % of imports). The European component of the trade balance has been increasingly negative in the last few years (see Figs. 3.44 and 3.45).

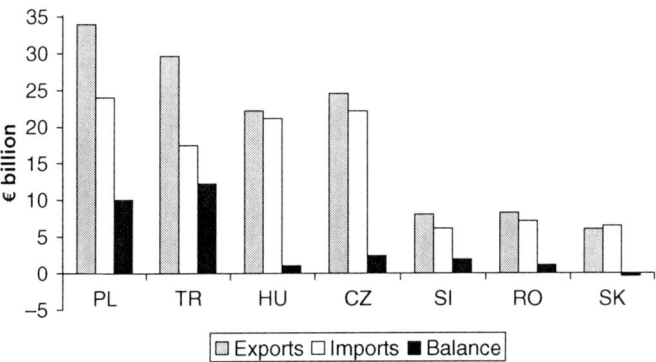

Fig. 3.42 Foreign trade between the EU and the major accession and candidate countries in 2000 (billion euro). In the year 2000, Slovakia has again a slightly positive balance with the EU
Source: European Commission 2001.

The decline in Italian exports to European countries is due to a reduced demand in Germany and France and to the loss of the competitive edge of Italian products. In 2004, Mr. Berlusconi, the then Italian prime minister, presented the euro – dollar strength as the scapegoat for Italian economic weakness. His argument, however, proves to be wrong, because it is precisely the trade towards the Euro Area that was declining (see Fig. 3.46). This phenomenon became more striking in 2004: Figure 3.46 shows shares of total imports and exports originating from trade with countries contributing more than 1.5 % to total Italian trade. It is apparent that the export shares to Germany and France are less that the corresponding import shares. In Chapter 4, "Competitiveness and Quality", we shall see that measures to help allow the Italian economy to recover include: to position itself in the top-

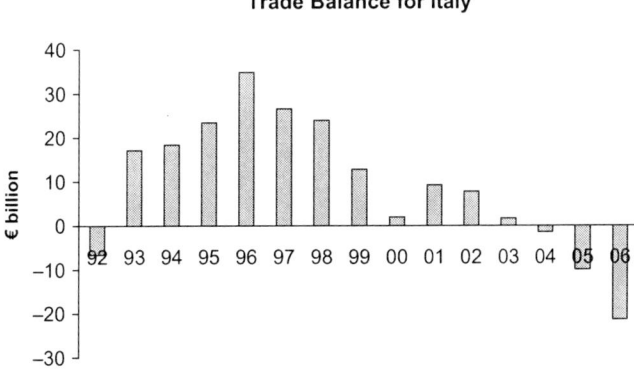

Fig. 3.43 Trade balance in Italy from 1992 to 2006 (billion euro). Further details are in Fig. 3.45
Source: IsTAT 2001–2007.

Foreign Trade in Italy in 2003 and 2006

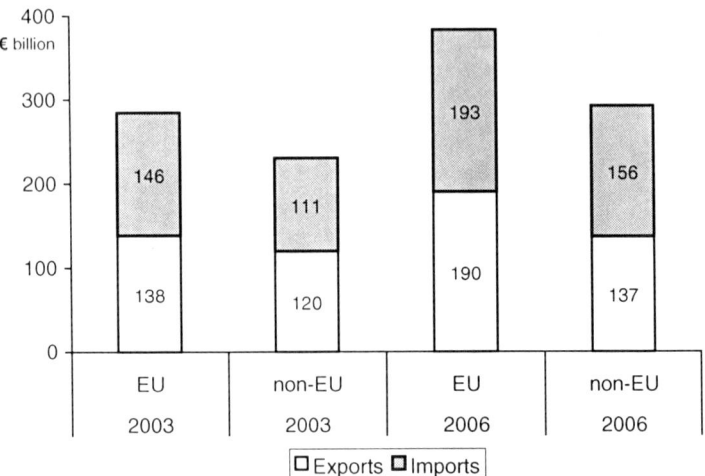

Fig. 3.44 Foreign trade in Italy in 2003 and in 2006. In 2003, the overall balance was favourable, but the balance within the EU was not favourable. In 2006, both the EU and the overall balances were unfavorable
Source: ISTAT 2005, 2007.

quality/high-price markets, to acquire market shares in the high-tech sectors and to address the growing market of budget products with satisfactory quality.

The 10 highest-ranked product categories for Italian exports in December 2004 are listed in the Table 3.10. Collectively, they make up one-fourth of Italian exports.

The Italian sectors that struck a favourable or unfavourable trade balance in the period 2003–2006 are shown in Fig. 3.47. The heaviest dependency is on energy,

Fig. 3.45 Balance of trade in Italy in the period 2000–2006. The white bars represent the trade balance with the whole world, including trade within the EU, which is represented by the grey bars. The overall balance was favourable until 2003, but the balance within the EU was unfavourable during the whole period
Source: ISTAT 2005.

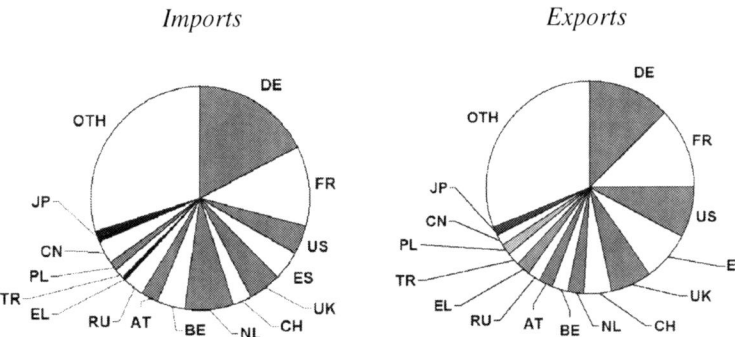

Imports   *Exports*

Fig. 3.46 Foreign trade in Italy during December 2004: shares of exports and imports to countries contributing to trade by more than 1 %. The balance was negative by € 1 billion during that month (1 % of exports)
Source: ISTAT 2005.

Table 3.10 The 10 highest-ranked product categories for Italian exports in December 2004

Top Italian Exports		
Sector	2003	2004
Vehicles	3.8	3.5
Spare parts for vehicles	2.8	3.4
Special-purpose machines	3.0	3.2
Pharmaceuticals	2.2	2.7
General-purpose machines	2.1	2.7
Footwear	1.8	2.1
White goods	1.9	2.0
Petroleum refined products	1.7	2.0
Steel products	1.2	2.0
Garments	1.4	1.8
Other sectors	78.0	74.7
Total	100	100
Total of top 10	*21.9*	*25.4*

Source: ISTAT 2005.

raw materials, and chemicals but the food and beverages sector is also unfavourable. The most favourable sector is engineering products, which bears witness to the excellence of the Italian precision mechanics manufacturing.

3.12 EU Foreign Trade in the Textile Industry

Emerging economies, such as Turkey and China, are severely challenging western economies in low-cost labour-intensive sectors. Textile trade competition (see Figs. 3.48 and 3.49, already anticipated in Section 3.6, "Emerging Economies") is particularly interesting when studying the European position in the globalized

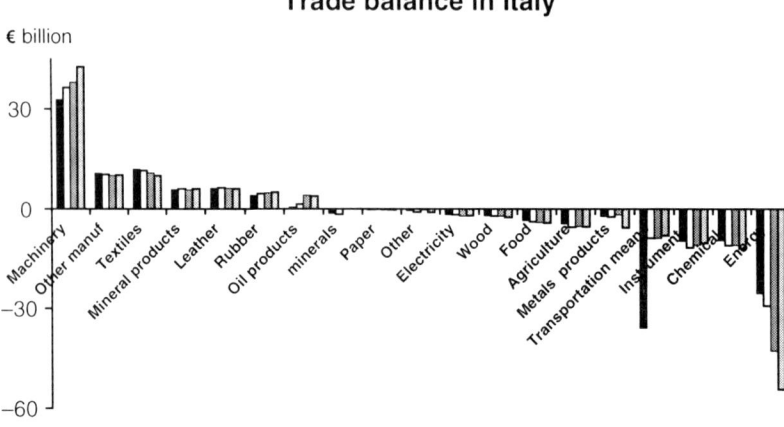

Fig. 3.47 Trade balance in Italy in various sectors (2003–2006). The best performing sectors are machinery and mechanical equipment and other manufactured products. The highest deficits in 2004 to 2006 are in energy and chemicals
Source: ISTAT 1999.

market, considering the aggressive presence of emerging economies. The evolution of the textile market supports our assertion that the only way to safeguard growth and employment in Europe is through positioning in the quality market: European expensive products can prosper in the globalized economy despite the aggressive competition from the emerging economies. In Europe, the textile industry lost five

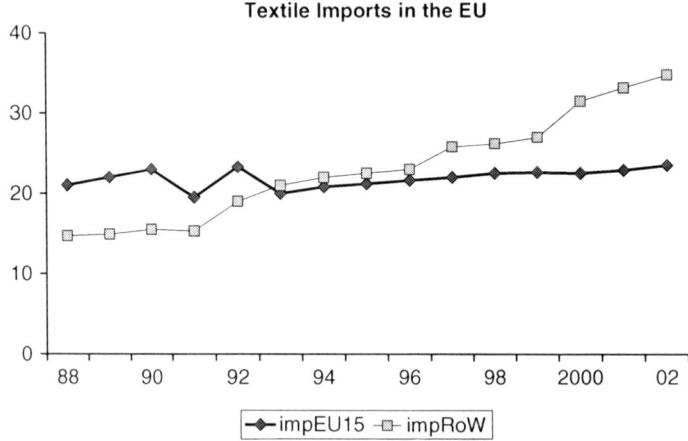

Fig. 3.48 Textile imports (as a percentage of consumption) in the EU from intra EU trade and from the rest of the world in 1988–2002
Source: European Commission [EC 2007-10].

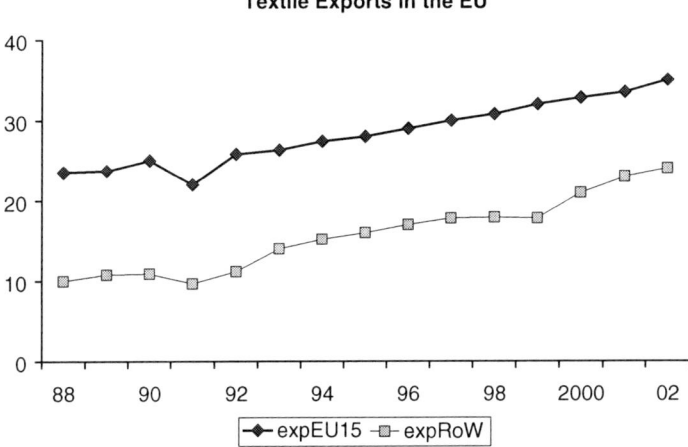

Fig. 3.49 Textile exports (as a percentage of production) from the EU15 countries intra EU and to the rest of the world in 1988–2002
Source: European Commission [EC 2007-10].

million jobs between 1988 and 1994. This situation improved at the turn of the century and then fell back from 2004 because of booming Chinese exports. Europe will be safe only if it manages to promote its new, high-quality textiles. Indeed, a new competitive textile business is expected to grow in Europe in the next few years, and the European Commission wrote (before the 2004 fallback) that this business might result in thousands of jobs being created every year in Europe [EC 2003-9].

Textile imports in EU15 countries from within Europe stabilized during the period 1988–2002, while textile imports from the rest of the world grew (Fig. 3.48). Textile exports within Europe of EU15 manufacturers grew during the period, as well as their exports to the rest of the world, despite of the aggressive price competition from emerging economies (Fig. 3.49). Economic performance in 1991 was affected by the transition in Europe following the collapse of the Soviet Union and the unification of the Democratic Republic of Germany with the Federal Republic.

3.13 Foreign Trade and Competition with Emerging Economies

In Section 3.6, "Emerging Economies", we saw that the Triad's ranking in the global market may change in the long term because of the aggressive presence of the BRIC*K* economies. In this section, we shall look at foreign trade with the emerging economies in order to understand to what extent they challenge the European Union's competitiveness and EU international trade.

3.13.1 China

Foreign trade between the European Union and China has boomed over the 1990s and it has become even stronger since China joined the World Trade Organization (WTO) in 2001. Membership in the WTO forced China to adopt legislation patterns more similar to western ones in key sectors like industrial and intellectual property protection. China's merchandise exports to the EU registered an average annual growth rate of 23% from 1990 to 2002, and of 25% in 2004–2006 (see Fig. 3.50). According to forecasts, if they kept on growing at the same pace, they would overtake U.S. exports by 2010.

China's boom reminds us of Japan's big expansion in the second half of the last century. However, China's economy [Wolf 2003] is more oriented towards the outside, compared with a more insular economy such as Japan's. The Chinese demand for natural resources has triggered a global commodity boom; China's role in stimulating the economy in the rest of the world could soon be stronger than the role played by Japan in the last decades of the 20th century, given the greater interconnection between the Chinese economy and the rest of the world. The growth of the Chinese economy has had a positive impact on the economies of other countries exporting basic and consumption goods (concrete, steel, machinery, and means of transport, but also agricultural produces such as soya) and of countries exporting sophisticated products and services. Among the leading countries are Argentina, Brazil, and Chile, followed by the U.S., Japan, and many EU countries. The industrialization of the BRICKs is doubling the world's workforce and China accounts for more than half of this increase. The world's most populated country, with one billion three hundred million inhabitants, has finally become a large consumption market willing to accept foreign goods as well. Hence, its imports in 2002 grew by 45%. The Chinese middle-upper class loves all things foreign, so China is becoming a huge market for luxury cars, mobile phones, and the most sophisticated equipment

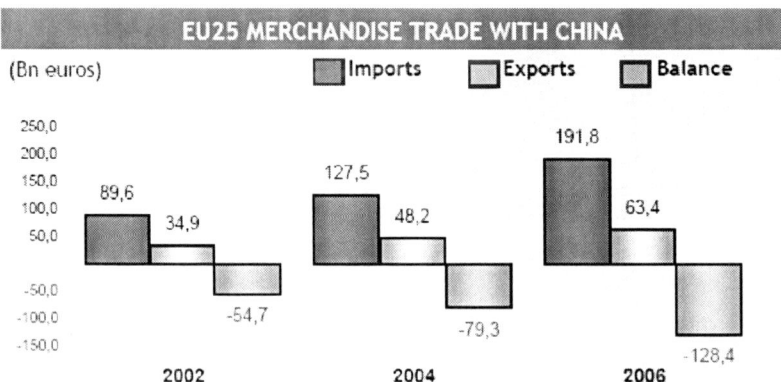

Fig. 3.50 Growth of merchandise trade between the EU and China from 2002 to 2006. Total trade (imports plus exports) was only € 4.1 billion in 1980 and € 16.2 billion in 1990
Source: European Commission 2007.

[Rampini 2003-2]. In the first half of the twenty-first century, China could be the driving engine of Asian and the world economy, having the same role the U.S., Japan, and Germany played in the past.

In 2001, international trade amounted to 26 % of GDP in China [EC 2004-1] and to 10 % in Japan. In addition, Chinese exports are dependent on foreign investments. In 2000, half of Chinese exports were produced by offshore companies based in China, taking advantage of the favourable conditions of the labour market, which offers a nearly unlimited resource of labour at a low cost. Chinese authorities estimate that in the next 10 years 150 million citizens will leave the countryside to find jobs in the industry or services sectors. By 2025, the Chinese rural population is expected to drop by 300 million people. This projection confirms that an important component of the competitive edge of China's economy will continue to be for some time the low cost of labour. China is a formidable competitor in markets based on price competition. Conversely, the Union can still have a leading role in those markets where competition is based on quality. Chinese investments in R&D are well below the average in industrialized countries, as they only account for 1.23 % of GDP, or half the EU average. However the Chinese Ministry of Science and Technology reports that in the period 1999–2004 S&T research increased in China by 10 % [RTD Info 2007-1] and China proved to be able to act fast and effective. The Chinese commitment to space activities or to a moon exploration project with a human landing by 2010 should not be deemed particularly significant with regard to technological fall-out, because, although this project is likely to be accomplished in due time, we cannot expect the same boost to technological innovation in the Chinese industry as in the U.S. during the 1960s and 1970s following America's space programs [Kroeber 2004]. The Chinese industry has never had a strong technological base, and in China there has never been a private high-tech industry, as opposed to the U.S., where a private industrial base of 20,000 enterprises participating in the moon exploration program managed to take advantage of the technology implemented in space programs, thereby making the U.S. the world leader in technological development. However, the picture might change: Rampini reports [Rampini 2005-2] that Chinese R&D investments increased by 25 % per year in the period 2000–2004.

Wang Hui, manager of the S&T Dept of the Chinese Trade Ministry [Hui 2003], maintains that the limit to Chinese export growth is the quality of Chinese products, which do not meet the safety, quality, and environmental standards required under the legislations in force in the importing countries. Joint ventures with leading innovative industries from Taiwan and from other countries investing in China might help the Chinese industry to bridge the technological gap with the most advanced economies.

China is well integrated in the world economy from a monetary perspective. It plays an essential role in supporting consumption in the U.S. through a regular and substantial purchase of dollars, namely, $ 10 billion a month, aimed at sustaining a fixed exchange rate of the renminbi, the national Chinese currency (also referred to as yuan). A virtuous cycle has been created between the U.S. and China: American consumers can change their cars or buy a house because the Chinese savers (who save up to 28 % of their income) give them credit [Rampini 2003-1].

3.13.2 India

Foreign trade between the European Union and India (see Fig. 3.51) is, in figures, a fourth of the trade with China, but it has deep roots established over the past decades and is slowly, yet constantly, growing. Moreover, Europe's trade with India features a strong services component. The Indian merchandise exports to the EU registered an average annual growth rate of 14 % from 1990 to 2002, and of 19 % in 2004–2006.

In figures, services account for 18 % of merchandise trade in India, while they only account for 6 % in China. As we have already pointed out, India offers high-quality services, particularly in the information technology and communications sector. It was precisely because of the magnitude of Indian exports in the ICT sector that the U.S. and Europe decided to introduce tariff barriers [Luce 2003].

It is not easy to draw conclusions and make any reliable forecasts for the long term, given the fact that unexpected events could bring uncertainty and instability in international relations and in the political and social structure of the emerging countries. However, it seems quite certain that the BRIC*K*s, particularly China and India, will play a primary role in the economy of the next few years. The Union will still be a leading player at the international level if it manages to strengthen its scientific and technological base and if it aims for a supremacy based on quality and not on price. However, in some sectors, such as software products, ICT services, and health care, India is already a formidable competitor with European providers, thanks to its quality products and services at unbeatable prices.

Fig. 3.51 Growth of merchandise trade between the EU and India from 2002 to 2006. Total trade (imports plus exports) was only € 4.4 billion in 1980 and € 11.2 billion in 1990
Source: European Commission 2007.

3.13.3 Brazil

The Brazilian trade balance is positive at 5 % of total trade in 2005. Imports are machinery, electrical and transport equipment, chemical products, oil, automotive parts, and electronics; main partners are U.S. (17.5 %), Argentina (8.5 %), Germany (8.4 %), China (7.3 %), and Japan (4.6 %). Exports are originated by the natural resources of the country (iron ore, soybeans, coffee, and ethanol) and by her manufacturing capabilities (transport equipment, autos). Main partners are U.S. (19.2 %), Argentina (8.4 %), Netherlands (4.5 %), Germany (4.2 %), and - increasingly - China (5.8 %). Merchandise trade with the EU increased by 68 % from 2002 to 2006 up to € 26 billion with a trade balance favourable for Brazil, see Fig. 3.52 [WorldFactbook 2007].

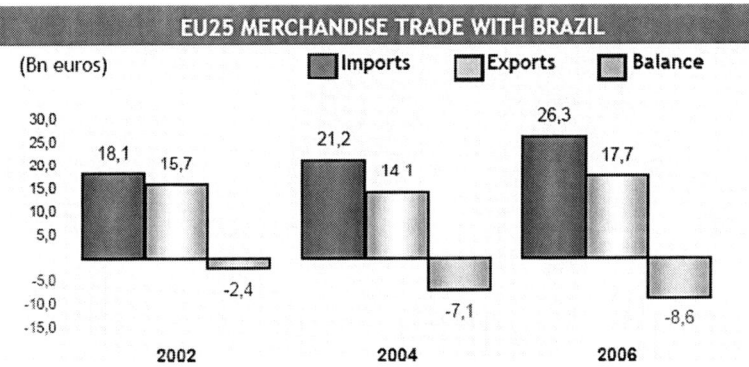

Fig. 3.52 Growth of merchandise trade between the EU and Brazil from 2002 to 2006
Source: European Commission 2007.

3.13.4 Russia

The Russian trade balance was positive at 30 % of total trade in 2005. Oil, natural gas, metals, and wood and wood products account for more than 80 % of exports and 32 % of government revenues, leaving the country vulnerable to swings in world commodity prices. Other exports include chemicals, and a wide variety of civilian and military manufactures. Main export partners are The Netherlands (10.3 %), Germany (8.3 %), (Italy 7.9 %), China (5.5 %), Ukraine (5.2 %), Turkey (4.5 %), and Switzerland (4.4 %). Main imports are machinery and equipment, consumer goods, medicines, meat, sugar, semifinished metal products; import partners are Germany (13.6 %), Ukraine (8 %), China (7.4 %), Japan (6 %), Belarus (4.7 %), U.S. (4.7 %), Italy (4.6 %), and South Korea (4.1 %). Merchandise trade with the EU increased by 88 times from 1980, and by 14 times from 1990 up to € 389 billion in 2006 with a trade balance favourable for Russia, see Fig. 3.53 [WorldFactbook 2007].

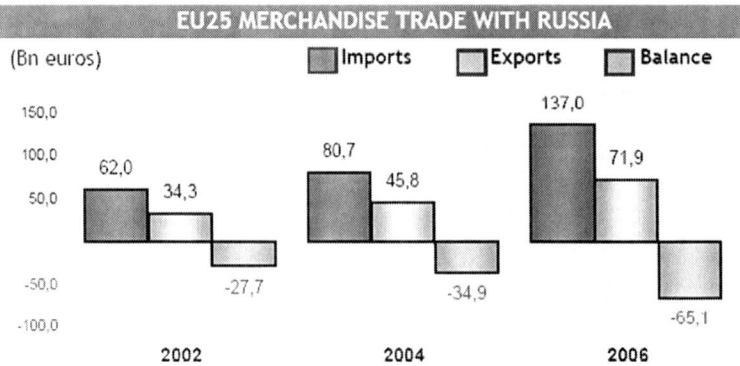

Fig. 3.53 Growth of merchandise trade between the EU and Russia from 2002 to 2006. Total trade (imports plus exports) was only € 4.4 billion in 1980 and € 11.2 billion in 1990
Source: European Commission 2007.

3.13.5 South Korea

Korea has achieved an incredible record of growth. GDP growth in 2002 was 7 %, and it stayed above 4.5 % since. A downturn in consumer spending was offset by rapid export growth. Moderate inflation, low unemployment, an export surplus with a positive trade balance at 2.7 % of total trade, and fairly equal distribution of income characterize this solid economy. The ratio of exports to GDP rose 15 times from 1953 to 2000, up to 45 %. High level and quality of R&D and high number of international patents position Korea in the international high-tech market.

Korean exports include semiconductors, wireless telecommunications equipment, motor vehicles, computers, steel, ships, petrochemicals; main partners are: China (21.8 %), U.S. (14.6 %), EU (11.7 %), Japan (8.5 %), and Hong Kong (5.5 %). Imports include machinery, electronics and electronic equipment, oil, steel, transport equipment, organic chemicals, plastics; main suppliers are Japan 18.5 %),

Fig. 3.54 Growth of merchandise trade between the EU and Korea from 2002 to 2006
Source: European Commission 2007.

China 14.8 %), U.S. 11.8 %), EU (7.1 %), and Saudi Arabia 6.2 %). Merchandise trade with the EU increased by 43 % from 2002 to 2006, to € 61 billion (see Fig. 3.54) [Mahlich et al., ed.s 2007].

3.14 Services and Infrastructures

We have already introduced the topic of the cost of basic services and infrastructures. It is worth studying the difference in costs within the Triad.

We can observe (see Fig. 3.55) that in the Union in 1996, some basic costs (oil for heating, natural gas, diesel, electricity, and the Internet) were far higher than in the U.S. It should be noted that the lower costs in Europe in 1996 were relative to dedicated and switched telephone communications, an area in which the Union had implemented deregulation and competition policies.

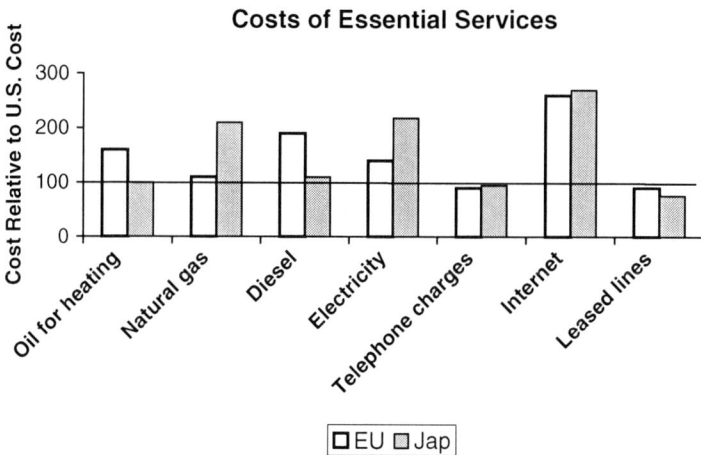

Fig. 3.55 Impact of the costs of essential services on production costs for the enterprise. Industrial prices in the EU (*white bars*) and Japan (*grey bars*) for the year 1997, including taxes, and normalized to purchasing power. Values relative to U.S. = 100
Source: OECD quoted by [EC 1998-3] Data from the year 1997.

The swift establishment of the new economy can bring dramatic changes in the market positioning of some countries whose economy is based on traditional industries and mature markets. Later on, we shall discuss on what terms the Union's industry will be able to maintain its market shares in the new economic environment.

3.15 Comparing Countries' Competitiveness

The World Economic Forum calculates a Growth Competitive Index (GCI) for 102 states [WEF 2003, WEF 2004, WEF 2005, and WEF 2006]. It is centred on the idea that economic growth can be assessed by taking into account three categories,

namely, macroeconomic environment, quality of public institutions, and technological progress. An index is calculated for each one of these categories and is later
synthesized in the GCI.

The macroeconomic environment index (MEI) describes economic stability, the
curb on inflation, the functioning of the banking system, the tendency of state
budgets to balance out, the cancellation of public debt, and the waste in public
spending.

The Public Institutions Index (PII) is used to assess the role played by the
institutions in the promotion and support of private business prosperity: guaranteed
property right, judicial system efficiency, control over organized crime, curb on
corruption, and the integrity of civil servants.

The Technology Index (TI) measures the extent of innovation brought about by
technological progress. It is based on the assumption that technological progress
can lay the ground for long-term economic growth. The situation is not the same
in countries whose economies are at the frontier of technological innovation and in
countries with less advanced economies, in which growth can only be guaranteed
through the adoption of stable and sound technologies that have already been tested
in other contexts. For this reason, the WEF divides economies into two groups,
namely, the core innovators and the others. The threshold is fixed on the basis of
the average yearly number of utility patents granted during the 1980s in the United
States per million population. The ranking of the core innovators' group performance in granted patents is shown in Table 3.17. The group was initially defined
as the set of countries with more than 15 patents granted on average in the 1980s
per million population. The composition of this group is, however, changing with
time. We see in Table 3.18 and in Fig. 3.58 that, during the last two decades, *new
entries* have been aggressively coming into the competitiveness arena, displacing
the established leaders.

In Table 3.11, the first 70 countries are presented with their GCI ranking from
2002 to 2004. They are ordered according to their ranking in 2004. For ease of
reference, Table 3.12 presents a subset of Table 3.11 considering only EU27 member
states and Croatia and Turkey, two countries for which accession negotiations are
ongoing.

Some countries considered by WEF in 2003 and 2004 had not been taken into
account in the previous year. In order to be able to make consistent comparisons
between ranking positions, the data in Table 3.12 have been normalized by the
author in Table 3.13 by recalculating the ranking considering only the countries
for which WEF data have been published throughout the reference period.

Some countries in Table 3.11 have a different weight than others in terms of
their economic power. Table 3.14 presents the ranking in the period 2001–2004
according to GCI for the 13 countries with the highest GDP in the world in the year
2000. The Russian Federation has also been added. It ranked 18th in the year 2000 in
terms of GDP. The business downturn in the United States in 2001 affected Canada's
economy due to its strong interdependency with the U.S. Canada competitiveness
lost three positions in the ranking during the second part of the period. However her
economy, the world's ninth-largest, recovered in the following years. Japan climbed

Table 3.11 GCI Index ranking of the first 70 countries in the years 2004, 2003 and 2002

Country	2004	2003	2002
Finland	1	1	1
United States	2	2	2
Sweden	3	3	3
Taiwan	4	5	6
Denmark	5	4	4
Norway	6	9	8
Singapore	7	6	7
Switzerland	8	7	5
Japan	9	11	16
Iceland	10	8	12
United Kingdom	11	15	11
The Netherlands	12	12	13
Germany	13	13	14
Australia	14	10	10
Canada	15	16	9
United Arab Emirates	16	—	—
Austria	17	17	18
New Zealand	18	14	15
Israel	19	20	17
Estonia	20	22	27
Hong Kong	21	24	22
Chile	22	28	24
Spain	23	23	20
Portugal	24	25	18
Belgium	25	27	21
Luxembourg	26	21	—
France	27	26	28
Bahrain	28	—	—
South Korea	29	18	25
Ireland	30	30	23
Malaysia	31	29	30
Malta	32	19	—
Slovenia	33	31	26
Thailand	34	32	37
Jordan	35	34	44
Lithuania	36	40	39
Greece	37	35	31
Cyprus	38	—	—
Hungary	39	33	29
Czech Republic	40	39	36
South Africa	41	42	34
Tunisia	42	38	32
Slovakia	43	43	46
Latvia	44	37	43
Botswana	45	36	35
China	46	44	38

Table 3.11 (continued)

Country	2004	2003	2002
Italy	47	41	33
Mexico	48	47	53
Mauritius	49	46	41
Costa Rica	50	51	49
Trinidad and Tobago	51	49	42
Namibia	52	52	47
El Salvador	53	48	60
Uruguay	54	50	40
India	55	56	54
Morocco	56	61	52
Brazil	57	54	45
Panama	58	59	51
Bulgaria	59	64	58
Poland	60	45	50
Croatia	61	53	48
Egypt	62	58	—
Romania	63	75	67
Colombia	64	63	61
Jamaica	65	67	67
Turkey	66	65	65
Peru	67	57	55
Ghana	68	71	—
Indonesia	69	72	69
Russian Federation	70	70	66

Source: Author's editing of data from World Economic Forum.

the ladder in the period. Following a volatile performance in the 1990s, from 2003 the Japanese economy entered its longest post-war recovery; the expansion continues in 2007, with annual GDP growth likely to reach 2.3 % [Economist 2007-20] (Table 3.15).

The U.S., Japan, and Germany are either stable or improving their GCI index position. Other countries with the highest GDPs are losing ground in the ranking as far as the indexes are concerned. South Korea shows a notable improvement in 2003 and a setback in 2004.

We will go now into some details concerning GCI ranking. Italy's ranking is discouraging. It undermines Italy's position as one of the world's most industrialized and prosperous countries, and damages economic growth, a favourable balance of trade and the creation of new jobs. According to the 2004 World Economic Forum, Italy's ranking is plummeting. On the basis of the Growth Competitiveness Index (GCI), Italy (see Table 3.13) ranked 33rd in 2002, 39th in 2003, and 42nd in 2004 and in 2006. Italy lags behind all the other G7 countries, the EU15 countries, the EFTA countries, nine out of the 10 European Union 2004 enlargement countries, the majority of Asia's emerging economies and a few of Africa's emerging economies. The loss of six positions in the general index from 2002 to 2003, followed by a

Table 3.12 Ranking according to the GCI index for EU25 and candidate countries in the years 2004, 2003 and 2002

Country		2004	2003	2002
Finland	FI	1	1	1
Sweden	SE	3	3	3
Denmark	DK	5	4	4
United Kingdom	UK	11	15	11
The Netherlands	NL	12	12	13
Germany	DE	13	13	14
Austria	AT	17	17	18
Estonia	EE	20	22	27
Spain	ES	23	23	20
Portugal	PT	24	25	18
Belgium	BE	25	27	21
Luxembourg	LU	26	21	—
France	FR	27	26	28
Ireland	IE	30	30	23
Malta	MT	32	19	—
Slovenia	SI	33	31	26
Lithuania	LT	36	40	39
Greece	EL	37	35	31
Cyprus	CY	38	—	—
Hungary	HU	39	33	29
Czech Republic	CZ	40	39	36
Slovakia	SK	43	43	46
Latvia	LV	44	37	43
Italy	IT	47	41	33
Bulgaria	BG	59	64	58
Poland	PL	60	45	50
Croatia	HR	61	53	48
Romania	RO	63	75	67
Turkey	TR	66	65	65

Source: Author's editing of data from World Economic Forum.

further three positions in the following year, was due, according to the WEF, to a decline on all fronts and, particularly, to a drop in the MEI index concerning the macroeconomic environment. Italy received lower and lower marks due to her poor economic stability, worsening public debt, decline in savings, increased waste in public spending, worsening perception of the services provided by the public administration, and less strict control over corruption and organized crime. As for technological progress, Italy got lower marks because of the lower priority level attached by her government to ICT and its successful promotion.

Germany gained one position in the GCI index in 2003 and maintained her position in 2004. WEF reported improved quality of public administration services, including the fight against corruption. However, some negative elements, such as state aid undermining genuine competition and an increase in the budget deficit to 3.6 % of GDP, offset the positive judgments.

Table 3.13 Normalized ranking according to the GCI index for EU25 and candidate countries. Normalization of ranking in Table 3.12: only the countries for which data are available during the whole period are ranked. Data for Luxembourg, Cyprus, and Malta are not available

Country		Normalized		
		2004	2003	2002
Finland	FI	1	1	1
Sweden	SE	3	3	3
Denmark	DK	5	4	4
United Kingdom	UK	11	15	11
The Netherlands	NL	12	12	13
Germany	DE	13	13	14
Austria	AT	16	17	18
Estonia	EE	19	22	27
Spain	ES	22	23	20
Portugal	PT	23	25	18
Belgium	BE	24	27	21
France	FR	25	25	28
Ireland	IE	27	29	23
Slovenia	SI	29	29	26
Lithuania	LT	32	38	39
Greece	EL	33	33	31
Hungary	HU	34	31	29
Czech Republic	CZ	35	37	36
Slovakia	SK	38	41	46
Latvia	LV	39	35	43
Italy	IT	42	39	33
Bulgaria	BG	54	62	58
Poland	PL	55	43	50
Croatia	HR	56	51	48
Romania	RO	57	73	67
Turkey	TR	60	63	65

Source: Author's editing of data from World Economic Forum.

France gained three positions in the GCI index in 2003 and maintained her 25th position in 2004. It scored high under the PII and TI indexes, which offset a low score in the MEI caused by a long-drawn-out budget deficit exceeding 3 %.

The United Kingdom lost four positions in 2003 and recovered in 2004. WEF relates the setback in 2003 to problems in public administration: control over organized crime, biased civil servants, worsening of public finances, and curb on inflation. The recovery was helped by higher scores under the TI index due to the higher priority level attached by the government to ICT and her promotion.

The United States maintained their number two ranking after Finland. However, their performance was jeopardized by an increase in the federal deficit, a reduction in savings, a perception of bias in civil servants' decisions, and less control over organized crime. In 2004 the U.S. was still the world leader in technology, yet their number of registered patents had decreased.

Table 3.14 Ranking in 2001 to 2004 of the 13 largest world economies according to the GCI index. The 13 economies with the largest GDP in the year 2000 are taken into account together with the Russian Federation

	Competitiveness ranking of the largest economies			
	2004	2003	2002	2001
1	U.S.	U.S.	U.S.	U.S.
2	Japan	Japan	Canada	Canada
3	UK	Germany	UK	UK
4	Germany	UK	Germany	Germany
5	Canada	Canada	Japan	France
6	Spain	Korea	Spain	Japan
7	France	Spain	Korea	Spain
8	Korea	France	France	Korea
9	China	Italy	Italy	Italy
10	Italy	China	China	China
11	Mexico	Mexico	Brazil	Mexico
12	India	Brazil	Mexico	Brazil
13	Brazil	India	India	India
	Russia	Russia	Russia	Russia

Source: Author's editing of data from World Economic Forum.

Japan gained five positions in 2003 and two more in the following year, thanks to good performances under the technological indexes, a considerable increase in patent registration, and a favourable perception of the government's promotion of ICT. The budget deficit in 2004 was still high, despite having decreased. The judgment about waste in public spending had worsened.

To complete this overview of the G7 countries, let us consider Canada. Canada lost seven positions in 2003 because of the deteriorating perception of public services, ranging from justice to corruption control and the public's lack of confidence in politicians. Canada improved by one position in 2004: it showed the highest macroeconomic stability in the G7 group. It supported the spreading of new technologies, although patent registration decreased.

Among the 10 EU accession countries, Lithuania and Slovakia have performed well—and Estonia extremely well, positioning herself in the 2004 ranking according to GCI in 19th position, in front of eight EU15 countries. Slovenia improved from 2002 to 2004 and positioned herself as the second of the accession countries. Hungary and Poland lost five positions from 2002 to 2004. The Czech Republic gained one position in the same period. Latvia improved by eight positions in 2003 but had a setback in 2004. Of the two Mediterranean accession countries, Malta in 2004 was positioned just below Ireland, and Cyprus below Greece. Comparison data about performance in previous years are not available for these two countries. The candidate countries and the 2007 accession countries are placed at the bottom of the ranking in Table 3.13: Bulgaria improved by four positions from 2002 to 2004. Turkey improved by five positions during this same period. Croatia lost eight positions from 2002 to 2004. Romania improved by 10 positions from 2002 to 2004.

Table 3.15 Ranking according to the Growth Competitiveness Index for 2003 and ranking according to its components: MEI (Macroeconomic Environment Index), PII (Public Institutions Index), and TI (Technology Index). The first 42 countries are shown. The complete table is in [WEF 2004]

Country	GCI	MEI	PII	TI
Finland	1	2	2	2
United States	2	14	17	1
Sweden	3	8	7	4
Denmark	4	5	1	8
Taiwan	5	18	21	3
Singapore	6	1	6	12
Switzerland	7	6	8	7
Iceland	8	16	3	15
Norway	9	4	16	13
Australia	10	7	4	19
Japan	11	24	30	5
The Netherlands	12	9	11	18
Germany	13	21	9	14
New Zealand	14	13	5	23
United Kingdom	15	12	12	16
Canada	16	11	24	11
Austria	17	10	14	27
South Korea	18	23	36	6
Malta	19	29	18	17
Israel	20	44	15	9
Luxembourg	21	3	13	42
Estonia	22	34	28	10
Spain	23	17	31	25
Hong Kong	24	15	10	37
Portugal	25	31	22	22
France	26	20	23	28
Belgium	27	19	27	29
Chile	28	35	19	31
Malaysia	29	27	34	20
Ireland	30	22	25	38
Slovenia	31	37	35	24
Thailand	32	26	37	39
Hungary	33	38	33	32
Jordan	34	42	20	—
Greece	35	33	42	30
Botswana	36	30	26	—
Latvia	37	36	45	26
Tunisia	38	32	32	—
Czech Republic	39	39	47	21
Lithuania	40	41	41	36
Italy	41	28	46	44
South Africa	42	40	43	40

Source: Author's editing of data from World Economic Forum.

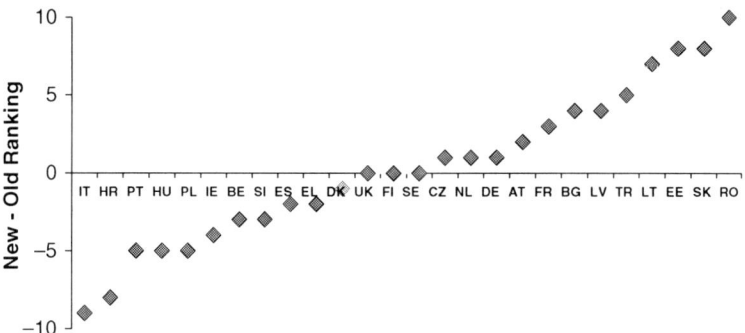

Fig. 3.56 Change in ranking of European countries according to the GCI Index from 2002 to 2004. Data for Cyprus, Luxembourg, and Malta are not available
Source: Author's editing of data from World Economic Forum.

Figure 3.56 shows the changes in GCI ranking of the EU member states and of the candidate countries from the year 2001 to 2004. Denmark, Sweden, Estonia, and Lithuania climbed the ladder; Italy, Ireland, and Poland went down. Finland's mark is set on zero, indicating no change in its leading position. Figure 3.57 plots the score of the GCI Index in 2004 versus the change in positioning in GCI ranking from 2002 to 2004. The candidate countries and the 2007 accession countries are, with Poland, at the lowest score levels, but Romania is advancing while Croatia is

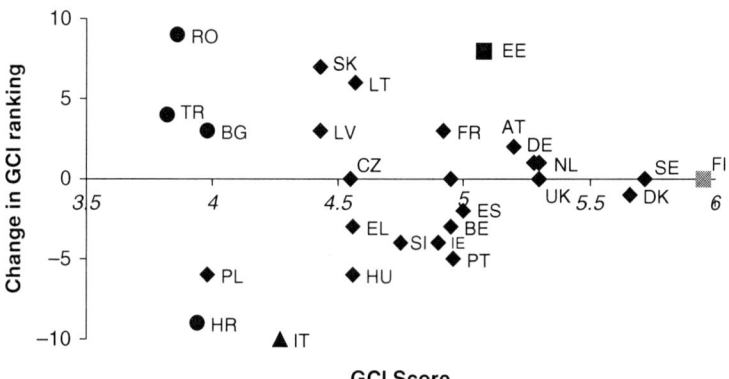

Fig. 3.57 Score of the GCI Index in 2004 versus the change in positioning in GCI ranking from 2002 to 2004 of European Union members (*black diamonds*) and candidate countries (*black circles*). Three extreme cases are identified with special signs: the worse EU15 performer, Italy (*black triangle*); the best performer among accession countries, Estonia (*black square*); and the best overall performer, Finland (*grey square*). Data for Luxembourg, Cyprus, and Malta are not available
Source: Author's editing of data from World Economic Forum.

in a losing position. The performance of Estonia is impressive, far ahead of all the 2004 accession countries, and well ahead of half of the EU15 countries.

Table 3.16 describes the GCI, MEI, PII, and TI variation for the world's 13 largest economies from 2002 to 2003. The Russian Federation was added (its GDP was in the 18th position in 2000). The countries are ranked by decreasing GDP in the year 2000. The second column shows how these countries would rank if they were ordered by GDP ppp (purchasing power parity) and not by GDP. In this case, China would be second after the U.S. We saw earlier in this chapter that China is becoming a very visible presence in the world economy under many aspects. Since the year 2000, its GDP has been larger than that of Italy.

The core technology-innovation country group is ranked in Table 3.17 according to the number of U.S. patents granted in 2002. This group is defined by the WEF according to a criterion established on the basis of figures dating back to the 1980s. The core innovating countries are those that in the 1980s were granted at least 15 patents in the U.S. per million population. These countries were 18 in number in 1990 (see Table 3.17) and at that time Italy ranked 17th, but in 2002 it ranked 25th because a number of new, aggressive actors came into play (see Table 3.18).

The new entries in the core technology-innovation country group are four Asian *tigers*, namely, South Korea, Singapore, Taiwan, and Hong Kong, and two European countries, Iceland and Ireland. Figure 3.58 shows on a logarithmic scale the increase

Table 3.16 Change of the GCI, MEI, PII, and TI in 2003 with respect to 2002 for the 13 largest economies (G7 countries in *italics*), ordered by decreasing GDP. The Russian Federation has been added (its GDP was in the 18th position in 2000). The second column shows the ranking according to GDP ppp (purchasing power parity), as opposed to GDP. GDP and GDP ppp refer to the year 2000 and are in trillion dollars

Ranking			Increase 2003–2002				Year 2000	
GDP	GDP ppp	Country	GCI	MEI	PII	TI	GDP ppp	GDP
1	1	*U.S.*	0	3	0	0	9.6	9.8
2	3	*Japan*	5	0	−3	0	3.4	4.8
3	5	*Germany*	1	−7	5	2	2.0	1.9
4	7	*UK*	−4	−1	−6	0	1.4	1.4
5	6	*France*	4	−1	8	1	1.4	1.3
6	2	China	−4	0	−10	0	5.0	1.1
7	8	*Italy*	−6	−6	−5	−3	1.4	1.1
8	12	*Canada*	−7	−3	−13	−3	0.8	0.7
9	9	Brazil	−7	−10	−4	1	1.2	0.6
10	11	Mexico	8	2	12	6	0.9	0.6
11	14	Spain	−1	1	−3	−1	0.8	0.6
12	13	Korea	7	8	−2	12	0.8	0.5
13	4	India	1	−3	8	−5	2.3	0.5
	10	Russia	1	5	−3	1	1.4	0.3
		average	0	−1	−1	1		
		G7 average	−1	−2,1	−2	−0,4		

Source: Author's editing of data from World Economic Forum for indexes and from [Economist 2003-6] for GDP and GDP ppp.

Table 3.17 Average annual utility patents granted in the U.S. to core technology-innovating economies in the 1980s, in 2001, and in 2002 (number of patents granted in the U.S. per million population). Countries are ranked by decreasing number of patents in 2002

Countries		U.S. Patents			Ranking		
		2002	2001	1980s	2002	2001	1980s
United States	U.S.	301	314	166	1	1	2
Japan	JP	273	261	101	2	2	3
Taiwan	TW	241	240	13	3	3	19
Sweden	SE	190	196	94	4	5	4
Switzerland	CH	189	196	190	5	4	1
Israel	IL	165	163	42	6	6	10
Finland	FI	156	140	37	7	7	12
Germany	DE	138	136	85	8	8	5
Canada	CA	110	116	50	9	9	7
Singapore	SG	98	72	2	10	14	25
The Netherlands	NL	87	83	52	11	11	6
Luxembourg	LU	83	—	—	12	—	—
Korea	KR	80	74	1	14	12	28
Belgium	BE	70	70	27	15	15	14
France	FR	68	68	43	16	16	9
Austria	AT	65	72	40	17	13	11
United Kingdom	UK	64	66	43	18	17	8
Norway	NO	54	59	23	19	19	15
Denmark	DK	53	90	32	13	10	13
Iceland	IS	46	63	9	20	18	21
Australia	AU	44	45	22	21	20	16
New Zealand	NZ	37	32	15	22	23	18
Ireland	IE	34	37	9	23	21	22
Hong Kong	HK	33	34	5	24	22	23
Italy	IT	30	30	17	25	24	17

Source: Author's editing of data from World Economic Forum.

Table 3.18 New entries in 1990–2000 among the core technology-innovating economies: countries that in the 1980s had fewer than 15 yearly granted utility patents in the U.S. per million population and passed this threshold before the year 2000. Ranking stands for ranking according to GCI

Country	U.S. Patents		Ranking	
	1980s	2001	1980s	2001
Taiwan	13	240	19	3
Iceland	9	63	21	18
Ireland	9	37	22	21
Hong Kong	5	34	23	22
Singapore	2	72	25	14
Korea	1	74	28	12

Source: Author's editing of data from World Economic Forum.

Patents Granted
Change of number of patents granted
100
10
1
0
CH UK FR DE AT DK NL US IT SE AU CA NO NZ BE JP IE IL FI IS HK TW SG KR

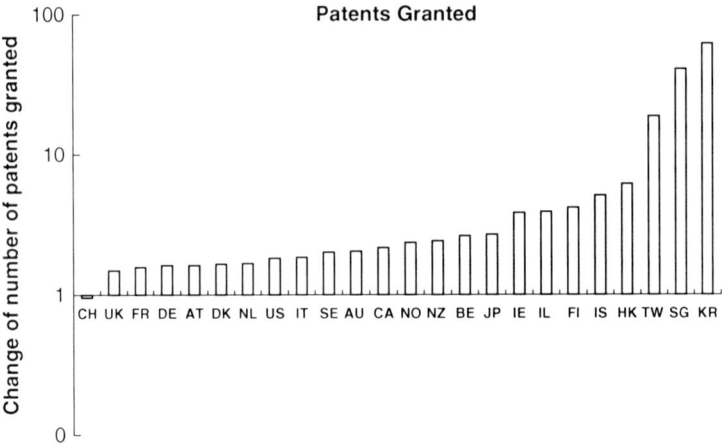

Fig. 3.58 Change from the 1980s to 2002 in the number of yearly granted utility patents in the U.S. per million population. The scale is logarithmic
Source: Author's editing of data from World Economic Forum.

in number of granted patents in 20 years. South Korea has jumped up by almost two orders of magnitude. Major increases are also seen for Singapore, Taiwan, Hong Kong, Iceland, Finland, Israel, and Ireland. We will see later in this book that these countries' performances are underpinned by major investments in research and technological development.

Even the daily newspapers have reported, starting in the year 2001, on the disappointing ranking obtained by Italy on competitiveness (see Fig. 3.59).

Ranking According to Competitiveness

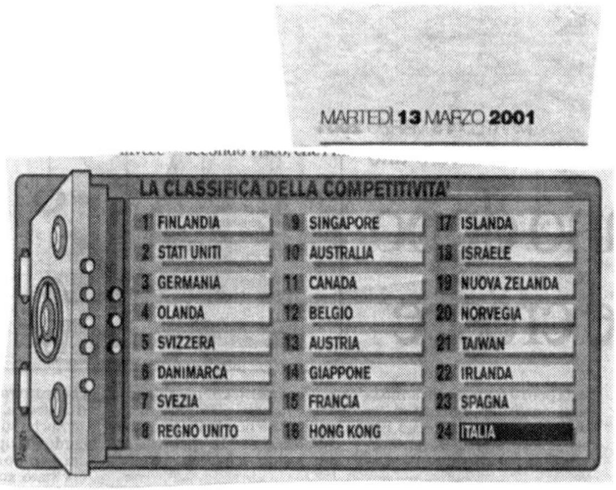

MARTEDÌ 13 MARZO 2001
LA CLASSIFICA DELLA COMPETITIVITA'
1 FINLANDIA
2 STATI UNITI
3 GERMANIA
4 OLANDA
5 SVIZZERA
6 DANIMARCA
7 SVEZIA
8 REGNO UNITO
9 SINGAPORE
10 AUSTRALIA
11 CANADA
12 BELGIO
13 AUSTRIA
14 GIAPPONE
15 FRANCIA
16 HONG KONG
17 ISLANDA
18 ISRAELE
19 NUOVA ZELANDA
20 NORVEGIA
21 TAIWAN
22 IRLANDA
23 SPAGNA
24 ITALIA

Fig. 3.59 Ranking according to competitiveness. In 2001 the daily newspapers started reporting on the disappointing ranking obtained by Italy for competitiveness
Source: *La Repubblica* 13 March 2001, reproduced with kind permission.

3.16 Factors Contributing to Growth in Competitiveness

We have examined the Union's situation in terms of competitiveness and productivity in comparison with the U.S. and Japan. Now we are going to see what contributes to growth in competitiveness and to a stronger presence in the new high-tech markets. The Union's future prosperity and the competitiveness of its economy depend on different factors. Let us list some of them:

1. research intensity and quality
2. development of human resources
3. ICT investments
4. business reorganization, and
5. access to financing

Let us take into consideration the first factor, namely, *research*. The level of *business* expenditure in research (as a percentage of GDP) in EU27 is almost half the level in the U.S. and Japan (see Fig. 3.61). Italy is well below half of the European average (see data for 2004 in Fig. 3.60). As to R&D in the *public sector*, the EU, the U.S., and Japan (Fig. 3.61) are more or less on an equal footing. Italy is below by almost 25 %. Korea had a total of 2.7 % of GDP in R&D in the year 2000, of which 80 % from private firms [Mahlich et al., ed.s 2007].

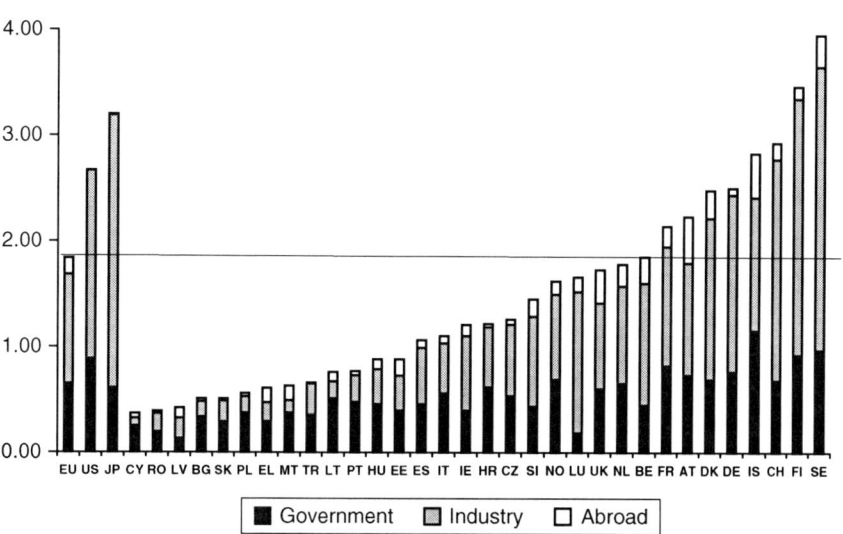

R&D Expenditure (% GDP)

Fig. 3.60 Investments in Research and Technological Development as a percentage of GDP in EU30, EFTA, the U.S., and JP. Government contribution is in black, industrial contribution in gray and contribution from abroad in white. Data are from 2004 or the latest year available. Countries are ranked by increasing total research investment
Source: Eurostat 2007.

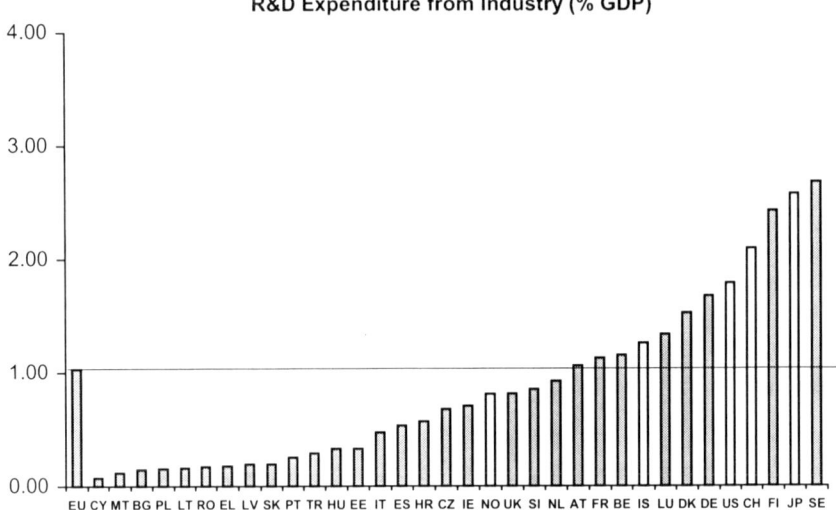

Fig. 3.61 Investments in Research and Technological Development from industry as a percentage of GDP in EU30, EFTA, the U.S., and JP. EU30 countries are in gray and EFTA, the U.S., and JP in white. Data are from 2004 or the latest year available. Countries are ranked by increasing industrial research investment
Source: Eurostat 2007.

When comparing investments in research, the European situation is even more serious than it might appear because of a phenomenon known as the *European paradox* [Caracostas et al. 1998]. Indeed, in Europe the return on investments made in research, in terms of increased scientific production, is the same or higher than in the other regions of the Triad (see Figs. 3.62, 3.63, and 3.64). However, the return in terms of technological or commercial impact, as measured by the number of

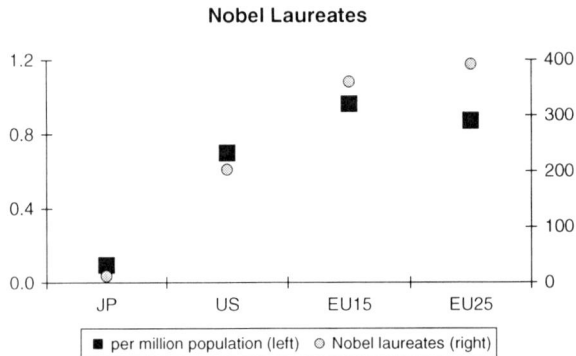

Fig. 3.62 Number of Nobel Prize winners in Japan, the U.S., EU15, and EU25. Absolute values up to 2004 are on the right scale; values per million population are on the left scale
Source: The Nobel Foundation, 2005.

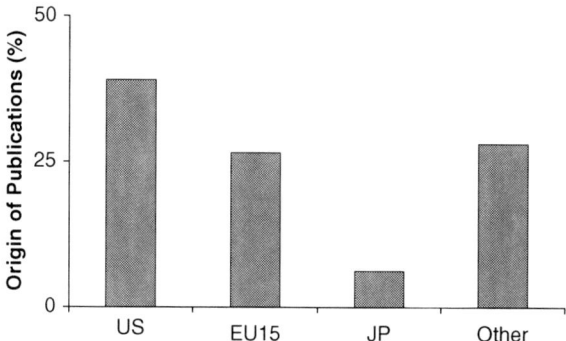

Fig. 3.63 Distribution (%) of geographical origin of scientific publications (all disciplines)
Source: [Caracostas et al. 1998].

patents filed, is much lower (see Fig. 3.65). The European paradox seems to point to a weaker connection between business and research in Europe compared with the U.S. or Japan.

Figure 3.60 shows different levels of R&D intensity in the member states. The picture is completed by Fig. 3.66, where R&D intensity is presented versus its annual growth, both measures referring to the EU average. Such averages define four quarters in the graph: the virtuous economies are in the NE quarter. Finland and Sweden have higher R&D investments than any other member state, and they also exceed the U.S. These countries are preparing a long-term future of prosperity and competitive presence in the world market. The member states placed in the NW quarter (Ireland, Portugal, Spain, Greece, and Austria) have lower-than-average R&D intensity, but a higher growth. They are well placed to bridge the gap with the

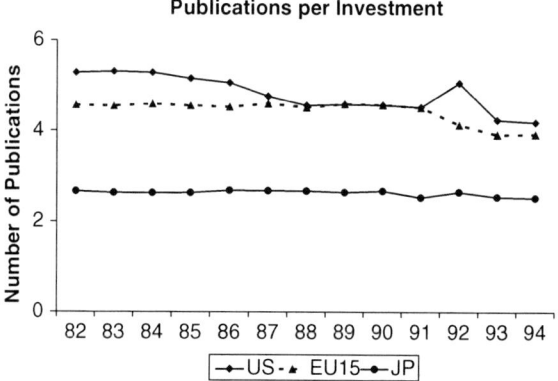

Fig. 3.64 Number of publications per unit research investment in the public sector
Source: [Caracostas et al. 1998].

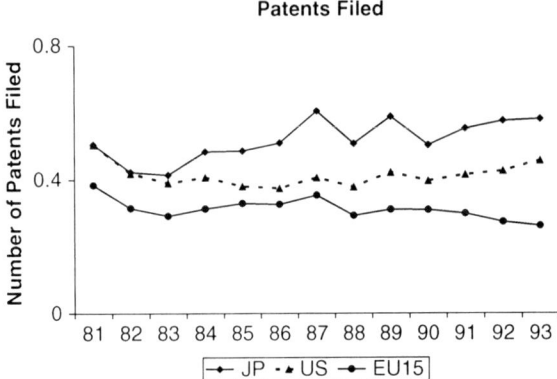

Fig. 3.65 Number of patents filed per unit research investment in the private sector
Source: [Caracostas et al. 1998].

most innovative EU economies. Italy is placed in the SW quarter. Its R&D policy is preparing a future of decline and is putting at risk its position among the largest industrialized countries. The Italian situation regarding R&D is particularly disturbing, its performance being positioned in the quarter of the chart of Fig. 3.66 where value and trend are both lower than the EU average: not only does Italy lag behind Europe, but it is also losing ground. The few countries (Greece, Portugal, and Spain)

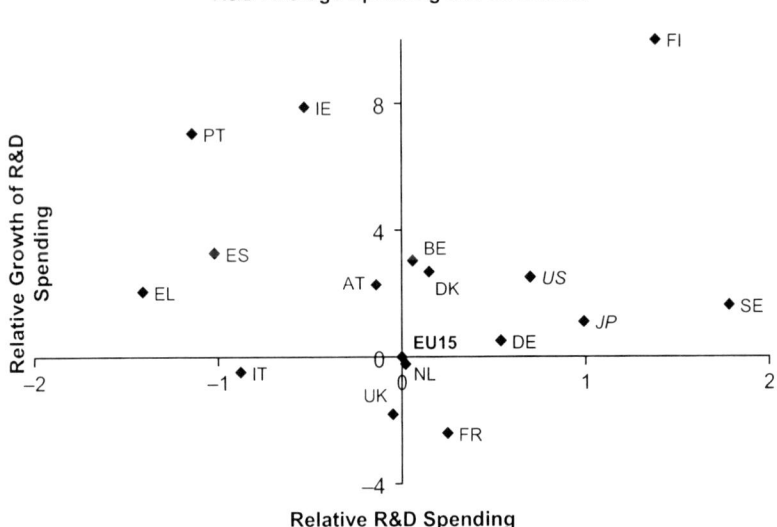

Fig. 3.66 Intensity of R&D as % of GDP (*horizontal*) versus its annual growth (*vertical*), relative to EU average (1999). Data for Luxembourg are not available
Source: Edited by the author from European Commission data [EC 2001-4].

that had a worse performance in R&D spending during 1999 have a positive trend and will leave Italy alone at the bottom of EU ranking.

On the occasion of the presentation of the first edition of this book at the University Roma-La Sapienza, Luciano Caglioti, professor of chemistry and Vice-provost responsible for research, made a statement that reflects his ideas as presented in a very stimulating book about Italy [Caglioti 2003]: "Saying that Italy is 42 positions behind Finland—he said—is based on the wrong assumption that Italy is lining with other countries trying to achieve the same goal, while Italy chose to go her own way, Italy is going somewhere else." Should you want to know where she is going, read about Italy's decline, starting.

Figure 3.67 shows the number of U.S. patents granted in the Triad countries during the year 2002. It is not surprising that the best performers in R&D (see Fig. 3.60) positioned themselves at the high end of the patent ranking (see Table 3.17 and Fig. 3.67).

Research and technological development (R&D) bear their fruits in the medium to long term. The time to market for an innovative idea ranges from 2 to 30 years, depending upon the industry sector: innovative software products may reach the market much earlier than, for example, pharmaceutical products or efficient exploitation processes for alternative energy sources. The economies lagging behind in R&D ranking will have a hard time catching up, because their position in high-tech international markets has already been compromised (see Figs. 3.68 and 3.69).

Figure 3.69 shows a correlation between intensity of R&D investments and trade performance in high-tech sectors. The best performers (Sweden, Finland) are among the countries that make the highest investments in R&D. Ireland is a case apart, because its highly favourable balance of payments in the high-tech sector does not originate from endogenous R&D but mainly from foreign investments. In Fig. 3.69, Sweden and Finland show a higher R&D intensity than the U.S. However, for a fair comparison, the U.S. should be disaggregated into its 50 states, in which case only Sweden would be posted among the 10 countries with the highest R&D intensity.

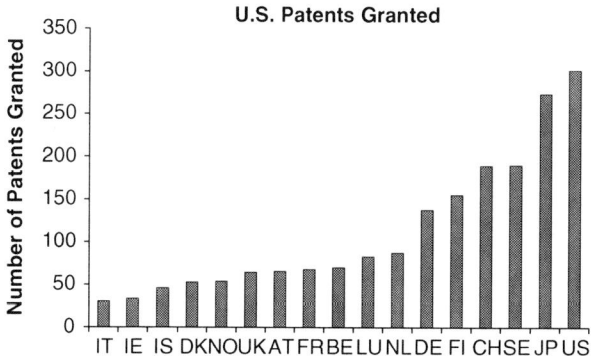

Fig. 3.67 Number of U.S. patents granted in 2002 in the Triad countries
Source: World Economic Forum 2003.

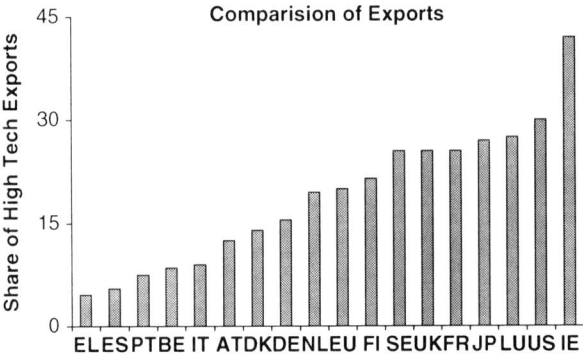

Fig. 3.68 Comparison of the percentages of high-tech exports in 2001 in the EU member states and in the Triad
Source: European Commission Scoreboard 2002.

Concerning the second factor contributing to growth in competitiveness, namely, *development of human resources*, our attention should focus on education, lifelong learning, and extent of participation in production activities. The percentage of citizens aged 25 to 64 having completed third-level education is low (see Fig. 3.70), as is the percentage participating in lifelong learning (see Fig. 3.71). Despite general acceptance that lifelong learning is of crucial importance, there has been no progress in the EU between 2000 and 2002 in improving participation. Only five member states (UK, FI, DK, SE, and NL) surpass the target of 12.5 % of the population set by the Lisbon agenda. The accession countries made progress from 2001 to 2002, but the level attained is quite low [EC 2003-6]. This problem, common to many European countries, is even more serious in those countries that experienced a drop

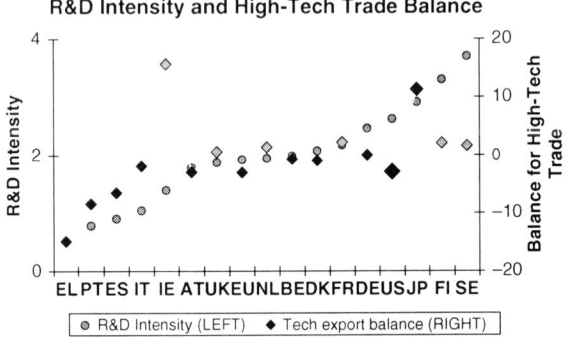

Fig. 3.69 Correlation between R&D intensity, measured in % GDP (grey dots) and trade balance for high-tech products (high-tech trade balance as % of total export). The European countries with a positive balance (grey diamonds: FI, FR, IE, NL, SE) are—with the exception of Ireland—among those investing the highest GDP percentage in R&D. EU stands for EU15
Source: European Commission, Second European Report on S&T Indicators. Key Figures (1999) and Towards ERA National R&D Policies indicators 2001.

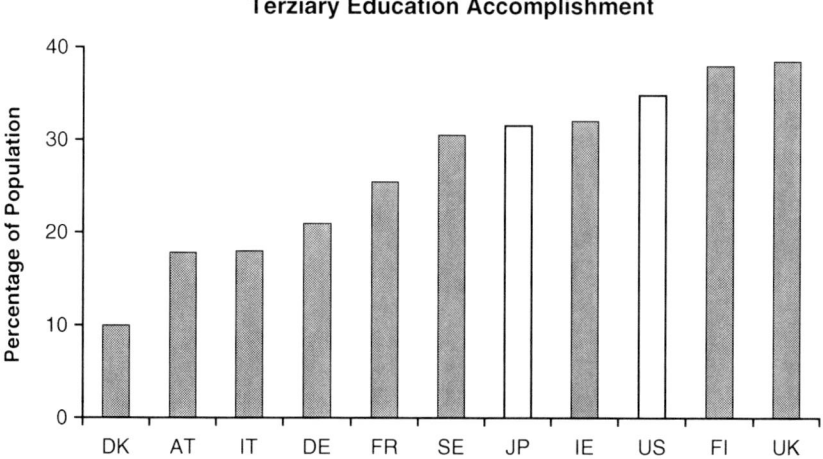

Fig. 3.70 Population of some countries of the Triad that has attained tertiary type education: percentage of graduates among the whole population at typical age of graduation (2000)
Source: European Commission, Benchmarking enterprise policy 2002.

in birth rate during the last decades of the 20th century, because an ageing workforce requires higher lifelong investments.

Concerning education and training, the outcome of a recent study conducted by the Shanghai Jiao Tong University is quite alarming [Shanghai Univ. 2004]. This

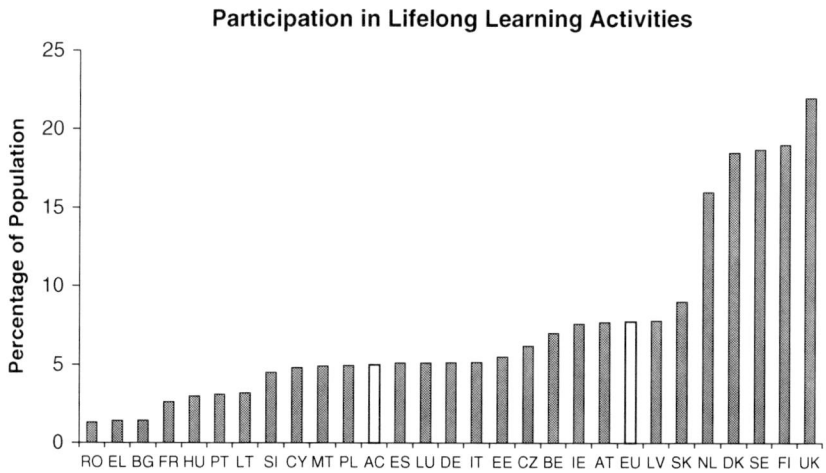

Fig. 3.71 Population of EU25 and candidate countries aged 24–64 participating in lifelong learning during the four weeks prior to the survey (as a percentage of the age group). Data for Croatia and Turkey are not available. AC stands for the average of the 10 accession countries, and EU is short for EU15
Source: European Commission, Scoreboard 2003.

The Best Universities

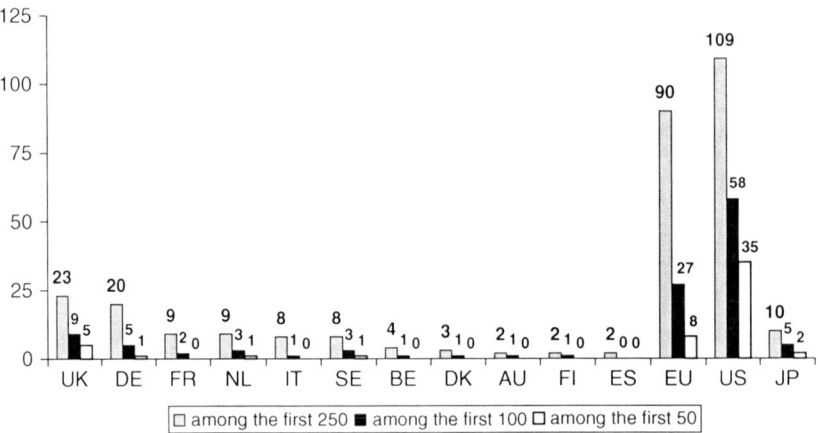

Fig. 3.72 Ranking of European Universities among the best in the world
Source: [Shanghai Univ. 2004].

study draws up a ranking of the world's top universities (see Fig. 3.72), based on the following criteria:

- Number of graduates who are now Nobel laureates in physics, chemistry, medicine, and economics
- Number of cited researchers, beyond a certain level, in the period 1981–1999 from 21 theme categories in life sciences, medicine, natural sciences, engineering, and social sciences
- Number of papers published on the world's leading scholarly science and technical journals from 2000 to 2002
- Number of citations on the Science Citation Index-expanded and on the Social Science Citation Index
- Performance of Academic staff with respect to the above indexes, normalized to the size of the academic staff

The results, presented below, are confirmed by another study conducted by the Higher Education Supplement of *The Times* [Higher 2004]. *The Times'* study is based on peer review and a study of the impact of R&D.

The U.S. has the lion's share with 35 universities among the first 50, followed by the EU with 8 and Japan with 2 (see Fig. 3.72). Five of the top 50 universities are outside the Triad, in Canada (2), Switzerland (2), and Australia (1). The UK has by far the best performance within the EU, followed by Germany, The Netherlands, Sweden, and France. On a per capita basis, the Scandinavian countries (Denmark, Finland, Norway, and Sweden) perform as well as the countries in the highest-ranked positions. Among the top 100 universities, there is not one belonging to an enlargement country. The only Italian University ranking among the top 100 is the University of Roma-La Sapienza, which ranks 70th. The best university from an enlargement country is the Hungarian University of Szeged, ranking 201st.

The other Italian universities, besides the University of Roma-La Sapienza, ranking among the top 250 are the University of Milan, University of Florence, University of Padua, Polytechnic of Milan, University of Bologna, University of Naples Federico II, and University of Pisa.

The 101 world's top universities according to [Shanghai Univ. 2004] are listed in Table 3.19.

Concerning the third point, *ICT investments*, we can observe in Fig. 3.73 that 2002 values are lower than 2001 values. This situation has been caused by a slowdown of the ICT market growth, following the collapse of the stock market bubble in the year 2000. In some states of the EU, the spreading of technology and computer literacy is not enough to give a significant boost to industry and services productivity. Five of the accession and candidate countries, together with five northern EU countries, rank above the U.S. in ICT expenditure as a percentage of GDP. Over half of EU30 countries lag behind the U.S. value.

The increased productivity registered by the U.S. in the 1990s is linked to ICT penetration in more than 50 % of businesses and households. Economists in the 1970s and 1980s used to say that computers, however ubiquitous, were not visible in productivity statistics. Recent economic studies [Economist 2003-2] have shown that a significant impact on productivity caused by the introduction of new technology can only be seen when technology penetration in companies reaches or exceeds 50 % of their employees and when penetration in households reaches 50 %. We can compare the delay in the visible effects of ICT on productivity with what has been observed about the structural changes in the economy at the beginning of the 20th century: the introduction of electric power, telephone communications, and motor vehicles into industry and society had a real impact on the economy only several decades after the first appearance of these technologies.

In 2002, the penetration of the Internet in business in the EU was on average above 75 %, with an increase over 2001 (see Fig. 3.74). The Scandinavian countries are positioned well at the top of the ranking. A significant increase was registered in Europe from 2001 to 2002 in the percentage of Internet users in the whole population (see Fig. 3.75).

Figure 3.76 suggests a correlation between pc GDP and investments in ICT. Sweden and the UK are at the U.S. level of investment in ICT as a percentage of GDP, and their position is well above the EU average. Greece, Spain, Italy, Portugal, and Ireland lag behind.

We shall discuss the fourth point, *business reorganization*, in detail in Chapter 7, "Entrepreneurship, Innovation, and Competitiveness" and in Chapter 8, "Competitiveness in the Knowledge-based Society". We shall see that for a company wanting to do business in the information society era, it is not enough to install a PC and an Internet connection on each worker's desk. To take any advantage of the introduction of technology, a change in mentality and attitude is needed as well as a new approach to the market and to production processes.

Table 3.19 The world's top universities according to a study conducted by the Shanghai Jiao Tong University [Shanghai Univ. 2004]. European universities are in *italics*. Those in *italics* with a * are European universities that are not based in a European Union member state (Switzerland and Norway). When the ranking algorithm generates the same score for two or more universities, the same ranking position is assigned to all universities sharing the same score value

101 universities in the top ranking of the world's universities	
1. Harvard Univ.	2. Stanford Univ.
3. California Inst. Tech.	4. Univ. California-Berkeley
5. *Univ. Cambridge*	6. Massachusetts Inst. Tech.
7. Princeton Univ.	8. Yale Univ.
9. *Univ. Oxford*	10. Columbia Univ.
11. Univ. Chicago	12. Cornell Univ.
13. Univ. California-San Francisco	14. Univ. California-San Diego
15. Univ. California-Los Angeles	16. Univ. Washington-Seattle
17. *Imperial Coll. Sci. Tech. Med.*	18. Univ. Pennsylvania
19. Tokyo Univ.	20. *Univ. Coll. London*
21. Univ. Michigan-Ann Arbor	22. Washington Univ.-St. Louis
23. Univ. Toronto	24. Johns Hopkins Univ.
25. *Swiss Fed. Inst. Tech.-Zurich*	26. Univ. California-Santa Barbara
27. Univ. Wisconsin-Madison	28. Rockefeller Univ.
29. Northwestern Univ.	30. KyotoUniv.
31. Univ. Colorado-Boulder	32. Vanderbilt Univ.
32. Duke Univ.	34. Univ. Texas-Southwestern Med Center
35. Univ. British Columbia	36. Univ. California-Davis
37. Univ. Minnesota-Twin Cities	38. Rutgers State Univ.-New Brunswick
39. *Karolinska Inst. Stockholm*	40. *Univ. Utrecht*
40. Pennsylvania State Univ.-Univ. Park	40. Univ. Southern California
43. *Univ. Edinburgh* Continued on next page	44. Univ. California-Irvine
45. *Univ. Zurich*	45. Univ. Illinois–Urbana
47. Champaign Univ. Texas-Austin	48. *Univ. Munich*
49. Australian Natl Univ.	49. Brown Univ.
51. Case Western Reserve Univ.	52. Univ. North Carolina-Chapel Hill
53. Osaka Univ.	53. Univ. Pittsburgh
55. Univ. Arizona	55. *Univ. Bristol*
55. New York Univ.	58. *Univ. Heidelberg*
59. *Uppsala Univ.*	60. *Tech. Univ. Munich*
61. Rice Univ.	61. Carnegie Mellon Univ.
63. *Univ. Oslo*	64. Tohoku Univ.
65. *Univ. Copenhagen*	65. *Univ. Paris 06*
67. Univ. Virginia	68. Nagoya Univ.
68. *Univ. Sheffield*	70. *Univ. Roma-La Sapienza*
70. Texas A&M Univ.-Coll Station	72. *Univ. Paris 11*
72. Univ. Rochester	74. *Univ. Helsinki*
75. *King's Coll. London*	75. Univ. Maryland-Coll Park
75. Univ. Florida	78. *Univ. Leiden*
79. McGill Univ.	80. Purdue Univ.-West Lafayette
81. Ohio State Univ.-Columbus	81. Univ. Utah
83. Tufts Univ.	84. *Univ. Vienna*

Table 3.19 (continued)

101 universities in the top ranking of the world's universities	
84. Univ. Groningen	86. McMaster Univ.
87. Michigan State Univ.	88. Univ. California-Riverside
89. Univ. Manchester	90. Univ. Iowa
91. Univ. Gottingen	92. Univ. Melbourne
93. Lund Univ.	94. Hebrew Univ. Jerusalem
95. Free Univ. Berlin	96. *Univ. Basel
96. Univ. Illinois-Chicago	98. Boston Univ.
99. Univ. Ghent	99. North Carolina State Univ.-Raleigh
99. Emory Univ.	

Source: [Shanghai Univ. 2004].

Concerning the fifth and last point, *access to financing*, in many European countries equity financing via the stock market is not very developed (see Fig. 3.77). Stock market capitalization accounts for 27 % of the financial market in Europe as compared with 34 % in the U.S. Enterprise borrowing is 37 % in Europe as against 20 % in the U.S. [Unice 2001]. In the EU, equity capital often comprises only capital injected by the entrepreneur. In addition, the tax system often provides incentives for credit financing rather than equity financing. Finland is an exception. Finland, in fact, in many ways is very much oriented towards a U.S. business model. Figure 3.77 shows a big swing in many countries between the minimum and maximum values of market capitalization during the period 2000–2003. Values at the end of the period

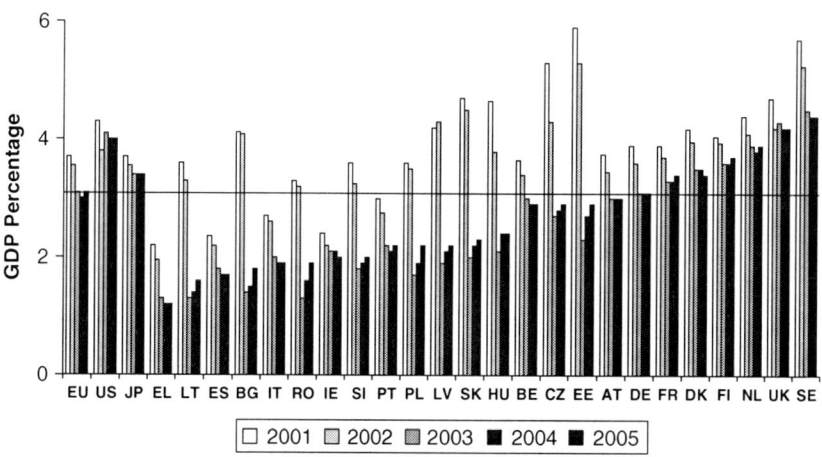

Fig. 3.73 ICT expenditure (% GDP) in EU27, the U.S., and JP. BE stands for Belgium + Luxembourg. EU is short for EU27. Data for Cyprus and Malta are missing
Source: European Commission, 2004, 2007.

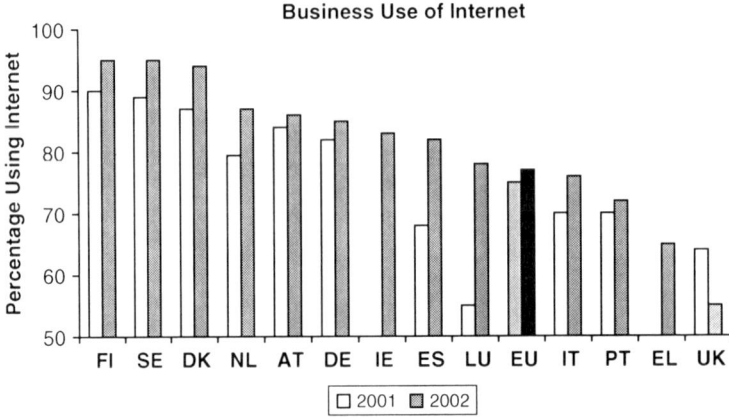

Fig. 3.74 Percentage of Internet penetration in business in the years 2001 and 2002
Source: European Commission, Scoreboard 2003.

are lower than the values during 2000, following the collapse of the stock market bubble and the slowdown in the economy.

Venture capital is equity investment made for the launch, early development, or expansion of a business. Venture capital is provided by specialized institutions or by specialized departments of commercial banks, and also by private investors, known as *business angels*. Business angels are capitalists (often wealthy retired business executives) who promote the development of innovative enterprises, at times with a spirit of generosity, unselfishness, and social liability. Venture capital is the key

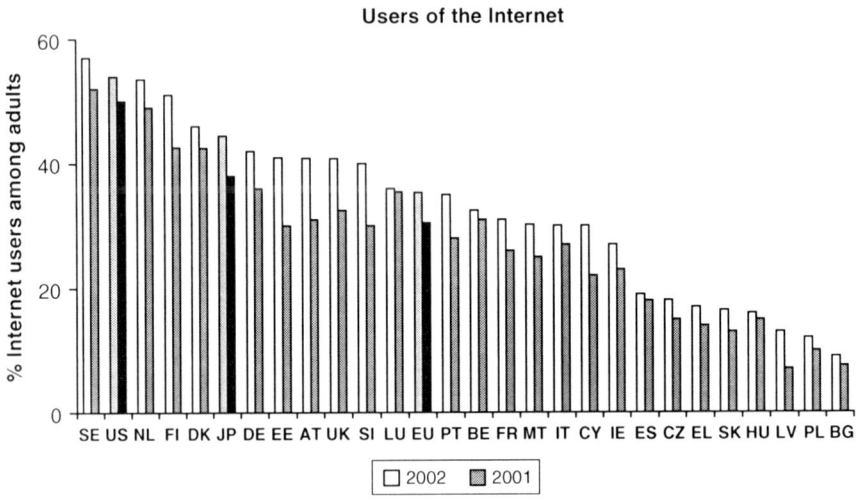

Fig. 3.75 Percentage of Internet users in the whole adult population in the years 2001 and 2002
Source: European Commission Scoreboard 2003.

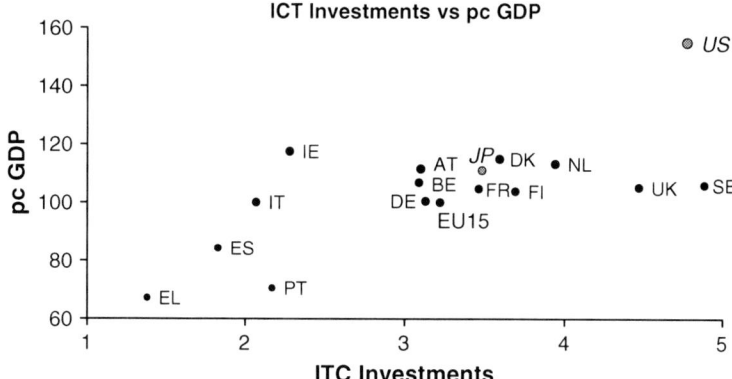

Fig. 3.76 Investments in information and communication technologies as a percentage of GDP (*horizontal*) versus pc GDP relative to EU average (*vertical*) in the year 2002. Data for EU15, the U.S. and Japan. As to ICT investments, Sweden and UK are at the U.S. level and well above the EU average. Greece, Spain, Italy, Portugal, and Ireland lag behind
Source: Edited by the author from data published in the EɪᴛO 2004 Report and from EC Facts and Figures 2004.

to the financing of innovative and fast-growing small companies that may not be able to finance their expansion through loans because of their high risk, limited capital, and lack of track record. Venture capital helps their development and may allow eventual access to the regulated stock market. Different stages of financing are included under venture capital: seed investments (funding the later stages of R&D and the development of initial prototypes), startup (financing product development

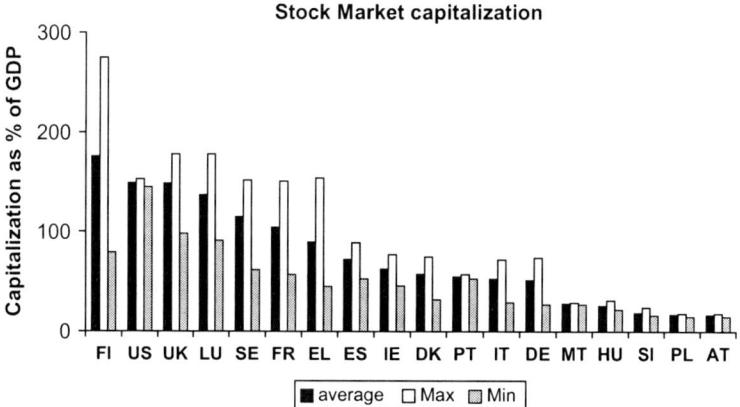

Fig. 3.77 Average market capitalization as percentage of GDP during the four years 2000–2003 (*black*) and maximum (*white*) and minimum (*grey*) values in the same period. The FR data in the figure add up the values for France, The Netherlands, and Belgium. Data for Bulgaria, Cyprus, the Czech Republic, Estonia, Croatia, Lithuania, Latvia, Romania, Slovakia, and Turkey are missing
Source: Edited by the author from European Commission, Scoreboard 2003.

and business development prior to market entry), and later stage (to provide financial resources for the expansion of production capacity and company growth). In some southern European countries, the high inflation experienced in the last decades of the 20th century suggested state-guaranteed bonds as a safe, highly remunerative investment for small savers—in the short term—because of the high interest rates that go along with high inflation. We will expand on this point later, considering the Italian case with some detail.

The habit of saving through equity investments came forward gradually in those countries at the end of last century with the big gains offered by a soaring stock exchange, but it was soon abandoned after the collapse of the stock market bubble. This situation is reflected by the level of capitalization of the stock exchange (see Fig. 3.77) and, more dramatically, by the size of the venture capital market (see Fig. 3.78) or the number of networks of capitalists promoting the development of innovative enterprises, known as business angels (see Fig. 3.79). The very idea of investing a part of one's portfolio in venture capital products is not familiar to the small savers and to some extent to the banking community in a number of southern European countries, where *venture* capital translates in the national languages as *risk* capital, and conjures up the dangers associated with the operation rather than the chances of high rewards.

The financial services available in Italy for business support are inadequate, especially for expanding SMEs (see Fig. 3.90). Stock market capitalization is low as is the small saver's propensity to invest in shares. This situation derives from the tendency to invest in treasury bonds guaranteed by the state. This type of investment became popular in the 1970s and 1980s, when runaway inflation led to a high

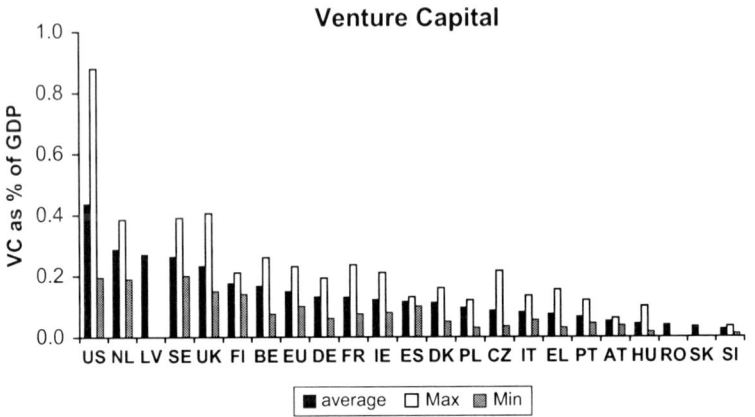

Fig. 3.78 Average venture capital (early and later stage) as percentage of GDP during the four years 1999–2002 (*black*), and maximum (*white*) and minimum (*grey*) values in the same period. The data for Luxembourg are missing and the EU15 value does not include them. Data for Bulgaria, Croatia, Cyprus, Estonia, Lithuania, Malta, and Turkey are missing
Source: Edited by the author from data published in the EITO 2004 Report and from EC Facts and Figures 2004.

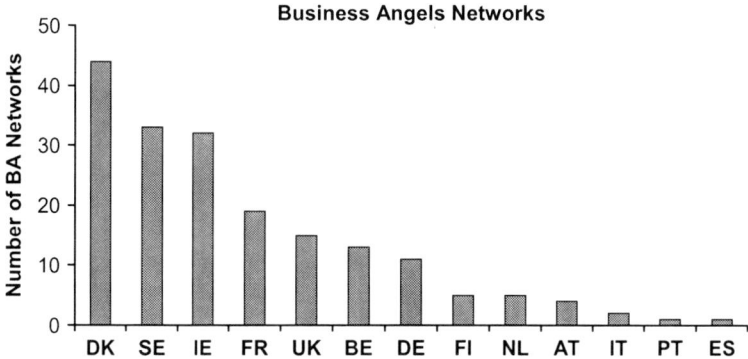

Fig. 3.79 Number of business angels networks per million enterprises in the EU (average number in the period 1999–2002). Data for Greece and Luxembourg are not available.
Source: European Commission, [EC 2001-3].

cost of money, which resulted in doubledigit returns on treasury bonds, which were therefore perceived as a good investment, at least in the short term. As a consequence, investing one's savings in shares has never been popular among the Italians. The above-mentioned gradual move towards the stock exchange at the time of the soaring stock market did not quite take hold in Italy. When the economic recovery and the curb on inflation, implemented in view of qualification requirements for the Economic and Monetary Union, contributed to a reduction in treasury bond interests rates, savings were diverted to fixed-interest investments, which proved to be more profitable (and more risky, too: see Argentinean bonds and Parmalat corporate bonds).

3.17 Small and Medium-Sized Enterprises

The European Union's business situation differs from country to country with reference to company size. Employment in Europe is almost equally divided into microenterprises (0–9 employees), small enterprises (10–49) and medium-sized enterprises (50–249 employees), and large-sized enterprises (250+), as shown in Fig. 3.80. The so-called SMEs (small and medium-sized enterprises) are those with less than 250 employees, i.e. SMEs include microenterprises in addition to small and medium-sized ones. Southern countries have an average enterprise size lower than the EU average (Fig. 3.81).

In 2001, there were 20 million enterprises in the European Union, only 0.2 % of which had more than 250 employees. In addition, only 34 % of workers were employed in the small percentage of large-sized companies, while the remaining 66 % were employed in microenterprises (0–9), small enterprises (10–49), and medium-sized enterprises (50–249), as shown in Fig. 3.80 and Table 3.20.

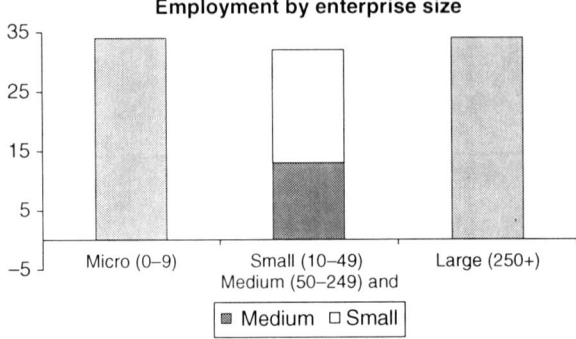

Fig. 3.80 Percentage of employment in EU enterprises of different size
Source: Edited by the author from European Commission data [EC 1997-3].

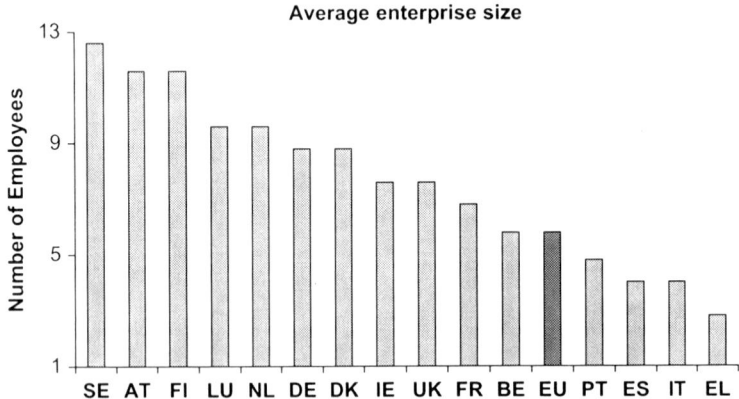

Fig. 3.81 Average size of the EU enterprise
Source: Edited by the author from European Commission data [EC 1997-3].

Table 3.20 Size of enterprises in the EU in the year 2001

	Dimension	% enterprises	% employment
Micro	0–9	93	34
Small	10–49	6	19
Medium	50–249	1	13
Total SME	0–249	99.8	66
Large	250+	0.2	34

Source: European Commission, Observatory of European SMEs [EC 2002-3].

In the regions of the Triad, there is a sensible difference in the average number of workers per company: 6 workers per company in Europe, 19 in the U.S., and 10 in Japan (see Fig. 3.82). Figures 3.83 and 3.84 provide a more detailed comparison between EU19,[4] the U.S., and Japan (year 1996).

In comparing the percentage of large-sized companies based in the U.S. with the percentage of large-sized companies based in the Union (see Fig. 3.84), it should not be forgotten that the U.S. large domestic market (with no customs barriers and a single currency) has been in place for over 200 years, while the European single market was established only a few decades ago and is still in the making. However, the mergers of European companies and the creation of transnational European companies is the beginning of a process that is going to reduce the company size gap in the future.

In countries with a prevalence of small enterprises, it is essential to ensure their growth, survival, financing, and development. But SMEs are actually very important even in those countries characterized by business giants or large conglomerates, because small and medium-sized enterprises can help future economic development; they can be a source of employment and stimulate the creation of new jobs. Figure 3.85 shows that in EU15 and EFTA, large companies created fewer net new jobs than SMEs, and among SMEs only microenterprises created more net jobs in the year 2000 than those created in 1988. Being more streamlined, SMEs can be rapidly transformed and can easily become more specialized. Synergy with larger companies and intense international business contacts, typical of the economy of the information society, make it possible for SMEs to be one of the players in the new globalized and networked economy, in which even small and medium-sized enterprises can operate in faraway markets.

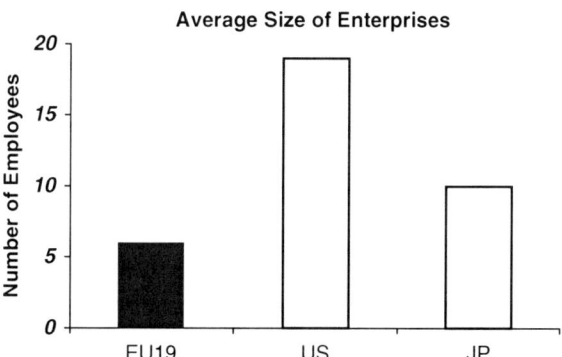

Fig. 3.82 Average size of enterprises in the Triad (1996). EU19 is the EU15 group of nations plus the EFTA countries, namely, Iceland, Liechtenstein, Norway, and Switzerland
Source: Edited by the author from European Commission data [EC 2002-3].

---

[4] EU19 stands for the EU15 group of nations plus the EFTA countries, namely, Iceland, Liechtenstein, Norway, and Switzerland.

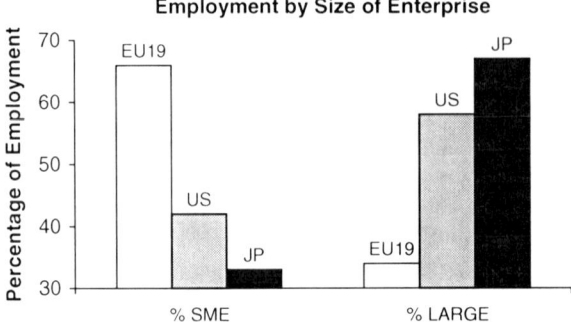

Fig. 3.83 Comparison of employment in small-sized and large-sized enterprises in the Triad. Data from 1996. EU19 is the EU15 group of nations plus the EFTA countries, namely, Iceland, Liechtenstein, Norway, and Switzerland
Source: Edited by the author from European Commission data [EC 2002-3].

A survey conducted by the European Commission in 2002 on the limiting factors for the development of SMEs reported the following results (see Fig. 3.86): 20 % of SMEs in 2002 had problems recruiting qualified personnel and 15 % of SMEs had problems accessing development financing. In the light of what has been previously reported (see Figs. 3.70 and 3.71, Fig. 3.77, Fig. 3.78, and Fig. 3.79), we can see that the Italian SMEs are facing a particularly difficult situation given Italy's position on issues like training and access to financing, compared with the other European partners.

European SMEs have increased their international business contacts. This trend increased between 1999 and 2001, particularly in Spain, Greece, Great Britain, and Belgium (see Fig. 3.87). In Chapter 5, "EU Research: Objectives and Results", we shall see that Community-financed research programs have played a fundamental role in the internationalization of European businesses.

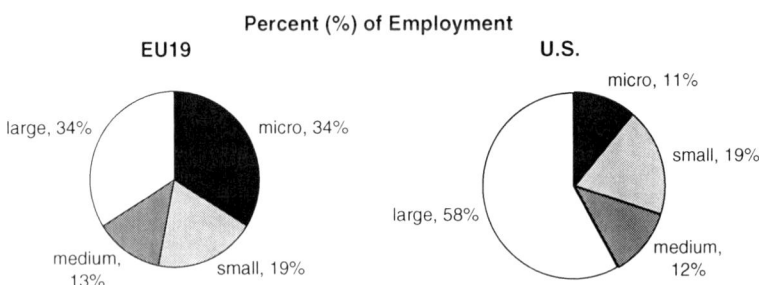

Fig. 3.84 Percentage of employment in the U.S. and in the EU versus size of enterprise. Data from 1996. EU19 is the EU15 group of nations plus the EFTA countries, namely, Iceland, Liechtenstein, Norway, and Switzerland
Source: Edited by the author from European Commission data [EC 2002-3].

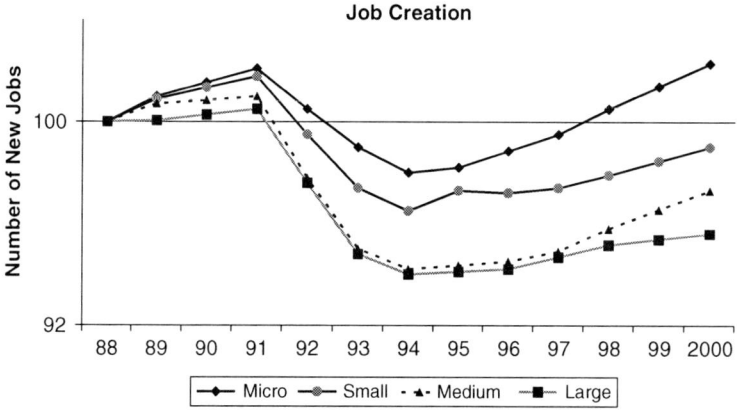

Fig. 3.85 Job creation in EU19 within companies of various sizes. New jobs created from 1988 to 2000 referred to the 1988 value (=100). Only microenterprises employed more workers in the year 2000 than those employed in the same size category in 1988
Source: Edited by the author from European Commission data [EC 2002-3].

SMEs grow and thrive in the market thanks to synergies and forms of cooperation with other enterprises. The nature of business contacts among SMEs differs in the various member states. Industrial clusters are an Italian feature (but not only Italian). In Italy, 43 % of the workforce is employed in industrial clusters. Some well-known, thriving industrial clusters in Italy are, for example, the production of eyeglass frames (Belluno area), silk (Como area), ceramics (Sassuolo area), biomedical (Emilia-Romagna region), and precision machine tools (Emilia-Romagna region).

Clusters are groups of interdependent companies based in special geographical areas. They belong to the same sector of business or to interdependent sectors and develop synergies by sharing technologies, business strategies, production processes, and competences. They pool procurement and different services, thereby

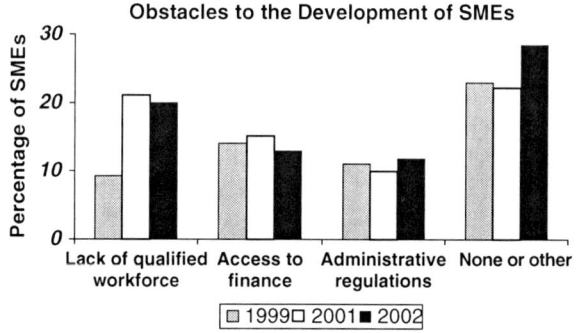

Fig. 3.86 Obstacles to the development of SME business in the EU in 1999, 2001, and 2002
Source: Edited by the author from European Commission data [EC 2002–3].

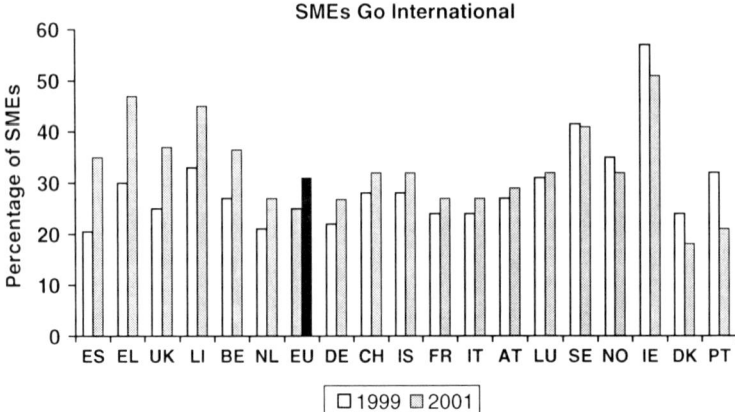

Fig. 3.87 Percentage of SMEs with more international business contacts in 1999 (white columns) and in 2001 (grey columns) than five years before. EU stands for EU15 and EFTA countries. Countries are ranked by decreasing difference of the 2001–1999 values
Source: Edited by the author from European Commission data [EC 2002–3].

generating economies of scale. The success of a cluster of SMEs depends on the presence of local traditions and assets, qualified human resources, entrepreneurship, education and training, qualification, and research. Sometimes, the success of an industrial cluster is so closely linked to the local know-how and skills that we tend to talk about *sticky knowledge*, meaning that it cannot be easily transferred somewhere else.

Public authorities, being aware of the great importance of clusters for economic development, try to support their growth through fiscal and legal incentives. Nevertheless, public authorities often are not perceived as favouring growth and development (see Fig. 3.86). In Italy, through the laws 317/1991 and 598/1994, authorities have tried to build a legal framework in favour of industrial clusters and their businesses. Such policies were aimed at creating research centres and other infrastructures for experimentation, pilot projects, and staff training, with the objective of improving the technological level and innovation capacity of SMEs. Thanks to the involvement of the local administration, decentralized initiatives linked to the local demand have been implemented.

Clusters originated well before the Information Society era. In the present context the geographical proximity is less relevant than it used to be because distance may be compensated by networking. The concept of clusters need revamped; the networks of excellence and the EU research area (described in Chapter 5, "EU Research: Objectives and Results" can contribute to the evolution of the idea of clusters.

SME competitiveness depends on the acquisition of state-of-the-art technologies. In Europe, the situation has improved considerably over the last few years, and even the countries that were lagging behind in the past in the use of Internet are now gaining ground (see Fig. 3.88). Since 2001, the 50 % threshold has been overcome (with the exception of France and Portugal; in France, the slow penetration of the

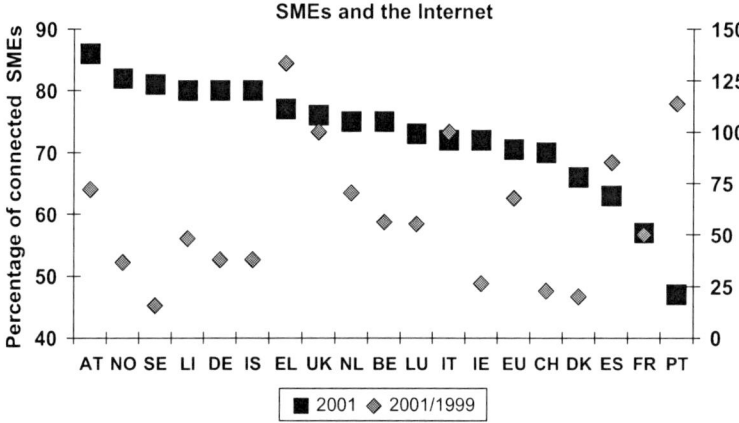

Fig. 3.88 Percentage of SMEs connected to the Internet in 2001 (*black squares, left scale*), and percent difference from the number of those connected in 1999 (*grey diamonds, right scale*). EU stands for EU15 and EFTA countries
Source: Edited by the author from European Commission data [EC 2002-3].

Internet may have been affected by the widespread use of an earlier technology, Minitel, that kept some users prisoners of an obsolete solution that offers only a small subset of the Internet functions). The implementation of ICT has a positive impact on productivity only if it is diffused over the entire industrial region and at least 50 % of companies have full access to such technology [Economist 2003-2].

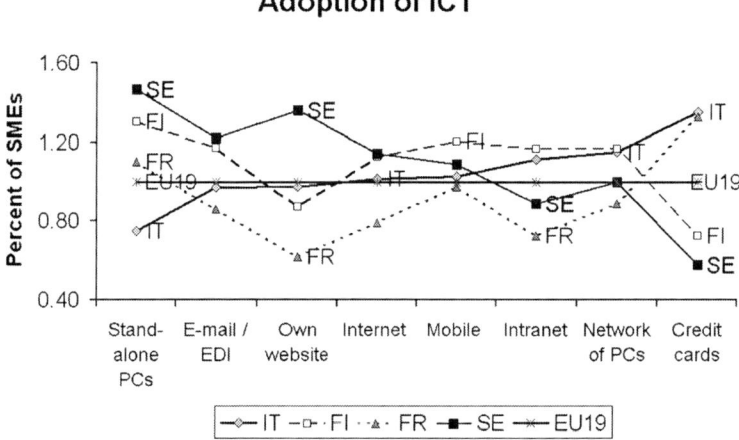

Fig. 3.89 Percentage of SMEs of five EU15 countries that have adopted some ICT technologies, with respect to EU15 and EFTA (EU19) average. Data from the year 2001. The lines in the figure have the sole purpose of facilitating the association of the points with the countries
Source: Edited by the author from European Commission data [EC 2002-3].

The Observatory of European SMEs has conducted a study on European companies about the acquisition of different types of ICT [EC 2002-3]. Figure 3.89 shows the percentage of SMEs (as against the EU15 and EFTA average) that have introduced different types of technologies, such as stand-alone PCs, e-mail, own website, Internet, mobile telephony, Intranet, network of PCs, and credit card payment. The lines in Fig. 3.89 have the sole purpose of facilitating the association of the points with the countries and make reading easier. As the figure shows, Italy is performing well and is ahead of the EU19, although some countries (Sweden, Finland) have registered higher rates. France is well behind the other countries.

SME growth (and competitiveness) also depends on access to financing, as Fig. 3.86 shows. SME satisfaction with the services offered by banks in different countries is shown in Fig. 3.90. The level of satisfaction among the Italian SMEs is the lowest, namely, 53 %, as against the EU19 average of 65 %.

Figure 3.81 about average company size clearly shows that Italy is completely different from the other most populated countries in the Union. This study on research and competitiveness highlights the peculiarity of the Italian situation, which cannot be neglected. It makes us aware of the need for special R&D activities to sustain the growth of Italian companies.

The majority of microenterprises do not participate in any national research programs, and even fewer participate in Community programs. For micro and small enterprises, the struggle for survival is often more of a priority than the need for innovation (see Fig. 3.91). Because of their small size, they cannot benefit from economies of scale, profit margins, and possible future technological innovations, which are the strategic and financial purposes of long-term investment in research. Those companies that do not actively participate in technological progress need specific tools to share the innovations derived from technology transfers or to benefit from consultancy contracts. They need to adopt materials, instruments, and processes of proven robustness and effectiveness. Research must bring benefits to the entire business fabric, and there should be more contacts between the top

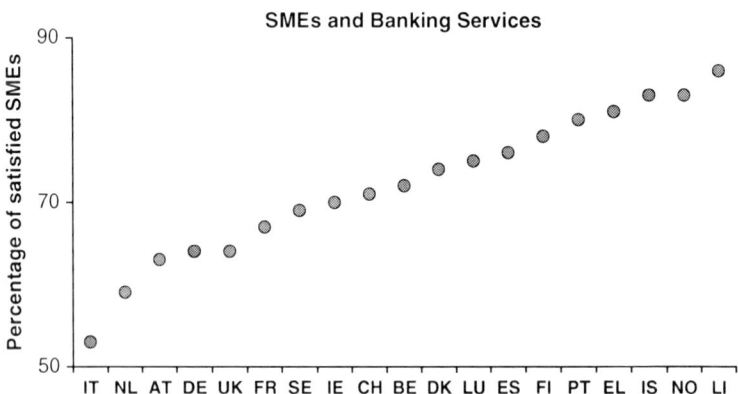

Fig. 3.90 Percentage of SMEs that declared themselves satisfied or very satisfied with the banking services available in their country. The EU15 and EFTA average is 65 %
Source: Edited by the author from European Commission data [EC 2002-3].

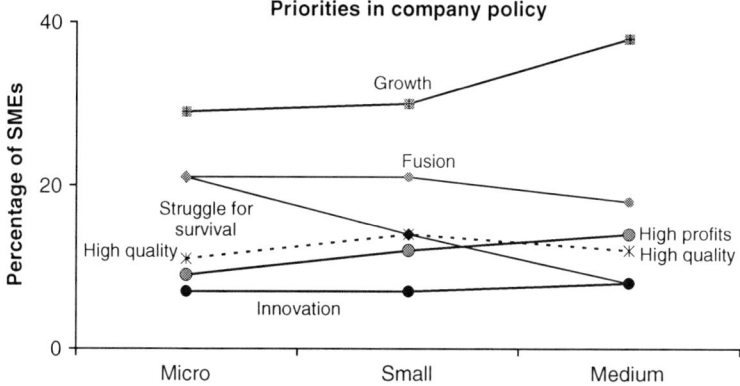

Fig. 3.91 Main priorities in company policy for European SMEs. Struggle for survival weights three times more heavily than innovation. The lines in the figure have the sole purpose of facilitating the reading
Source: Edited by the author from European Commission data [EC 2002-3].

innovators and the remaining, bigger share of the business world. However, the adoption of exogenous R&D results does not come without problems, and it is only effective if management and staff in the recipient enterprise understand and value technological innovation and are in command of it. Only under these conditions can they adopt the innovative breakthroughs and integrate them in the processes and products of their firm.

As we shall see in the second part of this book, Community R&D programs have been defined after broad consultation with thousands of operators from the different member states. The most active states in this phase of consultation will have more say in the definition of these programs, which will therefore mirror the research requirements of economies characterized by large industrial groups. The absence at these negotiating tables (or at best the scanty, not very qualified presence) of the Italian authorities, as well as of the Italian business and research world, is the main reason why such programs give more priority to the needs of economies with a bigger company size than that of the Italian companies. Consequently, Italian SMEs do not find any programs that meet their specific needs, and they do not take an active role in answering calls for research proposals. In addition, those organizations that do not directly and actively participate in scientific and technological progress may have difficulties in uptaking technological innovations generated by active participants in R&D programs.

3.18 Policy Guidelines

Every year, the European Commission issues policy guidelines [EC 2001-6, EC 2002-7, EC 2003-9, EC 2005-6, EC 2006-15, EC 2007-27] aimed at the attainment of the strategic objectives set by the European Council. The first European Council during the term of office of the 2000–2004 Commission, namely, the Lisbon Council in March 2000, defined an ambitious strategic plan for the creation, by 2010, of the most dynamic and competitive knowledge-based economy, capable of ensuring sustainable economic growth, higher employment rates, improved work conditions, and social cohesion. This plan, called the *Lisbon Agenda* or the *Lisbon Strategy,* is discussed in Chapter 8, "Competitiveness in the Knowledge-based Society". The policy guidelines mirror the Council strategy and are updated year after year. They include general observations and specific recommendations for each member state. They are concerned with issues such as budget policy, the labour market, productivity, entrepreneurship, and human resource management. The following considerations are derived from last years' guidelines, included in the recommended reading at the end of this chapter.

Economic policy recommendations are aimed at stimulating growth in a context of stable prices, while promoting savings and investments. Investment in training and research should increase productivity and, consequently, competitiveness and global market shares. Domestic demand should be stimulated so that GDP can grow faster than productivity in order to ensure higher employment levels. The fiscal policy must apply suitable structural measures in order to make member states' budgets balance out and to ensure that public debt is reduced and stays below 60 % of GDP.

Wage policies must be in line with price stability policies and reward productivity and qualification. Wage increments must be commensurate with productivity growth in order to support competitiveness and labour market development. Fewer rules, more flexibility, and opportunities for vocational training and retraining are recommended, as is the promotion of part-time work and flexible working hours. The Commission encourages the elimination of those rigidities that, instead of protecting workers, leave many on the sidelines of the labour market in a situation of long-term unemployment. A compromise between flexibility and stability can give more opportunities to people looking for a job, while providing companies with more labour. Public spending should be limited and prioritized in order to stop the squandering of resources and to ensure the development of human resources in terms of qualification and participation. The reform of the pension system should take into account demographic changes, reduce early retirement incentives, be economically sustainable, safeguard pension adequacy, and provide for special measures according to the different member states.

The diffusion and development of technologies has not been sufficient in most EU countries to progress along the lines of the ambitious objectives set in the Lisbon agenda, namely, the attainment by 2010 of the most competitive knowledge-based economy in the world. The 2002 guidelines suggest improvement in ICT infrastructures and commitment to education and lifelong learning. They suggest that, by 2010, up to 3 % of GDP should be invested in R&D, of which two-thirds should come from the private sector. The 2002 document is also concerned with the study of employment levels, productivity levels, and participation in production activities in the member states, based on 2000 and 2001 data (see Figs. 3.92 and 3.93).

In order to interpret the data in Figs. 3.92 and 3.93, it should be noted that in recent years economic growth in the Union has been linked more to the percentage of hours worked than to an increase in productivity (see Fig. 3.94).

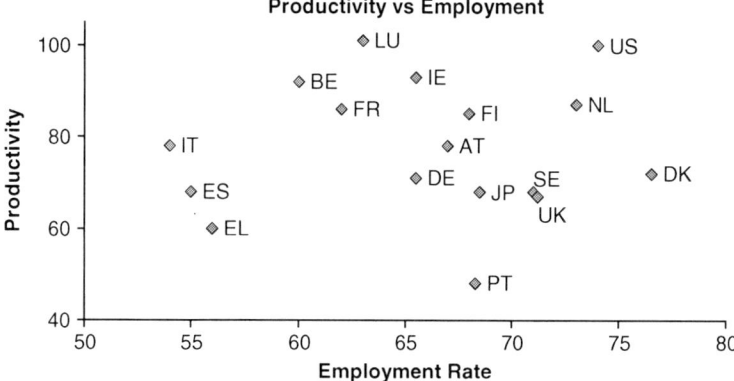

Fig. 3.92 Employment versus productivity levels in the EU15 member states. Productivity is relative to U.S. = 100. Employment is the rate of those employed in the working age population. Data from 2000
Source: Edited by the author from European Commission data [EC 2002–7].

In the light of these figures, we can say that the low rankings of Greece, Italy, and Spain in terms of participation are an indicator of a structural weakness that limits economic growth. These data reflect different lifestyle choices but also conceal anomalous levels of the shadow economy (see Fig. 3.21). As a matter of fact, these three states are those with the highest percentages of black-market work. The increase in participation must be stimulated with incentives to favour employment of the more disadvantaged categories, marginalized because of the structural shortcomings of society or owing to wrong welfare measures. Some good examples of measures or incentives to be implemented by public authorities in order to increase participation are the following: extended maternity leave, day nurseries,

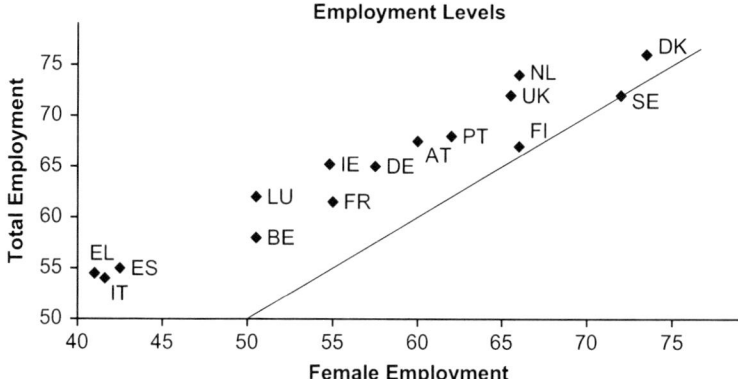

Fig. 3.93 Rate of participation to production activities of the total working-age population versus female employment. The diagonal line corresponds to equal male and female employment. Data from 2001
Source: Edited by the author from European Commission data [EC 2002-7].

Fig. 3.94 Productivity component of economic growth in the Euro Area and participation component. Data from 1996 to 2000
Source: Edited by the author from European Commission data [EC 2002-7].

convalescent homes, home help for the elderly and the disabled, a legal framework
promoting part-time and flexible jobs, education and lifelong learning programs,
time-limited unemployment benefits (dependent on the participation in qualifying
activities aimed at reintegration in the job market), tax incentives for businesses
favouring workers' reintegration, promotion of occupational mobility through lifelong learning programs, and promotion of geographical mobility through incentives
in the real-estate market and through special relief in real-property transfer tax when
property transfer is caused by mobility.

Unemployment is higher in economically depressed areas, where it goes hand in
hand with the demand for qualified workers. The labour market is uneven and badly
affected by the lack of geographical and occupational mobility, as well as by the
lack of salary differentials based on productivity levels.

The accomplishment of the Community objective set in Lisbon in 2000 would
create 15 million jobs by 2010 and presupposes the extension by five years of the
retirement age. Unemployment has decreased since 1997, but human resources are
not being exploited to their full potential in the European Union. An effective human
resource training policy is needed in order to create new competences and skills. In
addition, training should be provided to workers throughout their professional life.

Living conditions will improve only with an increase in participation in production activities and with more productivity growth, which, in turn, is linked to
effective investments and technology penetration.

3.19 Policy Guidelines for Italy

The rise in prices in 2002 and 2003 was higher in Italy than the European average
because of the inadequacy of checks and controls over dealers' speculation associated with the changeover to the single currency. The economic growth in Italy was
slower than the European average, and this has badly affected tax revenues. There
is still a wide gap between the North and the South of the country as far as growth
and employment are concerned. Italy failed to reach the objective of balancing the

state budget by 2003, and public debt is still huge - the highest in Europe both as GDP percentage and in absolute terms - and started to grow again in 2005. Thanks to the measures applied to reduce the budget deficit, Italy was able in 2003 to comply with the objective of containing the deficit within 3 % of GDP. However, such measures were not structural ones but rather one-off actions, based on pardon and securitization of public real property and other state goods. In addition, the laws on decriminalization of false accounting, and other *ad personam* laws proposed by the government chaired by Berlusconi and approved by the Parliament [Economist 2002-3, Economist 2004-2] and the justification of tax evasion by the institutions themselves [Luzi 2004, Bocca 2004] have contributed to weakening the constitutional state based on law and order and have increased the tendency to ignore the law and evade taxation. The reforms started in 1998 are bringing positive effects to the labour market, such as a reduction in unemployment to less than 9 %. However, the big gap between the North and the South is still there. Flexible work contracts have become increasingly common, but unlike permanent contracts in medium- and large-sized companies, they offer a relatively low level of protection and security, thus creating a two-tier job market, the lower tier having very limited security, insurance, and protection in case of pregnancy, serious illness, or unemployment.

The level of participation is still among the lowest in the Union and hides a prosperous shadow economy. The actions taken by the government to regain some illegally unreported economic activities have not produced the desired results. Foreign investments in Italy dropped dramatically in 2002, as reported in Chapter 14, "Economic and Monetary Union". Innovative programs still have many shortcomings, and corporate investment in R&D is very low (0.5 % in 2000). Since the Italian economy is mainly dependent on microenterprises, these data conceal a structural problem.

The specific recommendations for Italy included in the 2003 document concern the reorganization of public finances through the application of structural measures; employment growth, particularly the reduction of the North–South gap; help for women and older workers through the setting up of support facilities (such as nurseries) to favour their employment; the creation of a suitable environment for the development of a knowledge-based society, by increasing education and lifelong learning opportunities; and more investment in ICT, research, and innovation schemes. Other recommendations concern the need to increase the competitiveness of the energy and services sectors, to improve the business environment, and to accelerate the process of national implementation of Community directives for the accomplishment of the single internal market.

The Italian government is intolerant of the European Commission's criticism, and it acts as if it did not know that any criticism (and sanctions if needed) levelled by the European institutions has a legal basis in the Treaties. To compare Italy with other troubled countries (particularly with Germany and France), just to minimize the observations raised in Brussels, means forgetting that the Italian situation is very different because of the huge size of public debt. The prolonged budget deficit after the financial ruin of two major Italian companies in the food sector, Cirio and Parmalat, could downgrade Italy's credit rating because of the perceived insolvency risk on the international corporate bonds markets. A negative rating would further reduce investments in Italy, and the Treasury would be forced to increase interest rates on state guaranteed bonds to attract international investments, thereby generating a borrowing spiral.

3.20 Prospects

The European Union is successful in some high-tech industrial sectors and can compete in mature, quality-based markets, but it ranks behind the U.S. in terms of R&D investments and patent registrations. Some emerging economies have planned huge long-term investments in training, ICT, and research (South Korea, for example), and they are starting to compete in the international arena. Other countries can rely on a low-cost labour force. In some industrial sectors, the situation of some emerging countries is privileged on both fronts. Finland and Sweden are investing in research more than any other Union country (and more than the average of the U.S.), and in this way they are laying the ground for long-term prosperity.

The rapid transformations brought about by the new economy can radically change the competitiveness level of some countries whose economies are based on traditional industrial sectors and are therefore more competitive in mature markets. The capacity to innovate is the key to securing a long-term high standard of living and to reducing unemployment in Europe.

Italy is one of the European countries with the lowest percentage of workers with third-level qualification. It also has one of the lowest rates in terms of presence of top universities, ICT and R&D investments, patent registrations, computer literacy, and Internet penetration. The peculiar condition of the Italian business sector, characterized by microenterprises and an average company size that is half the European average, is the main reason why Italian companies are less keen to invest in research than European companies in general. Additionally, since Italy has not invested her best resources in the definition of Community programs on research, these programs are not fine-tuned to the Italian demand and therefore are either disregarded or ineffective.

Italy is paying a very high price in terms of the competitive loss of its economy (see Table 3.13) owing to, *inter alia*, an insane research policy. In only two years, Italy's competitiveness, according to the authoritative study conducted by the World Economic Forum, has dropped from the 33rd to the 42nd position. Italy's long-term prospects are quite alarming, as the new economy, widespread technological progress, and globalization will be affecting the competitiveness of those economies based on mature markets and traditional industries.

This rather gloomy picture of Italy's decline, which has been outlined above and elsewhere in this book, is corroborated not only by the comprehensive data reported here but also by the opinions of some authoritative experts. In January 2004, the second edition of Petrini's book [Petrini 2004-1] entitled *Italy's Decline* was published. Petrini writes that our economy is the strongest exporter in low-tech sectors, a heritage left by the old manufacturing industry. Today, however, our leadership is challenged by countries like Taiwan, South Korea, and China. Among the experts writing about Italy's decline, we find not only famous left-wing economists like Paolo Sylos Labini (who, in [Sylos Labini 2003], wrote: "two traditional industries of our country, namely, the textile and footwear industries, are doomed to become marginal"), but also the then Governor of the Bank of Italy, Antonio Fazio, who (in the final remarks of the 2003 annual report) spoke of a decline in competitiveness and in productivity [BdI 2003]), and the then president of Confindustria, Antonio D'Amato, who stated that "We risk lagging behind the rest of Europe" [Petrini 2004-2]. It is fair to say, however, that these opinions are not shared by all. The president of the Italian Republic, Carlo Azeglio Ciampi [Ciampi 2003-1], made the following statement:

> Having traveled Italy, I have seen many striking examples of vitality, creativity and flexibility everywhere, which make me disagree with the rhetoric of Italy's decline that is currently spreading and risks to prostrate our abilities and desire to improve.

It should be noted, however, that Ciampi made this statement at the bestowal of the 2003 Leonardo Quality Awards, coveted prizes conferred by the Presidency of the Republic to the best-performing enterprises. In 2003, Leonardo Prizes were awarded to Barilla (pasta), Benetton (clothing), Bucellati (jewels), Frescobaldi (wines), and Nonino (spirits): five successful companies far from the high-tech industry. By the way, also the following president of the republic, Giorgio Napolitano in 2006 awarded Armani (fashion), René Caovilla (shoes), Safilo (spectacles), and Ferrari (wine). This confirms that the best part of the Italian industry stays away from those high-tech market sectors that are getting increasing shares in world trade.

The Italians' desire for improvement of their political leadership and of the economic performance of their nation lead the country to change course in the political elections of 2006; different priorities were set, including the realignment of public finances towards budget balance and debt reduction. However, the electoral law put in place during Berlusconi's 2001–2006 term in office—once that public-opinion surveys indicated a likely failure of the ruling government coalition—was meant to create an unstable majority in Parliament, which weakened the action of Prime Minister Prodi's 2006–2008 centre-left government. During his two years in office, an economic policy of rigor and austerity has reduced the budget unbalance well below the requirements of the Treaty and a full-blooded campaign against tax evasion has initiated bearing its fruits. However the struggle is still up-hill and the process of putting the economy on its feet should be accelerated because at the current pace no less than *50 years* would be required to bring the public debt at the reasonable level of 60 % of GDP. Of course no political leader dares say so during the ongoing electoral campaign.

3.21 Recommended Reading

Recommended reading for this chapter are listed in Appendix C and are available on the web site companion to this book, at the URL http://stajano.deis.unibo.it/RQC.htm

Chapter 4
Competitiveness and Quality

4.1 Cost of Labour

The European Union is a region with a high cost of labour and high health care and social security costs. Expenditures for wages, social services, training, health care, and protection of the environment can only be partly compensated for by high productivity, the elimination of customs barriers and currency exchange costs, and a high level of efficiency in public administration (where efficiency is the norm). Success in competition in worldwide markets depends on the ability to highlight the superiority of European services and products on the basis of their quality, design, innovative content, and capacity to meet the demand of an evolving market. Quality-based supremacy is a way of reconciling a high standard of living and high European wages with maintenance of competitiveness of European services and products.

The cost of labour in the manufacturing industry increased rapidly both in Europe and the U.S. in the 1980–1998 period. It is complex to compare absolute values of hourly labour cost in the Triad over a long period because of exchange rate fluctuations. A snapshot of 2001 is shown in Fig. 4.1 from data published by the Institut der deutschen Wirtschaft and reported by Salzburg Agentur [Agentur 2003], an investment promotion agency of the City of Salzburg, Austria. Hourly labour appears to be cheaper in the EU, but if it were calculated again in 2007, U.S. labour would appear cheaper owing to the change in the euro-to-dollar exchange rate.

Figure 4.2 presents the evolution of hourly labour costs in the Triad from European Central Bank data, expressed in U.S. dollars. EU data are derived from the labour costs in the four major European economies. The EU graph, since labour costs are denominated in dollars, reflects a combined effect of national dynamics and of the changes in the euro-to-dollar exchange rate (see Fig. 14.11), and highlights a problem European exports have been facing with the weak dollar since the beginning of the new century.

The hourly cost of labour in the manufacturing industry is higher in the G7 countries than in the economies of emerging countries in Asia and also higher than in central and eastern Europe. We will show, in Chapter 11, "Enlargement of the European Union", that the accession countries have a competitive advantage over EU15 with respect to labour cost. For the EU15 states, it is very important to establish

Hourly labor costs in 2001

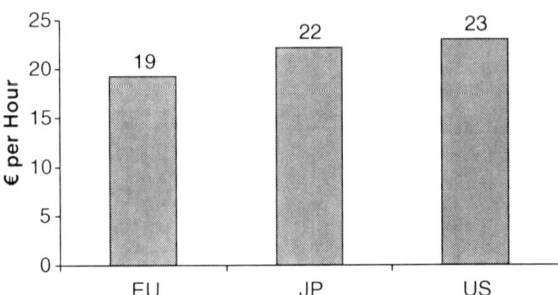

Fig. 4.1 Hourly labour costs in the Triad in the year 2001
Source: Institut der deutschen Wirtschaft reported by Salzburg Agentur (2003).

an advantage in terms of quality, because a comparison in terms of price would be unfavourable. It is in the interest of all 27 member states of the European Union to improve the quality of their products and to jointly face the challenge of the emerging Asian economies. Joint ventures between EU15 and AC12 are combining the advantages of lower labour costs and technologically advanced industrial processes, creating a synergy that can be favourable to both parties and that will improve working conditions, employment, and market shares. The challenge is to bridge the skills mismatch in the AC12 markets, to contain the rise in inflation in AC12, and to cope with the risk of unemployment in EU15 in case of major delocalization. Delocaliza-

Changing Hourly Labor Cost in the Triad

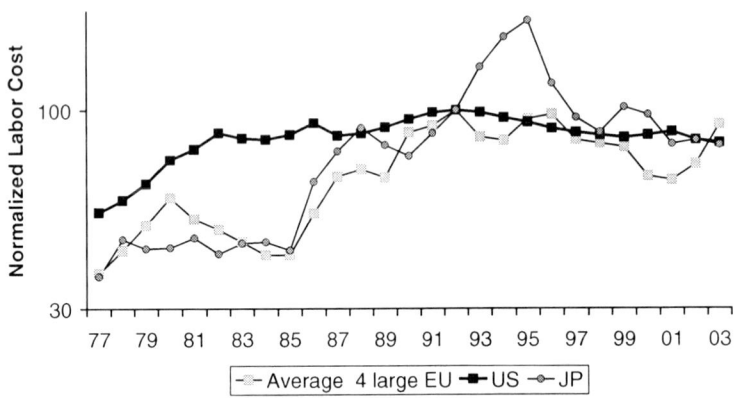

Fig. 4.2 Hourly labour costs (manufacturing industry) in the Triad in the period 1977–2003. The curve EU displays the average of the cost in the four major EU economies: DE, FR, UK, IT. Data are in U.S.$, normalized to 100 for the year 1992. The changing exchange rate $-€ and $-¥ increases the variance of non-U.S. hourly costs that would be more flat if expressed in the national currencies
Source: European Central Bank (2005).

tion should be supported by national authorities in the delocalizing as well as only to the extent that it addresses a demand in the country where the foreign investment is placed, while it should not be encouraged if it is exclusively a quest for cheap labour. We will see later in this chapter that a case of virtuous delocalization in Poland is that of Fiat, an Italian car manufacturer.

The EU15 countries have a positive trade balance not only with the accession countries but also with the U.S., many emerging economies (e.g. India), and the rest of the world. This can be attributed, among other things, to the EU15 capacity to sell high-quality products.

The trade balance between the European Union and the U.S. has been positive since the end of the 1990s, in favour of the Union (see Fig. 3.35), and in 2002 it was still positive by a value of € 65 billion [EC 2003–3]. The highest contribution comes from quality-based competition. In 2001, the trade balance was positive between the countries of the Euro-zone and the U.S. but was negative for the Union as a whole because of the negative contribution of the United Kingdom.

The trade balance between the European Union and Japan is unfavourable for the Union, (see Fig. 3.36). The highest negative contribution comes from price-based competition in both mature sectors (e.g. motor vehicles) and high-tech consumer goods sectors (e.g. audiovisual). For the Union and for Italy in particular, quality is the key to commercial success worldwide.

For example, the demand for luxury and high-quality cars is still higher than production capacity even after the depreciation of the dollar. In the U.S. or in China, owning a Ferrari is a symbol of social success and people are willing to pay any price for it (those who can afford it!). The same can be said for high fashion: Chinese or Turkish products can be bought at lower cost, but if style and fashion are what count, then French or Italian products are most sought after.

Quality and design make up for the lack of raw materials. When Piero Fassino was foreign trade minister in D'Alema's cabinet (1998–2000), he stated, "MADE IN ITALY is our black gold."

4.2 Measuring Quality

What is quality? How can it be defined and measured? There are many different definitions of quality, and the difference in these definitions originates from the context in which quality is considered. For example, in banking-software development, quality can be defined as the unique two-way correspondence between project specification and code development. This definition does not fit our case: in the context of a quest for a strategy to increase industrial competitiveness, we define quality of goods and services as the mix of characteristics appreciated by buyers, which attracts them and induces them to pay an extra charge. Said characteristics can be intangible (e.g. design, elegance, popularity) or measurable, though with different accuracy levels for the various attributes (e.g. availability, certification, compatibility, customer care, dimensions, flexibility, heat dissipation, lifetime, possibility of customization, reliability, resistance to adverse environmental conditions, speed, usability, versatility, wear, weight).

Shortly, we will see how to measure quality. Quality indexes are direct or indirect evaluations of some of the above-mentioned characteristics. At times these are qualitative and approximate, just like quality indicators, which are indirect evaluations of the conditions in which a quality product can be produced.

The indicators taken into account here are related to the resources of competence and innovation, which create the conditions necessary to produce quality goods and services in a company, namely,

- high intensity of R&D, and
- highly qualified workers

or are subsequent measurements of the characteristics of goods and services, market conditions, and position of the company in relation to the competition, namely,

- high market share
- high unit value, and
- limited production in countries with low wages

At the outset, it is important to note that *income elasticity* measures the increase in propensity to purchase by buyers with growing income, that *price elasticity* measures the variation in propensity to purchase depending on product price changes, and that *quality elasticity* measures the increase in propensity to purchase based on the perception of a high level of quality of the product. Furthermore, *vertical differentiation* of products indicates a wide range of products with similar functions and that consumers accept a wide range of prices for products of different quality (e.g. measurement instruments). In contrast, *horizontal differentiation* of products indicates a variety of preferences among customers or that the single consumer prefers to buy two different versions of the same product rather than two units of the same version (e.g. ice cream flavours).

Given the above, we can specify that the quality product market is characterized by

- high income elasticity
- low price-elasticity
- vertical product differentiation
- production in companies and states with a high cost of labour
- limited number of competitors, and
- high profits

In a state with an economy based on quality, the conditions for development are political and economic in nature. *Economic* factors for the development of a high-quality product market include high average income, high income dispersion, information campaigns on quality, certification, entrepreneurship, and access to financial markets. *Political* factors for the development of a quality product market include private and public investments in education, lifelong learning, technological research and development, open markets, liberalization, mobility, and organization of companies and states promoting excellence, creativity, and innovation.

At the company level, competitive strategies based on quality include focusing on improving product quality rather than reducing prices; gaining a quality-based advantage over competitors; increasing the willingness of customers to spend the extra money for quality; and offering only high range products. The instruments that a company can adopt to upgrade quality are research and development, qualification of personnel, use of new materials, use of new processes, study of market demand, functional improvement, improvement of customer care, company reorganization, strategic alliances and focusing only on high quality products and services with the view to creating an implicit association of the company's brand to quality. Marketing can increase the willingness of buyers to spend more for quality. Certifications, standards, and benchmarking increase the quality of products and processes and its visibility.

We will concentrate on analyzing the contribution to competitiveness that derives from product quality. Before examining in detail the criteria for measuring and improving quality, it is important to remember from the previous Chapter 3, "Competitiveness of the European Economy", that EU future competitiveness depends on five factors: research intensity and quality, development of human resources, ICT investments, business reorganization, and access to financing.

We propose three indexes to measure quality (please remember that these are qualitative and approximate measurements):

1. Unit value of goods (€ / Kg)
2. Share of exports in industrial sectors characterized by quality competition
3. Share of exports in high-price/high-quality segments in each sector

4.3 Unit Value

The *unit value* of goods (€/Kg) is the value of goods divided by their weight.

For example, the unit value in €/Kg of a supply of ironmongery articles is 1.00 €/Kg and is lower than the unit value of a supply of watchmaking articles at 10,000.00 €/Kg.

With homogeneous goods such as cars, by organizing various products according to the unit-value index, it is possible to see empirically that they are classified in ascending order of quality. For example, see Table 4.1.

The unit value of exports of a country is higher if the goods are more sophisticated and of higher quality. The unit value of exports of the member states of EU15 [Aiginger (ed.) 2001] varies between 0.43 €/Kg in Greece and 5.50 €/Kg in Ireland. The average unit value of exports from the European Union is 2.5 €/Kg, and Italy's value is 3.1 €/Kg.

Table 4.1 Unit value of some cars

Car model	Cost K€	Weight Kg	Unit Value €/Kg
Fiat Panda	6	715	8
Alfa 166 3 liter	46	1490	30
Ferrari 550	188	1690	111
Rolls Corniche	355	2735	129

Source: Quattroruote, 2003.

4.4 RQE Index

The RQE index (Revealed Quality Elasticity) [Aiginger (ed.) 2001] is linked to the share of exports in those industrial sectors characterized by quality competition. It is measured by taking the share of exports in *quality elastic* industrial sectors in relation to the share in *price elastic* sectors. The index is based on the fact that industrial sectors are price elastic when export unit values, which are higher than import unit values, are associated with lower export quantities. In contrast, sectors are quality elastic[1] when the difference between import and export unit values has the same *sign (+ or −)* as the difference between import and export quantity. The index is calculated taking into consideration bilateral trade between the EU15 states and 30 trade partners (the other 14 states of EU15, the U.S., Japan, eight emerging economies, and six of the AC10 countries). For each industrial sector and state, the series of signs is calculated in relation to the difference between import and export unit values for each of the 30 trade partners and in relation to the difference between import and export quantities for those partners. The RQE index is the normalization on a 0–100 scale of the number of equal signs *(+ or −)* in both series (a number that may vary between 0 and 30). Empirically, the normalized RQE index ranges between 25 and 53.

This index reveals the intrinsic characteristic of an industrial sector, unchanged over time or across countries. Countries that are active in industrial sectors with a high RQE concentrate their exports on quality elastic goods. The RQE is a quality index that can be defined on an aggregate level for a state or separately for each industrial sector in each state.

---

[1] To better explain how the RQE index of a given industrial sector in a given economy can provide an indication of quality elasticity, that is to say, the willingness of the market to purchase on the basis of quality and not price, let's take into consideration two extreme examples and let's evaluate intuitively the signs of the differences between export and import prices and quantities. Let's refer to the Italian economy. For the first case, let's take T-shirts; for the second case, let's take luxury sedans. The T-shirt sector is price elastic: the unit value of exported goods (e.g. Benetton) is higher than the unit value of imported goods (e.g. T-shirts made in China); the quantity exported is lower than the quantity imported. The luxury car sector is quality elastic: unit values of exported cars (e.g. Lancia) are lower than the unit values of imported cars (e.g. BMW series 7); the quantity exported is lower than the quantity imported in this sector.

4.5 PPS Index

Another quality index is the PPS index (Position in Price Segments), which is related to the share in exports in high price/high quality segments. It is measured by taking the difference between the percentage of exports in a given industrial sector that fall in the high-price segment and the percentage of exports that fall in the low-price segment. The PPS index is defined as follows. For each state of EU15, all the imports of each industrial sector from 30 countries (the other 14 states of EU15, the U.S., Japan, eight emerging economies, and six of the AC10 countries) are taken into consideration. For each sector and each state, the 30 import unit values of goods are registered. Import unit values related to each sector are divided into three parts that define the high/medium/low segment, respectively, of prices in that sector of industry. Then, for each state, exports to the 30 countries are taken into account, and their positions are analyzed in relation to the price segments defined for the respective industrial sectors on the basis of imports. The PPS index is the difference between exports that fall in the high segment and those that are part of the low segment.

Countries that are active in industrial sectors with a high PPS concentrate their exports on high-priced goods. PPS is a quality index that can be defined on an aggregate level for a state or separately for each industrial sector in each state.

4.6 Exports and Quality

Figure 4.3 shows the export unit value of each state in relation to the pc GDP in that state. Enlargement countries, which are characterized by a pc GDP lower than that of EU member states, are positioned in the low-export unit-value area. The linear trend line suggests a correlation between export unit value and pc GDP, which can be interpreted as the driving effect of domestic demand, where the demand for quality and performance increases as the pc GDP grows. Figure 4.3 shows this trend and some interesting exceptions. Belgium, The Netherlands, Finland, and Denmark have less favourable rankings in export unit value compared to other countries with the same pc GDP level, indicating the high share of capital-intensive industries in these countries. Greece is ahead of many accession countries in terms of pc GDP thanks to its income from tourism, but it is behind them in the unit value of exports in the manufacturing industry. Ireland has the highest export unit value due to her export of high-tech manufactured goods. The U.S. and Japan are among the top countries for pc GDP but rank low in export unit value. This ranking reflects the intensity of their exports to geographically close markets where there is a high degree of price elasticity.

The positive European trade balance arises from quality-elastic sectors. By examining the percentage of exports in high-quality sectors (measured using the RQE index in Fig. 4.4), it is possible to see that Germany and France are top ranking, while Portugal, Finland, Spain, Italy, and the United Kingdom have a percentage

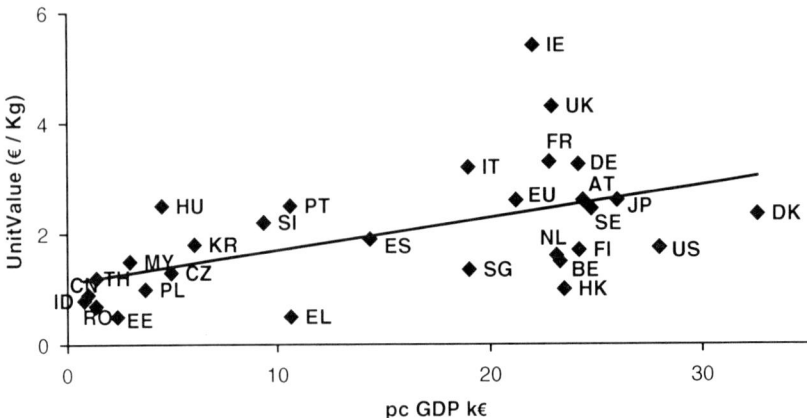

Fig. 4.3 Unit value to export in €/Kg (vertical) versus pc GDP in thousand euro (*horizontal*) for the Triad countries and a number of emerging Asian economies. A linear trend line is shown
Source: European Commission, "Europe's position in quality competition" [Aiginger (ed.) 2001].

equal to the European average. In Germany the sector[2] that gives the highest contribution is *machinery*; in France, *aircraft* and *beverages* dominate. Some of the countries ranked at the bottom, for example, Belgium, The Netherlands, and

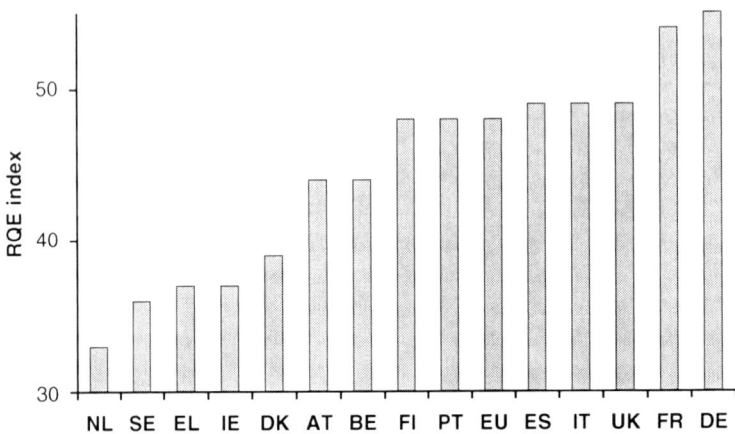

Fig. 4.4 Percentage of high-quality exports from the EU, according to the RQE index. Data for Luxembourg are missing. Data from 1998
Source: European Commission, "Europe's position in quality competition" [Aiginger (ed.) 2001].

---

[2] The classification used is the one adopted by Karl Aiginger [Aiginger (ed.) 2001] in his study, where 93 industrial sectors are taken into consideration.

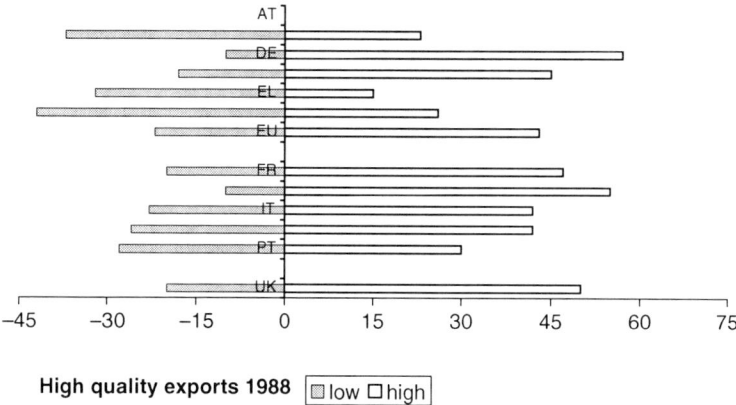

High quality exports 1988   □ low □ high

Fig. 4.5 Percentage in 1988 of high-quality exports (*white bars*) and low-quality exports (*grey bars*) according to the PPS index. Data for Luxembourg are not available
Source: European Commission, "Europe's position in quality competition" [Aiginger (ed.) 2001].

Denmark, have a positive trade balance but specialize in industries with medium or high price-elasticity. The United Kingdom has a negative trade balance despite its specialization in quality-sensitive industries. In the years around the turn of the century, the overvaluation of the pound sterling contributed to this position.

Figures 4.5 and 4.6 show the percentage in 1988 and in 1998 of the exports falling in the low- and high-quality segments, measured using the PPS index. There

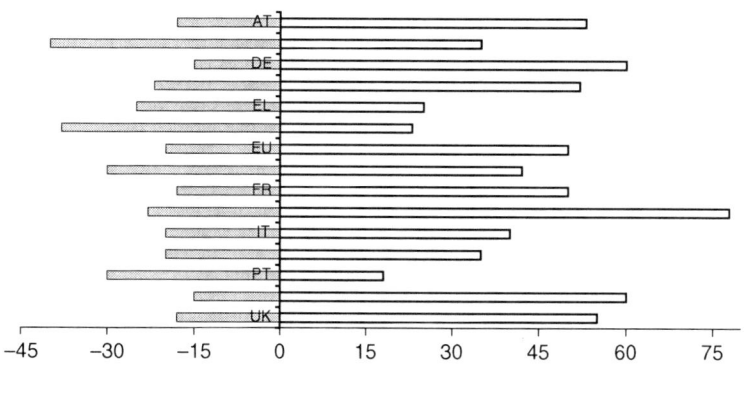

High quality exports 1998   □ low □ high

Fig. 4.6 Percentage in 1998 of high-quality exports (*white bars*) and low-quality exports (*grey bars*) according to the PPS index. There is a visible move towards high quality from the values of 10 years earlier, displayed in Fig. 4.5. Data for Luxembourg are not available
Source: European Commission, "Europe's position in quality competition" [Aiginger (ed.) 2001].

is a visible move towards higher quality during the 10 years from 1988 to 1998. Ireland was top ranking in 1998, followed by Germany. For the states that were not members of the EU before 1995, the data relating to 1988 are missing.

It is important to note that quality measures are imprecise and indicative. This is reflected in the difference between the data in Figs. 4.4 and 4.5, resulting from the use of different indexes for the appreciation of quality. However, the picture is qualitatively consistent. This can be seen in detail by examining once again the analysis related to Table 4.2, already given from another point of view in Chapter 3 "Competitiveness of the European Economy", and by taking into consideration data related to the specializations of member states. Table 4.2 lists the top five industries that in each state have generated within the European Union the highest percentage of added value. The states are ordered on the basis of the number of high-tech industries among those industries in which the states specialize, according to the definition given above. It can be seen that Belgium, Greece, Italy, Portugal, and

Table 4.2 Specialization in the EU: The first five industries that in each state have generated the highest percentage of added value. High-tech industries are in *italics*. This table contains the same data as in Table 3.7, but here the states are ordered on the basis of the number of high-tech industries among those industries in which the states specialize

Country	Specialization
Germany	*Electricity distribution and control; industrial process control equipment; motor vehicles; machine-tools; electrical equipment*
Ireland	*Recorded media;* jewellery; *medical equipment;* other chemical products; *office machinery, computers*
The Netherlands	*TV;* jewellery; *medical equipment;* other chemical products; *office machinery and computers*
Sweden	Sawmilling; pulp and paper; *weapons and ammunitions; TV, radio, and telephony;* first processing of iron and steel
Denmark	*Toys;* fish; meat; *ships and boats;* other transport equipment
United Kingdom	Agro-chemical products; *aircraft and spacecraft;* grain mill products; processed fruits and vegetables; *ships and boats*
Austria	Knitted crocheted fabrics; sport goods; *machine-tools;* made-up textile articles; *TV and radio recording apparatus*
France	*Steam generators; watches;* wooden containers; sport goods; Detergents and perfumes
Italy	Ceramic tiles; tanning and dressing of leather; leather clothes; *motorcycles and cycles;* luggage handbags and footwear
Finland	Pulp and paper; sawmilling; wood and panels of wood; leather clothes; *TV, radio, and telephony*
Portugal	Articles of wood and cork; footwear; knitted crocheted fabrics; other wearing apparel; tanning and dressing of leather
Belgium and Luxembourg	Jewellery; other textile; made-up textile articles; first processing of iron and steel; glass and glass products
Greece	Cement, lime, and plaster; textile fibres tanning and dressing of leather; other wearing apparel; knitted crocheted fabrics
Spain	Fur; cutting and shaping stone; ceramic tiles; vegetable and animal oils and fats; cement, lime, and plaster

Source: European Commission [EC 1998-3].

Spain—unlike Germany, Ireland, and The Netherlands—are characterized by a specialization in low-tech industries. Also, the other states that are ranked higher than Italy in Fig. 4.6 in quality exports (Sweden, the UK, Denmark, Austria, France) in Table 4.2 have more sectors specializing in high-tech (in *italics*) than Italy does.

The position of Ireland in Fig. 4.3 is apparently incompatible with its position in Fig. 4.4. While it is important to remember that indexes of quality are imprecise and indicative, it is possible to give an explanation for this apparent contradiction if we consider Table 4.2 and Figs. 4.15 and 4.16 together. Ireland specializes in the export of high-tech products with a high unit price, but these are mainly price-elastic high-tech commodities (computers, media).

Figure 4.7 shows the EU trade balance in 1998. The states are in the order of trade balance, and the value is presented together with the trade balance of exports related to sectors with a high RQE index. The overall trade balance of the EU in 1998 was favourable, with Germany and Italy at the top of the ranking. The UK had an unfavourable trade balance in both high and low RQE sectors, but the balance of high RQE sectors was less negative and reduced the total imbalance. The same consideration applies to Greece, Spain, Portugal, and Austria, where the reduced imbalance in the high-quality sectors alleviated the deficit of the overall trade balance. France and Germany had a trade surplus only in the high-quality sectors, but this was enough to get an overall favourable balance.

In the 10 years between 1988 and 1998, a significant increase (Fig. 4.8) in export unit value was registered in Ireland, which was by far the leader (it had been second to the United Kingdom in 1988). The situation in Ireland was characterized by heavy foreign investments (in particular from the U.S.) in high-tech sectors. Sweden also took a significant step forward, going from the 12th to the eighth position. Over these 10 years, Sweden doubled its export unit value. In Belgium, the export unit value decreased.

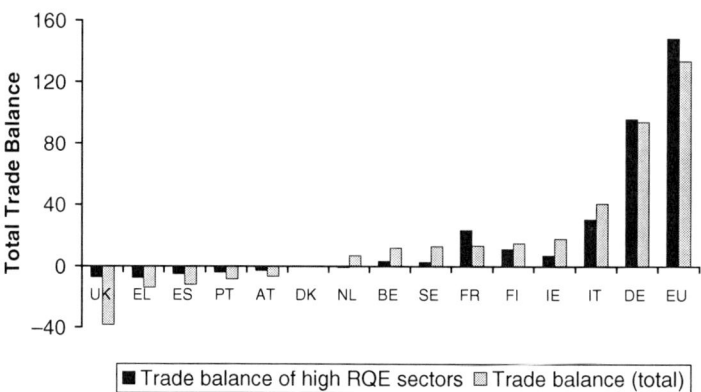

Total Trade Balance vs Balance in Quality Sectors

■ Trade balance of high RQE sectors ☐ Trade balance (total)

Fig. 4.7 Trade balance in 1998 (billion euro) and trade balance for exports in high-RQE sectors. Data for Luxembourg are not available
Source: European Commission, "Europe's position in quality competition" [Aiginger (ed.) 2001].

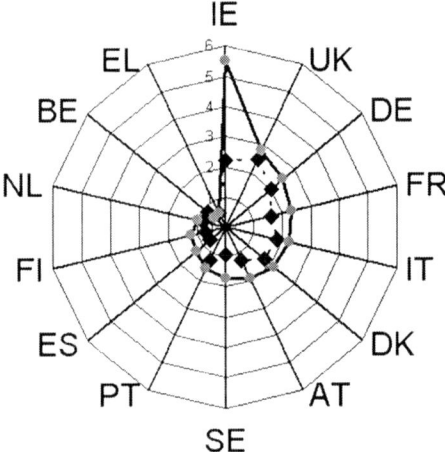

Fig. 4.8 Change in the unit value of export from 1988 to 1998 in the EU. Data for Luxembourg are not available. The lines in the figure have the sole purpose of facilitating the reading by joining points that belong to the same year
Source: European Commission, "Europe's position in quality competition" [Aiginger (ed.) 2001].

The 93 sectors into which industry is divided in Aiginger's study are ranked according to quality elasticity, as measured by RQE. Figure 4.9 shows that the top five sectors are other general-purpose machinery; agricultural and forestry

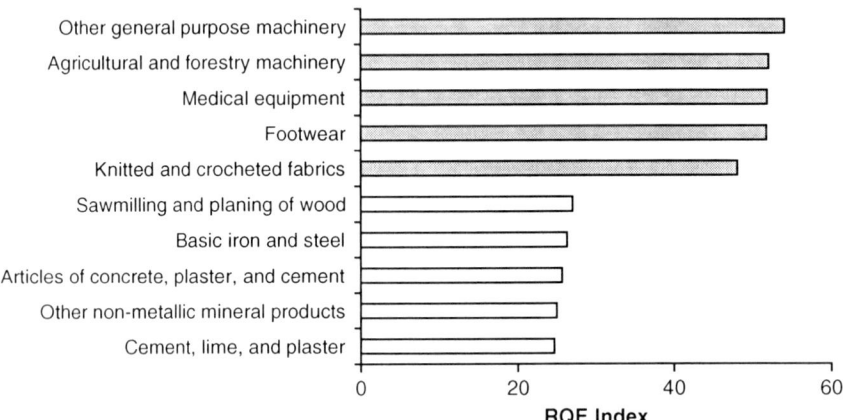

Fig. 4.9 Top and bottom ranking of industrial sectors according to Quality elasticity in 1998
Source: European Commission, "Europe's position in quality competition" [Aiginger (ed.) 2001].

machinery; medical equipment; footwear; and knitted and crocheted fabrics. In these sectors, competition is quality based.

The last five sectors in terms of quality elasticity are sawmilling and planing of wood; basic iron and steel; articles of concrete, plaster, and cement; other non-metallic mineral products; and cement, lime, and plaster. In these sectors, competition is price based.

4.7 Quality Premium

Quality premium can be calculated for a state or for a business sector: it is the difference between the real value of exports and their hypothetical value if they were priced at the price of imported products in the same state or sector. For example, the quality premium of French wine is positive because the price of wine exported from France exceeds the price of wines imported in France. The EU as a whole had a quality premium of € 161 billion in 1998. The highest contributions (see Fig. 4.10) came from Germany, Italy, France, and the United Kingdom. The largest part of the quality premium was accrued in the chemical sector. Machinery, food and beverages, motor vehicles, and textiles followed at a distance.

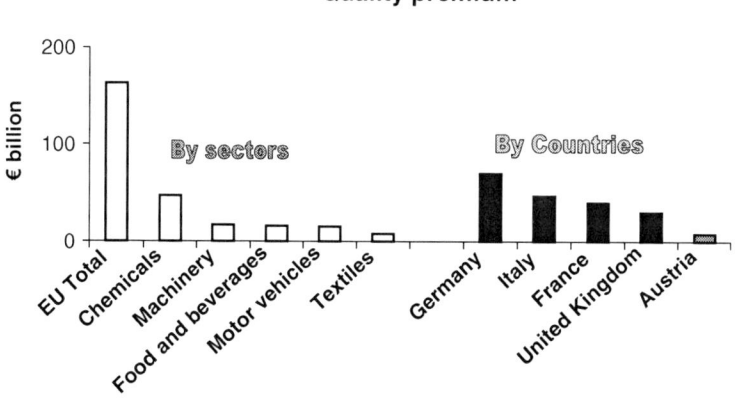

Fig. 4.10 Quality premium by sectors and by countries in € billion (1998)
Source: European Commission, "Europe's position in quality competition" [Aiginger (ed.) 2001].

4.8 Classification of Industrial Sectors in Terms of Quality

The 93 industrial sectors taken into consideration in the study by Aiginger are ordered in terms of the value of the RQE index, which measures quality elasticity.

Eleven of these sectors are classified as *technological*. Eight of these (see Fig. 4.11) fall into the category of high RQE. The sectors not included are computer,

Elasticity of industrial sectors

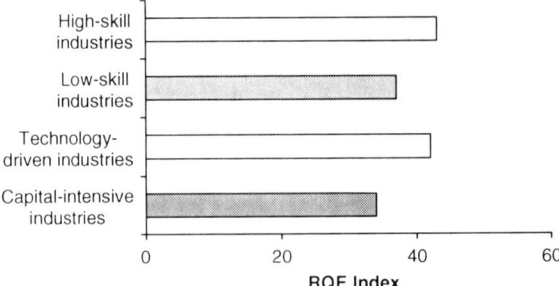

Fig. 4.11 Average quality elasticity of groups of industrial sectors, as measured by the RQE index (1998)
Source: European Commission, "Europe's position in quality competition" [Aiginger (ed.) 2001].

audio and video, and electronic components. These sectors have reached a development phase in which production has moved to regions with a low cost of labour. In these sectors competition is price based (see Fig. 4.12). On this matter we discussed earlier in this chapter the position of Ireland (see Fig. 4.3).

Quality is of the utmost importance in the following sectors: footwear, electronic games, toys, tobacco, and watches. Figure 4.13 shows three groups of industrial sectors: research intensive, advertising intensive, and capital intensive. In each group the turnover is disaggregated across sectors with a high/medium/low RQE index.

Exports in price segments

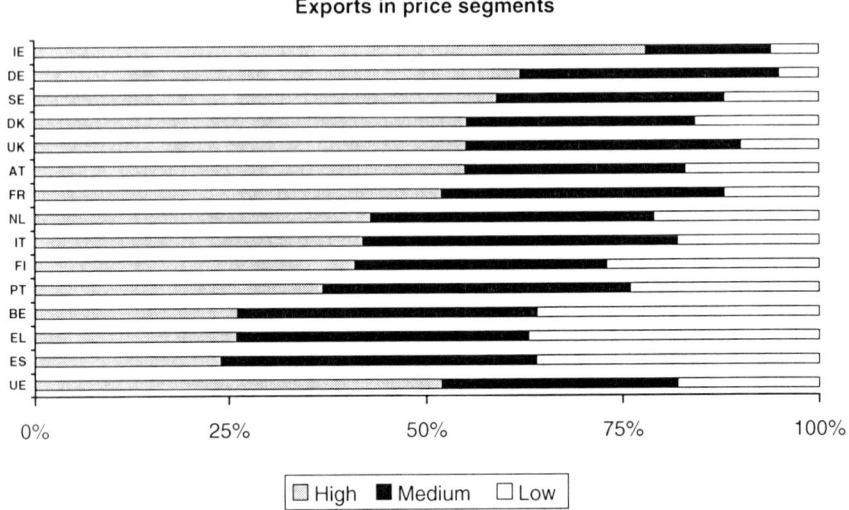

Fig. 4.12 Percentage of exports in different price segments (1998). Data for Luxembourg are not available
Source: European Commission, "Europe's position in quality competition" [Aiginger (ed.) 2001].

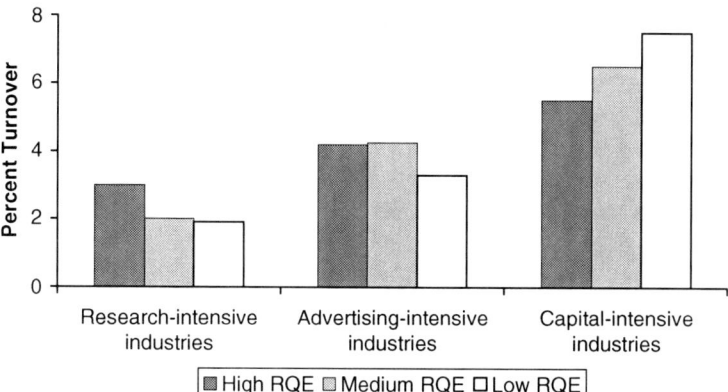

Fig. 4.13 Percentage of turnover in industrial sectors with different structural characteristics in three quality groups
Source: European Commission, "Europe's position in quality competition" [Aiginger (ed.) 2001].

Within the sectors with a high research content those with a high RQE prevail indicating that R&D intensive industries perform better in the high-quality sectors. In advertising-intensive sectors, the highest turnover is in those with a high or medium RQE. On the other hand, in capital-intensive sectors, those with a low RQE have the highest turnover. Figure 4.14 shows that in labour-intensive sectors, those with a medium or low RQE generate the highest added value.

Figure 4.12 shows the percentage of exports in different price segments for EU15 (except Luxembourg) in 1998. Ireland and Germany had the highest percentage of exports in high-price segments (78 % and 62 %, respectively), and both countries exported less than 10 % in the low-price segment during 1998. Sweden, Denmark, and the United Kingdom followed. Italy was below the European average

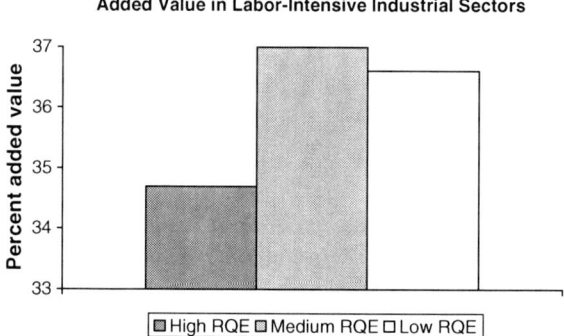

Fig. 4.14 Percentage of added value in labour-intensive industrial sectors in three quality groups
Source: European Commission, "Europe's position in quality competition" [Aiginger (ed.) 2001].

for percentage of exports in high-price sectors. Spain, Greece, and Belgium held the relative majority of shares of exports in the low-price segment.

In 1998, the main export sectors in Ireland were chemical, data processing systems and software, machinery, zootechnics, and products for zootechnics. The main export sectors in Germany were machinery, vehicles, chemicals, metals, manufacturing, food, and textiles. The main export sectors in Italy were industrial engineering, textiles and clothing, machinery, vehicles, transport equipment, chemicals, food, beverages, and tobacco. The main export sectors in France were machinery, transport equipment, aircraft, plastic, chemicals, pharmaceuticals, steel, and beverages. The main export sectors in Belgium were iron and steel, transport equipment, tractors, diamonds, and oil products.

Competitive success in terms of quality is characterized by

1. a high unit value of export goods (see Fig. 4.15)
2. a presence in high-price sectors (see Fig. 4.15), and
3. advanced status of a state's economy as regards the RQE index (see Fig. 4.16)

Figures 4.15 and 4.16 help us see the positioning of EU15 countries in the quality arena, and Fig. 4.17 shows us the change in quality of their exports between 1988 and 1998.

In 1998, Germany was in an advanced position in all three areas (Figs. 4.15 and 4.16). The main contribution to the positive German balance of payment was provided by the quality of the industrial production of machinery and motor vehicles. Greece exported in sectors with price-based competition and in Fig. 4.15 is positioned in the low-price segment.

**Unit Value of Exports
vs Position in Price Segments**

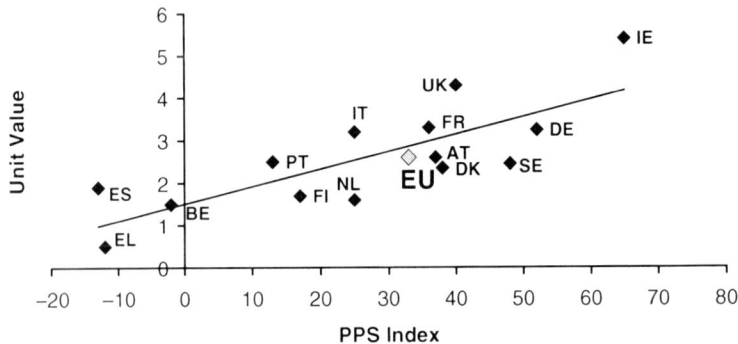

Fig. 4.15 Unit value of exports (€/Kg) versus the position in price segments as measured by the PPS index. The EU15 average is represented by a grey diamond. Data from 1998
Source: European Commission, "Europe's position in quality competition" [Aiginger (ed.) 2001].

Positioning of Exports According to PPS vs RQE

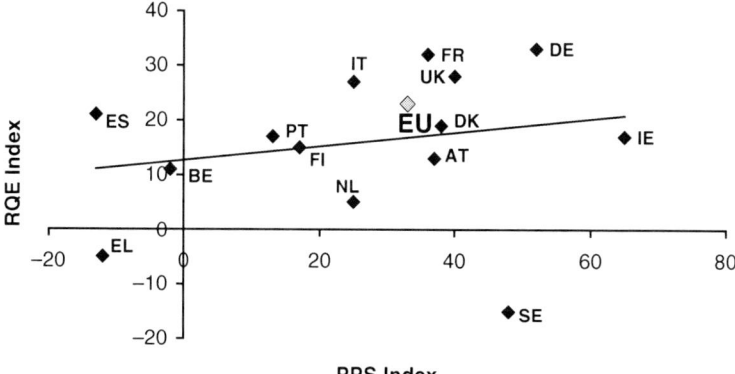

Fig. 4.16 Positioning of exports according to competitive mode, as measured by RQE index versus position in price segments as measured by the PPS index. The EU15 average is represented by a grey diamond. Data from 1998
Source: European Commission, "Europe's position in quality competition" [Aiginger (ed.) 2001].

The three criteria to evaluate the success of quality-based competition are usually correlated as can be seen in Figs. 4.15–4.17, with the exception of Spain, Ireland, and Sweden.

In 1988-1998, Spain made progress regarding quality-sensitive competition, followed by Germany, France, and the United Kingdom. Italy lagged behind, well below the European average. Spain had a high percentage of quality-based industries (49 %), but its industries were positioned in low-price segments—for example, Seat,

Variations of the Positioning of Exports

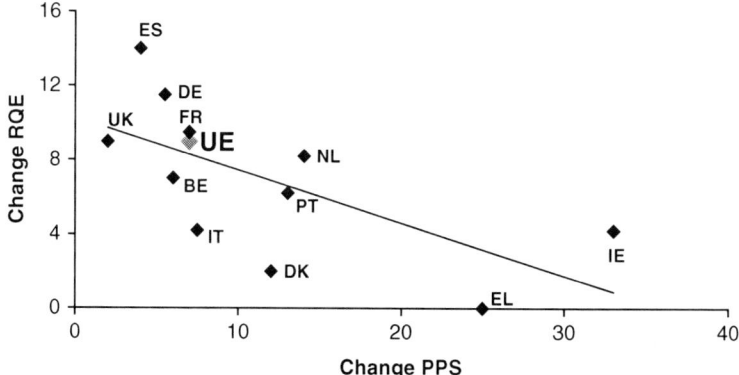

Fig. 4.17 Variations in the period 1988–1998 of the positioning of exports according to competitive mode, as measured by changes in the RQE index versus position in price segments as measured by changes in the PPS index
Source: European Commission, "Europe's position in quality competition" [Aiginger (ed.) 2001].

a motor vehicle manufacturer, specializes in budget cars. Ireland's exporting sectors were positioned in high-price segments, while it was positioned in the medium sector in relation to the quality index because electronic components and audio and video equipment were becoming highly price-elastic commodities. Sweden exported goods in high-price segments, but it was characterized by capital-intensive industries (paper, wood, steel, basic chemicals), which are price elastic.

Ireland and Greece moved forward in the 1988–1998 period, improving their positions in high-price segments, but they did not improve their positions at all (or not very much) in terms of quality. Portugal and The Netherlands improved in both areas.

The positioning of exports of member states in relation to unit value and presence in high-price sectors in 1998 (see Fig. 4.15) was characterized by a strong correlation between the two indexes, with the exception of Ireland, which had an exceptionally high export unit value justified by the specialization of Irish exports, as we saw in Fig. 4.3.

Germany was in a leading position, followed by France, the United Kingdom, and Italy.

EU – U.S. trade performance in the 1988–1998 period in high-tech sectors is shown in Fig. 4.18. In that period, the EU was well positioned in exports with reference to quality, but the U.S. was gaining ground and moving into less price-elastic sectors. Imports in the U.S. were moving towards high-quality segments more rapidly than in the EU or Japan. This shift indicates an evolution of demand. Imports of high-tech products from the U.S. to the European Union were higher than exports, but the latter increased in the 1988–1998 decade. The EU export unit value increased, and so did the EU import unit value, but by a smaller amount.

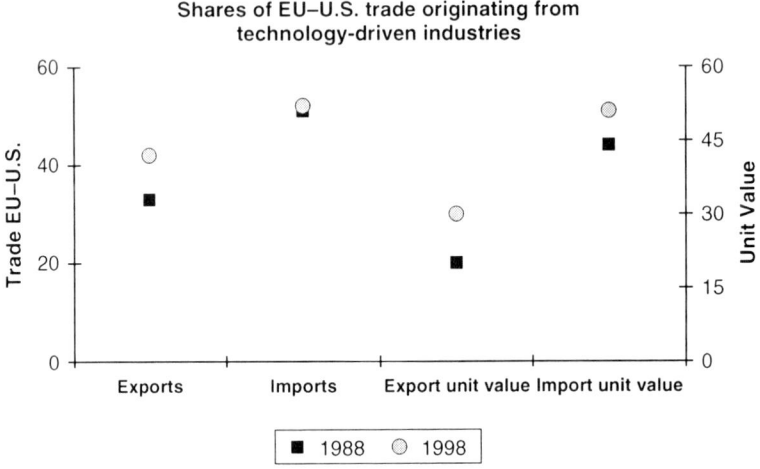

Fig. 4.18 Percentage and Unit Value of EU – U.S. trade in the period 1988–1998, originating from technology-driven industries. This figure presents exports from the EU and imports into the EU. Unit values are in €/Kg

Source: European Commission, "Europe's position in quality competition" [Aiginger (ed.) 2001].

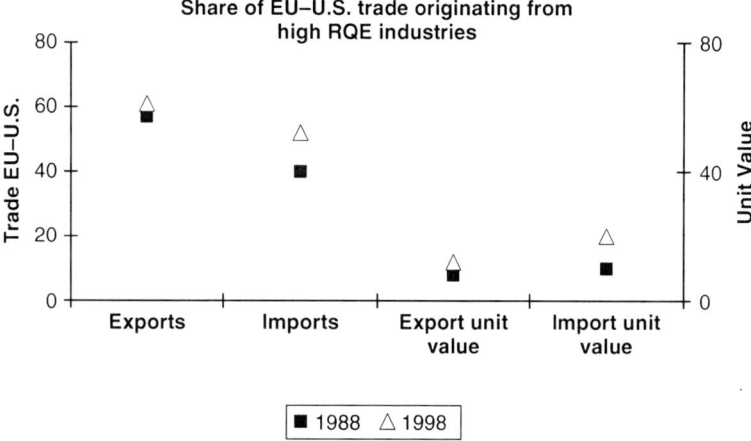

Fig. 4.19 Percentage and Unit Value of EU – U.S. trade in the period 1988–1998, originating from high RQE industries. This figure presents exports from the EU and imports into the EU. Unit values are in €/Kg
Source: European Commission, "Europe's position in quality competition" [Aiginger (ed.) 2001].

Figure 4.19 shows aggregate data of EU – U.S. trade in relation to production, exports and imports. The EU had the highest share of high-quality industries in the Triad. This positioning improved slightly in the 1988–1998 period. From a quantitative point of view, there was a surplus of exports from the European Union to the U.S., but the positive trade balance in the sector decreased in the 1988–1998 period. Both export and import unit values increased in that same period.

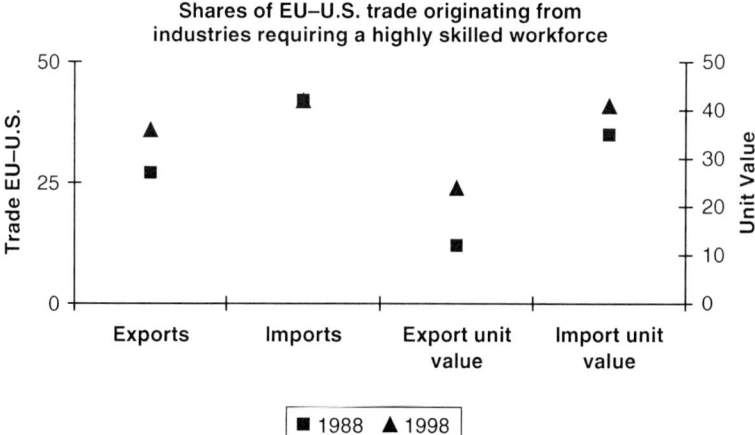

Fig. 4.20 Percentage and Unit Value of EU – U.S. trade in the period 1988–1998, originating from industries requiring a highly skilled workforce. This figure presents exports from the EU and imports into the EU. Unit values are in €/Kg
Source: European Commission, "Europe's position in quality competition" [Aiginger (ed.) 2001].

The trade balance as related to the products of high-skill industries was persistently unfavourable to the EU (see Fig. 4.20), but the U.S. – EU difference decreased; the increase in export unit value was higher than the increase in import unit value.

The sum of trade shares belonging to the three categories of industries presented in Figs. 4.18, 4.19, and 4.20, exceeds 100 %, because the three sets of industries are not disjoint and several industries are considered in more than one category.

4.9 System to Monitor the Quality Position of National Economies

Aiginger's study [Aiginger (ed.) 2001] proposes considering a group of indicators to evaluate and monitor the position of the states of the Union in relation to quality competition. There are 16 indicators related to production, quality, and imports/exports, and they are listed below. By *share* we mean the percentage in the European market.

 1. Share of quality-intensive industries in value added
 2. Share of quality-intensive industries in exports
 3. Share of exports in high-quality sectors of industries
 4. Export unit value
 5. Import unit value
 6. Relative unit value (export/import)
 7. Share of value added in technological industries
 8. Share of exports in technological industries
 9. Share of value added in skill-intensive industries
10. Share of exports in skill-intensive industries
11. Share of value added in industries with a high content of knowledge-based services
12. Share of exports in industries with a high content of knowledge-based services
13. Share of value added in industries with high product differentiation
14. Share of exports in industries with high product differentiation
15. Share of value added in globalized industries
16. Share of exports in globalized industries

The position of each state, for each indicator, is evaluated in relation to the European average. Ireland and France lead this ranking with 16 indicators out of 16 above the European average. The United Kingdom is below the average of EU15 only as regards the "Relative unit value (export/import)" indicator.

Compared with the European average, Germany is above average or aligned as regards 10 indicators and is ranked behind as regards six indicators:

• Share of value added in technological industries
• Share of value added in industries with a high content of knowledge-based services
• Share of value added in industries with high product differentiation
• Share of exports in industries with high product differentiation

- Share of value added in globalized industries, and
- Share of exports in globalized industries

Italy is ranked above the European average as regards nine indicators out of 16:

- Share of quality-intensive industries in value added
- Share of quality-intensive industries in exports
- Export unit value
- Relative unit value (export/import)
- Share of value added in skill-intensive industries
- Share of exports in skill-intensive industries
- Share of value added in industries with high product differentiation
- Share of value added in globalized industries, and
- Share of exports in globalized industries

Italy is in line with the European average as regards

- Share of exports in high quality sectors of industries

Italy is ranked below the European average as regards the following six indicators:

- Import unit value
- Share of value added in technological industries
- Share of exports in technological industries
- Share of value added in industries with a high content of knowledge-based services
- Share of exports in industries with a high content of knowledge-based services, and
- Share of exports in industries with high product differentiation

According to these monitoring criteria, Italy would not appear to be ready to face the challenges presented by the knowledge-based society.

4.10 Is "Good Enough" Better Than "Best"?

When studying quality-based competition, it is necessary to take into account a phenomenon described by Clayton Christensen [Christensen 2002, Christensen 2003] according to which a *good enough* product that satisfies the demand for quality can replace the *best* product already present on the market and providing better performance but at a higher price. Emerging countries could seize a share of the quality market in which European industry is established as soon as they are able to offer products *good enough* to attract the attention of consumers who seek quality, but are also price conscious. In the top-range motor vehicle field, Asian manufacturers now offer products *good enough* to seriously challenge the top German brands, which until now had strongly held a dominant position in the segment of *best* vehicles. In the photography sector, digital equipment cannot compete in terms of image quality with traditional cameras using chemical films but digital cameras are starting to be *good enough* to seriously challenge Kodak, the market leader in the manufacture of film. In the air transport field, we will describe in Chapter 8, "Competitiveness in the Knowledge-based Society" the case of Ryanair, a company that offers low-cost, no-frills travel, which is *good enough* to win out over traditional transport companies that offer *best* services, but are considered to be too expensive by the customers. Consequently, Ryanair is successful at a time when air transport is in crisis and long-standing traditional companies are going bankrupt. Among emerging economies, China is already able to offer *good enough* products in mature-technology sectors (e.g. household appliances, television). In 2002, the Chinese group TLC bought for € 8 million the German company Schneider Electronics AG, which had become insolvent [People's Daily 2002]. TLC, the largest Chinese manufacturer of televisions, with a turnover of 28 billion renminbi ($ 3.5 billion) has announced a 35 % increase in profits for the year 2003 [CNN 2004]. TLC has reduced by five times the price of televisions, which are sold on the Chinese market at € 300, compared with the € 1,500 they would have cost had they been imported from Germany by Schneider. Similar success stories have occurred in other sectors, for example, in the refrigerator compressor sector.

4.11 Outlook

Competitiveness of the European economy can only be achieved as a result of excellence in sectors where price is not the only factor defining the competitive edge. However costs must still be reduced through the improvement of infrastructures and the full realization of the internal market and of the Economic and Monetary Union. These measures are certainly necessary, but they are not sufficient to ensure competitiveness.

To maintain their competitive advantage, the states of the European Union must increase the quality of their export products. But it is of the utmost importance that the EU industry cover the whole quality spectrum of the internal market in order to face world competition on all fronts, including the less-than-best but good-enough products, in order not to be invaded by foreign low-cost imports. To this end, EU member states economies, while targeting the market demand for products and services of excellence—for which they have a competitive edge—and positioning themselves in the high-price segments of every industry, must not neglect the demand of price conscious consumers in their own markets.

In the next decades, the now favourable position of the EU15 in the quality-based competition arena may be compromised by three phenomena:

1. a convergence towards the standards of EU15 of accession countries and other economies, over which today the EU15 can boast a quality premium
2. the erosion of high-quality-goods markets by operators that can produce, at competitive costs products *good enough* to attract a portion of the demand for high-quality goods

3. competition on the innovation and quality frontier by new, relentless players who combine a low cost of labour with large investments in infrastructures, training, and research

To attain the objectives of the Lisbon Agenda, the European Union needs to accelerate the development process of science, technology, training, and innovation and plan a different allocation of financial resources. Rumours in January 2004 [Parker 2004–1] indicated that this acceleration was part of the plans of the European Commission. The final agreement in 2006 [EC 2006–8] although 17 % short of the initial EC proposal, assured a substantial increase in the Community budget for the period 2007–2013, as shown in Chapter 12, "The Budget of the European Union". The Lisbon Agenda will be presented and discussed in detail later, in Chapter 8, "Competitiveness in the Knowledge-based Society".

As we wrote at the end of Chapter 3, "Competitiveness of the European Economy", Italy is one of the lowest-ranked countries in Europe (and at a great distance behind the U.S.) in terms of ICT and R&D investments per capita, registration of patents, computer literacy, and diffusion of the Internet.

In terms of quality and positioning in high-price segments where the high cost of labour is sustainable, Italy is close to or above the European average (see Figs. 4.15 and 4.16). However, the loss of ground is serious (see Fig. 4.17): Italy is significantly below the European average in terms of variation of the RQE index. The outlook for Italy (which at present, in terms of GDP, is the seventh-ranked economy worldwide and had been in sixth position until the year 2000, as shown in Table 3.14) promises to be alarming as regards long-term evolution, when the new economy, the pervasiveness of technologies, and globalization will have changed the positioning of economies based on mature markets and traditional processes. The scenario is particularly serious because Italy is not in the bottom position only in terms of strategic investments in relation to GDP but also, following the reduction in investments for research created by the Financial Acts since the year 2002 in terms of the annual variation of said investments.

Prospects for a recovery may arise from the inversion of the trend; from an increase in private and public investments in education, training and research; from the alignment of national research policies to the research policies of the most advanced countries of the Union; and from the recovery of the delay in the accomplishment of the e-Europe Community program, which aims to diffuse the Internet and to *electrify* companies, trade, and public administration.

4.12 Fiat Auto: A Case Study

The case of Fiat, a major Italian industrial group, is a case of great interest for this study on industrial competitiveness. Here we are not able to give the matter the attention it deserves, and we limit ourselves to making a few comments supporting our arguments: there is a correlation between research, quality, and competitiveness;

and enterprises need a reorganization to face the challenges of the new globalized economy and exploit its opportunities.

The history of Fiat is a significant part of the history of Italian industry. Fiat had a driving role in the postwar industrial reconstruction; it employed hundreds of thousands of people, and it changed the lifestyle of Italians by producing affordable cars. The history of Italian industrial development includes significant achievements of the Fiat group, among which are the first diesel-electric locomotive (1922), the first high-speed train with variable attitude (1969), the first bodywork welding robot in the world (Comau, 1970s), innovations in the supply system for small diesel engines, known under the name of "Unijet common rail technology" (1990), the positioning of New Holland as a leading manufacturer of agricultural machinery, Ferrari's Formula 1 championship victories, of which six consecutive ones in the years 2000s, and 12 *Car of the Year* awards for a Fiat Auto motor vehicle (from 1967 to 2008). This list of achievements illustrates the excellence and potential of Fiat but it supports also the idea of the European paradox (discussed in Section 3.14, "Services and infrastructures") that points to the European weakness in turning great ideas into success stories in the market: the invention of the common rail (awarded with The Economist Innovation Award in 2002 [Economist 2002–1] was sold by Fiat to Bosh in 1994 and was consequently adopted worldwide by most car manufacturers instead of becoming a unique feature of Fiat's highly performing small Diesel cars.

Before the creation of the internal market of EU15, Fiat prospered in Italy; thanks to its status of national champion, protected by state aid, public investments that favoured the development of private transport, customs duties, import charges imposed on Japanese cars, and recent measures to support demand by offering scrapping incentives. In fact, in the first decades after the war, Fiat enjoyed a situation of monopolistic privilege in Italy, which ensured profitable growth but did not stimulate the pursuit of quality and excellence.

When *Avvocato* Gianni Agnelli took over the management of the group in 1966, Fiat had a 70% share of the motor vehicle market in Italy. With significant state aid, Fiat acquired Lancia and Alfa Romeo, its two major Italian competitors, and expanded in southern and eastern Europe, in South America and in the world, attaining global leadership positions in the 1970s and achieving the largest (30%) share of the European motor vehicle market in the 1980s.

Beginning in the second half of the 1990s, Fiat Auto went through a very deep crisis that compromised the position of the entire Fiat group, of which it made up 50% in terms of capital. The diffusion of Fiat Auto vehicles in northern Europe and the United States was precluded due to the lack of reliability, an average life cycle much shorter than vehicles produced locally and vulnerability to rust. By the beginning of the 2000s, the Fiat Auto share of the European market had been reduced to a third of the share it had in the 1980s. The Italian market share had slipped to 28% in 2003 down from 52% at the beginning of the 1990s. A serious problem of liquidity forced the group to sell some family jewels, including 50% of Ferrari. In the general worldwide crisis of the car sector, the situation of Fiat appeared to be particularly serious [Confapi 2002]. In fact, Fiat was the only

European manufacturer with negative profitability in 2001 and 2002 and with a 6 % reduction in turnover in 2003. Car models in the product line were short lived, contrary to competitors that evolved a model over decades to build a brand out of a product, like Opel with Kadett, Ford with Escort, or Volkswagen with Golf. Fiat instead seemed to be erratic in its moves from one model to the next as it was with its logo that changed no less than 14 times (Fig. 4.21).

In the year 2000, a contract was signed with General Motors (GM); John Griffiths comments on The Financial Times [Griffiths 2007] as follows:

> Fiat was close to broke and seeking salvation through an alliance with General Motors—'a bit like being rescued by the Titanic,' in the words of auto industry expert Professor Garel Rhys.

With that contract, GM acquired 20 % of Fiat Auto with 5.7 % of GM shares, undertaking to purchase, following a request by Fiat, the remaining 80 % at a fair market price in the five years starting from 2004. In 2002, Fiat Auto was [Martin 2002] the sixth-ranked worldwide manufacturer of motor vehicles, with over two million vehicles produced per year and a turnover of over € 30 billion. However, Fiat reported a financial loss of over € 400 million in 2001, and stayed in the red figures up to and including 2005, ranking as the weakest of the surviving European mass-market carmakers. In the three years from 2001 to 2004, the Fiat stock quotes on the New York Stock Exchange dropped by 70 %, while the worldwide crisis of the car sector has resulted in a fall of this industry index by 16 % (see Fig. 4.18). This situation would mean that if the put option of the contract with GM had been exercised, the fair market price of the remaining 80 % of Fiat Auto would have been evaluated at half or one-third of its value at the time the contract was signed. In December 2004, The Financial Times wrote [FT 2004–2]:

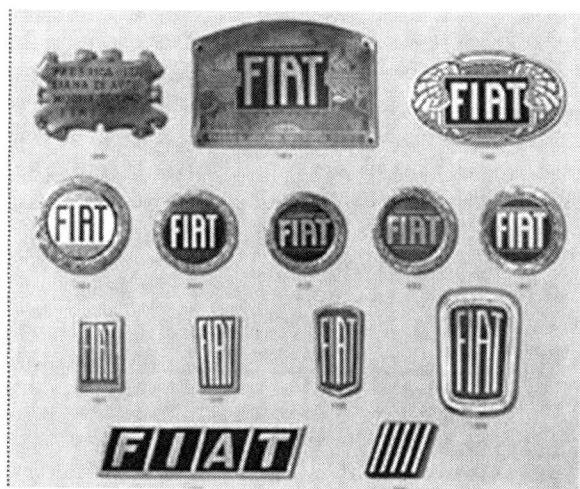

Fig. 4.21 Fiat changing logo
Source: Fiat, 2007.

The Fiat group would love to dump Fiat Auto on GM. But it is hard to see a foreign takeover of such an icon as Fiat winning political or public acceptance, especially if GM then shuts most or all of it down. More likely would be a payment of some kind from GM that would be part of yet another rescue of Fiat.

In 2005, Fiat obtained € 1.55 billion to relinquish the put option, and the Fiat – GM honeymoon was interrupted on Valentine's Day 2005 [Economist 2005–2]. This was one of the decisive changes introduced by the new Fiat CEO, Sergio Marchionne, who was in office since June 2004 and started the recovery of a company believed to be heading towards bankruptcy.

A comparison presented by Eurobusiness [Rubython 2002] on Fiat's decline is enlightening: between 1995 and 2001 (see Table 4.3), Fiat Auto spent about one-fifth of the amount spent by Volkswagen, its main European competitor, for research and development of new models. In the meantime, Volkswagen became the leading European manufacturer and established itself in the profitable American and Chinese markets.

It is significant to note that when Fiat Auto started to dismiss employees and to announce the closure of factories in Italy, for the first time the trade unions claimed, in addition to the traditional demand for job preservation, also the need for investment in research. What had been for decades an isolated claim of intellectuals and scientists became a request familiar to the man in the street: research for competitiveness. In 2003, the Italian government started to talk about research as a priority, but if we read the Financial Acts of 2003–2007, we realize that research, universities, training, and lifelong learning actually continue to be relegated to a Cinderella role.

The 14 May 2007 issue of Fortune has Mr. Marchionne on the cover and the feature article is titled "The Turnaround" [Faris 2007]. Fortune's previous front page on Fiat dates back to 1992 and reads: "Arrivederci, Fiat. Will Fiat Exit the Car Business?" Faris writes: "Marchionne has engineered a remarkable recovery and has accomplished the impossible; he resembles less a car executive than an orchestra conductor". Two months after being appointed, he announced a restructuring plan that should take the losing Fiat group to a € 1.4–1.8 billion net profit in 2007, a target that he might achieve, since he already flipped red to black from − $ 332 million in 2005 to + $ 384 million in 2006. He wiped away dozens of non performing directors and reorganized Fiat's car business. The production cycle was reduced from 36 to

Table 4.3 R&D investments (billion euro) between 1995 and 2001 for the development of new car models for Fiat and for a few European Fiat competitors

Firm	R&D
Volkswagen	20.9
Mercedes	13.1
Renault	10.4
BMW	9.9
Fiat Auto	4.5

Source: Eurobusiness [Rubython 2002].

18 months and research investments increased by 10 % a year, up to € 1.6 billion
(3.2 % of turnover after tax) in 2006 [Michellone 2007]. It is expected that the profit
after tax in 2007 will be over € 1.8 billion.

The financial markets have acknowledged the changed perspectives for Fiat.
Fiat share on the Milan stock exchange was, on 1 November 2007, 65 % up from
52 weeks earlier at € 18.55 (see Fig. 4.22). Fiat placed € 1 billion 10 years bonds
in June 2007, priced with a risk premium of less than 1 per cent above the European "risk-free" rate for the first time in more than two years. Fitch's upgrade by
two notches from BB to BBB-, could reduce further the cost of Fiat's borrowings

Fig. 4.22 Value of Fiat shares at the New York Stock Exchange (*lower curve*) during 2001–2004,
compared with the automobile industry average and the S&P500 index (*light upper curve*)
Source: ©*Financial Times*, 3 March 2004:, reproduced with kind permission.

[Michaels 2007–1]. Mr. Marchionne said: "The decision by Fitch gives us great satisfaction because it is recognition of the speed with which we have managed a fundamental change at the group."

The origins of this crisis and the miracle of the recovery are complex and cannot be exhaustively analyzed in the space we are dedicating to it in this context. It seems, however, that four factors undisputedly contributed to the crisis:

1. diversification of the Fiat Group in other sectors, different from the car sector, which it identified as more profitable and in which the Group concentrated its energy and investments
2. insufficient R&D investments to improve the auto product and produce a line of products that would be competitive on the international market
3. excessive outsourcing (about 75 %) of its production, identifying its suppliers on the basis of price rather than quality, in contrast to its German competitors, which had gone as far as to support research and development of associated or subcontracting companies
4. fifty years of relying on state aid, which reduced the need to confront the market in terms of fair competition, thus weakening the stimulus for innovation and the quest for quality

The factors that contribute to the recovery appear to be the following:

a. flush out non performing management, starting from the top
b. give executives freedom of action, encourage them to take risk, and supervise closely their performance
c. reduce the development cycle by a factor two, down to 18 months
d. streamline the design and share platforms and subsystems across product lines
e. increase R&D investments and internationalize research cooperation
f. establish strategic industrial alliances worldwide, in particular in emerging markets (China, India, Russia, Poland, and Turkey—in addition to the well established presence in Brazil)
g. analyse carefully and constantly customer demand and satisfaction
h. delocalize in European countries where the manufactured product meets a market demand and the local workforce is qualified, low-cost, and highly motivated

At the end, this short analysis we expand on the last point, to mention Fiat's investments and delocalizations in Poland and in Turkey, as we will refer in Section 11.5.8, "Poland". In its plant in Tychy, Poland now possessing a workforce of 2,700 qualified, reliable, and dedicated workers, Fiat is producing its award-winning new Panda. This industrial project would not have been possible in the Mirafiori plant in Turin, Italy, let alone in the Termini Imerese plant in Sicily: in both cases, the cost of labour would have imposed a price on the final product that would not have been competitive. The delocalization to Poland is a positive move because it rescues a viable industrial project, matches the availability of qualified and affordable workforce, and can answer a demand in the local market for budget cars. In Bursa, in its Turkish joint venture Tofa:, Fiat manufactures the budget multipurpose award-winning vehicle Doblò. It is sold internationally as well as locally: you can see Doblò as a taxi or a police car on Turkish roads. This is also a healthy case of delocalization where the product meets a local demand, to the advantage of Fiat, Tofa:, and the Turkish workers.

4.13 Recommended Reading

Recommended reading for this chapter are listed in Appendix C and are available on the web site companion to this book, at the URL http://stajano.deis.unibo.it/ RQC.htm

Part II
Research and Technology Policy in the European Union

Chapter 5
EU Research: Objectives and Results

An overview of the history, objectives, participation methods, results, and open problems of EU research is presented below [Guzzetti 1995, Caracostas et al. 1998, Muldur et al. 2006]. By EU research we mean the research and technological development activities undertaken under the control and with the financial support of the EU, as dictated by the Treaty on EU. These include a series of strategic R&D actions that comprise only a minor part of the overall R&D undertakings in the EU countries. EU research has the objective of achieving industrial competitiveness and contributing to other EU policies. It does not include R&D aimed at enlarging the knowledge of mankind or R&D actions that, in the spirit of subsidiarity, are conducted under national budgets, and address specific country objectives. Chapter 6, "Framework Programs" describes the domains addressed by EU research.

EU-funded research programs have played an important role in the implementation of EU policies and in the process of integration of the member states. While the milestones of the new treaties marked the institutional evolution of the European Economic Community towards the European Union, European citizens, and particularly researchers, scholars, industrialists, and students, became more and more aware of being part of a borderless community much wider than their own country. This community has been working in an active and peaceful environment for Europe's economic, social, and political integration. EU research programs have facilitated the mobility of scientists, researchers, industrialists, and students throughout Europe, thus contributing to the creation of a feeling of belonging to the European Union. EU research programs have also promoted industrial partnerships between European companies, which have led to greater competitiveness and cohesion.

A real EU technology policy was formalized only in the mid-1980s [Stajano 1999-1] with the reform of the Treaty of Rome through the Single Act (1987). However, the first signs of a technology policy can be traced back to the 1950s even though the policy was limited to agriculture and the sectors linked to postwar industrial reconstruction.

A. Stajano, *Research, Quality, Competitiveness,*
© Springer Science+Business Media, LLC 2009

5.1 From Euratom to the Framework Program

Today's European collaboration in the research sector can be traced back to the glorious century-old traditions of the great European universities such as: Bologna (Italy), Cambridge and Oxford (UK), Coimbra (Portugal), Heidelberg and Lubecca (Germany), Leuven (Belgium), Paris (France), Praha (Czech Republic), Salamanca (Spain), Utrecht (The Netherlands), and many others. Over the centuries, European universities have been the nodes of a network of exchanges and of philosophical, legal, and scientific collaborations that went on even when Europe was ravaged by wars that left nothing but death and destruction in their wake. Therefore, it is not surprising that, soon after the end of the Second World War (1940–1945), some European countries joined forces to create CERN (1953), the international laboratory for high-energy physics research, and that in 1957, at the same time as the Treaty of Rome, the six founding members of the EEC created Euratom for the peaceful use of nuclear energy for the power plants required by the industrial reconstruction of Europe.

These two organizations were very different and had quite different, or rather opposite, destinies [Guzzetti 1995]. CERN is a nonprofit scientific organization, located in Geneva, Switzerland, that does not aim at any industrial fallout effect. Research aims at increasing human knowledge in elementary particle physics, a field to which European scientists had made great contributions even before the war. This research field requires complex and expensive experimental equipment that no member state alone can afford. Over the years, important discoveries have been made at CERN, which has greatly contributed to the promotion of scientific cooperation and the development of an international scientific culture among the communities of European researchers. Within the research field, CERN has been compared to a *beating heart* able to accelerate the circulation of ideas and people within the high-energy physics sector. In addition to many scientific results, some important technological results have been achieved as well, which have had a great impact in the market and on civil society—such as the world wide web, which resulted from the need to exchange scientific information between lab networks in different countries.

Euratom was a research centre of several thousand scientists located in Ispra, Italy. Euratom, in contrast to CERN, cannot be considered a success, and it is important to understand why. The founding idea was to increase energy production to satisfy the energy requirements for European reconstruction and development. Electricity generated through nuclear fission was seen as the cheap and safe energy source of the future, which was perhaps in keeping with the knowledge of the day, but later proved wrong. In a spirit of *nuclear optimism* and especially under the pressure of the lobbies of U.S. technology providers, the nuclear industry was presented as a mature sector requiring only the construction of nuclear power stations rather than preliminary research on plant safety and waste disposal. European policymakers did not take into account the energy market trends, the prospecting for new oil fields, and research on new energy sources, or consult the European companies that would build the nuclear power stations or the utility companies that would manage the distribution and supply of energy to the end user. The combination of the reduction in oil prices owing to the discovery of new oil fields, a new environmental awareness, environmental movements, and nuclear safety problems led many European governments to reduce or ban nuclear production programs in the decades following the start of Euratom. This situation caused a long period of decline for the organization. Unlike CERN, Euratom, which represented a top-down action taken by politicians under U.S. pressure without any expert advice or market research, was a stronghold of highly specialized and highly paid officials with open-term contracts and limited professional mobility. When, after a few years, it became clear that the project was no longer meeting EEC priorities, the European Council could not dismantle Euratom, due to pressure from the trade unions and the Italian government. So, Euratom survived as a plethoric, sclerotic, expensive, rigid, and obsolete structure searching for its raison d'être. When at last most of its personnel reached retirement age (in the 1990s), Euratom was rejuvenated and focused on new goals. Its structure was changed, and the Joint Research Centre (JRC), described in Chapter 6, "Framework Programs", was formed as an organization that carries out research to support community policies, providing the impartial technical – scientific advice and know-how necessary to carry out institutional activities of the Union.

5.1.1 Targeted Shared Cost Research

A new approach to technology policy and to the role of research in industrial development was suggested in 1979 by Etienne Davignon, the European Commissioner for Industry. He entered into negotiations with the "Big 12 European information and communication technology (ICT) companies" to see whether it would be possible to ensure a world market share and competitiveness for the European ICT industry. Davignon's vision started a synergic process between research and technology policy and generated a new spirit of participation and cooperation within Europe.

Learning from the experience of Euratom, Commissioner Davignon unified under single management the EEC-funded industrial research and the industrial policy and negotiated with the Big 12 companies, which were fiercely competing against each other in the market, to create a common space of precompetitive and prenormative cooperation in order to

- reduce the technology gap and prepare future industrial developments,
- stop the brain drain from Europe to the U.S.,
- overcome the *national champions* policy of the member states, which gave privileges to national leading companies to the detriment of the competition policy and the creation of an internal market, and
- create synergies between European research and European industry.

Commissioner Davignon's great contribution was twofold: on the one hand, he acknowledged that the European users of ICT technologies were the captives of proprietary systems binding them to their U.S. providers; on the other, he managed to reconcile the competition policy with the public financing of industrial cooperation activities by defining a space of precompetitive cooperation. He aroused the European industry's interest in creating, together with its European competitors, the conditions for being successful in a market dominated by U.S. oligopolies while preserving the identity and independence of every single actor and promoting competition among these actors in the market.

The Big 12 companies were asked to draw up, together with the Commission, a strategic R&D program that would associate them with each other, with other European industries, and with European universities in order to achieve three objectives:

1. The development of standards of European origin to allow new actors to enter the market, to offer open structures and interoperable systems, to overcome the domination of the U.S. technological oligopolies, and to exploit the potential capabilities of the European industry, setting it free from U.S. dependence
2. The development of a technological base for the growth of European industry
3. The creation of synergies between industries, universities, and research centres throughout Europe

The Big 12 companies were Aeg (DE), Bull (FR), Cge-Alcatel (FR), Gec (UK), Icl (UK), Nixdorf (DE), Olivetti (IT), Philips (NL), Plessey (UK), Siemens (DE), Stet (IT), and Thomson (FR).

Davignon showed the industrialists that they had long-term common interests and objectives and that, compatibly with the competition policy, there were opportunities for collaboration in the development of standards and basic technological research that would have led the individual industrial groups to independently integrate new technological developments in their product lines. The member states' governments, however, expressed reservations about Davignon's program. They had been used to following a policy protecting *national champions*, namely, national leading companies, and they prevented the full accomplishment of the internal market by hindering cooperation and creating obstacles to standardization processes that were seen as attacks on their national leaders. It took Davignon four long years of discussions and negotiations to overcome the reservations of the boards of directors of the Big 12 companies, which used to look daggers at their competitors. In the end, the Big 12 companies accepted with conviction the EEC proposal and lobbied their governments, which passed in the Council the Commission proposal for a pilot project whose costs would be shared between industry, universities, and the EEC.

The above process shows the role of the European Commission in the planning and implementation of policies. The Commission acts like an entrepreneur [Lembke 2002] that combines, with a specific purpose, several actors and interests, promoting an agreement between them and offering a regulatory framework for the realization of such an agreement. In this regard, the author would like to mention the briefing he received when he joined the EC from Horst Hünke, his first boss and his master: "The Commission should limit its action to serving good coffee in order to attract industrialists around a table. It's then up to them to write the industrial strategy." This statement was very inspiring, although deliberately incomplete, since it did not mention the cookies served with the coffee: a 50 % contribution to R&D expenditures.

The consultation with the Big 12 marked the beginning of the European Strategic Program for R&D in Information Technologies (*Esprit*). Esprit was a 50–50 shared-cost program with industrial partners from at least two member states (see Fig. 5.1) in which both small and medium-sized enterprises and universities or research centres could participate. The intellectual property regulations issued within the research contracts allowed each partner to economically exploit the results achieved and to have access under fair conditions to the background information on the research subject under study that, before the beginning of the collaboration, belonged to the other partners. In fact, in all fairness to Mr. Hünke, the author would like to mention that he was not only making good coffee but also feeding the EC legal service with drafts of the Esprit research contract, a masterpiece that became the model for publicly funded R&D also for other initiatives within the Triad.

These rules were the incentive to face the intrinsic complexity of international collaborations and led to an increase in the scientific, cultural, and technological level of the participants. The Esprit model was studied and applied in the U.S. and in Japan for their national research programs funded by public money.

For the next 15 years after 1983, Esprit was the main industrial research program funded by public money. Together with Acts and Telematics (two other ICT

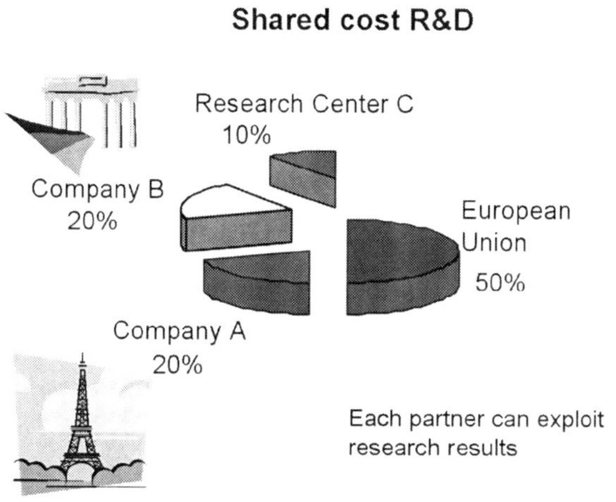

Shared cost R&D

Research Center C
10%

Company B
20%

European
Union
50%

Company A
20%

Each partner can exploit
research results

Fig. 5.1 The European Union funds industrial research aiming at competitiveness. The Community contribution adds up to 50 % of the total costs, including overhead. Consortia must include at least two industrial enterprises from two distinct member states. This figure shows a case where the consortium is made of two industries (from two different countries) and one research centre. In the example, Company A and B each contribute to 20 % of the costs, and both have access to 100 % of the intellectual property of the results

programs on, respectively, communications and applications) and Brite (a program on industrial and materials technologies), Esprit has created a European industrial research community. It is a very widespread and top-quality community, proven and reinforced by hundreds of thousand of person-years of transboundary research collaboration, which has allowed scientists, industrialists, researchers, and Ph.D. students of different EU countries to get to know and respect each other and to exchange new ideas and experiences. Research programs have also played an important role in the institutional transformation and cohesion between member states, and contributed to the development of the European Economic Community into the European Union.

The industrial research community has created innovations not only in the technological base, which is essential for competitive growth, but also the world of information and communication technology *users*. It has prepared the Union for the information society and has accelerated the convergence processes in communication, electronics, information processing, and the media [Forrester 2004].

Esprit has achieved important technological results that have been integrated in systems and applications and that have created the conditions for the development and success of new companies or the transformation of other companies. Commissioner Davignon's successful action led to an important transformation of the role of research within the European Community, with the consequent inclusion of technological research in the Treaty amended by the Single Act (1987).

Over its long and successful life the Framework Program was refocused to best support the industry competitiveness [Bianchi 1999]. From bridging the technology gap and diffusing technologies (1983) the program moved to aggregation of firms and to coping with Japanese competition (1987); then the focus was on user-supplier collaborations, best practice and dissemination; and contribution to other Community policies than competitiveness, such as internal market and employment (1994). The following refocusing was on socio-economic values and support of SMEs (1998); the establishment of the European Research Area (ERA) and the building of the knowledge-based society (2002); until the present seventh Framework Program, planned as the research instrument aligned with the 2007–2013 financial perspectives that supports the revamped Lisbon Strategy, relying on a 65 % increase of R&D resources to unify competitiveness and cohesion policies towards the achievement of sustainable growth.

Associating over the years competitiveness with other EU policies put R&D centre stage among the Union policy making and justified a significant increase in Community funding [Peterson et al. 1998].

5.1.2 Framework Program

The Treaty on European Union provides (in the articles from III-248 to III-255) for the legal basis for the research programs funded by the European Union. An excerpt follows:

Article III-248

1. The Union shall aim to strengthen its scientific and technological bases by achieving a European research area in which researchers, scientific knowledge and technology circulate freely, and encourage it to become more competitive, including in its industry, while promoting all the research activities deemed necessary by virtue of other Chapters of the Constitution.

2. [...] the Union shall encourage undertakings, including small and medium-sized undertakings, research centers and universities in their Research and Technological Development activities of high quality. It shall support their efforts to cooperate with one another, aiming, notably, at permitting researchers to cooperate freely across borders and at enabling undertakings to exploit the internal market potential, in particular through the opening-up of national public contracts, the definition of common standards and the removal of legal and fiscal obstacles to that cooperation. [...]

Article III-249

[...] the Union shall carry out the following activities, complementing the activities carried out in the member states: (a) implementation of research, technological development and demonstration programs, by promoting cooperation with and between undertakings, research centers and universities; (b) promotion of cooperation in the field of the Union's research, technological development and demonstration with third countries and international organizations; (c) dissemination and optimization of the results of activities in the Union's research, technological development and demonstration; (d) stimulation of the training and mobility of researchers in the Union. [...]

Article III-251

1. European laws shall establish a multiannual framework program, setting out all the activities financed by the Union. [...]

3. A European law of the Council shall establish specific programs to implement the multiannual framework program within each activity [...].

Since 1982, all research activities have been coordinated through a five-year Research and Technological Development *Framework Program (FP)*. Starting in 2007, FPs span over seven years, to synchronize with the timing of financial perspectives. The decision-making process leading to the adoption of a Framework Program is defined in the Treaty and summarized in Fig. 5.2. A key feature of the process is an open and extended consultation lasting several years and involving many thousand experts from industry, academia, member state administrations, and user groups. A feeling of ownership of the research policy is created in those who participate in consultations. They understand the ultimate objectives of the policy and are brought to refrain from participating in the program for the sole sake of grabbing the public funding.

The planning and finalization of the Framework Program, from the presentation of the proposal by the Commission up to its approval and implementation, takes over three years during which Commission, Council, Parliament, the Economic and Social Committee, the Research Group, the Committees appointed by the Council, commissions of experts and working groups, and research users play, at different stages, their respective roles. The implementation of a five-year FP takes seven to eight years, taking into consideration that each program covers a five-year period during which calls for research proposals are announced whose execution lasts one to three years.

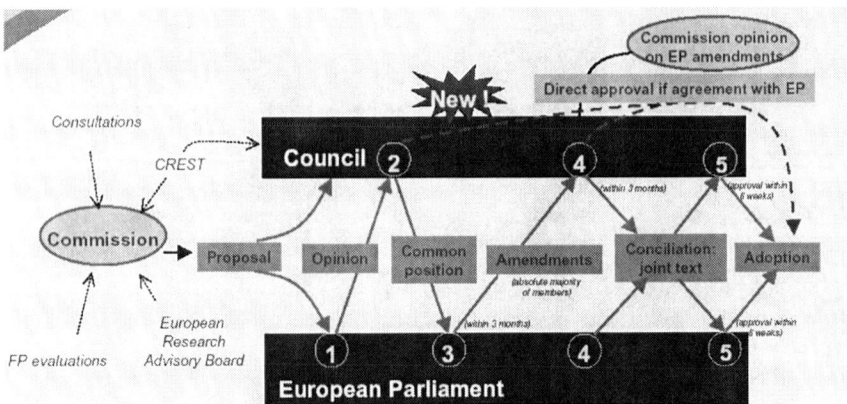

Fig. 5.2 A schematic representation of the decision process leading to the adoption of the Framework Program for Research and Technological Development. The time elapsed from initial consultation to adoption is over three years. (European Commission 2003)

The extent to which a given FP is able to meet the expectations of a member state depends on the capacity of that member state to participate in the preparation and discussion of the program's strategy, structure, objectives, and contents. A positive effect on that country during the implementation of the Framework Program is possible only if the orientation, objectives, and contents of the specific programs implementing the Framework Program cover issues already defined as national priorities. To be effective, the FP must be in line with national, industrial, and academic research policies and satisfy the national requirement for innovation in industry and services. The most active countries strongly intervene in the preliminary stage of definition of the orientation of the FP. During the definition of the contents and objectives of the framework programs, the Commission is and has always been very open, inviting experts from all member states. While some member states accepted these invitations with qualified, informed, and efficient participation, others, including Italy, did so with less enthusiasm.

The EU Research and Technological Development Framework Program is not a plan for the redistribution of financial resources, but rather an instrument to promote the research actions required to increase the competitiveness of European products and services. The selection of the proposals received is carried out, regardless of their geographical origin, on the basis of their quality and their compliance with the admission and selection criteria in order to ensure the feasibility of the overall objectives of the plan. Even though there is a quest for the widest and most balanced participation by organizations from all member states, an even redistribution of returns in proportion to the contribution to the Community budget is not required.

Research project contracts are not structured so as to favour accounting aggregation of payments on a national level. However, it is possible to calculate the national distribution of the financial resources in the proposals considered for the negotiation of contracts. Even if it is not straightforward to take into consideration the redistribution effect during contract negotiation or execution and any transnational subcontracts, a clear picture can be drawn with sufficient approximation about where the financial resources go.

The comparison between Italian participation in the program and Italian contribution to the Community budget would be meaningless if the resulting difference were not as significant and persistent as to be considered symptomatic of a structural gap. After the EU enlargement to the EU15 in 1995, the Italian contribution to the Community budget accounted, until 2004, for about 12–13 %. The Italian participation in the Framework Program has always been lower than its contribution rate to the Community budget by several percentage points.

The ex post recognition that the results of resource allocation did not reflect the potential contribution of a certain member state has led, over the years, to the implementation of some corrective actions, involving, in some cases, Italy, too. If, on the one hand, these actions have ensured a somewhat partial and temporary accounting correction, on the other, they have not filled the structural gap relating to insufficient participation in the definition of program objectives, content, and methods. The program is therefore conceived on the basis of the requirements of other member states and does not take into account the peculiar features of Italian industry, characterized by microenterprises or small and medium-sized enterprises rather than big multinationals and conglomerates.

5.1.3 R&D Planning Together with Industry

In the years following the *Esprit* pilot phase, consultation and planning also involved other operators in addition to the Big 12:

- small and medium-sized enterprises (SMEs)
- technology providers
- users
- system integrators
- service providers
- researchers

In this way, over the years, the number of subjects involved in R&D planning has grown from the initial 12 participants to some thousands. This orientation can be seen in the funding distribution (see Fig. 5.3) and in the type of collaborations (see Fig. 5.4). While at the beginning collaborations involved mainly big companies, with the passage of time, an increasing number of partnerships have been formed between big and small companies (see Fig. 5.5); also thanks to the SMEs' greater flexibility and capacity to rapidly focus on the main issues of technological evolution.

If, for big companies, SMEs are the ideal partners to cover development border areas, on the other hand, big companies can offer SMEs marketing infrastructures and distribution and assistance channels, thereby allowing SMEs to operate in a wider market than they could have reached with their own limited sales and assistance networks.

Fig. 5.3 Percentage of funding in the Esprit program during the period 1994–1998 to various classes of partners: large enterprises, SMEs, universities, and other undertakings. (European Commission 1999)

Distribution of funding in FP4

Fig. 5.4 Percentage of FP4 projects (1994–1998) with the participation of large enterprises and SMEs associated with providers and users of technology. (European Commission 1998)

Industrial Participation in FP4 Consortia

Fig. 5.5 Change in the number of collaborations of large enterprises with other large enterprises, SMEs, universities, research centres, and other undertakings between FP3 (1990–1994) and FP4 (1994–1998). The lines in the figure have the sole purpose of facilitating the reading by joining points that belong to the same category. (European Commission 1999 [EC 1999-3])

Collaborations Among Partners in FP3 and FP4 Consortia

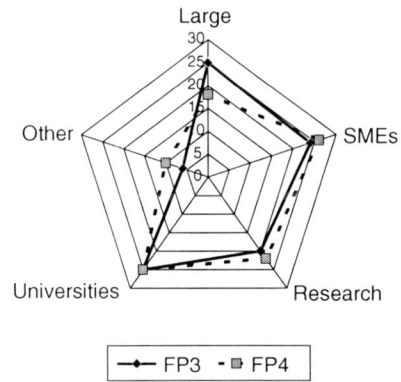

5.2 Results of EU Research

In this section, we provide a general description of the results of EU research without referencing any specific scientific, technological, or application result. Some specific results will be presented in Chapter 6, "Framework Programs".

The results of EU-funded research are beyond counting and of very different nature: ideas, standards, patents, components, materials, processes, systems, applications, etc. As we will see, many of them—and not the least—are intangible and have to do with the transformation of the people, enterprises and countries involved. Results are integrated into solutions to the problems of society in any sector: aerospace, agriculture, banks, communications, distribution, e-commerce, emergency management, energy, environmental protection, the food industry, health care, industrial planning, insurance, logistics, media, microelectronics, packaging, the pharmaceutical industry, public administration, tourism, and transport, to name a few.

The implementation of the research policy has anticipated institutional changes and widened the EU's scope. As an example, in the quest for strategic technological development in the interests of the Union, beyond the specific interests of individual countries, decisions on specific programs were at times made with qualified majority voting prior to the enforcement of this rule following the Treaty of Amsterdam. Research widens the EU's scope: as an example, since the 1980s, the association to R&D programs between undertakings from future enlargement countries and EU15 firms has favoured the integration of those countries. Firms participating in the Framework Program can achieve great results such as: uptake new leading-edge technologies, processes, materials; improve capabilities, behaviour, and ability to compete; realize increased turnover and profitability; enhance productivity; improve market share; access new markets; reorient the commercial strategy; improve the competitive position; enhance their reputation; and reduce the commercial risk. Research generates high social intangible benefits and opens up for a possibly very high return on investments (ROI) in the long term. ROI can be the highest in risky exploitation of the leading edge technological breakthroughs; firms must be encouraged to take risks and the public funding can help the private enterprise to adventure in explorations that it could not afford with its sole resources.

We will show in the following pages that R&D collaboration has led to integration processes among enterprises that go beyond research alliances and that research and technology policy contributes to the realization of the internal market and to EU cohesion and enlargement while using only a small part of the EU budget.

5.2.1 Overcoming the Technological Gap and Dependence

One of the most important results of EU research over the past 25 years has been the creation of open architectures and the overcoming of the technological gap between the EU and the U.S. EU action to define interoperability platforms and create open systems has allowed new operators, particularly SMEs, to enter the ICT market, thus overcoming the dependence of the European information technology industry and information users on the American oligopolies that dominated the European market until the end of the 1970s.

Standardization in data transmission, document architecture, image coding, and the standardization of physical and software interfaces has opened and widened the market. Standardization supported the competition policy by opening markets to new actors and helped creating new business that generated jobs by the million in mobile communication and services, content industry, and multimedia.

The technological gap with respect to the other regions of the Triad has been bridged at least in some sectors. For example, in the EU, there are some leading undertakings in the sectors of aerospace (e.g. Airbus, ESA, Galileo); production and design of microprocessors (e.g. ST-Microelectronics, ARM); microprocessor cards (e.g. Gemplus); software development (e.g. SAP); media and content industry (e.g. Philips, Giunti Interactive Labs); and of telecommunication technologies and services (e.g. Nokia, and numerous mobile service providers).

The standards originating in Europe protect the potential of the European industry and represent one of the necessary elements for the creation of the internal market.

5.2.2 Industrial Partnerships

Research collaborations have led to mergers, acquisitions, and partnerships, thus creating an industrial structure suitable for the EU single market. Vertical collaborations between operators acting at different stages of the value chain started from research collaborations, and often created stable industrial partnerships and synergies. Some successful multinationals (e.g. ST-Microelectronics) are the result of research collaborations; many other precompetitive partnerships have created a suitable space for the growth of the European industry.

5.2.3 Support to SMEs and the Sharing of Experiences

SMEs are a vital and essential part of the economy of almost all EU countries. Even though most of them are unable to directly invest in research because of their reduced dimensions, their competitiveness must be assured through appropriate actions to ensure their initiation into technology and the sharing of innovative experiences. In some cases, SMEs that have specialized in fast-developing sectors are at the cutting edge of research and, through industrial collaborations and subcontracts, provide the skills required for the innovations of big companies. Synergies among SMEs and collaborations between large and small companies can be important steps in the quest for competitiveness, as shown by the outcomes below:

- Collaborations between large and small companies have offered the latter new outlets as well as the chance to enter new markets. Moreover, small companies that are financially unable to invest directly in research have had access to the results of EU-funded research through the specific actions of technological initiation and transfer. On the other hand, big companies have exploited small companies' flexibility and agility to satisfy new requirements in planning, safety, and advanced applications.

- Sharing of successful experiences (best practice) increases industrial culture and contributes to the cohesion of the national and European industrial system.
- Regular evaluation of one's own performance using reference standards and structured comparisons (benchmarking) maintains the quality level required to compete in international markets.
- Initiation into technology: Technology is so widespread that it is necessary to introduce it into the industrial and service sectors that are the least technologically advanced. Specific EU actions have already satisfied this need.
- Other actions at both the EU and national levels involve SMEs in the fight against the digital divide, namely, the danger of marginalization of people, enterprises, or public services that have not been able to keep up with the changing business environment. Among the Italian initiatives to eradicate the digital divide, the author takes pride in citing a project managed by his daughter Cecilia in Consorzio Gioventù Digitale, a nonprofit organization founded by the City Council of Rome and supported by industrial sponsors, committed to the digital inclusion of Roman citizens. The project, called *Nonni su Internet*[1] (*Pops on Internet*), introduces the elderly to ICT through training provided by supervised high school youngsters.

5.2.4 The Power of Networking

Research programs have created a community of scientists, industrialists, and researchers united in temporary and voluntary associations based on common development objectives. These associations have often defined the rules for integration on an industrial scale rather than just at the level of research. Working together and with the help of EU funding, some partnerships have been created from the bottom up and have contributed to the implementation of an EU-wide process of innovation and cohesion that has proven to be much more effective than any top-down action. The knowledge network and the awareness of the existence of qualified subjects in several industrial sectors have been the basis for a structural change in EU industry and have contributed to the creation of the internal market.

Figure 5.6 shows the contribution of the research programs funded within the fourth Framework Program (FP4, 1994–1998) to the internationalization of industrial partnerships. For each EU and EFTA state, the rate of national and transnational partnerships is indicated. The first shaded area in each row represents the percentage of research alliances within the same country. As can be seen, even in the biggest European economies (Germany, France, the United Kingdom, and Italy) the rate of national partnerships is at most 13 %, which means that at least 87 % of research partnerships involve industries or universities or research centres of another state. The following areas in each row represent the percentage of research contracts between undertakings in the country named in that row and undertakings in other

---

[1] URL: http://nonnisuinternet.gioventudigitale.net

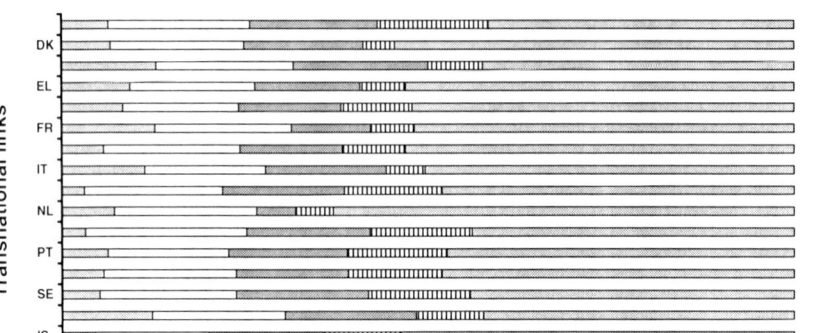

Fig. 5.6 Transnational alliances initiated by R&D contracts in FP4. This figure presents for each EU and EFTA state the percentage of research contracts with undertakings in the same country and of contracts with foreign ones. (European Commission Science and Technology indicators 1997)

EU or EFTA countries. Transnational partnerships represent a good opportunity to start collaborations and processes of mutual knowledge, and sharing of experience and skills that can change the European industrial framework and narrow the gaps between the member states. Links formed in projects persist beyond the initial R&D co-operation and may move from the lab to the board room, creating alliances and mergers.

The success of Esprit (information technologies) and of other programs like Race/Acts (communication technologies), Telematics (applications), and Brite (industrial and materials technologies) has been widely recognized as a key factor in the achievement of the strategic objectives of the Community, that is, the internal market, cohesion, and enlargement.

Figure 5.7 shows which countries have been in the first and second most frequent transnational research partners in Fig. 5.6 in the contracts originated by R&D in FP4. The UK has been able to take advantage of the research and industrial networking opportunities offered by the FP more than other economies with comparable or larger GDP. Italy lags far behind.

In the 1990s, Esprit focused on collaborations between technology providers and users, creating what are called *vertical* partnerships to distinguish them from what are called *horizontal* partnerships between companies of the same sector. Only horizontal partnerships concerning prenormative R&D processes prior to standardization continued to be promoted. On the other hand, vertical partnerships were greatly promoted, as they generated synergies and complementarities. In addition, they could lead to technological transfers and industrial exploitation of results at the end of the funded projects without any problem of competition between research program participants and without the need for any participant to look for

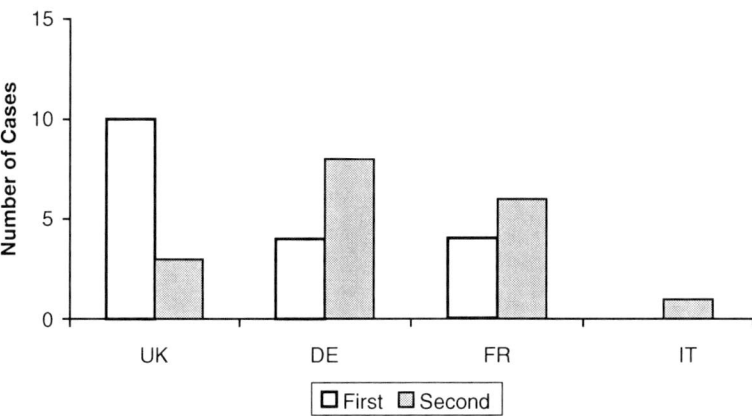

Fig. 5.7 Number of cases of positioning as most frequent and second most frequent transnational partner of EU and EFTA states in the FP4 research contracts. (European Commission Science and Technology indicators 1997)

independent access to the market according to its specific company strategies and range of products. To give an example, consider a partnership between a company designing low-consumption microprocessors, another company producing micro-processor cards, and a mobile telephony company, each with its own specific skills, market, and demand for technological solutions. The activities of the three partners follow one after the other in cascade; therefore useful synergies can be created without any interference between the industrial and commercial operations of the different partners.

By involving ICT users, this new orientation of the program has promoted the spread of automation processes in public administration and industrial sectors out-side the ICT core business (such as aerospace, agriculture, banks, communications, distribution, e-commerce, emergency management, energy, environmental protec-tion, the food industry, health care, industrial planning, insurance, logistics, media, packaging, the pharmaceutical industry, public administration, tourism, and trans-port, to name a few). Furthermore, it has contributed to achieving competitiveness and growth objectives by accelerating innovation. Best-practice sharing, first-users actions, training, diffusion of results, and technological transfer have marked the passage from technology push to user pull, in response to the demands of a society in transition, where ubiquitous ICT technologies require the availability of infras-tructures and solutions for the emerging knowledge-based society.

5.2.5 Research and the Internal Market

Research collaborations have contributed to the removal of some nontariff bar-riers that prevented the realization of the internal market and have promoted a dynamic adjustment of the European industry towards the highest quality levels.

These changes have occurred both through research collaborations and through the above-mentioned best-practice sharing and structured comparisons with reference standards and between different competing solutions (benchmarking). A bottom-up transformation of the European industrial sector, which is characterized by the following three elements, has thus occurred:

1. The pace of innovation, the de facto standards, and the competition rules are set by the most innovative countries and companies.
2. Competition is no longer based on price but on quality.
3. Market segmentation is no longer based on the geographical origin of products but rather on their functionality, quality, and performance.

Other factors resulting from research programs that accelerated the creation of the internal market have been:

- the mobility of workers, particularly of researchers and Ph.D. students
- the harmonization of standards in design, workers' performance and safety, and, above all, in interfaces
- the diffusion of open and compatible architectures in which heterogeneous multivendor systems can successfully work together

Open systems not only accelerated the creation of the internal market as far as the ICT industry is concerned but also have been instrumental in overcoming or attenuating the technological dependence on U.S. oligopolies that conditioned the spread of information technologies, in particular in the European countries with limited endogenous innovation capacities.

5.2.6 Research, Cohesion, and Enlargement

Research programs have triggered transnational industrial partnerships. R&D programs have involved and still involve participants from countries that are about to become members of the EU: for example, Portugal and Spain at the beginning of the 1980s, the member states of the European Economic Area (EEA) at the beginning of the 1990s, the East German regions after the fall of the Berlin Wall (1989), the central and eastern European countries (CEEC) since the mid-1990s, and currently the three candidate countries.

5.2.7 Summary: Unsolved Problems

Despite the great number of successes obtained by EU research programs, there are still many unsolved problems:

- *Intellectual property protection.* The sharing of the results obtained by research programs funded by public money implies the risk of losing the intellectual property of inventions. University and research culture, which is intrinsically open to the sharing of research results and to their publication, must be combined with the requirements of protection of confidential information in business. Moreover, the digitization and convergence processes have created some very serious cloning problems as regards information products, which require new safety technologies as well as production and distribution processes to protect intellectual property.

- *Law on patents.* Because of market globalization, national patents are not enough to protect inventions. On the other hand, the complexity and high costs of patents in all EU member states have often discouraged intellectual property owners to embark on a complex and expensive process. Up to now, owing to the complexity of patent coverage in the European single market [EC 1997-2], the position of European researchers has been less favourable than that of U.S. researchers. The European Commission has been working since the 1970s for the adoption of a European patent that overcomes the language and legal barriers burdening patent procedures. A historic outcome for the completion of the internal market was the European patent agreement signed in 2003 [Council 2003]. It remains for the Court of Justice to define the application rules and the judicial procedures. In the knowledge-based society new instruments are required both technological and legal to address the protection of intellectual and industrial property more effectively than what copyright and patents can do.

- *Access to funding* for the creation and development of a company. Even though great progress has been made over the past 20 years, the Union, and particularly some EU member states do not offer appropriate financial support for the creation of new companies and the funding of innovation.

- *Support for the creation of startup companies.* Research programs sometimes create the conditions for turning an idea into a process or product with great market potential. Researchers sometimes have the necessary information and know-how to industrialize their inventions, but they often lack the entrepreneurial culture and the capacity to draw up a business plan and to collect the financial resources required to start a company. Even though some universities and public corporations have taken several actions to create a suitable climate for the creation of new companies, the situation in most of the EU territory is still inadequate except for some areas (e.g. in the Cambridge area in the UK [SQW 1985, SQW 2000]).

- *Promotion of the valorization of results.* Many research results age in researchers' cupboards or in mass storage devices and become obsolete before being industrialized and launched in the market. The know-how developed within research programs is not duly valorized. Within actions aiming at intellectual property protection, the Framework Program should invest more resources for the technological transfer and diffusion, valorization, and industrialization of research results.

- *Diffusion of innovation* in companies not taking part in research. In some EU member states, and particularly in those, like Italy, where there are mainly microenterprise, small-sized, and medium-sized companies, it is necessary to ensure that certified technological innovation results are spread throughout the whole production community. In order to maintain its market share, a company not investing directly in R&D must be able to use the instruments and infrastructures developed, verified, and adopted by industrial operators that, thanks to their size and farsightedness, have been able to invest in research. If innovation is not spread throughout the whole production sector, there will continue to be a loss of competitiveness and a reduction in international market shares, which will be increasingly acquired by companies and countries where industrial policy supports the diffusion of technology and the creation of infrastructures. We explained in Chapter 3, "Competitiveness of the European Economy" that the adoption of exogenous R&D is not a small challenge, because innovation is not a commodity one can buy off the shelf. The recipient enterprise must understand and value technological innovation and be able to integrate it in its products and processes while remaining in full control of them.

5.3 Research Funding

Europe is the second region in the world for R&D spending (Fig. 5.8) with 29 % of the total world expenditure, which is concentrated in the Triad at 94 %. In terms of per capita R&D spending, Europe is at 60 % of the U.S. and at 35 % of Japan [Economist 2007-26] and in terms of researchers as a percentage of the workforce, the EU is at 0.5 %, namely 35 % below American and Japanese figures. EU15 position is worse in the strategic sector of ICT where EU15 invests only 17 % of the

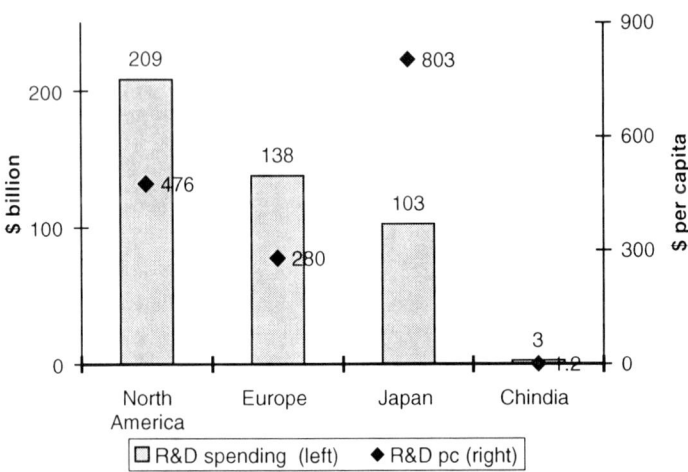

R&D spending in the world

Fig. 5.8 R&D spending in the world: absolute values left scale, grey bars: R&D spending in the world in $ billion (year 2006). Per capita R&D spending: right scale, diamonds: in $ (year 2006). Chindia stands for China and India
Source: [Economist 2007-26].

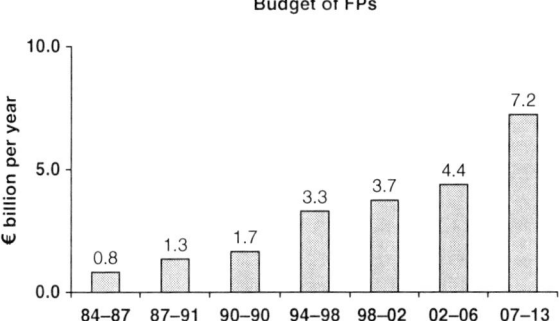

Fig. 5.9 R&D in the EU: R&D yearly funding from 1984 to 2013 (billion euro). The figures indicate the yearly Community contribution. An almost equivalent amount is contributed by the participants
Source: European Commission 2002.

Triad's ITC R&D spending [EC 2006-11]. Multiannual R&D framework programs have been increasingly funded (see Fig. 5.9) and have shown a strategic orientation in line with the requirements of a changing society. While China and India are at 0.6 % of world's R&D spending and practically invisible in Fig. 5.8, other Asian economies (Korea, Singapore, and Taiwan) spend 4.8 % of the world's R&D expenditure. The case of Korea's R&D was addressed in Section 3.6.4, "Korea". Some EU27 countries (Sweden and Finland) are above the U.S. in terms of R&D spending (see Fig. 3.61). However, if countries were compared with the U.S. states considered separately, Sweden would rank sixth and Finland would not be within the first ten.

Another major difference between Europe and the rest of the world is on the percentage of R&D financed by private firms. This different attitude of the European firms reduces the power of intervention of the member states administrations and of the EU. Conditions can be put in place as incentives to R&D private spending, but if the business culture is not receptive only limited results can be expected. As it says, you can bring the horse to water, but you can't make him drink. In Europe there is also a gender gap which is not reported in the U.S. Higher business R&D investments in the U.S. attract outsourced research also from European firms; they attract also international Ph.D. students and researchers and contribute to brain drain from several EU countries.

Investing in research pays, although it is quite difficult to quantify the return. A study by Muldur et al. estimates direct ROI three to fifty times after 25 years and enumerates many important intangible results such as: new jobs; increased turnover and profitability; enhanced productivity and competitive position; and improved market share. According to this study the € 2.7 billion yearly increase in community R&D funding from 2007 might generate € 60 billion increase in EU GDP by 2030 [Muldur et al. 2006]. However R&D money might not suffice if there are not enough scientists and engineers and if the society is not supportive of innovation.

Financial commitments passed from € 3 billion in 1984–1997 to € 6 billion in 1994–1998 and were equal to

- € 13 billion for the 4th Framework Program, 1994–1998
- € 15 billion for the 5th Framework Program, 1998–2002
- € 17.5 billion for the 6th Framework Program, 2002–2006
- € 50.5 billion for the 7th Framework Program, 2007–2013

Increased funds should go along with increased efficiency and reduced red tape (a recurring complain form the research groups that keeps many SMEs out of the EU research and advantages the very large multinationals that participate in hundreds of project and can justify the cost of EU admin experts among their staff). Increased funds is also instrumental to a complete integration of the accession countries, however this should not happen at the cost of compromising on quality.

The Commission's initial proposal [EC 2005-1] for the seventh Framework Program (that spans seven years from 2007 to 2013) follows the European Council recommendation for a substantial effort to meet the Lisbon objectives. The proposal suggested a maximum amount of commitment credits of € 75.8 billion, which is $2\frac{1}{2}$ times the amount of the sixth FP on a yearly basis. A final agreement was reached at € 50.5 billion, as shown in Fig. 5.14 and in detail in Chapter 6, "Framework Programs" In 2007, a new source of R&D financing was introduced that should mobilize additional € 30 billion for loans to research undertakings, as described in Section 5.4, "R&D Funding from the Financial Market ,".

It is worth noting that the financial resources of the Framework Program are about twice as much as the above figures, considering that the contribution by participating corporations usually amounts to 50 %. Thanks to these resources, it is possible during the execution of FP7 to finance every year an order of magnitude of 10^5 full-time researchers, taking into account not only their salaries but also any indirect expenses, overhead expenses, and the depreciation of the costly infrastructures and equipment required for research. These facts may help the reader to appreciate the potential impact of the Framework Program in Europe.

Budget structure, namely, the funding of various R&D domains, has changed over time, according to changing strategic priorities. We will analyse in detail the research contents of the fifth, sixth, and seventh Framework Programs in Chapter 6, "Framework Programs". While at the beginning the focus was on energy, the focus later passed (Fig. 5.10) to enabling technologies; technologies, infrastructures, and services for the knowledge society; support for sustainable development; improvement in the quality of life; valorization of results and of the human potential in R&D; and lately on new ideas to support fundamental research. The naming of the R&D budget lines has changed over time, so that the grouping underlining Fig. 5.10 is to some extent dependent on the interpretation of how old denominations fit into the new headings. Like in the case of the changes of EU budget over the years (see Chapter 12, "Budget of the European Union") the author suspects that changing names sometimes is used to pretend more profound transformations than those implemented. Figures 5.11–5.15 show the sector distribution of EU contributions in the last three framework programs.

FPs Budget Structure

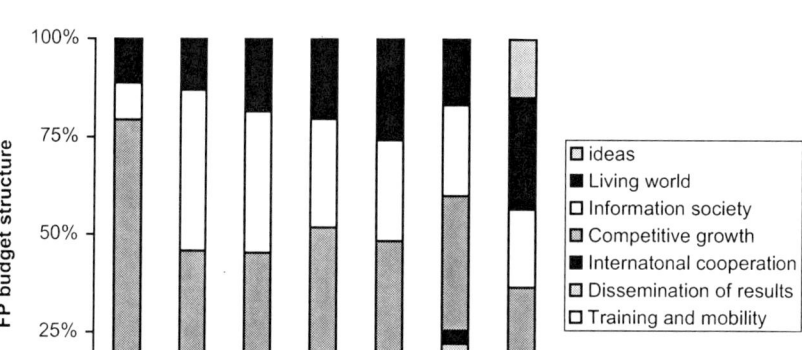

Fig. 5.10 R&D in the EU: Changing priorities are reflected in the changing percentages in the various sectors of the Framework Program from 1984 to 2013
Source: European Commission 2003 S&T Indicators 1999–2003

FP4 1994–1998

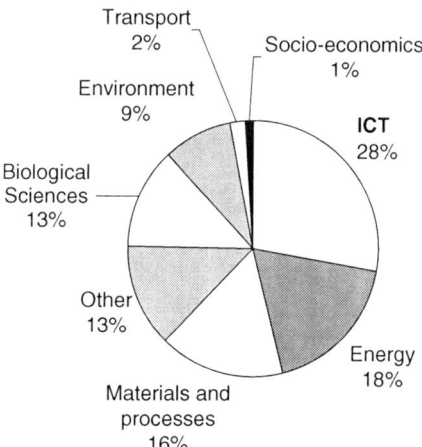

Fig. 5.11 Funding of specific programs in FP4 (1994–1998) adds up to € 12.3 billion
Source: European Commission 2003

In 1998, the increase of FP5 funding over FP4 only slightly exceeded the compensation for inflation (see Fig. 5.16). Subsequent FPs enjoyed a real increase in funding on top of compensation for inflation.

In the passage from the fifth to the sixth Framework Program, the sectors that benefited from an increase in funding higher than inflation were linked to small

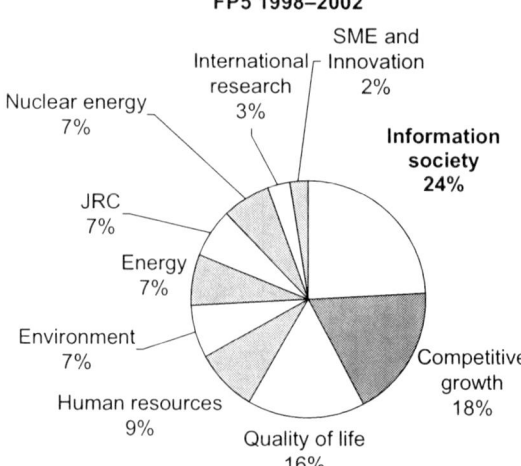

Fig. 5.12 Funding of specific programs in FP5 (1998–2002) adds up to € 15 billion. (European Commission 2003)

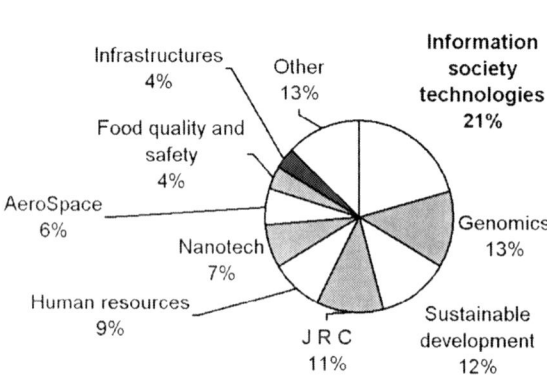

Fig. 5.13 Funding of specific programs in FP6 (2002–2006) adds up to € 17.5 billion
Source: European Commission 2003

enterprises and innovation, human resources and the European Research Area. From the sixth to the seventh Framework Program higher than inflation increase was across the board with special focus on frontier research and all the areas connected with the revamped Lisbon strategy and the building sustainable and environment-friendly growth in the knowledge-based economy. In Chapter 12, "Budget of the European Union", we saw that the EU budget (equal to about € 112 billion in 2006) accounts for 2.06 % of the consolidated public spending of all EU member states. EU-funded research in the sixth R&D Framework Program (equal to about

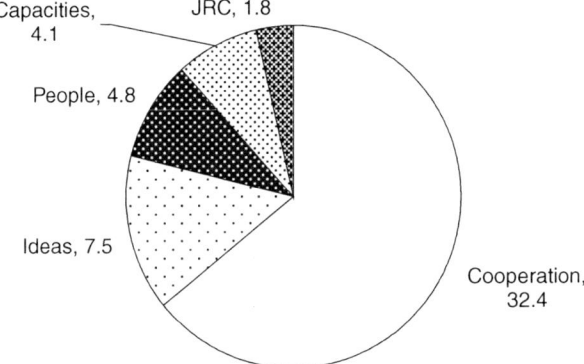

Fig. 5.14 Funding the components (€ billion) of the specific programs in FP7 (2007–2013). FP7 adds up to € 50.5 billion. (European Commission 2007)

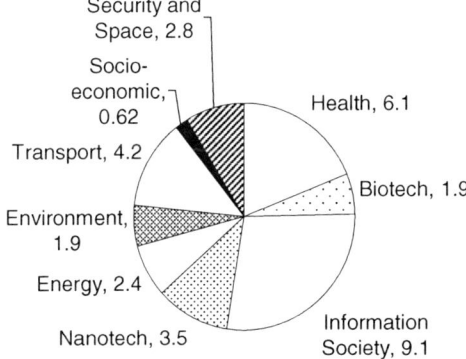

Fig. 5.15 Funding the components (€ billion) of the specific program Cooperation in FP7 (2007–2013). The program Cooperation adds up to € 32.4 billion. (European Commission 2007) *Source*: European Commission 2007.

€ 4 billion/year) accounted for about 4 % of the EU budget, while spending in FP7 reaches 6.6 % of the EU budget. The current FP accounts for about 0.14 % of the consolidated EU member states' public spending. The total R&D spending in the EU (including public and private spending) is 1.8 % of the GDP.

Total public R&D funding (EU and member states) amounts to about € 80 billion per year. Figure 5.17 shows that the percentage of public funding to international research actions reached 22 % in 2007. The first area at the bottom of the graph represents the Framework Program. The second zone represents a number of cooperative programs not under the European Commission's jurisdiction (CERN, high energy physics; COST, S&T co-operation among 19 countries; EMBL, molecular biology laboratory; EMBO, molecular biology organization; ESA, European space

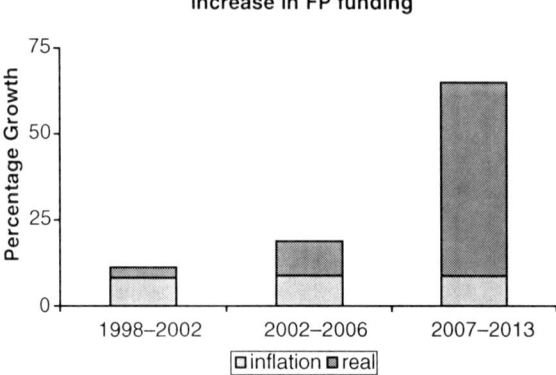

Fig. 5.16 Percentage of growth of FP5, FP6, and FP7 funding, with respect to the previous FP. Inflation compensation and real growth are shown. (European Commission 2003)

agency; ESF, European science foundation; ESO, European astronomical observatory; ESRF, European synchrotron radiation facility; ILL Langevin institute for physics, chemistry). The third zone represents the Eureka program. Only 11 % of total public R&D funding is financed by the EU budget and falls under the EU's direct responsibility and coordination. Even though the Framework Program only accounts for a small percentage of the expense of publicly funded research, it has a great impact on the growth and competitiveness of the European industry because of the strategic nature of the funded topics, the strict monitoring of the execution of the works, the management of the intellectual property of the resulting inventions and patents, the multiplying mechanism of the valorization of results, and the financial contribution from research participants (usually 50 %).

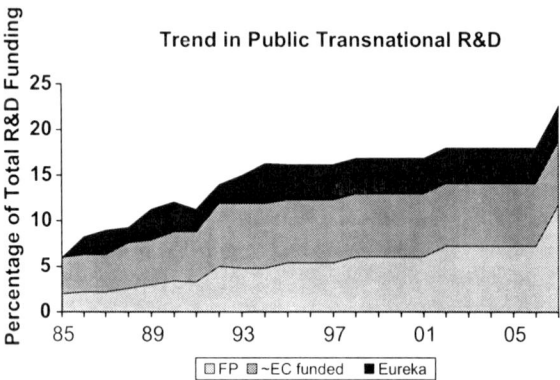

Fig. 5.17 Trend in public transnational R&D funding in Europe (1985–2007) as a percentage of total public funding. The first zone from the bottom represents the Framework Program. The second zone is a number of cooperative programs not under the European Commission's jurisdiction. The third zone represents the Eureka program. (European Commission 1998, 2007)

R&D Expenditure

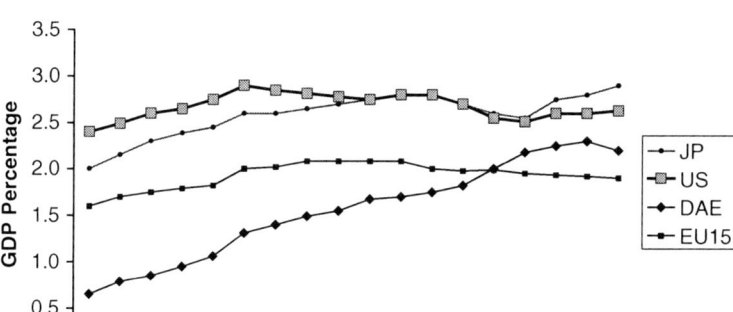

Fig. 5.18 Public expenditure for research as a percentage of GDP in EU15, the U.S., Japan, and the DAE countries (dynamic Asian economies: Korea, Singapore, Taiwan) in the 1980–1997 period. Consolidated data for DAE available up to 1997. (European Commission 2003)

Figure 5.18 shows total research spending in the EU in comparison with total research spending in the U.S. and in Japan during the 1980–2010 period. Another group of countries has made an aggressive entry into the technological research arena: a number of dynamic Asian economies (Korea, Hong Kong, Singapore, and Taiwan). Research spending in the European Union is, on average, at 72 % of the U.S. level. In 2003, the objective of the EU was to raise R&D spending to 3 % by 2010. Even if this objective is achieved, which is not certain, the situation would remain very serious and would have long-term effects since research benefits on competitiveness emerge with a delay of up to 20 years or more. In particular, the new discoveries in biotechnology research produce results much later than ICT research, where the lag time between new ideas and market launch varies from two to ten years. In the field of biotechnologies, the transfer of an innovative idea from the research lab to the hospital wards and the chemists' shelves may easily take over 20 years.

Details on funding, content, and results of FPs are presented in Chapter 6, "Framework Programs".

5.4 R&D Funding from the Financial Market

The seventh Framework Program plans to generate an additional € 30 billion within 5 years through a new mechanism, the Risk Sharing Finance Facility (RSFF), drawing resources from the European Investment Bank and from the private financial market. The RSFF is a new financial tool proposed by the European Investment Bank (EIB) and the Commission to encourage the banking sector to take more risks in financing technological enterprises. A total of € 1 billion is provided by the Community budget (provisioning) to which € 1 billion is added by the EIB, that is, a total amount of € 2 billion per year, allowing a volume of € 10 billion in loans over 5 years. Since the EIB finances an average of one-third of the total cost of the project, an additional € 20 billion will be provided by private finance. Therefore, in total, this will make € 30 billion which should be mobilized for R&D [RTD Info 2007-1].

5.5 Participating in Research Activities

The activities funded within the Framework Program aim at achieving the objectives indicated in the Union treaty for research actions: the strengthening of the technical and scientific base of the European industry, the promotion of its competitiveness, and the support of EU policies.

In the previous sections of this chapter, we have outlined the long process leading to the definition of the strategic content of the Framework Program through several consultations with thousands of operators in industry, the economy, the scientific community, and national or local administrations.

The Framework Program is not an indiscriminate, across-the-board research funding plan but rather a program specifically aimed at supporting EU development policies and especially at increasing the European industry's competitiveness through focused R&D actions.

Participation in the Framework Program research activities funded by the EU budget is open to consortia formed among organizations located in the member states, in the states involved in the EU enlargement, and in states associated with the EU by cooperation agreements. In the last case, participation and funding procedures are different from those applied to the EU member states or to applicant countries. The undertakings admitted are production companies in the industrial, trade, and agricultural sectors; service providers; small enterprises, groups of small enterprises; research groups from universities or research institutes; and public administrations.

Participation in the program offers European organizations the chance to join the network of the most advanced industrialists and scientists in Europe, to prepare for market globalization, to benefit from the European internal market, and to exploit the synergies between technology providers, integrators, researchers, scientists, users, and public administration.

To consider the Framework Program only as a source of financing would be a mistake. Participation is justified and is a source of development and success only if the research topics are included in a company's development strategy and only if the suggested research would be carried out even without the EU contribution. A company must consider the Framework Program as an opportunity to collaborate with excellent partners and share knowledge and complementary skills. The EU financial contribution should allow companies to explore more advanced innovation areas and more risky ventures than the ones that they could face with their own skills and financial resources. Through EU-funded R&D, the research partners should move towards the most advanced technological and application developments in order to prepare the ground for great innovations able to increase the company's competitiveness in the market over the long term.

Participation in the program is quite expensive due to the complexity of transnational partnerships and collaborations, the burden of EU bureaucracy, the mobility requirement, and the need for documentation in a language common to all partners, of the whole R&D process: from the preliminary stage of proposal definition to reporting, in the operative stage, on work progress, results, and failures. These expensive activities are compensated for by the EU contribution and are often the occasion for improvement in the company processes and preparation for the international industrial, trade, and scientific collaborations typical of the globalized and networked market of the knowledge-based society.

A company, a research centre or a university department wishing to participate in the research activities of the FP will find it easier to do so if the administration of its country has duly and promptly performed all its tasks during the planning, orientation, and definition of the FP and of the specific programs implementing it. As already mentioned in a previous section of this chapter, the preparation of the Framework Program is a very complex activity taking up to three years and ending with the definition of the strategic areas identified by the community of industrial and scientific operators as the objectives to be pursued over the period covered by the FP. The European Council, together with Parliament, makes the final decisions on research priorities and financial coverage. Conflicting strategies may be presented due to the industrial structure of the member states and the peculiar features of the areas of excellence of the different countries. These strategies may be aimed at supporting the conflicting interests of one of several groups: the large industrial conglomerates, the basic industry developing enabling technologies, the small enterprises, the applied industry, or the end users. A member state's administration and its industrial and scientific communities must ensure their informed and qualified participation in the complex stages of study, planning, and proposal of the FP in order to focus the objectives on supporting the sustainable development of that same country. On the other hand, if an administration deals with the FP only upon attribution of the research contracts, with a view to earmarking available funds, it is impossible to ensure participation that, in addition to the financial contribution, also provides the partnerships, synergies, and innovation required for the growth of the national economy.

Unfortunately, Italy has been less active than almost any other member state in defining and protecting the strategic interests of its industrial and academic sectors. Industrial and university researchers have underestimated the strategic importance and professional worth of taking part in work groups for the definition of programs and the evaluation of proposals. Therefore, the request for innovation of the Italian production system has not found in the FPs that same incentive of growth and competitiveness from which many other member states have benefited.

At the October 2002 annual meeting of the Società di Politica Industriale (Industrial Policy Society) held in Palermo, Italy, the then president of the European

Commission, Mr. Prodi, stated that in the study groups defining EU intervention strategies (not only in the research area) in Brussels, Italian interests were not represented as much as they should have been considering Italy's contribution to the EU budget. This situation is not because of secret plots but rather because of the fact that Italy seems impervious to any invitation to participate in activities defining the European industrial and economic future. This is one aspect of a broader phenomenon: while in Italy a higher percentage of people than in other EU states express favourable feelings towards the Union, the Italian government has for decades considered Community institutions with disdain, and the appointment to the Commissioner post has at times been used to get rid of some awkward presence on the national political scene. Conversely, in 1972, the offer of a post in the national government was sufficient for Franco Maria Malfatti to resign as Commission president after less than two years in office. Franco Frattini, the Italian vice-president in office, might be doing the same in view of a minister post in Italy after the elections due in April 2008.

It is therefore not surprising that the success rate of the research proposals presented by Italian coordinators, that is, the rate of proposals turning into real contracts, is low in comparison to the proposals coming from other countries. This situation further discourages Italian participation which ends up often being characterized by proposals aimed at collecting financial contributions rather than achieving strategic objectives.

The drawing up of a proposal for an EU research project is a very complex and expensive task, which requires a long period of preparation for the creation of the consortium and the definition of the roles and work program. In order to understand how to draw up a potentially successful proposal, it may be useful to analyse the evaluation criteria applied by the European Commission.

5.6 Evaluation Criteria of Research Proposals

At the URL http://www.cordis.lu/, the European Commission publishes the evaluation criteria applicable to any research proposal. Every call for proposals for the assignment of research contracts during the implementation of the Framework Program is also published at this URL, along with specific evaluation criteria applicable to the call in question, according to the type of contract and the nature of the specific programs. We analyse below some general criteria [EC 2003-2] applicable to industrial research programs, recommending, however, that any interested reader should visit the above-mentioned URL and carefully review the documents relating to the relevant call for proposals.

Calls for proposals are usually announced four times a year, except for some activities for which the call for proposals is always open. Each specific program calls for proposals for certain domains within its scope, with a frequency of its own. Each call contains a work program, which represents one of the stages of implementation of the specific program and includes a list of the research areas for which proposals are called, and an indication of the financial resources allocated to projects in that area.

Proposals are presented by consortia, which, in most cases, must be formed by at least two industrial organizations from two different member states and must have a suitable size for achieving the research objectives. Each proponent must ensure a specific contribution, thereby justifying its membership in the consortium.

Research proposals are evaluated by a panel of independent experts according to specific guidelines and ethics rules. The proposals selected for the negotiation of a research contract must have high scientific, technological, and management quality in relation to the objectives of the relevant specific research program.

Analysis of the criteria used by the European Commission to evaluate proposals is very useful for drawing up a proposal highly likely to be selected for the negotiation and funding of a research contract. The following six criteria apply:

1. Relevance
2. Potential impact
3. Scientific and technological excellence
4. Quality of the consortium
5. Management quality
6. Mobilization of resources

A brief explanation and interpretation of each of the above criteria are provided below:

> *Relevance:* the extent to which the proposal meets the work program and correctly interprets its objectives.
>
> *Potential impact:* the extent to which
>
> - the proposal has a strategic impact on competitiveness and contributes to the achievement of EU social objectives or to the implementation of EU policies,
> - the proposal includes the valorization and diffusion of results in order to ensure effective technological transfer and the use of project results even outside the consortium,
> - the proposal highlights the added value resulting from research carried out at the European level rather than within research programs of the individual member states,
> - EU support contributes to the extent and results of the suggested action,
> - the proposal mobilizes greater European resources than the critical mass needed to ensure the achievement of significant objectives,
> - the proposal faces and solves problems linked to standardization.
>
> *Scientific and technological excellence:* the extent to which
>
> - the proposal clearly defines some specific objectives,
> - the objectives represent a significant development with respect to the state of the art,

- the scientific and technological approach ensures that the consortium can achieve its research and innovation objectives.

Quality of the consortium: the extent to which

- both the role and funding of each participant and the ways in which participants form, on the whole, a top-quality consortium are clear and justified,
- each participant is qualified and involved in the activities assigned to it,
- participants complement one another well and there is a good integration between their skills and experiences, which creates synergies and added value with respect to the research programs of the single member states,
- there is good participation from small and medium-sized enterprises.

Management quality: the extent to which

- it is possible to demonstrate the high quality of project management and the latter is proportionate to the project complexity,
- there is an appropriate plan for the management of knowledge, intellectual property, and other aspects linked to innovation,
- the coordination methods proposed are strong enough to support both the actions proposed and the synergy within the consortium,
- the ethical and safety problems linked to the proposed research have been duly identified,
- risks have been duly evaluated,
- clear and significant milestones, documenting partial results during the execution of the project, have been duly defined, as well as success indicators together with criteria, measurement methods, and expected values in order to evaluate the achievement of the objectives and partial results.

Mobilization of resources: the extent to which

- the proposal implies an appropriate use of human, infrastructural, and financial resources for the success of the project,
- these resources are duly and effectively integrated,
- the financial plan is proven to be appropriate,
- the gender problems linked to research have been duly faced.

5.7 Recommended Reading

Recommended reading for this chapter are listed in Appendix C and are available on the web site companion to this book, at the URL http://stajano.deis.unibo.it/RQC.htm

Chapter 6
Framework Programs

This chapter covers the contents and presents some results of the research activities funded under the European Union's Framework Programs for Research and Technological Development. Other R&D activities supported by Community funding and relevant for achieving competitiveness are described in Chapter 7, "Entrepreneurship, Innovation, and Competitiveness". An extended survey of results stemming from projects of ongoing Framework Programs is presented in the recommended reading associated to this chapter.

6.1 Fifth Framework Program

The fifth Framework Program (1998–2002) [EC 1998-2, EC 2000-9] had an allocated budget of € 15 billion as shown in Table 6.1 below.

The last call for proposals of the fifth Framework Program (FP5) resulted in research contracts that began in 2002 and continued until 2005. In December 2002, the first call for proposals of the sixth Framework Program (2002-006) was published, and it led to contracts that began in the second half of 2003 to end by 2006. Other calls for the sixth Framework Program (FP6) followed, until 2006, and the execution of the relative contracts will continue until at least 2009. Consequently, for a certain amount of time FP5 and FP6 coexisted, as currently FP6 coexists with FP7.

FP5 included thematic programs that covered the following research sectors (not including the Community Joint Research Center):

1. Information society [EC 2000-1]
2. Quality of life [EC 2000-2]
3. Competitive and sustainable growth [EC 2000-3]
4. Sustainable development in terms of: environment, energy, nuclear energy [EC 2000-4]

Alongside these sectors there are the following "horizontal" activities:

1. International research [EC 2000-5]
2. Innovation and SMEs [EC 2000-6]
3. Human research potential [EC 2000-7]

A. Stajano, *Research, Quality, Competitiveness*,
© Springer Science+Business Media, LLC 2009

Table 6.1 Funding of the fifth framework program FP5 (1998–2002) in million euro

Fifth Framework Program (1998–2002)	
Information Society	3, 600
Competitive growth	2, 705
Quality of life	2, 413
Human resources	1, 280
Environment	1, 083
Energy	1, 042
JRC	1, 020
Nuclear energy	979
International research	475
SME and Innovation	363
Total	€ 15 billion

Source: European Commission 2003.

6.2 Sixth Framework Program

The sixth Framework Program (FP6, 2002–2006) [EC 2002-4, EC 2003-7] had a budget of € 17.5 billion as shown in Table 6.2 below.

Table 6.2 Funding of the sixth Framework Program FP6(2002–2006) in million euro

Sixth Framework Program (2002–2006)	
Information society technologies	3, 625
Genomics	2, 255
Sustainable development	2, 120
JRC	1, 990
Human resources	1, 580
Nanotechnologies	1, 300
Aerospace	1, 075
Food quality and safety	685
Infrastructures	655
Other	555
SMEs	430
ERA	320
International research	315
Innovation	290
Governance	225
Science and society	80
Total	€ 17.5 billion

Source: European Commission 2003.

6.3 Funding for the Seventh Framework Program

An agenda for the next ten years was drawn up in Lisbon, Portugal, during the European Council of 23 March 2000. The *Lisbon Strategy,* presented in Chapter 8, "Competitiveness in the Knowledge-based Society", is an all-encompassing umbrella that covers many EU policies, with the view to preparing Europe for the knowledge-based economy.

In March 2003, the European Council in Barcelona, Spain, set the target of raising the EU research effort to 3% of the GDP, two thirds of which should come from private investment. The March 2005 Council confirmed this target. The major financial effort in R&D envisaged by the Lisbon Strategy suggested that the Framework Program should be synchronized with the period of time covered by the financial perspectives for the EU budget, that is, not four years but rather seven years: 2007–2013.

The Commission announced [EC 2005-1] proposals for measures to create the conditions for increasing private commitment to research investments. Such measures include fiscal incentives for research and innovation, revisiting state aid rules, adapting intellectual property rights regimes, facilitating venture capital operations, and strengthening links between universities and industry. At the same time, the EC is putting the knowledge economy centre stage in many EU financial programs, in particular in the Structural and Cohesions Funds. These are undergoing a major redirection following the enlargement, and might be a relevant source of financing for R&D with Community funding under national planning.

While the member states' administrations can in principle decide to increase public R&D funding, raising *business* investments in R&D—which is what makes the difference for achieving industrial competitiveness, as the case of Korea epitomizes—can be recommended and facilitated with incentives, but cannot be imposed. In many member states, in particular in those characterized by a majority of SMEs, propensity to business R&D is low and there are minimal investments by SMEs in long term innovation because the main concern for their CEOs is survival in the next six months. However, research at the European level has some attraction compared with national programs, at least for the most open-minded entrepreneurs: they are induced to invest in R&D at EU level more than in national programs because results stemming from cooperative European research are supposedly applicable to a larger market and the risk of failure in exploring the frontiers of technology is reduced by the strength of the industrial partnerships.

The seventh FP has been announced as a program to build the European Research Area (ERA) of knowledge for growth. It has been structured with four specific programs that are described in some detail in next Section 6.4, "Implementing the Seventh Framework Program":

1. *Cooperation*: to gain leadership in key scientific and technological areas by supporting cooperation between universities, industry, research centres, and public authorities across the EU as well as with the rest of the world

2. *Ideas*: to stimulate creativity and excellence addressing leading edge domains with a long term vision
3. *People*: to develop human potential and support mobility and career development in European R&D
4. *Capacities*: to enhance research and innovation capacity with the infrastructures and means to allow the emergence of *poles of excellence,* namely, regional research-driven clusters bringing science and industry together, supporting research for the benefit of SMEs and of the society at large

Table 6.3 presents the allocation of resources as initially proposed by the European Commission after extended consultations and as finally adopted by Council and Parliament; it does not include € 2.75 billion for Euratom 2007–2011. Despite the reduction of the final financial allocation with respect to the Commission proposal, FP7 grants a 65 % per year increase over FP6. The research budget in the 2007–2013 Financial Perspectives has a larger increase than the overall EU budget, as we saw in Chapter 12, "The Budget of the European Union".

Table 6.3 Allocation of financial resources (billion euro) for the seventh Framework Program, 2007–2013. Commission Proposal of April 2005 and Adopted Decision of December 2006

Specific Program Thematic areas	Proposal Commission 2005	Funding Adopted
Cooperation	*44.4*	*32.4*
Health	8.3	6.1
Biotech	2.5	1.9
Information Society	12.7	9.1
Nanotechnologies	4.8	3.5
Energy	2.9	2.4
Environment	2.5	1.9
Transport	5.9	4.2
Socio-economic	.79	.62
Security and Space	4.0	2.8
Ideas	*11.9*	*7.5*
People	*7.1*	*4.8*
Capacities	*7.5*	*4.1*
Infrastructures	4.0	1.7
SMEs	1.9	1.3
Regions of Knowledge	.16	.13
Research Potential	.55	.34
Science in Society	.55	.33
Coherent Dev. of Res. Policies	–	.07
International Cooperation	.36	.78
JRC	*1.8*	*1.8*
Total EC	72.7	50.5

Source: European Commission 2005, 2007 [EC 2005-1, EC 2007-11]

6.4 Implementing the Seventh Framework Program

The seventh Framework Program presents significant changes with respect to the previous ones and relies on a budget allocation per year that grants a 65 % increase over the previous one. The first major innovation in FP7 is the institution of the European Research Council (ERC), a funding agency for frontier research. Long term research was already funded under FPs since 1989; the FP6 basic research initiative, still ongoing, is called Future and Emerging Technologies. The term "frontier research" reflects a new understanding of basic research. On the one hand it denotes that basic research in science and technology is of critical importance to economic and social welfare, and on the other that research at and beyond the frontiers of present knowledge of mankind is an intrinsically risky venture, progressing on new and most exciting research areas, and is characterized by an absence of disciplinary boundaries. The proposal selection and the project monitoring for frontier research is conducted by ERC with minimal red tape, simplified procedures, and reduced boundary conditions and constraints, as explained in Section 6.4.2, "Ideas". Another innovation of FP7 is the Joint Technology Initiatives (JTI) mainly concerned with areas of research that require considerable investments to ensure long-term success. The JTIs rationalize the procedures for participation and financing by combining private investment with public financing. JTIs are implemented by *sui generis* structures that bring the Commission and the private sector together; support activities for SMEs, Marie Curie mobility grants, and certain logistical and administrative aspects will become the responsibility of an executive agency. This evolution can be explained both by the increase in funding for the Framework Program, without a corresponding increase in the Commission's manpower, and willingness on the part of the Commission to concentrate on political and legislative tasks [RDT Info 2007-1].

6.4.1 Cooperation

The specific program "Cooperation" assures continuity with the previous collaborative cross-border R&D initiatives. The instruments to implement "Cooperation" are as follows:

Integrated projects create the prerequisites for the realization of the priority thematic areas of the Framework Program by grouping together a critical mass of activities in the areas of research, demonstration, training, innovation, and resource management (human resources, skills, infrastructures, and equipment).

Networks of excellence group together and interconnect centres of excellence in a specific thematic area in order to overcome the European fragmentation of research activities. In a sense they are, in the knowledge-based society, comparable to the clusters of the old economy.

Coordination of national research programs supports the synergy and collaboration between the national research programs of the member states.

Specific Targeted Research Projects (STREP) are shared-cost projects that implement research activities aimed at developing new products, processes, or services (or improving existing products, processes, or services) to satisfy society's needs and implement EU policies. STREPs are regulated by new provisions protecting the intellectual property of inventions.

Coordination actions aim at promoting and supporting the interconnection and coordination of research and innovation activities.

Supporting actions integrate research activities and aim at supporting the implementation of the FP. They include surveys, presentations of research results, seminars, and study groups.

The first call of the Cooperation Program was opened on 22 December 2006 in the area Information and Communication Technologies with a budget allocation of € 6 million.

6.4.2 *Ideas*

This is the first time that the EU has a dedicated mechanism to fund great ideas coming directly from its "brightest" and the best researchers. Never before has there been a research funding mechanism at European level that funds pure, investigative research at the frontiers of science and technology, independently of thematic priorities. As well as bringing such research closer to the conceptual source, the flagship FP7 program "Ideas" is recognition of the value of basic research to society's economic and social welfare [EC 2007-12]. The implementation of this program is under the responsibility of the European Research Council (ERC), a funding agency for frontier research, described in the next Section 6.5.

The first call *"Starting Independent Researcher Grant"* was published in December 2006 and closed on 25 April 2007. Its allocation was € 289 million.

6.4.3 *People*

The "People" Specific Program acknowledges that one of the main competitive edges in science and technology is the quantity and quality of its human resources. To support the further development and consolidation of the European Research Area, this Specific Program's overall strategic objective is to make Europe more attractive for the best researchers.

Abundant and highly trained qualified researchers are not only a necessary condition to advance science and to underpin innovation, but also an important factor to attract and sustain investments in research by public and private entities. Against the background of growing competition at world level, the development of an open

European labour market for researchers free from all forms of discrimination and the diversification of skills and career paths of researchers are crucial to support a beneficial circulation of researchers and their knowledge, both within Europe and in a global setting. Special measures to encourage young researchers and support early stages of scientific career, as well as measures to reduce the brain drain, such as reintegration grants, will be introduced.

Instruments for mobility and training (Marie Curie actions) offer researchers several transnational mobility opportunities at different stages of their careers and support institutions that host researchers who are participating in mobility programs. The "Marie Curie Actions" have long been one of the most popular and appreciated features of the FPs. They have developed significantly in orientation over time, from a pure mobility fellowships program to a program dedicated to stimulating researchers' career development. The "Marie Curie Actions" have been particularly successful in responding to the needs of Europe's scientific community in terms of training, mobility, and career development. This has been demonstrated by a demand in terms of highly ranked applications that in most actions extensively surpassed the available financial support [EC 2007-13]. Calls for Marie Curie Actions are recurrently announced on the CORDIS URL.

6.4.4 Capacities

The "Capacities" specific program is designed to build the new European economy by strengthening and optimizing the knowledge capacities that Europe needs if it is to become a thriving knowledge-based economy. By strengthening research abilities, innovation capacity, and European competitiveness, the program is stimulating Europe's full research potential and knowledge resources. The program embraces six specific knowledge areas, including [RDT Info 2007-1]:

Research Infrastructures promote the creation of a top-quality infrastructural research system and its effective use within Europe

Research for the benefit of SME, include two types of instruments:

1. subcontracting to external corporations (research centres, universities, or others) by a group of small and medium-sized research enterprises
2. research projects promoted by associations or industrial groups with the commissioning of research activities on behalf of SME consortia

Regions of Knowledge bring together the various research partners within a region. Universities, research centres, multinational firms, regional authorities, and SMEs can all link up and strengthen their research abilities and potential.

Research Potential creates synergies across various partners within the same knowledge area

Science in Society develops a dialogue between the scientific community and society at large on issues including ethics in science, science education, and scientific advice to get more people and organizations involved in shaping EU policy

International Cooperation activities ensure the balanced participation of third countries in collaboration with European partners

A call on "Supporting the coordination of national policies and activities of member states and associated states on international S&T cooperation" was published on 2 October 2007 with closing date of 12 February 2008 and a budget of € 11 million.

6.5 The European Research Council

The European Research Council (ERC) is a funding agency for frontier research established to implement the Ideas Program. ERC will select the most promising and interesting ideas. Decisions on what is promising and interesting will be taken autonomously by the scientific community, without that red tape that has at times discouraged submissions to the Framework program calls by European scientists.

Ernst-Ludwig Winnacker, the German molecular biologist and Professor of Biochemistry at the University of Munich, speaks in his new Brussels office of ERC director about the ERC's plans to make a big name for itself in the research world by funding outstanding younger and more experienced researchers from potentially any field in scientific disciplines [RDT Info 2007-1]:

> The rationale is to fund frontier research, investigator-driven research identifying people and projects in any field of science one can think of, including engineering, humanities, everything. The hope is that, through serendipity and hard work, results are achieved that can eventually flow into the innovation process. Scientific excellence is the criterion. The only other condition is that you have to work in Europe. If you are in Tallinn, and can convince the peer reviewers that this is the place to be for your work, then that is fine. Or you could move to somewhere else in Europe if the project so requires. You could also be a Chinese scientist in Shanghai who wants to work in Europe. If this person presents their case well enough, basically it can be done. And there are lots of European researchers working in the U.S., of whom some could be encouraged by the scheme to come back. So the conditions are to work in Europe and to be good. Anybody who feels courageous enough to say "yes, I am the best, I would win here" should apply. The prospective grantees decide whether they send in an application, and they have to convince the review panels that they are good and that the host institution is the place to do the proposed research. There is no political reason why you have to do it in a particular place. You have to convince the panels that you and the host institution together are the best thing that can happen. We have set up 20 peer review panels to study project applications covering the whole spectrum of science from A for archaeology to Z for zoology.
>
> We want to competitively fund frontier research and to this end envisage two funding streams/instruments. The first provides for ERC 'Starting Grants' to permit early independence for young scientists between two and nine years after their PhD. A first call for proposals has already been announced with a deadline for applications of 25 April 2007. Starting Grants will be anything up to € 400 000 for whatever the researchers need–their salary, their supplies, everything, so that they will be totally independent. The second funding stream relates to ERC 'Advanced Grants' for more established investigators in the later stages of their profession, where a call for proposals will be announced in spring or summer 2007.

6.6 The European Institute of Technology

The Commission put forward on 18 October 2006 a proposal establishing the European Institute of Technology (EIT) as an integral part of the revised Lisbon Strategy
for growth and jobs. The aim is to reinforce Europe's capacity to transform education and research results into business opportunities. This proposal was built on
extensive consultations with European stakeholders, member states, and the general
public. The financial envelope for the implementation of EIT during the period from
1 January 2008 to 31 December 2013 is set at € 309 million.

The EIT will contribute to bridge the innovation gap between the EU and its
major competitors by promoting further the integration of the three sides of the
knowledge triangle (education, research, and innovation) in a mutually-supportive
manner and providing a world-class innovation-oriented critical mass at the EU
level. In particular, the EIT will facilitate innovation partnerships, translate research
and technological results into business opportunities, promote entrepreneurial initiatives, and enrich higher education through the most up-to-date, directly applicable
knowledge.

The EIT is composed of two levels:

- A governance structure which is based on a Governing Board (GB). The GB will
  be responsible for steering the activities of the EIT. It will also take charge of
  selection, designation and evaluation of the Knowledge and Innovation Communities and for all other strategic decisions. It will comprise a balanced, representative group of high-profile people from business and academia, supported by a
  small number of administrative staff.
- Knowledge and Innovation Communities (KICs): The KICs are the defining
  characteristic of the EIT; they are autonomous partnership of universities,
  research organizations, companies, and other stakeholders in the innovation
  process. They will undertake: innovation activities; cutting-edge innovation
  driven research in areas of key economic and societal interest; education and
  training activities at master and doctoral levels; and dissemination of best
  practices.

The EIT shall give an answer to the need to support higher education as an integral, but often missing, component of a comprehensive innovation strategy. The
agreement between the Governing Board and KICs should provide that the degrees
and diplomas awarded through the KICs should be awarded by participating higher
education institutions, which should be encouraged to label them also as EIT degrees
and diplomas. Through its activities and work, the EIT should help promote mobility within the European Research Area and the Higher Education Area as well as
encourage the transferability of grants awarded to researchers and students in the
context of the KICs.

6.7 Projects and Results in the Areas of the Ongoing Framework Programs

The thematic areas of the ongoing Framework Programs (the sixth and the seventh) are grouped into the sectors illustrated below. FPs also finance the Joint Research Center (JRC).

1. Health
2. Biotechnology
3. Information Society
4. Nanotechnologies
5. Energy
6. Environment
7. Transport
8. Socio-economics Research
9. Security and Space
10. Joint research center

Alongside the *vertical* research sectors listed above are the *horizontal* activities listed below.

1. International cooperation
2. Innovation and SMEs
3. Human research potential

For specific initiatives addressed to small enterprises and international cooperation, besides the financing indicated in Table 6.3, under "Capacities", additional budget is earmarked from the allocations of the vertical sectors.

The *Health* sector includes research aimed at combating diabetes, degenerative diseases, poverty-linked diseases, and includes research to improve the health and well-being of European citizens through improved food quality; and to improve methods of analysis and control of food production.

The *Biotechnology* sector includes genomics, biotechnology, and their applications in medical diagnosis and in drugs development.

The *Information society* sector includes research to support the Lisbon Strategy objectives and the creation of a competitive society based on knowledge management. Research will lead to the integration of computers and communication systems in all environments, making a multitude of services, and applications accessible thanks to interfaces that can be easily used by everyone.

The *Nanotechnologies* sector includes research for the creation of a new approach to engineering, industrial processes, and the science of materials and for the development of knowledge-based multifunctional materials.

The sectors on *Energy, Environment,* and *Transport* include research to create sustainable development by means of energy and transport systems that respect the environment and promote economic growth; research to understand climatic changes and deal with problems related to the greenhouse effect; and research to develop water, land, and natural disaster management systems. These sectors also include actions of waste disposal and recycling and actions to safeguard and assure sustainable management of cultural heritage. These actions are linked to other community initiatives financed by other budget items that are mainly part of the Culture 2000 program [EC 2001–9].

The *Security and Space* sector finances research for the attainment of new levels of technological and safety excellence in the aeronautics sector, while respecting environmental constraints; for the development of the Galileo system for satellite localization; and for the development of satellite telecommunications.

The *Socio-economic* sector includes research in economic and social sciences in relation to the knowledge society, relations between citizens and institutions, European cohesion, and preservation of local cultural identities.

JRC concerns activities of the European Commission's research centres and includes consulting the EU institutions and performing research that contributes to the achievement of community policies in the context of the Framework Program. In particular the sectors of competence of the JRC include food safety, biotechnology, chemistry, quality and reliability of medical equipment, ecology, nuclear safety, and management of nuclear waste.

In the Recommended Reading of this chapter (see Appendix C, Section A3.8.6.6), we have examined each sector in detail, illustrating research scenarios, upcoming results from completed and ongoing projects, and the objectives of recently started projects. The domains addressed include: health; biotechnology and genomics; information society; nanotechnology; energy; environment; transport; socio-economics; security and space; and the Joint Research Center.

6.8 Recommended Reading

Recommended reading for this chapter are listed in Appendix C and are available on the web site companion to this book, at the URL http://stajano.deis.unibo.it/RQC.htm

Chapter 7
Entrepreneurship, Innovation, and Competitiveness

7.1 Community-Funded Research and Enterprise Innovation

Innovation policy tends to maximize the return, in terms of economic development, on investments made in research [Branscomb 1998]. Innovation is the most important component in economic development in today's society. The feedback between market-related and technology-related phases of innovation plays a central role and calls for networks of co-operation between firms and research institutions and for knowledge spillovers across them [Fischer 2006]. The EU Framework Program, by creating networks of enterprises with horizontal and vertical links, and networks between industrialists and scientists is a very powerful and effective catalyser of innovation. Knowledge spillovers are generated by the exposure to best practices and by bringing together the best scientists and industrial researchers from the most advanced undertakings and by uniting their skills and efforts towards the achievement of common strategic objectives. Networks are generated by the very nature of the instruments through which the Framework Program in implemented; in a sense, they are the new form of industrial districts fitted to the needs of the knowledge society, once the need for geographical proximity of those involved is overcome by the communication infrastructures of the networked economy.

Community-funded research does produce results, as explained in Chapter 5, "EU Research: Objectives and Results" and in Chapter 6, "Framework Programs". These results contribute, among other things, to narrowing the technological gap with the most advanced economies, building industrial alliances, and providing support to SMEs. In practice, community-funded research aimed at industrial competitiveness produces the following results: investigation of future technological scenarios that can open up new markets or help in the acquisition of shares in already existing markets; prenormative activities in view of the definition of standards in support of European industry; sharing of experiences on enterprise innovation through the introduction of new technologies, processes, and materials; interconnection across the most advanced and innovative sectors of the European economy; filing of patents; networking and strengthening the community of scientists, engineers, and researchers in academia and in industry; and acquisition by enterprises of new know-how and experience in terms of technology, organization, experience

A. Stajano, *Research, Quality, Competitiveness*,
© Springer Science+Business Media, LLC 2009

of new markets, new trade practices, and more familiarity with the different market demands in other regions within and beyond the European internal market.

Based on his own experience in Community research, the author believes that the process started by the Framework Program for research can act as a catalyst for enterprise innovation. The new ideas that introduce changes into organizations, initiate new processes, or put forward innovative solutions that lead to market success often come from outside the workplace, away from its routine. Indeed, the right conditions for innovation and the invention of new solutions are to be found outside day-to-day life, far away from everyday habits, in a favourable multidisciplinary context where people think outside the box and are encouraged to take risks. A recent study by Michel Syrett and Jean Lammiman [Syrett et al. 2002] on the right conditions for enterprise innovation brings more evidence in favour of this assumption.

Thanks to Community research, industrialists, and researchers are induced to leave their everyday environments and are given the opportunity to meet their most competent European counterparts in a context of technological and intellectual competition far away from the day-to-day routine. In this way, they are encouraged to take risks that they could not handle by relying on their own resources and without the Union's financial support. They are induced to think outside the box and share their views openly, away from the direct influence of the everyday working environment. They have to document their goals, their work in progress, their failures, and their ideas. They have the chance to meet people with different qualifications and experiences and to interrelate new types of knowledge, new situations and facts, which they had never taken before into consideration in the same context. These opportunities create the best conditions for innovative ideas. Nevertheless, the innovation cycle is not over, as ideas are like diamonds hidden in the dust. Turning a great idea into a great business is perhaps as complex as conceiving the idea; a long development process is necessary, often requiring investments ten or a hundred times bigger than those needed during the research phase.

Therefore, Community research is important not only for its visible achievements, but - perhaps most of all - for the process of transformation it brings to the people and into the enterprises that participate in it.

7.2   A New Entrepreneurship in a Knowledge-Based Economy

A basic element to introduce innovation is the creation of a new entrepreneurship capable of preparing enterprises and individuals for changes, in a dynamic context characterized by lifelong learning and constant evolution. In order to allow enterprises to prosper in a knowledge-based economy, it is necessary to make good use of the research and technology available and to promote competition. The European Union endeavours to promote innovation culture in order to make sure that enterprises can actually exploit the results of innovation and turn them into a competitive advantage.

The Information Society, as we will see in the next Chapter 8, "Competitiveness in the Knowledge-based Society", requires for new forms of entrepreneurship that empower each unit in the enterprise with the corporate knowledge of the firm. A new entrepreneurial culture enables the firm to outsource some of its R&D requirements because management and staff are capable to uptake the research results, value, and internalize them.

7.3 Context and Objectives of Innovation Policy

Enterprises play a central role in the Union's economic growth. In order to build a strong economy, the Union must encourage the entrepreneurial spirit and provide an appropriate legal and normative context, as well as suitable infrastructures in order to promote new business practices, favour the creation of new companies, and support their development.

The Union's enterprise policy aims at promoting and facilitating the setting up of new enterprises, and increasing their growth, financing, and innovation capacity. It aims at creating a dynamic business environment by ensuring to these enterprises and to their products and services real access to markets both within and outside Europe.

7.3.1 The Entrepreneurial Spirit

The Union has the objective of stimulating business activities in Europe and will pursue this goal by contributing to the creation of a favourable business framework for enterprises. Rewarding and supporting those who are willing to take risks and facilitating their access to financing are ways to favour the setting up of new enterprises. It is also necessary to introduce profound changes in the present education and training systems in order to develop the necessary know-how and skills. ICT can be the driver of productivity and competitiveness only as far as its adoption is spread across the enterprise and the qualified workforce is literate enough to using it as an enhancing infrastructure as opposed to a straight jacket.

7.3.2 Profiting from the Internal Market

The internal market is the business arena in which European enterprises compete. Despite the great progress made so far, there are still obstacles to the free circulation of goods and services. In order for enterprises to exploit their full potential in the internal market, such obstacles must be removed and full market access must be guaranteed. The liberalization of sectors such as transport, telecommunications, and energy must be completed and the legislation must be streamlined. The application of standards on a voluntary basis and flexibility in regulation will be increasingly important factors to ensure that the internal market brings real advantages to European companies and, at the same time, to guarantee the application of high environmental and health standards.

7.3.3 Sharing Innovative Experiences

Enterprises and member states are being urged to share their experiences and to learn from each other by making use of a series of reference tools and joint initiatives. If they manage to work together with the purpose of improving their individual and collective results, they will eventually be able to improve the enterprise business scenario and obtain positive results in terms of increased competitiveness.

7.3.4 Promoting New Entrepreneurial Models

The e-economy is gradually finding its place and is introducing changes and new practices in different sectors, which lead to more efficiency and lower transaction costs. Online commercial transactions will have an increasingly high impact on all economic transactions in the internal market (between one company and another, between companies and consumers, and between companies and public administrations). The international competitiveness of European companies will depend to a great extent on their capacity to adopt new information and communication technologies.

7.4 Funding Innovation Policy

Enterprise policy is financed under different Community programs, the most important of which are the Program for Enterprises and Entrepreneurship [EC 2000-10] and the Competitiveness and Innovation Framework Program (CIP) [EC 2005-9]. All activities relating to innovation are generally financed through the program for innovation. Other financial tools specifically intended for enterprises include the Structural Funds and the European Investments Bank (EIB), which provides financing for projects and gives loans through financial intermediaries operating at the national, regional, and local level. Financed projects include some training programs whose objective is to facilitate access to new markets, improve quality, safeguard the environment, and favour cooperation.

In Chapter 12, "The Budget of the European Union", we will mention that, at the beginning of 2004, the Commission proposed to raise the budget in the period between 2007 and 2013 to increase investments in research and innovation, with the goal of achieving the Lisbon 2000 objectives on competitiveness and employment. This Commission proposal [EC 2004-9] was eventually adopted by Council and Parliament in 2006 [EC 2006-8], with the significant increase of 75% over previous

R&D investments. Sustainable growth was at the heart of the policy supported by the Financial Perspectives. To reach this target, action is required across three key axes: making Europe into a dynamic knowledge-based economy geared towards growth; reinforcing cohesion; and ensuring the sustainable management and protection of natural resources. The first two of these axes will be addressed through competitiveness and innovation in the single market and strengthening research and technological development [EC 2004-12].

7.4.1 Competitiveness and Innovation Framework Program

The European Commission suggested in July 2004, in the context of its proposals for the next budgetary period, a Competitiveness and Innovation Framework Program (CIP) that would bring together into a common framework of specific support actions critical to boosting European productivity , innovation capacity, and sustainable growth, whilst simultaneously addressing complementary environmental concerns. The CIP will provide a significant and coherent basis for enhancing competitiveness and innovation, complementing the research-oriented activities promoted by the Community Framework Program on Research and Technological Development. The CIP will be composed of specific sub-programs: the Entrepreneurship and Innovation Program (EIP), the ICT Policy Support Program (Ictpsp), and the Intelligent Energy - Europe Program (IEEP). The financial resources to implement CIP are € 4.2 billion, over the period 2007–2013, of which € 2.6 billion are earmarked for EIP, € 0.80 billion for Ictpsp, and € 0.78 billion for IEEP.

The Entrepreneurship and Innovation Program will bring together activities on entrepreneurship, SMEs, industrial competitiveness and innovation. It will specifically target small and medium sized enterprises, from hi-tech "gazelles" to the traditional micro- and family-firms which make up the large majority of enterprises in Europe. It will cover both industrial and services sectors. It will also encourage entrepreneurship and potential entrepreneurs both generally and in particular target groups, paying special attention to gender issues. It will contribute to encouraging young people to develop an entrepreneurial spirit and promoting the emergence of young entrepreneurs. EIP will also be one of the instruments supporting the implementation of actions that aim at removing the obstacles so as to tap the full potential of technologies to protect the environment while contributing to competitiveness and economic growth, ensuring that over the coming years the European Union takes a leading role in developing and applying environmental technologies, and mobilizing all stakeholders in support of these objectives. The availability of venture capital is crucial for innovation, a business process connected with exploiting market opportunities closely related to the willingness to take risks and to test new ideas on the market. EIP will address persistent recognized market gaps leading to poor access to equity, venture capital and loans for SMEs, through Community Financial Instruments operated on behalf of the Commission by the European Investment Fund (EIF), the Community's specialized institution for providing venture capital and guarantee instruments for SMEs [EC 2005-9].

The *ICT Policy Support Program* (Ictpsp) is aimed at improving innovation performance and competitiveness via the uptake of ICTs by both the private and public sector. Ictpsp includes validation and deployment of trans-European ICT-based services. Ictpsp is part of the initiative "i–2010: European Information Society" targeted to revamping the Lisbon Strategy. It provides the backbone for the knowledge Economy and is a catalyst for organizational change and innovation. The program will strengthen the internal market for information products and services. It will aim to stimulate innovation through a wider adoption of and investment in ICTs to develop an inclusive information society, more efficient and effective services in areas of public interest and to improve quality of life [EC 2005-9].

The *Intelligent Energy – Europe Program*'s objective is to support sustainable development as it relates to energy and to contributing to the achievement of the general goals of environmental protection, security of supply and competitiveness. Energy efficiency and renewable energy sources are essential to meeting Kyoto requirements and to reducing Europe's growing dependence on energy imports, which could reach almost 70% of energy demand in 2030. The Union has been working towards the ambitious target of a 12% share of renewable energy in gross inland consumption by 2010 and to further reduce final energy consumption, but these targets will not be reached unless considerable extra action is taken at member state and Community levels. The Union has set herself clear quantitative targets for the uptake of intelligent energy to be achieved by 2010. These include doubling the share of renewable energy sources in EU energy consumption to reach a share of 12% of electricity generated from renewable sources in the internal energy market, and to increase the share of bio-fuels up to 5.75% in all petrol and diesel used for transport. A number of more qualitative targets are also to be achieved such as increased sales of energy efficient products/appliances and expanded highly efficient cogeneration, namely the simultaneous generation of both electricity and useful heat [EC 2005-9].

The CIP is complementary to, coordinated with, and mutually supportive with R&D under FP7, Lifelong Learning Program, and the European Social Fund (ESF). The CIP will address both technological as well as non-technological aspects of innovation. With respect to technological innovation, it will focus on the downstream parts of the research and innovation process. More specifically, it will promote innovation support services for technology transfer and use. For its part, the FP7 will continue and strengthen support of trans-national cooperation in research, technological development and demonstration, in particular between enterprises and public research organizations, of specific Research and Technological Development (RTD) schemes in favor of SMEs and of researcher's mobility between firms and academia. In doing so, it will focus more on the technological innovation needs of industry and introduce new actions, in the form of joint technological initiatives in key areas of industrial interest.

The European Commission has integrated its various educational and training initiatives under a single umbrella, the *Lifelong Learning Program (LLLP)* [EC 2007-23]. With a significant budget of € 7 billion for 2007 to 2013, LLLP enables individuals at all stages of their lives to pursue stimulating learning opportunities across Europe. It consists of four sub-programs addressing: schools; higher education;

vocational education and training; and adult education. Education and training are essential for ensuring that Europe's human capital is kept up to date with the skills and knowledge necessary for innovation. A highly skilled workforce responds better to the quickly changing demands of enterprise and finds it easier to move to new jobs. Education and training also contributes to the diffusion of knowledge and to the process whereby organizations learn from their experiences and improve their processes, products and services. Europe needs more and better investment in education and training, and the adoption of the proposed "Lifelong Learning Program" will help to promote entrepreneurship, support continuous vocational education and training and help organizations to become "learning organizations". The European Social Fund (ESF) will also support as a priority lifelong learning systems as part of its priority to increase adaptability of workers and enterprises, in particular by promoting increased investment in human resources by enterprises, especially SMEs, and workers. Actions to support digital literacy will take into account the policy work on basic competences being carried out under the auspices of "Education and Training 2010" and the support offered for digital literacy under the "Lifelong Learning Program" [EC 2005-9].

7.5 The Eureka Initiative

Created as an intergovernmental initiative in 1985, Eureka aims to enhance European competitiveness through its support to businesses, research centres, and universities who carry out pan-European projects to develop innovative products, processes, and services. By encouraging and assisting businesses to innovate, the Eureka initiative complements the European Union's Framework Program in working actively towards the common European objective of raising investment in R&D to 3% of GDP by 2010. Eureka currently counts 38 full members, including EU30, the European Commission, EFTA, the Russian Federation, and Israel.

More than € 24 billion of public and private funds have been mobilized through Eureka over the past two decades to fund some 1,800 completed projects. These have involved 11,000 partners from industry as well as research centres, universities, and national administrations. Although not itself a source of research funding, the initiative plays a catalytic role by awarding an internationally-recognized label to projects that meet its stringent evaluation criteria, facilitating applications for national public and private finance. Some 42% of participants in both innovative projects and strategic initiatives are SMEs, providing them with a global commercial environment and network. Partners in Eureka projects decide themselves on the content, duration, and the amount invested. Public authorities are asked to indicate their financial commitment on each project. This bottom-up, industry-led style, and flexible approach is meant to ensure that Eureka is particularly effective. However the multiple national funding sources may create a problem in the timely availability of development funds for the different project partners. As a consequence, each member of Eureka consortia is exposed to the risk that partners from other countries might not be regularly and timely funded by their national sources.

The Eureka initiative is committed to enhancing the competitiveness of European industry through the promotion of high-quality collaborative, market-led innovation. It enables industry, research centres, universities, and national administrations to join forces in near-market research and development through trans-national collaborative projects.

Eureka's organizational structure is composed of different bodies including:

- the Ministerial Conference, the political body of Eureka
- the rotating Chairmanship representing Eureka externally and implementing the Eureka Program
- the High-Level Group of representatives of the members is the key decision-marking body. It endorses new Eureka projects, takes decisions on the management of Eureka and prepares new Eureka policy
- the Eureka Secretariat (ESE) in Brussels is an international association which acts as the central support unit for the initiative
- the National Project Coordinators

Eureka projects are implemented via Clusters - and Umbrellas. *Clusters* are longer-term, industry-led strategic initiatives aimed at developing generic technologies of key importance for European competitiveness. They usually have a large number of participants, and aim to develop generic technologies. Because of their high profile, Clusters play a key role in promoting their industry sectors and in persuading national governments and financial organizations to support the clear objectives identified by industry itself.

Umbrellas - are thematic networks focusing on a specific technology area or business sector such as manufacturing, multimedia, food and biotechnology, fish breeding, laser technologies, transport, environmental research, and digital content. Their aim is to facilitate the generation of Eureka projects by offering partner search facilities, analysing R&D needs, and coordinating national and European R&D activities through a working group comprising public authorities and sectoral experts, and by a chair and secretariat hosted by one of the national members.

7.6 Obstacles to Innovation

The obstacles to enterprise innovation in Europe basically come from its institutional and legal framework and from the attitude of entrepreneurs. Institutions are bureaucratic and there are too many rules. Tax incentives should encourage entrepreneurs to taking risks in advanced research and to starting up new businesses. Moreover, public and – most important – private investments in R&D are quite low, and the labour market is too rigid. Entrepreneurs are not keen on changes and are often afraid of taking risks; they lack entrepreneurship and are not that willing to invest in lifelong learning.

7.6.1 Recommendations from Industrialists

The Union of the Associations of European ICT Industrialists (Unice) has drawn up recommendations [Unice 2001, Unice 2002] for possible solutions, policies, and initiatives.

Promotion of the development and dissemination of new ideas. The ICT sector in the U.S. is larger and more productive than in Europe because it invests more in new technologies and effectively disseminates the new knowledge. Conversely, in Europe, insufficient resources are invested in the development of new ideas. Additionally, the interconnection among enterprises, the academic and scientific world, and the public administration is weaker and less effective. Unice recommends the following measures: researchers mobility, increased public investments in R&D, streamlined procedures for enterprises willing to participate in Community research programs, adoption of the European patent, and enhanced intellectual property protection, particularly for software products.

Creation of a communication infrastructure capable of effectively supporting the development of the information society. Unice has empirically demonstrated that there is a link between the Internet connection cost and the number of Internet hosts. In order to bridge the gap with the U.S., Europe must follow the example of Finland and Sweden, which have lower Internet connection charges and a higher number of Internet hosts. Unice recommends the following measures: completion of the liberalization of the telecommunications market, thereby providing the appropriate level of regulation to safeguard European competitiveness in the global market; promotion of standardization through the involvement of European market players; promotion of wider best-practice sharing; and enhancement of the use of the broadband telecommunications infrastructure.

Increase in consumers' propensity to accept innovations in products, services, sales methods, and new technology. Widespread technological illiteracy, fear of fraud, lack of privacy and security, the cost of transactions, and insufficiently qualified personnel make people more reluctant to use e-commerce and, more in general, to adopt new technologies. Unice recommends the following measures: governments, together with management and labour organizations, should take measures to overcome consumers' reluctance; elimination of unnecessary regulations affecting the development of new markets; out-of-court dispute settlement systems for e-commerce; a legal and fiscal framework in support of e-commerce; and promotion of computer literacy.

Improvement of labour market flexibility. Unice recognizes the need to reward qualifications through appropriate wage differentials; provide incentives for higher education and for lifelong learning; ensure geographical and professional mobility to the labour market; and cut social security contributions. Unice recommends the following measures: bringing public education into line with the Lisbon 2000 objectives, thereby ensuring that, at the end of their studies, all young people should have basic training in ICT; increasing investments in lifelong learning; and facilitating qualified labour mobility within and across the states, including immigrants into the European Union from third countries, in order to cover the need for a qualified workforce in the short term.

Creation of a European financial market. The size of the European financial market is € 15 trillion, less than half of that of the U.S. Moreover, stock market capitalization accounts for 27% of the financial market in Europe as against 34% in the U.S. Enterprise borrowing is 37% in Europe as against 20% in the U.S. The European financial market needs to become more integrated and new banking products for company financing need to be developed, most of all to support fledging companies. Unice recommends the following measures: enhancement of the European venture capital market; enabling the financial market to have controlled access to pension funds; and accelerating the process towards the full accomplishment of the internal market for financial services.

Encourage entrepreneurs to take risks. The return on the capital invested in young, innovative enterprises of the new economy is higher in the U.S. than in Europe, thanks to the implementation of fiscal measures encouraging enterprises to take risks and facilitating the setting up of new companies. Unice recommends the following measures: a lowering of the tax burden on equity investment; reformation of stock option taxation; and change in the labour taxation system to adjust it to the needs of the new economy.

7.6.2 Recommendations from Industry Ministers

In June 2000, the Italian government (prime minister Giuliano Amato, Industry minister Enrico Letta) and the Organization for Economic Cooperation and Development (OECD) held an international ministerial conference in Bologna entitled *Promotion of SMEs Competitiveness*, in which the representatives of 48 countries (including all EU15, EFTA, and G8 countries) drew up a CHART OF POLICIES IN FAVOR OF SMES. This chart contains important observations and suggestions that have many points in common with the industrialists' recommendations.

The ministers agreed that SMEs competitiveness would benefit from the following:

- Regulations that do not impose unnecessary constraints on SMEs but rather favour entrepreneurship, innovation, and growth by making public administration accountable and effective, by pursuing a fair and transparent competition policy, by fighting corruption, and by implementing a stable, transparent, and nondiscriminatory tax system
- Education and training policies that support innovation and entrepreneurship
- Promotion of mobility
- Access of SMEs to financial markets
- Technology development and dissemination
- Social dialogue between entrepreneurs and institutions
- Measures in support of SMEs in line with the other national and supranational policies

The ministers agreed that innovation is vital to SMEs competitiveness and recommended the following:

- Better conditions should be provided to SMEs so that they can hire and train qualified staff in order to foster relations between enterprises and university research.
- Financial obstacles should be removed and enterprises encouraged to take risks through fiscal incentives.
- Access of SMEs to public research programs should be made easier.

The ministers agreed that industrial clusters and enterprise networks can give a boost to innovation and recommended the following:

- The creation of partnerships between private operators, nongovernmental organizations (NGOs), and public administration sectors should be supported as part of a series of strategies to stimulate the development of clusters and enterprise networks.
- The private sector should guide cluster-building initiatives, leaving the public sector with the task of acting as a catalyst on the basis of national priorities.
- The construction of clusters should be stimulated by improving transport and communication infrastructures and by fostering relations between the university and the business world.

The ministers agreed that e-commerce generates new opportunities and new challenges for SMEs and recommended the following:

- The standpoint of SMEs should be taken into consideration in the drawing up of directives and regulations relating to ICT and e-commerce.
- SMEs should be made aware of the advantages offered by the knowledge-based society; they should get rid of useless paperwork; a competitive market for network equipment should be promoted; and an effective e-government service should be ensured that should not be unfair towards SMEs.
- SMEs should become more familiar with e-commerce through pilot projects and demonstration and training centres; standard systems should be provided in order to ensure secure and certified transactions as well as privacy and intellectual property right protection; and a clear and consistent legal framework for e-commerce should be developed in order to settle e-commerce disputes outside the courts, thereby avoiding complications and excessive costs.

There is no substantial contradiction between the position of the European industrialists and that of the industry ministers at the OECD Conference. The solutions are therefore well known, but their actual implementation has been inhibited or delayed because it requires political will, change in habits, and availability of financial resources – priorities that, as a matter of fact, do not appear to be at the top of the governments' list, despite official declarations to the contrary.

7.7 Innovation in the U.S.

The U.S. Government and Congress favour innovation in different ways, among which are intellectual property laws, the promotion of industrial development of the outcomes of funded research, and incentives to investors [Branscomb 1998, Stajano 1999-2].

In the U.S., the laws in favour of innovation cover different areas, such as competition, intellectual property protection, a patent system that suits the needs of research and business, bankruptcy laws, labour mobility, and accessibility of financial markets.

Intellectual property laws are at the core of the innovation policy pursued by the U.S. administration. They are based on the Bayh/Dole Act (1996) [COGR 1999], by which all beneficiaries of Federal funds for research shall share the intellectual property rights to inventions resulting from research funded by taxpayers money. This Act also facilitates the concession of development and industrialization rights, and – where required by the nature of the industry – of exclusive exploitation rights for a defined goal and a predefined period of time.

In the U.S., financing innovation also means providing support for the setting up of new enterprises and risky initiatives: the financing of start-ups, provision of venture capital, and financing by the so-called business angels, who invest in innovative businesses. Moreover, the capitalization of the stock market, through family savings and pension funds, contributes to the financing of enterprise development.

Another important factor contributing to the promotion of innovation in the U.S. is the link between universities and the business world, which is characterized by communication, mobility, and synergy. In American society, an achievement in business is not less valuable than an academic success in the CV of a scientist.

The *U.S. recipe* to enhance the value of inventions and industrialize them can be summed up in five points:

1. access to the financial market is easier
2. risk-taking is encouraged and initial mistakes are excused
3. it is easy to set up a new company
4. mobility between research and business is encouraged
5. *business* contents are part of the education and training of technologists

A European observer may be surprised to realize that U.S. head hunters may take the mention of one business failure, along with a number of success stories, as a positive point in the CV of a young entrepreneur. They would consider that a business person in his or her late 30s had not taken enough risks, if not one failure was reported.

A report comparing innovation performance in the U.S. and in the EU published in 2005 by the Sant'Anna School of Advanced Studies, Pisa, Italy [Dosi et al. 2005], comes to conclusions in support of the new orientation of Community-funded research under FP7 (2007–2013):

> Certainly one observes significant differences across scientific and technological fields, but it happens that Europe has structural lags in top level science and innovative performance vs. the U.S., together with some points of strength in physical sciences and engineering. At

the same time, one also finds ample evidence of a widespread European corporate weakness, notwithstanding major success stories. We suggest that effective European catching up would require much less emphasis on various types of networking, interactions with the local environment, attention to user need—current obsessions of European and national policy makers—and, conversely, much more on policy measures aimed to both strengthen *frontier* research and, at the opposite end, strengthen European corporate actors.

During a study on Technology transfer in the U.S. conducted by the author in 1998 on behalf of the European Commission, he was impressed by the difference of Technology Licensing offices in the U.S. compared to their equivalent services in Europe. In the U.S. they were staffed by very knowledgeable, senior, highly-paid full-time staff that was competent in the research domains of the university departments they served and in intimate contact with the sectors of the business community that might be interested in the uptake of research results.

As an example, the operations of the Office for Technology Licensing (OTL) at Stanford, CA had a staff of 21 at that time. The Office was supported financially by a fraction (varying over the years between 5 and 15%) of the royalty income generated by the licenses, patents and copyrights that it had established and managed. The prime objective of the Office was the protection the intellectual property of the researchers' inventions in order to attract private investment and to sow the seeds of the products and services of tomorrow. An important role of the tech transfer offices was the raising of a new awareness in the researchers so that they would appreciate the obligation they have and the benefits they could earn out of exploiting their results. A new culture was created where publication in prestigious journals was not the only ultimate reward. This was music to the author's ears. The establish practice in Europe was to appoint to the management of a licensing office a senior professor without discharging her or him of the teaching duties; the focus on the personal intellectual interests would bias her/his support action, limited anyway by the teaching duties, the incomplete familiarity with the market and the business community, and the understaffing of the office.

Stanford OTL's support in going from invention to market typically used to begin by reviewing an invention with its inventor to learn about potential applications. The OTL staff was so competent and respected that they would have access to any researcher, who knew that they would only be of support and might generate visibility and financial profits. OTL was looking for inventions that at some stage (may be after 15 years) might generate at least $ 0.1 million. They then would develop a licensing strategy, consider the technical and market risks, decide whether to patent the invention, actively recruit companies that might be interested in the invention and seek a product champion within a company before negotiating a licensing agreement. If they decided to file for a patent, outside patent attorneys would have been involved. Fifteen percent of the royalties was deducted to support OTL's operating costs (the difference, if any, between the 15% deduction and OTL's budget was directed to the OTL Research Incentive Fund under the auspices of the Dean of Research). Any direct expenses, such as patent costs, were then deducted. The remaining net royalties were divided on the basis of this rule: one-third to the inventor(s); one-third to the inventor(s)' department; and one-third to the inventor(s)' school.

When the author presented these findings to some greedy European researchers, they objected that they would not be satisfied with the idea of getting only one third of the net royalties; they had to admit eventually that one third of a big sum may end up to be more than 100% of zero.

7.8 Technology Transfer in the European Union

Community-funded research is aimed at supporting the competitiveness of European businesses. Therefore, the outcome of research must also be transferred to the public administration and to enterprises and not only to those directly involved in research programs. Transferring technological innovation into the European Union's productive fabric is the final stage, and may be the most important one, of adjusting the business structure to the needs of the knowledge-based society. This means not only acquisition of new technologies, new materials, new processes, and new communication and knowledge-management infrastructures. It also – and above all – means that enterprises must adjust themselves, from a structural and cultural point of view, to the new environment and the new challenges.

Technology transfer in the Union is delayed owing to difficult access to financing, the dichotomy between academia and business, and the inadequacy of intellectual property laws. This topic is dealt with in more detail with reference to the difference between the U.S. and EU in [Stajano 1999-2] and with reference to the difference between the U.S. and Germany in [Abramson et al. 1997]. The Union's technology policy tried to bridge the gap starting from the early 1980s, when it facilitated access to financing and technology transfer. In the 1990s, the EC initiated a number of initiatives to enhance technology transfer, foster innovation, and help enterprises, in particular SMEs, to find their way to the financial market. Financing Innovation was one of the services provided by the *Innovation and Small Enterprise Program*. It was a tool to turn researchers' creativity into business. The Financing Innovation program offered:

- Business plan preparation support
- Business plan diagnostics
- Discussion forums
- Industry events
- Investor identification and matching
- Service provider directory
- Seminars
- Workshops
- Investment forums
- Peer group clubs, and
- Access to networks of local intermediaries

Under the Financing Innovation program, companies submitted their innovation proposals, which were examined and evaluated on the basis of their maturity and business prospects. Additionally, participating companies were helped in the preparation of a technology business plan and in finding potential sources of financing. Companies' expectations were identified, and the most readily available sources of financing were indicated to them.

Other Community actions include investment forums, stock market regulation, and the creation of new markets for startup companies and for companies committed to innovation.

In 2005, Financing Innovation evolved into the Gate-to-Grow Initiative [EC 2005-2], supported by the European Commission under its Innovation and SMEs program [EC 2005-3]. Gate-to-Grow is an online service centre where investors and entrepreneurs can access a complete range of services, resources, industry foresight, and networking opportunities. The initiative serves a variety of actors:

Entrepreneurs seeking financing
Investors
Technology incubator managers
Knowledge transfer officers
Academic researchers seeking to be networked with financial institutions and private investors
Academic researchers in quest for innovation
Academic researchers seeking to develop entrepreneurial capabilities
Innovative companies seeking expert service providers

Users of the initiative are assisted by a network of regional contact points in the European countries and find help and support in a number of events initiated by the Commission and by a group of private enterprises participating in the initiative including: venture capital networks, consultants, associations, university departments, incubators, and business innovation centres.

The Gate2Growth Initiative initiated *ProTon Europe,* a pan-European network of Technology Offices linked to Public Research Organizations and Universities. It addresses the significant amount of world-class research undertaken in universities and research institutions in Europe, which has actual or potential commercial relevance. To capitalize fully on the potential of these public research organizations, it is essential that commercialization becomes an integral part of the research process and alternative approaches to the ownership and exploitation of intellectual property rights are suitably explored. One of the main focuses of ProTon Europe is the identification of good practices that be applied in a practical way to improve the day to day performance in the enterprise.

7.9 Recommended Reading

Recommended reading for this chapter are listed in Appendix C and are available on the web site companion to this book, at the URL http://stajano.deis.unibo.it/RQC.htm

Chapter 8
Competitiveness in the Knowledge-Based Society

"It may have been a British scientist, Sir Tim Berners-Lee, working at a laboratory in Switzerland, who invented the world wide web, but America is the home of the Internet and all the business sectors it has spawned" [Economist 2007-27]. In several European countries, the spread of the Internet and the benefits from its use in business are low and EU average Internet diffusion is below the one in the U.S. A new impetus is required if the EU wants to stay in the competitive game of the knowledge-based society. In some European states, the access to and use of the Internet are still limited, and there are delays in Internet access in schools and in the training of workers. The diffusion of online services is slowed down by their non-ubiquitous availability, their cost, as well as by problems of confidentiality, safety, and fraud. The technological infrastructures do not cover evenly the whole territory: fibre optic cables, radio connections, and digital and cable TV must be everywhere and become affordable in order for the EU to compete in the arena of the globalized and networked society.

8.1 The Third Industrial Revolution

At the end of the 18th century, the first industrial revolution created the pre-conditions for economic growth and generated added value in sectors other than the agricultural one. For 200 years, the development limits were marked by the availability of energy. The second industrial revolution started at the beginning of the 20th century with the industrial and household use of electricity, cars, and telephones.

Over the last two decades of the 20th century, information and communication technologies and the availability of fast, reliable, and cheap means of transport created a platform that led to a radical change in society. From the 1990s onwards, the framework has been modified by a relentless process of convergence of information and telecommunications technology initiated by the digitalization of telecommunications. The spread of the Internet and of the world wide web, the globalization of the economy, the free movement of persons, and the agreements of the World Trade

Organization (WTO) contributed to create the third industrial revolution. There is now a new context in which new development limits add to the previous ones:

1. environmental sustainability of development
2. the capacity to manage the complexity of systems and the overload of information
3. the capacity to capitalize on corporate information and structure it into corporate knowledge
4. the gap between life conditions in the North and South of the world, and
5. the capacity to face the demographic pressures of emerging countries and the ageing of the population in the richest countries

As we said at the very beginning of this book, energy and technology continue to be essential for economic growth, but growth is now conditioned by two other factors: information and knowledge. While energy is limited and can be used only once, information is widely available and overabundant and can be used by several users at the same time. The new challenge is managing and exploiting information and structuring it into knowledge that can support a new approach to sustainable development and trigger an improvement in the quality of life.

Knowledge transcends the specific context and the time dependency of information and can be applied to a variety of situations by analogy to the cases that have contributed to generate it [Fischer 2006]. This is why corporate knowledge may result in the competitive leading edge of an enterprise thus able to capitalize on the experience gained in a variety of sites and domains and to reuse it facing new challenges.

The first commercial application of ICT dates back to 1850, when Julius Reuter created a commercial communication service between Brussels (Belgium) and Aachen (Germany, 200 km away) using carrier pigeons. Today the Reuters agency replaced pigeons with optic fibres and satellites but, practically, works in the same industrial sector, which in the meantime has greatly grown in terms of speed, reliability, availability, capacity, security, and added-value services.

The European Union has played a major role in the promotion of the economic advantages resulting from the information society with the opening of the telecommunication markets between 1996 and 1998. In all member states, there are now new operators and ICT service providers that have led to a greater availability and quality of services and to costs reduction.

The term *information society* was coined in Europe at the end of the 1980s; terms *knowledge society* or *knowledge-based society* have been recently used as synonyms for information society, but they indicate in facts a higher level concept, going beyond the information society's essentially technological substratum, which in turn is the main connotation of the reductive term *information superhighways* used in America in the late 1980s. Society has undergone a radical technological, economic, social, and cultural change. The knowledge society is characterized by a new science, knowledge management, which involves the whole fabric of society and particularly the world of education, training, and lifelong learning.

The EU technology policy and EU-funded research programs have played an important role in the definition and growth of the knowledge society and in its planning for sustainable and competitive growth of the economy while maintaining a high quality of life and increasing employment. Within the European Union, the technology policy in the knowledge society is addressed by the Lisbon Strategy, presented in Sections 8.4–8.7 starting, after introducing e-business and e-government.

8.2 e-Business

The term e-business includes buying and selling on line (e-commerce) and the restructuring of business processes induced by ICT. In detail, e-business encompasses:

1. relationships between companies and consumers (B2C: business to consumer). In its turn, B2C includes several types of commercial operations

    - direct e-commerce (e.g. music, software, databases, online newspapers, and information)
    - indirect e-commerce of goods (e.g. books, audio-CDs, audio-books, DVDs, flowers, and video on demand)
    - indirect e-commerce of services (e.g. banking services, tourism, and travel)

2. relationships between companies (B2B: business to business): the supply of goods and services and the outsourcing between companies through e-commerce
3. relationships between consumers (C2C: consumer to consumer): including both online direct sales and online auction sales based on catalogues or ads
4. relationships between consumers and companies (C2B: consumer to business): the demand for transport and tourist services at minimum prices
5. last but not least, enterprise reorganization and the restructuring of business processes to make the best use of digital technologies

The traditional methods of production, sale, and distribution are transformed by e-commerce in many different ways, since e-commerce replaces or expands the traditional sale channels:

- it offers consumers new solutions and a wide range of options
- it enables suppliers to reach new customers in new sectors or remote locations
- it reduces transaction costs
- it creates competition between suppliers
- it rationalizes the purchasing, production, and distribution processes

A Forrester Research report, referred to on the *Financial Times* [Malkani et al. 2003], estimated that the scope of e-commerce in the U.S. was equal to $ 100 billion in 2003 (1 % of GDP), while in Europe it was equal to € 15 billion in 2001, € 30 billion in 2002, and € 48 billion in 2003 (0.6 % of the GDP). According to another estimate [JMM 2002], B2C in the U.S. in 2003 was equal to $ 64 billion with an annual increase of more than 30%. The percentage of enterprises (with 10 or more persons employed) purchasing on-line – based upon the proportion of sales (turnover) that is realized via the Internet, EDI, and alternative networks – rises among larger enterprises, which may be due to their investment in more advanced networks and their more frequent purchases, which promote the use of systems such as electronic data interchange, often linked to logistical and stock control [Eurostat 2007-2]. The increased use of e-commerce is reported in Section 8.8, "8.8 Monitoring the Progress".

Being aware that e-commerce is an important factor for increasing the competitiveness of European industry, the European Commission started to finance a great number of R&D projects in this sector already in the 1990s [EC 1999-4].

8.2.1 e-Commerce on the Internet

Industry passed from e-commerce on proprietary networks for data transmission (EDI: electronic data interchange) to e-commerce on the Internet. EDI had the advantage of eliminating communication errors or mistakes made by operators and ensured documented, fast, accurate, and complete transactions. EDI represented great progress in terms of cost reduction, rapidity, and effectiveness of the whole production chain. The passage to the Internet represented, in contrast, a real revolution in the way of doing business because it opened up the market by introducing innovative approaches in terms of accessibility, reduction of transaction costs, competition between suppliers of goods and services, interactivity, and online auctions.

Trade relationships, which in the beginning were limited to relationships on private networks between companies that mutually trusted each other, have been opened up by e-commerce on the Internet, thus ensuring unconditional accessibility to companies, public administration, and end consumers. e-Commerce on the Internet has marked the passage from *closed circles* over proprietary networks, which were often restricted to a specific sector and with a limited number of known and reliable partner companies, to a *global market* open to everyone, with an unlimited number of partners. This shift required a new level of safety and privacy: safety became an integral part of the planning of an open and unprotected network with both known and unknown partners, which requires the confidentiality, safety, authentication, and certification of transactions.

Access to the global market through the Internet creates new demands:

1. a demand for new services: an efficient series of distribution channels and a transeuropean network for the physical delivery of goods; broadband Internet access; and telephone connection at reduced prices
2. a demand for technology: interoperability between marketing and sale systems and between distribution and payment systems; the spread of reliable technologies for electronic signatures, safe methods of payment and electronic

certificates, and anticloning systems to protect the intellectual property of contents (music, images, information, news)

3. a demand for awareness campaigns: to make consumers and companies prepared to take advantage of the opportunities offered by e-commerce through training, information, and demonstration projects

4. a demand for new regulations: a legal framework supporting the development of new production and trade settings; a legal, fiscal, and institutional framework allowing for the application of the new technologies without excessive costs and problems; regulations and controls for the protection of intellectual property; and coordinated protective actions against computer crimes

Europe has certain strong points for access to the global market:

1. the largest single market in the world
2. industrial capacities in the technology sector
3. advanced position in the mobile telephony sector
4. a great tradition in the creation of contents
5. a great variety of languages and cultures, which prepares European actors to face the rest of the world with its countless differences
6. a single currency in the biggest market in the world, which represents a great incentive for the adoption of e-commerce in Europe

The acceptance of the euro by the general public may improve thanks to the advantages experienced by using e-commerce with the single currency irrespective of the source of the provider. e-Commerce can contribute to the overcoming of problems linked to the geographical position of areas far from markets and to the reduction in demand for geographical mobility of workers. Moreover, e-commerce has already led to scale economies through virtual groupings.

8.2.2 e-Commerce in Italy

In 2002, the Italian government issued law no. 388/2002 to promote the use of e-commerce in Italy. This law provided for the expenditure of € 100 million to promote and assist in the use of telematics. Over 13 000 companies applied for the financing within a few days from the publication of the call (February 2003).

National newspapers and the financial press drew attention to the event (see Fig. 8.1), telling their readers "Now it's for real" because financial incentives had become available. In January 2003, 6,114 Italian enterprises were in e-commerce [Il Sole24Ore 2003]; in Lazio (the Italian region around Rome), according to data from the regional trade observatory, there was the highest concentration of e-commerce companies (891 out of 6,114). But in 2002, Italy was at a distance from the other large European economies in e-commerce spending; although Italian online spending is expected to grow, at the same time Italy is expected to continue to lag behind (see Fig. 8.2).

Fig. 8.1 Incentives to enterprises for e-commerce. An Italian national newspaper announces the government regulation implementing the law n. 388/2002
Source: http://www.repubblica.it 27 January 2003, reproduced with kind permission.

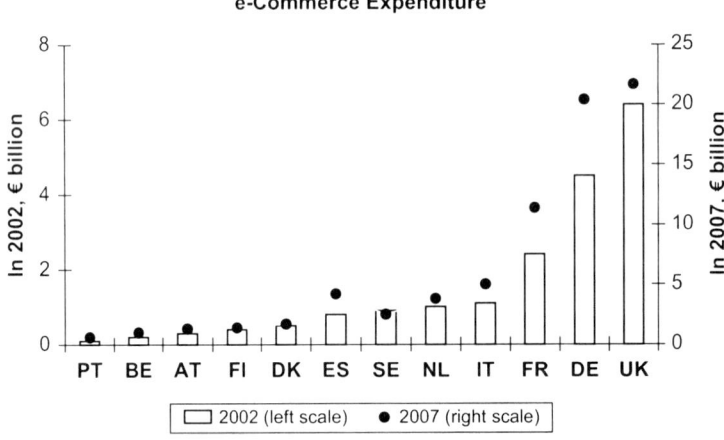

Fig. 8.2 Estimates of e-commerce expenditure (billion euro) in the EU in 2002 (bars, left scale) and 2007 (dots, right scale). Online spending will grow throughout Europe, with above-average increases in Spain, France, and Germany. Data from Greece, Ireland, and Luxembourg are not available
Source: *La Repubblica*, 27 January 2003.

8.3 e-Government

e-Government is introduced in the Commission's Information Society web site as "the use of Information and Communication Technologies to make public administrations more efficient and effective, promoting growth by cutting red tape. This is something which anyone who has spent hours waiting in line in a government building can appreciate." e-Government streamlines bureaucracy by increasing efficiency and offering online public services like information, certifications, collection of income tax returns, payments, submission of tenders, etc. e-Government ends up to reducing the cost of public administration, and delivering quality services to businesses and citizens. By connecting government departments, companies, and citizens, e-government-powered public services become faster and more personalized, allowing citizens and companies to get on with their lives, and build their businesses rather than waiting in line in front of government counters. e-Government can also strengthen democracy by improving two-way communication between the citizens and their government. While there is much research still to be done, this is not just a technical issue – technology must be combined with organizational change and new skills to fulfil the e-government promise. Moreover, national e-government solutions must not lead to new barriers within the European internal market: if national electronic identities are not interoperable, for example, both companies and people will face new barriers to working and living in other countries.

Modernizing Europe's public administrations therefore means helping researchers, companies, and public administrations work together across Europe to develop the technologies, exchange best practices, and forge a coordinated approach. We must avoid reinventing the wheel 27 times across the EU, learn from each other's successes and failures, and win the economies of scale that only a European approach can provide [EC 2007-25].

e-Government is more than the use of IT to reduce the lines in the public offices: to make real progress on transforming government services it is necessary to positively transform the relationship between government and citizens by enhancing the democratic relationship, and building better democratic dialogue between citizens and their administrators, which then may enhance the practice of citizenship within society [EC 2007-26]. e-Government actions include: covering the development of better services in health care; streamlining the administrative requirements to obtain official documents, licenses or to start a new business; reducing the dangers of marginalization of the computer-illiterate citizens; improving the protection of privacy in citizens' life.

By building a climate of cooperation and trust between the public administration and citizens e-government actions may achieve far reaching goals, establish a correlation between what citizens expect from the government, and what they are prepared to give in return, in terms of availability, participation, information, and loyalty. The ultimate objective includes a more equitable and fare taxation system. Section 8.8, "Monitoring the Progress" will show the level of member states' harmonization as revealed by e-government indicators.

8.4 The Initial Formulation of the Lisbon Strategy

From the beginning of the 1990s, the relationship between the European Commission and the whole ICT industry (a relationship created and reinforced also thanks to research programs) has enabled the EU to know and evaluate the difficulties, challenges, and potential opportunities of the knowledge society. In this respect, an important step was the drawing up, publication, and discussion of the White Book on Development and Competitiveness [EC 1993] as well as the later position of the European Commission within the G7 as a key partner on the information society and e-commerce.

In the information society, technologies are not the exclusive concern of the computer, telecommunication, aerospace, or biotechnological industries: ICT applications are beyond counting, practically in all sectors. The convergence of the digital technologies of computer, telephone, and television leads to a convergence of markets and business. Globalization of markets does not imply their homogenization; on the contrary, it is an opportunity to enhance the differences and peculiar features of the different languages, cultures, and business practices. The use of the primary resources of the information society, namely, information and knowledge, by several users at the same time changes the rules of the game. Industrial operators must learn to protect the intellectual property of inventions and organize their companies in a suitable way to structure information into knowledge, face change, promote innovation, and invest in personnel training and in lifelong learning.

Because of the complexity of the new technologies and their rapid evolution not every enterprise can be equipped with in-house advanced competence in all innovation fields. This situation results in research being often outsourced. However, integrating outsourced R&D results into the culture of a company is a very complex and challenging task. It requires a high education and training level in the receiving organization, the capacity and desire to keep on learning during one's whole professional life, and the willingness to change one's habits. The management of the receiving organization must be able to understand and appreciate the results that they acquire and to stay on top of the development processes as they are transformed by those results.

In March 2000, the European Council, following a communication by the newly appointed Commission, addressed the challenges that the knowledge-based society was determining for the European growth and committed the EU to become by 2010 "the most dynamic and competitive knowledge-based economy in the world, capable of sustainable economic growth with more and better jobs, greater social cohesion and respect for the environment." This commitment has come to be known as the *Lisbon Strategy* (LS), and it was implemented through a comprehensive interdependent series of actions both at the Community level and in the member states [EC 2002-9, EC 2005-11].

The Lisbon Strategy was seen by the EU in the year 2000 as the 1960s U.S. project *"the man on the moon,"* namely as a challenging and evocative mission that might mobilize energies and consolidate the enlarging European Union. The Lisbon Strategy became the umbrella under which all actions towards sustainable growth and increased competitiveness would fit, hopefully overcoming the shortcomings of horizontal cross-policy coordination, and of harmonization of vertical levels of governance. This caused a lack of focus whereby the implementation of the strategy suffered.

The strategy was formulated resting on three pillars:

- preparing the ground for the transition to a competitive, dynamic, knowledge-based economy
- modernizing the European social model by investing in human resources and combating unemployment and social exclusion

- reconciling economic growth with the environment-friendly use of natural resources

Research and development policy was put at the heart of the Lisbon Strategy to boost employment and growth in Europe. Research, with education and innovation, was meant to form the "knowledge triangle", which would allow Europe to maintain its economic dynamism and social model. R&D and innovation in the Union should be increased with the aim of approaching 3 % of GDP by 2010. Two-thirds of this investment should come from the private sector [EC 2002-9].

The author sees three major weaknesses in the Lisbon Strategy: the lack of involvement of the stakeholders in its dirigistic conception; the lack of focus because of non converging objectives of all the policies involved; and the lack of rewards and sanctions. The Commission might have made the same mistake with the Lisbon strategy that was made, 50 years earlier, with Euratom and that caused its failure (discussed in Chapter 5, "EU Research: Objectives and Results"). The Lisbon strategy seems to be a top-down project conceived by the Commission staff, without industrial consultation or involvement of all European social partners and, therefore, without a feeling of ownership by anybody and most importantly by no industrial group. The lack of a feeling of ownership is ever so much relevant because one cornerstone of the LS is the objective to increase technological R&D to 3 % of GDP, of which two-thirds should come from private firms. While member state administrations can inject financial resources into research centres and universities, they can only encourage private R&D expenditure with infrastructures and incentives. But private enterprises will not invest on a plan that they did not contribute to shape to their needs. The Lisbon Strategy might end up being only a recommendation without power of enforcement [Monti 2005] with no rewards for its followers and without sanctions against those that, at the receiving end, would not implement it. In a similar way, in another field, the economic ministers of defaulting member states tried to (and to some extend succeeded in) converting the Growth and Stability Pact into little more than a powerless recommendation.

The Eurostat Statistical Report on the information society in Europe [Eurostat 2003] stated that individual member states had made significant progress towards the targets of the Lisbon strategy, but that the EU as a whole was behind schedule to achieve the strategy within the proposed time frame. On 9 December 2004, Mr. José Manuel Barroso, the newly appointed president of the Commission, declared at an Unice meeting: "The Lisbon strategy must be a central priority [. . .]. There is much that remains to be done." In fact, the EU did not seem to be moving fast enough to reach its target. Time was running out and American and Asian competition was stronger than five years earlier. Little progress had been made from 1999 to 2003 in economic terms as well as in R&D intensity [Muldur et al. 2006]. Half way into the first decade of the new century, it appeared that the LS goals would not be met in the timeframe indicated: the EU would not become the most dynamic and competitive knowledge-based economy in the world by 2010, rather its positioning might even worsen, due to the burst of the Internet bubble, the aggressive international competition, the European decline in growth rates, and the flaws of the strategy. A rather gloomy picture was presented in 2004 by the Kok report [EC 2004-15], an independent review commissioned by the European Council to a high level group chaired by Wim Kok, former prime minister of The Netherlands. The report states that

> the growth gap with North America and Asia has widened, while Europe must meet the combined challenges of low population growth and ageing. The EU and its member states have themselves contributed to slow progress by failing to act on much of the Lisbon strategy with sufficient urgency. This disappointing delivery is due to an overloaded agenda, poor coordination and conflicting priorities. Individual states have made progress in one or more of these policy priority areas but none has succeeded consistently across a broad front. If Europe is to achieve its targets, it needs to step up its efforts considerably. The high level group recommends urgent action across five areas of policy:
>
> • *the knowledge society:* increasing Europe's attractiveness for researchers and scientists, making R&D a top priority and promoting the use of ICTs.
> • *the internal market:* completion of the internal market for the free movement of goods and capital, and urgent action to create a single market for services.
> • *the business climate:* reducing the total administrative burden; improving the quality of legislation; facilitating the rapid start-up of new enterprises; and creating an environment more supportive to businesses.
> • *the labor market:* developing strategies for lifelong leaning and active ageing; and underpinning partnerships for growth and employment.
> • *environmental sustainability:* spreading eco-innovations and building leadership in eco-industry; pursuing policies which lead to long-term and sustained improvements in productivity through eco-efficiency.

The Kok report contributed to a revamping of the Lisbon Strategy in 2005 and the replacement of the e-Europe projects with the i–2010 initiative.

8.5 The e-Europe Projects

The e-Europe projects, a Commission initiative active in the period 2000–2005 in support of the Lisbon strategy, were aimed at developing a competitive EU economy based on information and communication technologies and knowledge management.

Its first formulation, the e-Europe 2002 [EC 2001-8] plan included three lines of action:

1. make the Internet cheaper, faster, and safer
2. invest in human resources and training
3. promote the use of the Internet

The work method provided for by the plan foresaw

• the creation of a legal framework promoting e-Europe
• the creation of new services and infrastructures in Europe
• collaboration and benchmarking (comparison based on reference solutions)

The e-Europe 2002 plan had the aim of promoting the development of broadband telecommunication infrastructures and services and of applications and contents both for public services and for e-business. The partners provided for by the e-Europe plan are the European institutions, the EU member states, private enterprises, associations, and social institutions.

The e-Europe 2002 program was updated in 2002 with the e-Europe 2005 program [EC 2002-6] aimed at translating Internet connectivity into increased economic productivity and improved quality and accessibility of services for all European citizens (e-government, e-learning services, e-health services), based on a secure broadband infrastructure available to the largest possible number of people at competitive prices.

The main objectives of e-Europe 2005 were:

- modern online public services:

  - e-government services
  - e-learning services
  - e-health services

- a dynamic e-business environment
- a secure information infrastructure
- widespread availability of broadband access at competitive prices
- benchmarking and the dissemination of good practice

In detail, the target included the implementation by 2005 of: support for broadband access in less-favoured areas; providing broadband connections for all public authorities; ease of access for all citizens to public access-points to the Internet (PAPI); the launching by the member states, with support from the Structural Funds, of training activities to provide adults with the skills needed to work in a knowledge-based society; the establishment by the member states of health information networks between points of care (hospitals, laboratories, and homes); identification and, where necessary, elimination of factors which prevent businesses from launching into e-business; private sector development of interoperable e-business solutions for transactions, security, procurement, and payments; the establishment of a cyber security task force; secure communications between public services; and speeding up the transition to digital television.

8.6 i–2010 – A European Information Society for Growth and Employment

The e-Europe 2005 program was superseded in 2005 by "i–2010 – A European Information Society for growth and employment" [EC 2005-10]. Contrary to the initial steps of the LS, i–2010 results from a wide stakeholder consultation on previous initiatives and instruments. It proposes three priorities for Europe's information society and media policies:

i. the completion of a single European information space which would promote an open and competitive internal market for information society and media
ii. the strengthening of innovation and investments in ICT-research to promote growth and more and better jobs
iii. the building of an inclusive European information society that promotes growth and jobs in a manner that is consistent with sustainable development and that prioritizes better public services and quality of life

In 2005, large sales in systems software and e-business applications indicated that businesses were adopting new and more mature solutions, even if these new investments were limited to large companies or early adopters of advanced solutions. Users were quickly embracing new services brought about by convergence. Many member states achieved high levels of broadband adoption, which in turn stimulated the development of innovative advanced services. The transformation of the content market was already apparent in the growth of online music sales and new digital devices. Movie distribution and online TV were also advancing. The move from traditional content distribution to online availability was accompanied by an explosion of user-created content. In this context, the Commission proposed, in the framework of i—2010, to implement in 2007–2008 the following [EC 2005-10]:

- to review the regulatory framework for electronic communications
- to assess policy needs for media literacy and propose comprehensive approaches to radio-frequency identification (RFID) and to mobile TV
- to promote a comprehensive approach to the development of high quality innovative content
- to follow up on the security strategy with a communication on cybercrime

8.7 The Revamped Lisbon Strategy

Taking stock five years after the launch of the Lisbon Strategy, the Commission found, in 2005, the results to date somewhat disappointing and established that the European economy had failed to deliver the expected performance in terms of growth, productivity and employment. Job creation had slowed and there was still insufficient investment in research and development. "The Commission has therefore decided to focus attention on the action to be taken rather than targets to be attained. The date of 2010 and the objectives concerning the various rates of employment are thus no longer put forward as priorities" [EC 2006-14].

"The intergovernmental method without commitment, which constitutes the basis of the Lisbon strategy – declared the Belgian Prime Minister Guy Verhofstadt at the European Parliament [Verhofstadt 2006] – does not function correctly. We need an approach which is more restraining and Community-based." The weakness of governance was recognized a flaw of strategy as conceived in the year 2000. This assessment resulted in a new start for the Lisbon strategy focusing on growth and jobs, to be implemented via a reform in the method of coordination and in a revaluation of the role of the EU, influencing the debate on Community funded programs and opening the way for an ambitious seventh Framework Program for Research and Development [EC 2007-13, Muldur et al. 2006] and for other programs as presented in Chapters 11, "Framework Programs" and Chapter 7, "Entrepreneurship, Innovation, and Competitiveness".

The Lisbon strategy is revamped by restating policy priorities, particularly with regard to growth and employment [EC 2006-14]:

More growth: a new partnership for growth and employment is essential in order to give a fresh start to the Lisbon strategy. Accordingly, the Commission intends to:

- make the European Union (EU) more attractive to investors and workers by building up the internal market, improving the European and national regulations, by ensuring open and competitive markets within and outside Europe, and lastly by extending and improving European infrastructures

- encourage knowledge and innovation, by promoting more investment in research and development, by facilitating innovation, the take-up of information and communication technologies (ICT) and the sustainable use of resources, and by helping to create a strong European industrial base

More and better jobs: the Commission's new proposal concerning the financial framework for the period 2007–2013 reflects a switch of emphasis in favour of growth and employment. To create more and better jobs, the Commission intends to:

- attract more people to the employment market and modernize social protection systems. The member states and the social partners must implement policies to encourage workers to remain active and dissuade them from leaving the world of work prematurely. They must also reform the social protection system in order to achieve a better balance between security and flexibility;

- improve the adaptability of the workforce and business sector, and increase the flexibility of the labour markets in order to help Europe adjust to restructuring and market changes. Simplifying the mutual recognition of qualifications will make the mobility of labour easier throughout Europe. The member states should remove all restrictions in this area as quickly as possible;

- invest more in human capital by improving education and skills. The Commission intends to adopt a Community lifelong learning program.

Better governance: the Commission also stresses the need for responsibilities to be shared more clearly and more effectively. Overlapping, an excess of red tape and not enough political ownership are holding up progress. It will put forward a Lisbon action program in order to clarify what needs to be done and who is responsible.

The Commission will propose simplified coordination with fewer and less complex reports. It is also proposed that the national programs concerning the Lisbon strategy be presented in a format bringing together three coordination processes:

- labour market policies
- microeconomic and structural reforms
- macroeconomic and budgetary measures

The need for multiple redirections of the Lisbon Strategy results from the weakness of its initial conception; however at the end of 2007 the Lisbon Strategy appears to have been cured of most of the flaws that have characterized its slow start and its disappointing delivery. Not of *all* the flaws, since the objective of reaching R&D

investment at 3% of GDP of which two-thirds from private firms is still there as a blunt knife, supported only by the recommendation to the member states to define complementary measures to encourage private investment in ICT research and innovation; in addition LS still suffers from a nonetheless reduced lack of focus due to the vastness of its ambitions and the difficulty of cross-policy and multiple governance coordination. The merit of the revamped LS is to have confirmed the urgency of the challenges while moving away from the initially dirigistic approach. The Lisbon strategy relies now basically on strengthened shared cost cooperative research and innovation programs, a set of instruments that – having proved successful in the previous 25 years – were adopted and adapted to the new context: the Framework Program for research and technological development [EC 2007-13], strengthened and revitalized in its objectives, its instruments, and its financial coverage; the Competitiveness and Innovation Framework Program [EC 2005-9]; the e-Contentplus Program and the e-learning Program and their follow-up [EC 2005-12, EC 2006-15]. These programs are planned with full involvement of the stakeholders and reflect the perceived needs of the social actors that will ultimately implement them. In this way, rewards were introduced in the Lisbon Strategy for those member states and those social actors that would align towards the strategic objectives of the European Union. Another positive point of the revamped Lisbon Strategy is its link with the Structural Fund and Cohesion Fund programs, a link that might attenuate in the years to come the great divergence of member state's attainments documented by most information society indicators.

8.8 Monitoring the Progress

The Commission monitored over the years the position of the member states against a number of indicators relative to Internet penetration, ICT investments, e-business, e-government, and performance in education and research. It results that the member states are advancing at a different pace, that their achievements are in most cases below expectation, and that their performance is not converging.

8.8.1 PCs and the Internet

In the 1997–2003 period, there was an increase in Internet penetration in Europe, but investments, diffusion and use varied in different countries (see Figs. 8.3–8.6). On average, EU15 increased the number of Internet hosts[1] per 100 inhabitants by 250% during this period, but some countries lagged behind (e.g. Italy's hosts rose by 170%) and others (e.g. Finland and other EU and EFTA northern countries)

---

[1] 'Internet Hosting' means the provision of a variety of services including: e-mail hosting service, service that allows individuals and organizations to serve content to the Internet and to provide their own websites accessible; to provide backup services and various levels of technical support.

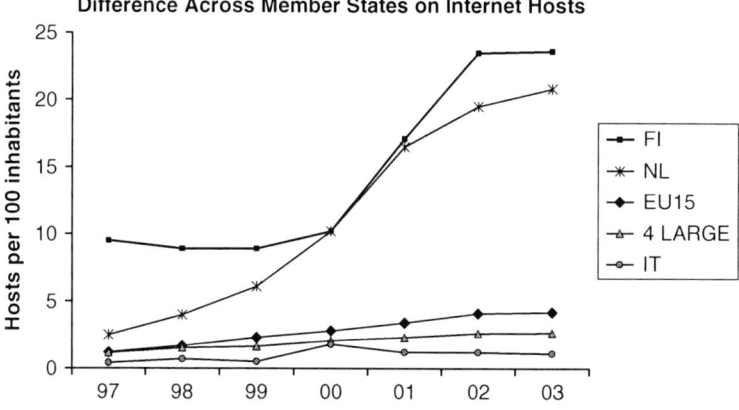

Fig. 8.3 Number of Internet hosts per 100 inhabitants in the years from 1997 to 2003. The lines show, from the bottom up, the values for Italy, for the four largest economies (DE, FR, UK, IT), for EU15, and for two well-performing countries, namely, NL and FI
Source: Eurostat statistics on the information society in Europe, 2003.

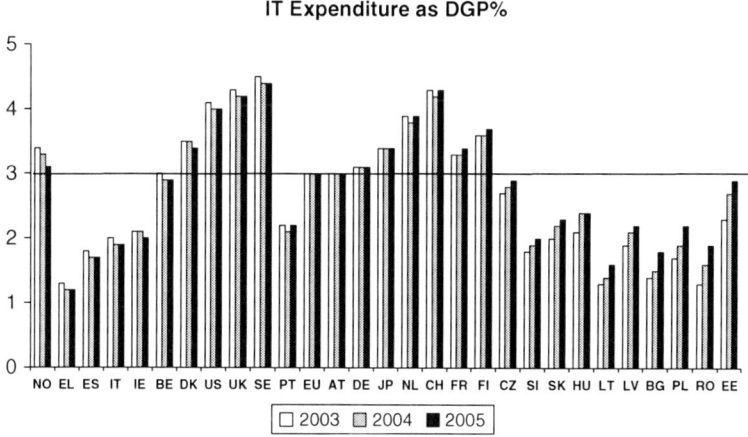

Fig. 8.4 IT expenditure in the member states, in EFTA and in the Triad, as a percentage of GDP, during the period 2003–2005. The horizontal line corresponds to EU25 average. Countries are ranked by increasing value of the difference of 2005 expenditure to 2003 expenditure. For countries on the left of Portugal in this chart, the expenditure decreased in 2005 with respect to 2003. Data for Croatia, Cyprus, Iceland, Luxembourg, and Malta are missing. EU stands for EU25
Source: Eurostat, 2007.

outperformed, ending the period with two to eight times more hosts than the EU15 average. Within the enlarged EU plus EFTA, in the year 2003 the number of Internet hosts per 100 inhabitants ranged from 0.2 in Romania to 34.8 in Iceland.

IT expenditure in the member states was on average at 3.0 % of GDP in the period 2003–2005 (Fig. 8.4), positioning the EU 25 % below the U.S. and 11 %

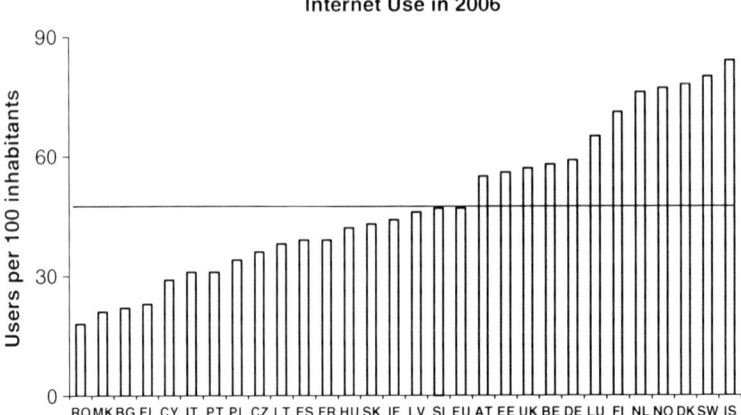

Fig. 8.5 Number of Internet users (percentage of individuals who accessed Internet, on average, at least once a week) per 100 inhabitants in the year 2006 in the EU and EFTA. Data for Malta are not available. The horizontal line corresponds to EU25 average. EU stands for EU25
Source: Eurostat, 2007.

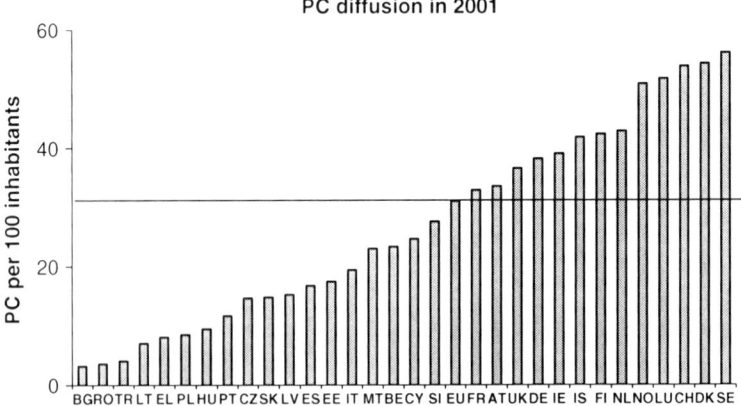

Fig. 8.6 PCs per 100 inhabitants in the year 2001 in the EU, candidate countries, and EFTA. The horizontal line corresponds to EU15 average. EU stands for EU15. Data for Croatia and Liechtenstein are not available
Source: Eurostat statistics on the information society in Europe, 2003.

below Japan; the wide spread across the member states ranged from 1.2 % in Greece to 4.4 % in Sweden. All the enlargement countries increased their investment in the period.

In the Union between 2000 and 2005, there was a great increase in the diffusion of PCs and in the way computers and the Internet are used (Figs. 8.5–8.7). In the Euro-Area, in 2005, half of the households had an Internet connection and 45 % of them had a broadband connection; of the individuals aged 16–74 living in a connected household over 75 % would use regularly the Internet if the connection is via

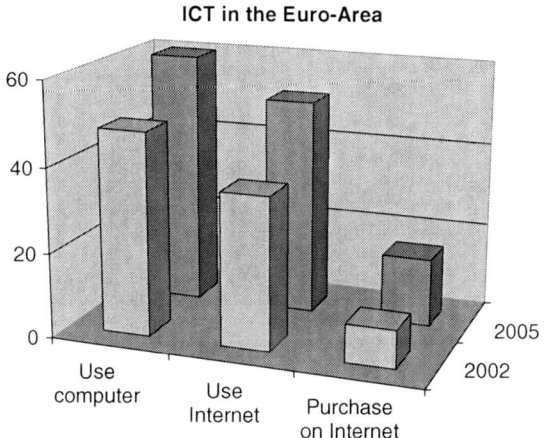

Fig. 8.7 Change in the use of computers and Internet in the Euro-Area between 2002 and 2005
Source: Eurostat Statistical pocketbook 2006.

broadband and 60 % if the connection is not broadband; according to Eurostat, in 2005, 64 % of individuals in EU27 have used a mouse to launch programs such as an Internet browser or word processor; 53 % have copied or moved a file or folder; 49 % have used copy or cut and paste tools to duplicate, or move information on screen; and 35 % have used basic arithmetic formulae to add, subtract, multiply, or divide figures in a spreadsheet; as to the enterprises: 92 % have access to the Internet, 64 % have a broadband connection, and 60 % have a web site [Eurostat 2007-2]. Younger people in the 16–24 age bracket use computers and the Internet twice as much as the older group aged 55–64 (Fig. 8.8) and the presence of children in the household is

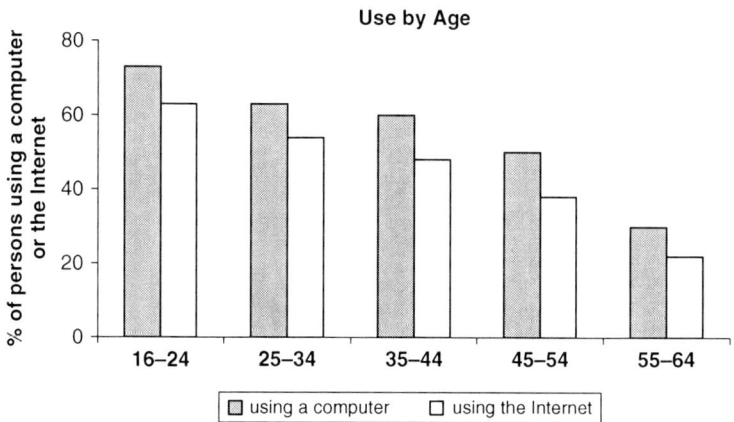

Fig. 8.8 Computer and Internet use by age group in EU15 in the year 2002. Data for Belgium, France, Ireland, and The Netherlands are not available
Source: Eurostat statistics on the information society in Europe, 2003.

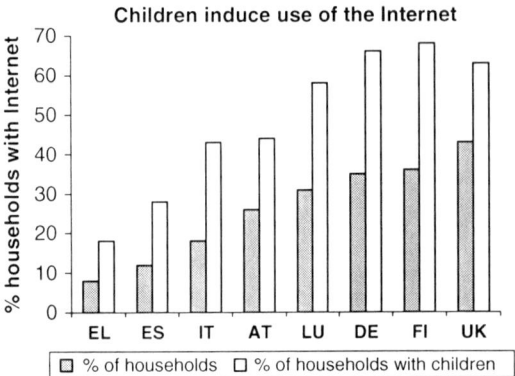

Fig. 8.9 Internet connection in households in general and in those with dependent children in EU15 in the year 2002. Data for Belgium, Denmark, France, Ireland, The Netherlands, Portugal, and Sweden are not available
Source: Eurostat statistics on the information society in Europe, 2003.

an incentive to the use of PCs and the Internet: households with dependent children show a significantly higher use of Internet connections than the households without children (see Fig. 8.9).

The costs of broadband Internet connections were rather uniform in the EU, in 2002, with the exception of Ireland (see Fig. 8.10). Broadband Internet is essential to ensure multimedia content with TV quality. Where cable TV or satellite receivers are very widespread, cable or satellite Internet competes with ADSL, and this contributes to reducing prices. Internet dial-up connection costs are relevant to household use (see Fig. 8.11). The large spread of costs in the year 2000 evened out from 2002.

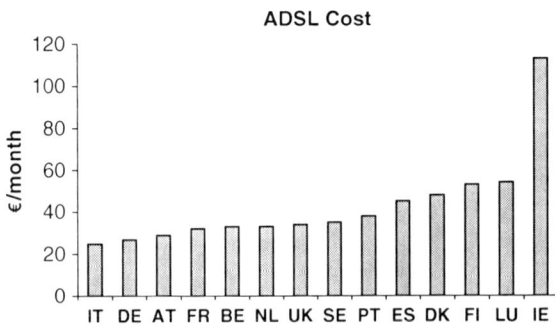

Fig. 8.10 ADSL cost (euro/month) in EU15 in 2002. Data for Greece are not available
Source: Eurostat statistics on the information society in Europe, 2003.

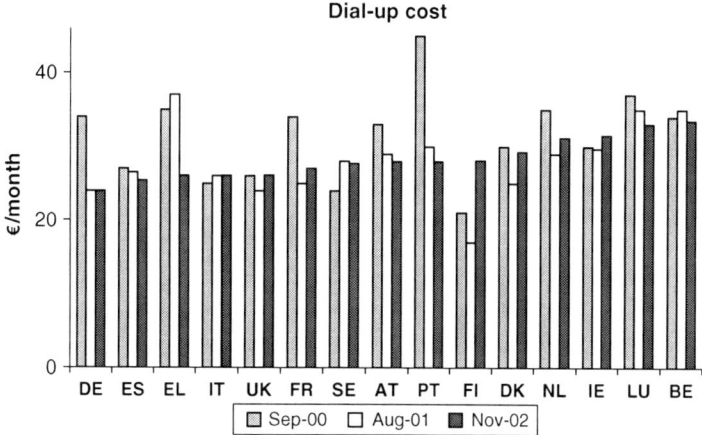

Fig. 8.11 Internet dial-up connection costs (euro/month) for residential users in EU15. The cost shown is of 20 hours per month off-peak time in 2000, 2001, and 2002
Source: Eurostat statistics on the information society in Europe, 2003.

8.8.2 Education and Training

In 2007, reaching the EU benchmarks and goals for 2010 continues to pose a serious challenge for education and training systems in Europe, except for the goal on increasing the number of mathematics, science, and technology graduates. Over 15 % of students aged 18–24 leave the school without attaining a diploma above lower secondary education; full participation in the knowledge-based society requires that each individual is equipped with at least basic education at upper secondary level. However, the share of young people who have completed upper-secondary education in the EU is below 80 % and has only slightly improved since 2000. Some countries with a relatively low share, notably Portugal and Malta, have made considerable progress in the recent past. Moreover, several new member states already perform above the benchmark of 85 %. In general women perform better they have a lead of about five percentage points in comparison to men [EC 2007-24].

The percentage of adults (25–64) participating in lifelong learning is at 9.6 %; lifelong learning is fundamental, not only for the competitiveness, and economic prosperity of the EU, but also for social inclusion, employability, active citizenship, and the personal fulfillment of people. Individuals must be able to update and complement their knowledge, competences and skills throughout life. The percentage of the working age population participating in education and training amounted to 9.6 % in 2006. Nordic countries, the UK, and the Netherlands currently show the highest participation rates. Additional efforts by many EU countries are needed to reach the benchmark of a 12.5 % participation rate in 2010 [EC 2007-24].

Increasing the capital investment in human resources includes – in addition to reducing early leaving of education, higher upper secondary attainment, and higher participation in lifelong learning - improvement of reading skills, of math and

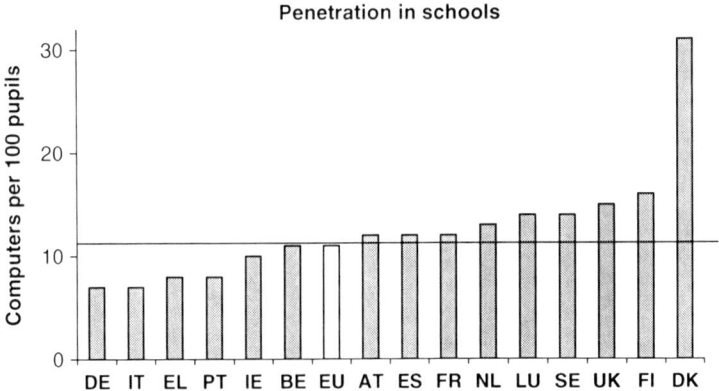

Fig. 8.12 PC penetration in schools: PC per pupil in EU15 in January–February 2002. About half of the PCs are less than three years old. The horizontal line corresponds to EU15 average. EU stands for EU15
Source: Eurostat statistics on the information society in Europe, 2003.

science abilities, provision for pre-school day-care, and increased public spending on education [EC 2007-24].

The number of PCs used during 2003 in EU15 for educational purposes in schools is shown in Fig. 8.12. However, equipping and wiring schools is not enough. Every student must have access to the Internet and to effective e-learning programs.

We mentioned in Chapter 3, "Competitiveness of the European Economy" that the use of ICT generates measurable benefits in term of productivity if the technology is widespread and reaches more than half the enterprises and involves more than 50 % of the individuals. It is also of the utmost importance that employees using computers and the Internet be trained so that they can take full advantage of the technology as opposed to being shy prisoners of processes and techniques that they do not fully understand and master.

Figure 8.13 gives the percentage of employees participating during 1999, in continuing on-the-job training on new technologies. The EU average is below 50 % and with a dispersed distribution, where a number of member states show too low a percentage of people attending formal technological education courses. Figure 8.14 shows the proportion of employees using a PC at work in 2002 and being trained on PC use: in three of the EU15 countries, the percentage of users is below 40 %, and in seven states, the percentage of those who received training is lower than the percentage of PC users – in one case, it is less than half. These are not the right conditions to generate productivity and competitiveness from the use of technology.

8.8.3 e-Commerce

Companies are transferring their selling and purchasing online, and increasingly so in the last few years, as shown in Figs. 8.14 and 8.15; on line transactions can be

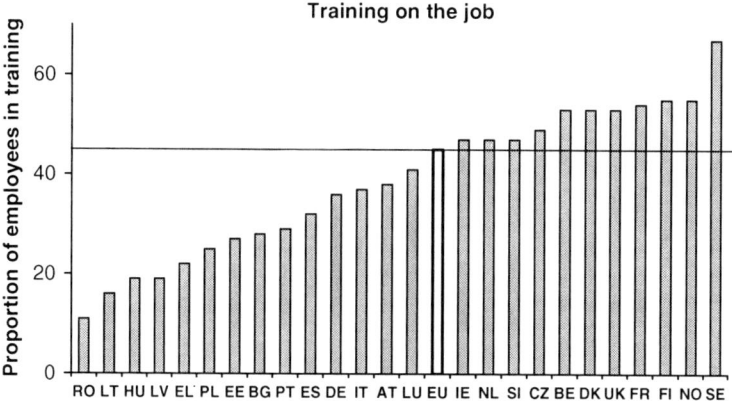

Fig. 8.13 Proportion of employees in enterprises with new technologies participating in continuing vocational training courses in the year 1999 in the EU, in candidate countries, and in EFTA. The horizontal line corresponds to EU15 average. EU stands for EU15. Data for Switzerland, Cyprus, Croatia, Iceland, Liechtenstein, Malta, Slovakia, and Turkey are not available
Source: Eurostat statistics on the information society in Europe, 2003.

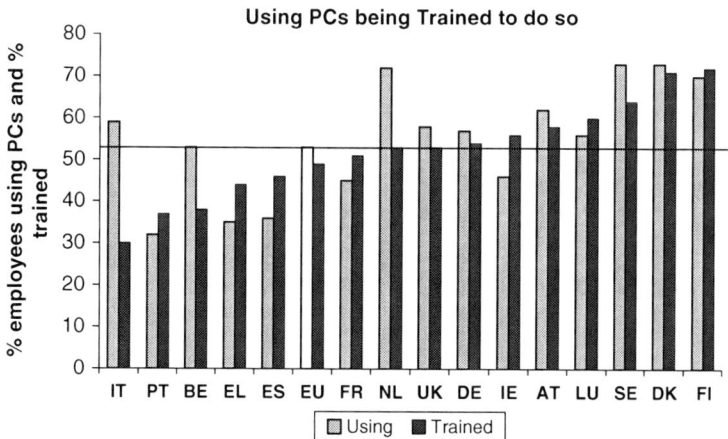

Fig. 8.14 Proportion of employees using PCs and having received training in EU15 (2002). The horizontal line corresponds to EU15 average. EU stands for EU15
Source: Eurostat statistics on the information society in Europe, 2003.

a source of savings in time and money and of quality improvement, as we show in Section 8.9, "Enterprise Strategy in the Knowledge-based Society".

While many enterprises have experienced electronic commerce, only about 1 % of EU enterprises sold in 2001 more than 50 % of their total sales via the Internet. Broken down by activity, EU enterprises in the hotel and accommodations sector made by far the most use of e-sales. Transport and communication sectors were also very active [Eurostat 2003].

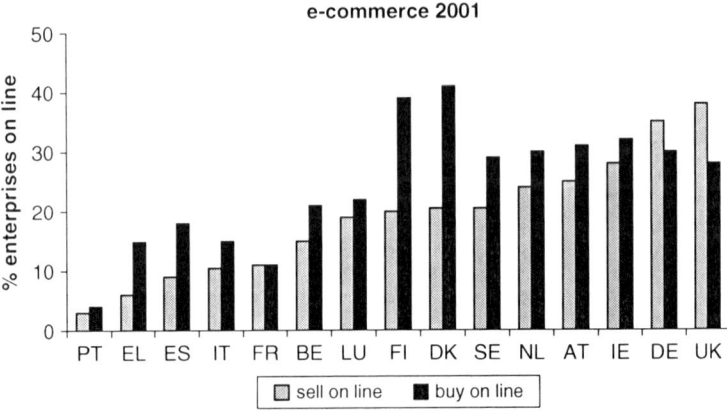

Fig. 8.15 Electronic commerce: Percentage of enterprises selling/buying online (2001) *Source*: European Commission e-Europe 2002 Benchmarking Report [EC 2002-5].

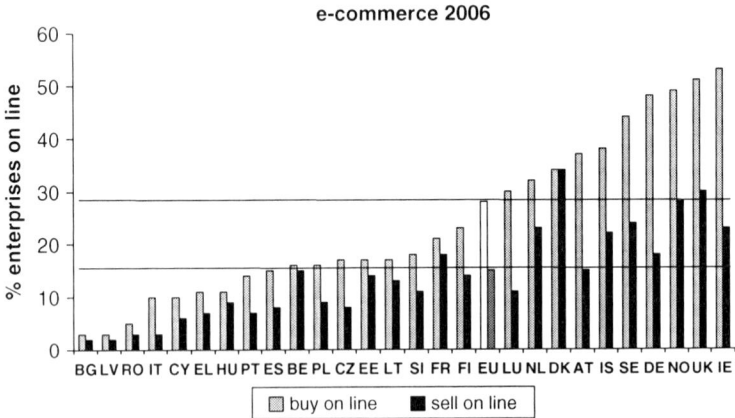

Fig. 8.16 Electronic commerce: Percentage of enterprises selling/buying online (2006). The horizontal lines indicate the EU25 average for buying (*upper line*) and selling (*lower line*). EU stands for EU25
Source: Eurostat 2007.

There are more companies buying than companies selling online (see Figs. 8.15 and 8.16): online sales require an updated web site, a safe connection, and, above all, distribution logistics. Buying online is easier because it only requires an Internet connection and a credit card. This is the reason why more companies are buying online than selling online.

But online buying is also developing at home. Figure 8.17 gives the ranking of online sales for different goods and services. The ranking is based on the number of purchase actions, not their value.

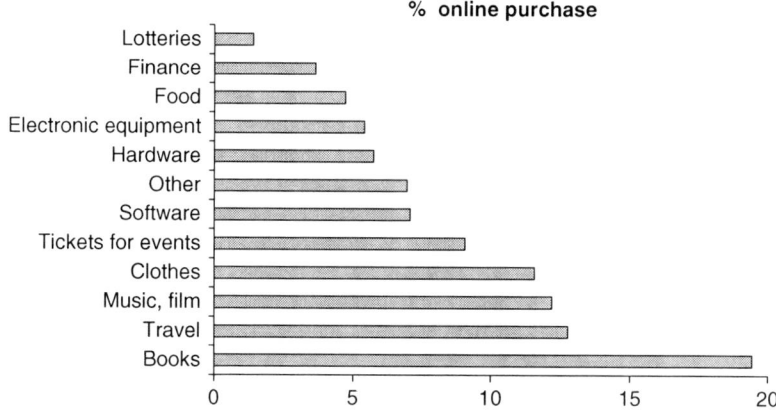

Fig. 8.17 Frequency of online purchases by individual persons for different classes of goods and services. Data from Denmark, Greece, Luxembourg, Austria, and Sweden were relative to the year 2002
Source: Eurostat statistics on the information society in Europe, 2003.

8.8.4 e-Government

The availability of on-line services from the administration, as perceived by the citizens, is very different in the various member states. Finland is, again, an example of excellence. The author was surprised by the official data that placed Italy higher in Fig. 8.19 than where his perception would have placed it. He received an interesting suggestion[2] from a political scientist working on the study of organizations: the red tape involved in the scrapping of old equipment could enable a number of obsolete (unused) dust collectors to contribute to the statistics of ICT equipment supposed to enable the Italian public administration to offer effective services to citizens and firms.

On the supply side (Fig. 8.19), most of the accession countries (with the exception of Malta and Estonia) are below the EU average, but significant progress is reported in the last few years. All EU member states made significant progress and Austria achieved the objective of covering all areas of administration-enterprise interaction with on-line services.

e-Government is a complex transformation of the relationship of the public administration with citizens and enterprises (Figs. 8.18 to 8.20). Quantitative indicators may not suffice to capture the process in its complexity. A study contracted by the European Commission, e-government unit, DG Information Society and Media [EC 2007-26], suggests that

> public value in e-government is not a process, but it is a relationship built on participation, organizational transparency and trust. A simple focus on the organization is not sufficient – it is the way in which the organization mediates a critical relationship between government and citizen that matters. It is not enough just to implement organizational change. Change in itself

---

[2] Marco Velicogna, private communication, April 2005

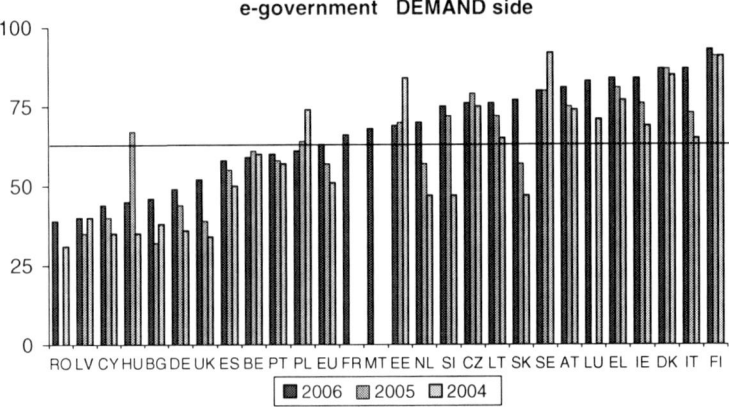

Fig. 8.18 e-government usage by enterprises. Percentage of enterprises using the Internet for interacting with public authorities demanding of online public administration services (2004–2006). EU stands for EU25. The horizontal line indicates EU25 average on 2006
Source: Eurostat 2007.

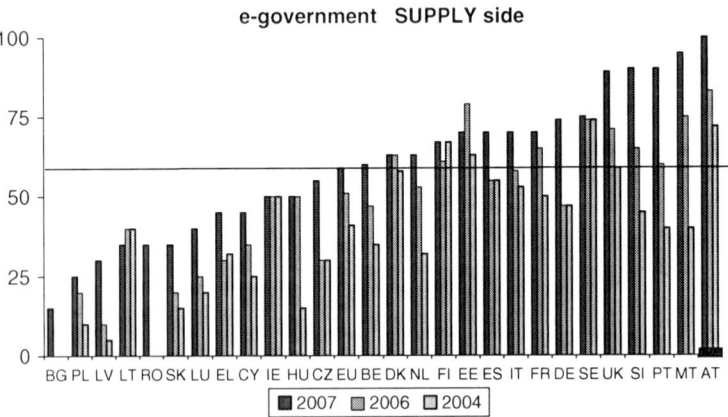

Fig. 8.19 e-government service offer by the public administration. Percentage of public administration services supplied online (2004–2007). EU stands for EU25. The horizontal line indicates EU25 average on 2007
Source: Eurostat 2007.

will not guarantee delivering services that provide public value. Progress in e-government comes through modernization and the effective use of IT that improve the trust of citizens in government. Efficiency is the operation of the governance process in a way that continues to demonstrate cost benefits; more for the same, the same for less. Effectiveness comes from the use of efficient processes to construct service portfolios that deliver individual and public value. Managing the transformation of efficiency into effectiveness involves flexible organizational behavior and relationship management with citizens. Consequently, the true measurement of the benefits of public service modernization cannot necessarily be found just in the traditional bottom-line financial approach.

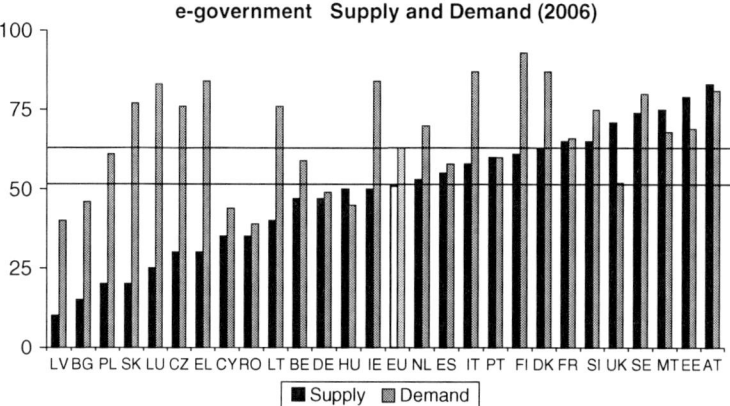

Fig. 8.20 e-government services in 2006. Comparison of Supply and Demand of online public administration services in EU27. The horizontal lines indicate EU25 average in 2006 for supply (*lower line*) and demand (*upper line*). EU stands for EU25
Source: Eurostat 2007.

8.9 Enterprise Strategy in the Knowledge-Based Society

e-Mail was created in 1969 within the ARPA network, a project of some of the major U.S. universities supported by funding from the U.S. Department of Defense. In the business field, e-mail was first used by big enterprises as a company communication system on proprietary and dedicated networks and then, at the end of the 1970s, on public networks.

In addition to increasing efficiency and productivity, e-mail has deeply transformed the relationships between company functions, as it has introduced a new level of informal communication and has reduced bureaucracy. The year 1982 is considered to be the first year of the Internet era, with international e-mail services between the United Kingdom and the other northern European countries. The spread of the Internet has been particularly fast: as a matter of fact, over a decade, the number of Internet hosts has exceeded one million units. In 1991, the CERN laboratory in Geneva, Switzerland, made public the first Internet browser and the HTTP protocol. In 1992, CERN made public the World Wide Web software. A new stage of transformation in the use of information and communication technologies thus began, which would change companies, services, the relationships between suppliers and customers, private communications, education, and little by little the whole society.

The www network has led to a great industrial revolution as important as the introduction of electricity, the car, or the telephone, and which has deeply changed production costs, working procedures, and trade practices, to the advantage of both enterprises and consumers.

The transformation of companies does not simply imply providing each desk with an interconnected personal computer. In order to get the most from the use of ICT and the www, it is necessary to reorganize the whole company according to a new strategy that will transform the production cycle and the relationships both within the enterprise and with its customers and that will increase the quality of products and services. Otherwise, the company will face failure and bankruptcy.

The Internet does not involve only new and small dotcom companies, since some big multinational corporations from the old economy, are also changing and getting great advantages from the web. Several cases of companies that have taken up the gauntlet of the new technology will be examined in this chapter. Six cases will be analysed:

1. General Electric, U.S.   industrial conglomerate
2. Siemens, Germany   industrial conglomerate
3. 7-Eleven, Japan   retail
4. Cemex, Mexico   concrete
5. Ryanair, Ireland   airline
6. Michelin, France   tyres

At the end of 1999, the following slogan appeared:

e-business or *out of business*

However, the burst of the speculative bubble of the *new economy* has reduced the growth expectations linked to e-business. It seems that the Internet has had the maximum impact on companies of the *old economy*. A certain number of dot companies have gone bankrupt. Some great successes in the application of Business-to-Consumer (B2C) and Business-to-Business (B2B) solutions have turned out to be a flash in the pan. On the other hand, some "old giants" have seized the opportunities offered by technology and Internet for a great company transformation [Economist 2001-1].

8.9.1 General Electric

Like Siemens, the enterprise we consider next, General Electric operates in a number of industrial sectors. GE and Siemens are very large undertakings, with hundreds of thousand of employees and a worldwide operation. The choice of both companies to be structured into single conglomerates rather than in holdings with a number of separate companies has been successful because they have learned how to live and prosper in the knowledge-based society and are able to capitalize on their ability to structure huge amount of information into a corporate knowledge base, which is their competitive edge. The size and diversity of the sectors in which both conglomerates operate make them highly complex businesses to run, which are only profitable because both undertakings are able to cross-fertilize one sector with the other and share valuable corporate knowledge throughout their worldwide operations. A conglomerate is not a means to reduce shareholders' risks by operating in different sectors so that if one sector faces a crisis another may balance it out: shareholders could achieve this with a balanced portfolio of investments in separate listed companies. A conglomerate is successful

Table 8.1 GE balance sheet

		2001	2006
Turnover	($ billion)	125	163
Operating profit	(%)	11.2	15.2
Employees	(thousands)	313	319
Number of countries		160	160
Number of shares	(billion)	10.0	10.5
Stock Market capitalization	($ billion)	360	390

Source: GE Annual Reports.

if it creates synergies by sharing the experience and knowledge accumulated in the diversified operations and if this advantage generates higher profits than the overhead costs generated by the increased complexity of the operations.

To describe General Electric (GE), *The Economist* wrote on 17 May 2001: "Elephants can dance." GE is the biggest industrial conglomerate in the world [Economist 2001-3]. Table 8.1 gives basic figures from the 2001 and 2006 GE balance sheets. GE's financial results highlight its ability to deliver. In 2006, GE generated double-digit earnings and revenue growth ($ 163 billion in revenues and $ 20.7 billion of earnings). Over the past five years, GE has grown its earnings an average of ten percent annually. In the *Fortune's* ranking of the most admired companies, GE has often been No. 1, and so was in 2006 for the second straight year; GE is one of only six "Triple A" rated U.S. industrial companies; and in 2007, it ranked No.4 in *Business Week's* "World's Most Innovative Companies." GE has *electrified* the market, that is, it has introduced some changes as radical as the passage from steam traction to electric traction in the railway sector.

Under the management of chief executive Jack Welch (who held office from 1981 to 2001 and then retired with a golden handshake, partly paid on the side, which has caused him some problems), in 2001 GE actions reached a value that was 62 times higher than their value in 1981. Jack Welch managed to achieve first or second world ranking in a series of industrial sectors:

1. Airplane engines, locomotives
2. Household appliances
3. Lighting
4. Supply of electricity
5. Generators and turbines
6. Nuclear reactors
7. Plastics
8. Image treatment for medical use

In addition to the above industrial sectors, we should mention GE Capital Services, GE's financial sector that contributes to 50 % of the group turnover and is one of the biggest financial companies in the U.S. GE also holds the NBC television network and has tried to take over Honeywell, an information technology company, but in 2001 its purchase offer was blocked by the European trust-buster.

If Bill Gates is the most famous American businessman, Jack Welch, GE's boss, was, in the 1990s, the most admired. For many years, Mr. Welch was for financial journals what Princess Diana was for romantic magazines.

The *first* merit of Mr. Welch was to create a conglomerate in which every unit was No. 1 or No. 2 in the world.

The *second* key to success was the ability to adjust to changes in the world economy and the speed in the adoption of innovations.

The *third* factor was the conversion of low-profit manufacturing processes into highly profitable services.

Under Jeffrey R. Immelt, GE's Chairman and CEO since 2001, GE is transformed into a more global, diverse and customer-driven enterprise [GE 2007]. GE has six strong businesses aligned to grow with the market trends of today and tomorrow. This is not by chance: it is a result of the considered and strategic investment in each business over time—and ahead of external realities.

1. GE Infrastructure is one of the world's leading providers of essential technologies to developed and emerging countries, including aviation, energy, oil and gas, rail, and water process technologies and services.
2. GE Industrial provides a broad range of products and services throughout the world, including appliances, lighting, and industrial products; factory automation systems; high-performance engineered plastics; security, and sensors technology; nondestructive testing; and equipment financing, management, and asset intelligence services.
3. GE Healthcare is a leader in the development of a new paradigm of patient care. GE Healthcare's expertise in medical imaging and information technologies, medical diagnostics, patient monitoring systems, disease research, drug discovery, and bio-pharmaceutical manufacturing technologies is dedicated to detecting disease earlier and helping physicians tailor treatment for individual patients.
4. NBC Universal is one of the world's leading media and entertainment companies in the development, production and marketing of entertainment, news, and information to a global audience.
5. GE Commercial Finance offers an array of services and products aimed at enabling businesses worldwide to grow. GE Commercial Finance provides loans, operating leases, financing programs, and other services.
6. GE Money is a leading provider of credit and banking services to consumers, retailers and auto dealers in countries around the world, offering financial products such as private label credit cards, personal loans, bank cards, auto loans and leases, mortgages, corporate travel and purchasing cards, debt consolidation, home equity loans, credit insurance, deposits, and other savings products.

GE is committed to the development of its human resources. GE strives to create a culture of opportunity where employees can challenge themselves to reach their full potential. The corporate commitments include:

- continue to provide a stable base of development, opportunities, jobs, and benefits
- implement best practices from U.S.-based pension plans to strengthen governance procedures globally
- focus on managing global diversity, in a range of different cultural settings
- continue to focus on individual competitiveness by improving employee tools and resources

Worldwide, GE invests about $ 1 billion annually on training and education programs to develop some of the best leaders and some of the most widely practiced business techniques. The centerpiece of its commitment to excellence in leadership development is the John F. Welch Leadership Center at Crotonville, New York the world's first major corporate business school.

GE is committed to R&D: it has more than 3,000 of the best and brightest researchers spread out at four multi-disciplinary facilities around the world. Headquartered in Niskayuna, New York, GE also has research facilities in Bangalore, India; Shanghai, China; and Munich, Germany. These facilities are delivering the innovations and breakthroughs that are driving growth for GE's businesses and revolutionizing markets. GE believes "what we imagine, we can make happen." GE filed 2,650 patents in 2006 and its research budget was of $ 3.66 billion, which by including the technology expenditures for improving existing products and services, and plant productivity brings GE's innovation investment in 2006 to $ 5.7 billion.

GE's business is online. Navigating in the GE portal, customers can buy goods and services from the vast GE catalogue. In 1997, GE Information Services (GEIS) was the only significant example of Business-to-Business e-commerce. Through an electronic data interchange system (EDI-B2B), it supplied 100,000 companies that, in total, carried out over one billion operations a year. In 2006, GE bought and sold $ 20 billion online, that is, more than the rest of the B2B market in the U.S. By eliminating human errors, EDI ensures savings without changing the usual working method. With the Internet, B2B transactions offer the opportunity of establishing new market relationships through auctions, direct sales without intermediation, and access to new customers. The Internet has thus led to savings and improved quality.

GE's e-business projects have introduced new sources of turnover:

1. services transforming every sale into an inexhaustible source of turnover
2. sales to new customers, for example, small and medium-sized enterprises that could not be reached before

Moreover, e-business widens GE's range of suppliers, thus increasing competition. For example, if you buy an airplane turbine, you are later approached with an offer to purchase an essential maintenance service ensuring long-term availability, reliability, and safety. GE has introduced diagnostic systems in the turbines that provide information about turbine working conditions while the plane is airborne. A satellite network collects the data about the turbine working conditions and transmits them to a GE processing centre. When the plane lands, the service personnel of the airport are informed about maintenance operations to be carried out before any anomaly that emerged during the flight degenerates into failure, grounding the plane or creating problems of reliability or safety.

8.9.2 Siemens

Siemens, headquartered in Berlin and Munich, Germany, is one of the world's largest electrical engineering and electronics companies. Siemens provides innovative technologies and comprehensive know-how to benefit customers in 190 countries. Founded 160 years ago, the company is active in the areas of Information and Communications, Automation and Control, Power, Transportation, Medical, and Lighting. With more than 62,000 patents (5,060 patent applications filed in 2007) and 32,500 researchers (with a € 3.4 billion budget), Siemens is one of the most innovative companies worldwide [Siemens 2007].

> After decades of under-performance, the giant German conglomerate has been shaken in the past few months by investigations into corruption scandals reaching back at least seven years. Its reputation and its business have suffered. Several executives have been arrested, and two convicted (so far). Most recently Siemens named a new chief executive, Peter Löscher, an Austrian poached from Merck, an American drugs company. The stock market has been delighted, pushing Siemens shares up to and beyond ten-year highs and valued it in the stock market at more than $ 110 billion. Peter Löscher, the first outsider to become chief executive of Siemens, spent his first three months at the firm traveling and listening. Then he announced plans to centralize the conglomerate and reduce its nine divisions to three: energy products (such as power turbines and transmission equipment); infrastructure (such as factories and trains); and health care (such as scanners) [Economist 2007-27, Economist 2007-28].

The Siemens' 2007 Annual report states [Siemens 2007]:

> Our Company's success depends on the answers we're giving to the toughest questions of our time—how to minimize resource consumption in industrial manufacturing, protect the environment and the climate, and improve healthcare for people everywhere. We can deliver answers because we're competitive, because we're active in nearly 190 countries around the world, because we waste no time in translating customer requirements into new products, systems, services and solutions—in a word, because we're highly innovative. Innovation is the springboard for the answers we provide. Only intensely committed, highly qualified people like ours can transform inventions into successful products and solutions. The growing value of our Company and the long-term successes we're scoring with customers around the world prove that we're finding the right answers to the questions of our time.

Siemens [Economist 2001-2] operations are coordinated and unified by a company-wide knowledge management system. The Siemens group [Economist 2001-3] has faced at the turn of the century the challenges of the knowledge society with a great change through an investment of € 1 billion for technological and organizational transformation into an e-company. Siemens became a company based on knowledge management and communication (Table 8.2).

The Siemens strategy has been evolving over the past 15 years to cope with a changing world, and that change accelerated with the dawn of the knowledge society era. The company's worldwide operations have been coordinated and integrated, so that each project contributes to the company's knowledge base, and each one of the 400,000

Table 8.2 Siemens balance sheet

		2001	2007
Turnover	(€ billion)	87	72
Gross profit	(€ billion)	2.7	5.1
Employees	(thousands)	470	398
Number of countries		190	190
Number of shares	(million)	888	915
Stock Market capitalization	(€ billion)	40	88

Source: Siemens annual reports.

employees has access to the accumulated experience of the enterprise as it builds up in its 190 national units throughout the world. The depth of the company's hierarchy was reduced and access to higher management enabled via informal communication channels.

In April 2005, a program was launched called Fit4More based on four pillars:

- Performance and Portfolio, to create a portfolio that would guarantee our future success, by realigning selected activities and expanding the strong core businesses (energy and environmental technologies; automation systems; public and industrial infrastructures; and healthcare solutions)
- People Excellence, to implement targeted measures, with a view to identifying and nurturing the young people who will one day lead the company
- Operational Excellence, to establish standardized processes and procedures throughout the company while concentrating intently on innovation, customer focus and global competitiveness.
- Corporate Responsibility, to focus on corporate values and assume societal responsibilities

The financial results demonstrated the effectiveness of the program's tools and levers. All Siemens' groups reached their margin targets by the second quarter of fiscal year 2007. Growth over the past two years totalled more than 15 %.

At the end of 2007, the new corporate program is called Fit 4 2010 and is meant to support a Siemens' growth twice as fast as the GDP growth. Its aims are:

- To attain high profitability targets, optimizing capital efficiency—with a target of return on capital employed (ROCE) of 14–16 % for the entire company by 2010
- To build on corporate strengths in order to capture and maintain No. 1 or No. 2 positions in attractive markets and further boost our businesses in the three sectors: Industry, Energy, and Healthcare
- To create a high-performance culture, strengthening our leadership development and promoting top talents around the globe
- To continue pushing innovation by applying proven methods and tools while sharpening Siemens' customer focus and enhancing its global competitiveness
- To align business performance with the corporate commitment to society and to the environment

Siemens is committed to training and lifelong learning: about 9,900 young people are at any time enrolled in Siemens' vocational training and work-study programs. Siemens offers a comprehensive range of training programs, preparing participants for new types of jobs in business administration, information technology, metalworking, and electrical engineering. In addition, it sponsors special programs for high school graduates and work-study programs for university students. Siemens is exporting its work-study system, which combines theory and practice, to many of its regional companies around the world. Outside Germany, about 3,100 young people are currently enrolled in Siemens apprenticeship programs, which have been adapted to country-specific requirements. These programs enable Siemens to maintain an outstanding workforce by constantly raising the educational standards of its employees, which in turn contributes to the high quality of its products and services. For university graduates just beginning their careers, Siemens offers more than direct entry opportunities. Siemens' two-year graduate program, which trains the managers of tomorrow, has three parts, each of which offers participants an insight into a different aspect of Siemens—for example, purchasing, product development, sales, and marketing. The program also includes an internship at one of the corporate locations outside the participant's home country. When people join Siemens, their education does not stop. In fact, it is just beginning. To continually broaden the basis of success, Siemens offers a wide range of training and continuing education programs—from management and language courses to seminars in technical fields—that enhance our employees' qualifications, foster their development, and equip them to excel throughout their working lives.

Siemens is committed to research. Its 32,500 researchers are working in laboratories throughout the world: in Germany (36 %), in the rest of Europe (28 %), in the Americas (23 %), in Asia-Pacific (12 %), and in Africa (1 %). R&D brings Siemens at the forefront of innovation. Three examples may help give an idea of the kind of breakthroughs achieved. Siemens' Industrial wireless LAN (Iwlan) technology makes it possible to precisely control industrial machines without using expensive cable connections. The solution increases factory flexibility and improves production quality. By guaranteeing data transfer rates in the millisecond range, special Iwlan software allows the effective monitoring and control of machines while eliminating data bottlenecks and costly error chains.

In Denmark, Siemens is using a patented single-cast process to manufacture turbine blades for wind power systems. Produced without joints, the blades are the toughest in the wind power industry. As the global market leader in the offshore sector, Siemens offers the world's most powerful wind turbines in series production. With a capacity of 3.6 MWs, these turbines generate over 100 times more power than the first wind turbines, which were marketed some 25 years ago.

Siemens' Somaton Definition has revolutionized computed tomography (CT) technology. Boasting two X-ray tubes and two detectors that rotate in synchrony, the innovative system provides twice the temporal resolution, operates at twice the speed, and delivers twice the power of conventional scanners—with substantially lower radiation exposure. The outcome: enhanced patient satisfaction and unprecedented image quality, so that diseases of the blood vessels, heart, and other organs can be detected earlier, diagnosed faster, and treated more precisely than ever before. The technology's wide popularity is yet further proof of its effectiveness: Since market launch, more than 250 Somaton Definition scanners have been installed worldwide.

Customer perception of Siemens as a single enterprise has been achieved by assuring that any customer has a single company contact even if he or she relies on goods and services provided by many different Siemens divisions. A full-page advertisement celebrated this achievement in several daily newspapers in 2003: "Simplify: one single business card suffices for the whole Siemens Group."

Siemens is an e-company that is online and whose target of 100 % of e-purchases was achieved by year-end 2005. The recipe for Siemens' success can be summarized in a few points:

- a corporate knowledge base that helps capture and share the experience of the whole group
- informal communications
- training and lifelong learning
- research and innovation
- presence only in industrial sectors where the group is or is equipped to become soon a world leader
- reorganization into small groups
- employee equity participation
- a completely online management with no paperwork
- compensation linked to the company's performance
- company standards
- cost reduction
- cohesion
- only *one* Siemens contact for Siemens' clients

Because of the great size of the Siemens industrial conglomerate, in the past the synergy between the different company divisions was so limited that it justified the motto "If only Siemens knew what it knows." The new company policy transformed such motto into "Siemens knows what it knows." The new orientation based on knowledge management generates cohesion, standardization, and synergy throughout the whole group, as well as a coherent and uniform relationship with the customers and the different operative divisions.

8.9.3 Cemex

Cemex operates in a traditional sector of the old economy, that is, concrete [Cemex 2007, Economist 2001-4]. The year 2006 was important for Cemex that commemorated 100 years of growth and the people who achieved it. The legacy of all of that hard work and dedication is a stronger and more solid company, better prepared for the challenges and opportunities that lie ahead. Cemex is a pioneer in e-business. Flexible and efficient, Cemex works all over the world from a developing country (Mexico).

Cemex is a leading global producer and marketer of quality cement and ready-mix concrete products, positioned in the most dynamic markets around the world: the Americas, Europe, Asia, Africa, and the Middle East. Founded in Mexico in 1906, Cemex has grown from a small regional player into a top global cement company.

Cemex customers' need for prompt on-site concrete delivery led Cemex to rethink and change the nature of the ready-mix business. Through a proprietary system for the dynamic synchronization of operations, Cemex was able to guarantee on-time delivery to its clients—versus the industry's standard four-hour delivery time—regardless of the weather or traffic.

In order to solve the traffic-jam problems in Mexico City and to deliver concrete on time, within 90 minutes from the loading of the concrete mixer, all concrete mixers have been equipped by the year 2000 with a computer and a satellite navigator. Once loaded, concrete mixers head for the town districts in which the building yards are located and are redirected in real time while on route towards the different customers according to the orders received via the Internet, thus guaranteeing delivery within 20 minutes from the order even in building yards that are more than a one-hour drive from the production centre.

Cemex covers the whole concrete cycle:

- extraction of limestone and clay
- fragmentation and transport
- pulverization
- homogenization
- storage
- hot calcinations
- grinding, and
- packing and shipping

Cemex represents a company operating in the old economy whose managers have understood at the right time the advantages resulting from the adoption of ICT and globalization (Table 8.3). Lorenzo H. Zambrano, Cemex chairman and chief executive, declared [Economist 2001-4]:

> Information technology allows CEMEX to eliminate repetitive activities, thus releasing the intellectual energies of our personnel and increasing competitiveness and profitability. Innovation is the driving force of Cemex's success, a digital enterprise. e-Enabling, the company information system, allows all our operators to have access to information and resources.

Table 8.3 Cemex balance sheet (million dollars) 2002–2004

	2002	2003	2004
Net sales	6,543	7,164	8,149
Gross profit	2,887	3,034	3,563
Operating income	1,310	1,455	1,851
Net income before income taxes	591	731	1,501
Activities in:	12 countries		

Source: http://cemex.com

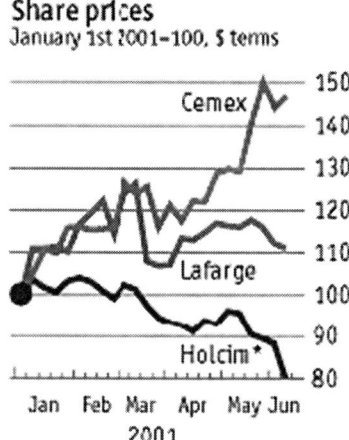

Share prices
January 1st 2001=100, $ terms

Fig. 8.21 Cemex grows in the stock exchange while its competitors decline
Source: [Economist 2001-4] "The Cemex way" © *The Economist* vol. 359, n. 8226 (16 June 2001),
reproduced with kind permission.

Cemex's success on the stock exchange is caused by its successful strategy (see
Fig. 8.21). Cemex's keys to success are:

1. standardization and quality
2. new technologies
3. new management criteria
4. introduction of innovative processes in developing countries
5. recognition of the information value
6. automated relationships in factories
7. automatization of operations
8. a satellite network to control worldwide operations since the 1980s
9. e-mails connecting the board of directors with executive managers

Cemex is committed to investing in people training and in capitalizing on corporate
knowledge. The 2006 report says [Cemex 2007]:

> Our primary asset is our people—motivated employees who, together, deliver positive results
> for our customers, our communities, our stockholders, and each other. From our special-
> ized training programs to our online learning system, we are committed to fostering our
> employees' continued professional and personal development. In addition to our professional
> development programs, we foster an open and responsive work environment in which all of
> our employees can achieve their full potential. To this end, we not only measure employee
> engagement, but also encourage employees to propose ideas as to how we can improve our
> business. In 2006, we received 2,683 ideas from throughout the country, with the three best
> ideas to be recognized at the 2007 annual employee meeting. We also promote employee
> health, safety, and wellness across our organization. We take significant steps to ensure that
> our people have the knowledge and tools to assure their own and their coworkers' safety. We
> provide comprehensive health-awareness and preventive-care programs. As our employees
> succeed in their professional and personal lives, they enable our company to make a greater
> contribution to society.
>   One of the qualities that set us apart is our unwavering commitment to continuous
> improvement. We are always looking for new ways to make our operations even better

by identifying, sharing, and implementing best practices across our global network of plants and facilities. We continually look for new opportunities to improve our performance by identifying, sharing, and implementing best practices across our global network of plants and facilities. Using knowledge shared from our German operations, we are aggressively expanding our alternative fuels program—particularly in Europe. This program uses more cost-effective and environmentally friendly alternative fuels, including biomass and solid waste, in place of traditional fossil fuels such as coal, fuel oil, and natural gas. As a result of our network collaboration, alternative fuels now comprise 25 percent of the total fuels used in our European cement plants, as well as a growing percentage of our diversified fuel structure worldwide.

Cemex is a pioneering innovator: In the 2006 report, we read [Cemex 2007]:

We began wiring Cemex in 1989, and we connected all of our facilities by 1992. Leveraging our early edge in technology, we now use the Internet to satisfy and fulfil our customers' needs through different electronic initiatives like our online storefronts and mobile sales technology. In this and many other ways, we set the industry standard with in-house technology resources and solutions, not off-the-shelf applications.

Upon acquisition of other companies, Cemex's efficient organization allows it to integrate the information system of the new company with that of the group in the record time of only four months.

Cemex is moving all its operations onto the web:

* finance
* purchasing
* sales
* relationships with customers, and
* relationships with suppliers

The savings thus achieved amount to $120 million a year.

8.9.4 7-Eleven

7-Eleven, Inc. is the world's largest operator, franchisor, and licensor of convenience stores with more than 31,600 units in 17 countries worldwide, of which 21,000 in the U.S. 7-Eleven also is one of the U.S. largest independent gasoline retailers. The turnover was of $27 billion in 1999. In 1927, 7-Eleven introduced [Economist 2001-5] a new retail system at Southland Ice Company in Dallas, Texas. In addition to selling ice bars, a resourceful salesman started to offer milk, bread, and eggs on Sundays and in the evening, when the other shops were closed. The trademark of 7-Eleven originated in 1946 when the stores were open from 7 a.m. to 11 p.m. Today, offering customers 24-hour convenience, seven days a week is the cornerstone of 7-Eleven's business. 7-Eleven, Inc. is privately held and became a wholly owned subsidiary of Seven-Eleven Japan Co., Ltd. in Tokyo, Japan, on 9 November 2005.

The secret of 7-Eleven's success was the innovative use of technology. In the 1980s, 7-Eleven replaced cash registers with point-of-sale (POS) computers monitoring purchasing habits and training personnel online. These systems are multimedia systems with moving images and sound, very easy to use even by temporary and unskilled personnel.

In the Japanese branch of 7-Eleven, an ad-hoc Windows operating system connects 61,000 computers in 8,500 shops in Japan, while Microsoft offers 24-hours-a-day, 7-days-a-week real-time support from Seattle, Washington, through a satellite connection with the whole network of shops. The whole hardware and software system is completely renewed every five years.

7-Eleven has created a *paperless business* through the elimination of all paperwork. The company monitors online the customer demand and preferences (range of products, presentation, prices), which allows it to get accurate demand estimates and adjust production and distribution (e.g. through the correlation between weather forecasts and consumer preferences).

7-Eleven carries out continuous quality control of sales and stock and adjusts product lines very often. A company road transport system guarantees maximum speed of purchasing and distribution, with a real time control of deliveries. 7-Eleven uses this company transport system to offer an efficient home delivery service, too.

7-Eleven has also extended its commercial operations to (Fig. 8.22):

- offer of services
- payment of bills
- other banking services, and
- order, payment, and collection for e-shopping

In this way, 7-Eleven can expand its market to increase profits without cutting personnel. The services offered include, without limitation, Internet workstations in supermarkets with the possibility of buying online, and asking for the delivery of the goods at the supermarket if nobody is at home during the day. This setup ensures another visit by the customer to collect the goods bought online, thus creating another occasion for shopping in the 7-Eleven store.

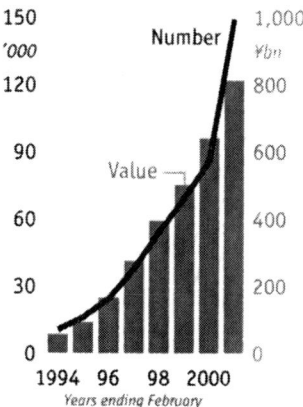

Fig. 8.22 A retail company, 7-Eleven, expanding its business into the financial sector
Source: [Economist 2001-5] "Over the counter, *e*-commerce" *The Economist vol.* 359, n. 8223, 26 May 2001, reproduced with kind permission.

8.9.5 *Ryanair*

Competition in the European skies started with the deregulation of air transport (1997) in Europe that followed the deregulation in the U.S. 20 years earlier (1978). Like Northwest Airways in the U.S., in Europe Ryanair, EasyJet, and then many others have bet on a new business model in air transport based on low-cost, no-frills flights.

Southwest Airlines is the model of a low-cost, no-frills airline. It has exploited the deregulation of air transport in the U.S. since 1978. With no meals or coffee, but fares at 20 % of the price applied by the other airlines, it offers short-distance point-to-point flights from secondary airports, sometimes one-hour drive from the city centre, using only Boeing 737 aircraft.

Southwest and Ryanair are two examples of the validity of Clayton Christensen's theory [Christensen 2002, Christensen 2003] according to which a fairly good service (in this case, low-cost, no-frills flights) is better than the best (flights with luxury service) if the best is too expensive.

Established in 1985, Ryanair started its activity with only one aircraft, a Bandeirante aircraft, and a staff of 25. Ryanair operated daily from Waterford in the southeast of Ireland to London Gatwick in the UK. Ryanair's first cabin crew recruits had to be less than 5 ft. 2 in. (157 cm) tall in order to be able to operate in the tiny cabin of the aircraft. In 1985, Ryanair transported 5,000 passengers a year. It worked to reduce transport prices (see Table 8.4), and to increase the number of passengers and the routes served. Over the years, Ryanair has built up a reputation for punctuality, has attracted millions of passengers; it is now ranked first among European airlines by number of passengers. The operating revenue in 2007 was 32 % up from the previous year to € 2.2 billion and net profit was up 42 % to € 435 million [Ryanair 2007].

Ryanair's primary commitment to safety begins with the hiring and training of Ryanair's pilots, cabin crews and maintenance personnel, and includes a policy of maintaining its aircraft in accordance with the highest European airline industry standards. Ryanair has not had a single incident in its 20-year operating history.

Ryanair endeavours to control its labour costs by continually improving the productivity of its already highly-productive work force. In the year ended March 31, 2004 productivity calculated on the basis of passengers booked per employee continued to improve, increasing to 10,049 passengers, a 21 % improvement on the previous year. Compensation for employees emphasizes productivity-based pay incentives; the pay

Table 8.4 The lowest fare offered by Ryanair for the Dublin–London and return flight. Occasionally special offers are advertised at even lower prices

Year	Price of Return Ticket
1989	£ 94^{99}
1990	£ 69^{00}
1994	£ 49^{00}
2002	£ 30^{00}
2005	€19.98
2008	€19.99

Source: http://Ryanair.com

of Ryanair staff is 10–20 % higher than that of Air France, British Airways, Iberia, and Lufthansa.

Ryanair was Europe's original low-fare airline and is Europe's largest low-fare carrier. In 1999, Ryanair had 35 routes and 6 million passengers. In April 2001, Ryanair established its first operating base on the European continent in Brussels–Charleroi. In 2005, Ryanair carried over 34 million passengers on 220 low-fare routes across 19 European countries. In 2007, the routes were 545 and the passengers over 42 million. Ryanair has 22 European bases and its fleet of 132 brand-new Boeing 737–800 aircraft grew by 29 % in 2007, with option for a further 42 new aircraft, which will be delivered over the next years to raise the capacity to over 70 million passengers.

Expansion of the fleet is the basis for the growth of Ryanair's business, and cost control is the key to the profitability of its operations, which achieve a record load factor (83 % in November 2004) and outstanding customer service (see Table 8.5). Ryanair's strategy is to deliver the best customer service performance in its peer group. Ryanair has achieved punctuality and baggage delivery records through the use of secondary, low-traffic airports, only one type of aircraft, the elimination of connections, and the introduction of simple and effective registration and boarding procedures.

While big airlines like British Airways and Air France went through a critical period and other big airlines like Swissair and Sabena have gone bankrupt, Ryanair successfully established itself in Europe on the basis of a new business model: low-cost, no-frills flights (Fig. 8.23). The stock quotes of Ryanair shares (see Fig. 8.24) rose by 30 % in the unfavourable period from February 2000 to March 2004, during which the stock exchange index of the airline sector fell by 30 %. In a Morgan-Stanley report on Ryanair in March 2005 [Morgan-Stanley 2005], a volatile short-term performance is expected, but in the long term "we can comfortably support"—they wrote—"a valuation 40–50 % ahead of today's levels (€ 6.15 on 25 February 2005). Ryanair is operating a superior low-cost business model in Europe." The quote reached € 8.25 in May 2007 but closed the year at € 5.

Even though traditional airlines are reducing their fares to face the competition of Ryanair, Ryanair's offer is still unbeatable because this airline has found the secret to reducing costs. Ryanair has cut booking costs by reducing them from about € 8 down to € 1 per booking. There is now no intermediation of travel agencies and no issue of tickets. The traveller chooses the flight and buys the ticket online. Ryanair's

Table 8.5 Outstanding customer service puts Ryanair first among major European airlines. Punctuality is the percentage of arrivals within 10 min of the scheduled time. Missing bags is the percentage of non delivered bags per thousand passengers

Airline	Punctuality	Missing bags
Ryanair	90.2	0.39
Alitalia	81.6	13.9
Lufthansa	79.2	17.7
Air France	75.7	16.5
British Airways	73.8	20.5

Source: Associations of European Airlines.

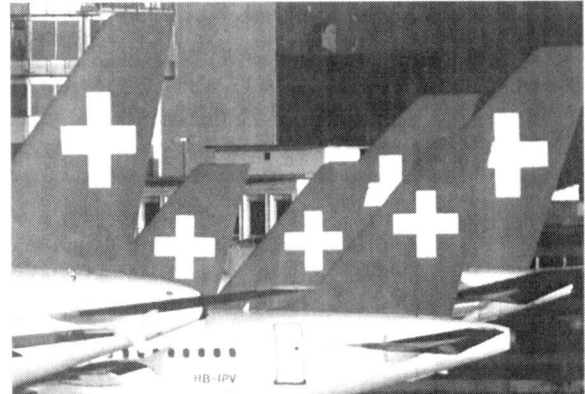

Fig. 8.23 Ryanair establishes itself while Swissair and Sabena go into bankruptcy
Source: "Black days" © *The Economist*, vol. 361, n. 8242, (4 October 4, 2001), reproduced with kind permission.

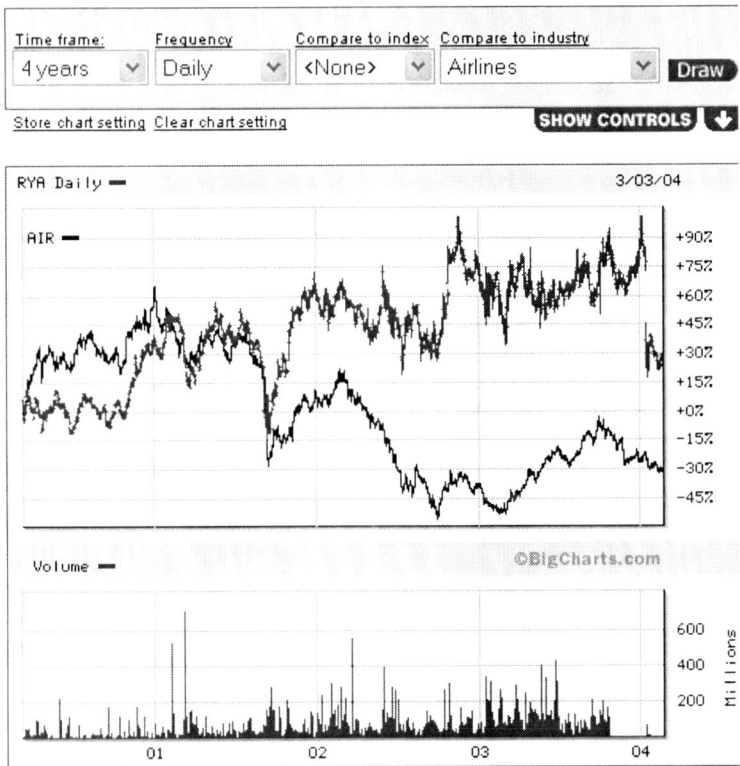

Fig. 8.24 Stock exchange quotes for Ryanair from 2000 to 2004, compared with the average of the airlines sector. A visible plunge after 9/11 recovers earlier for Ryanair than the airline sector's average. A price collapse of Ryanair in February 2004 was caused by allegations of illegal Belgian state aid in support of the Charleroi hub operations
Source: © *Financial Times*, 3 March 2004, reproduced with kind permission.

low fares are designed to stimulate demand, particularly from fare-conscious leisure and business travellers available to fly at odd hours, or in the middle of the week or who might otherwise have used alternative forms of transportation or would not have travelled at all. Ryanair sells seats on a one-way basis, thus eliminating minimum stay requirements from all travel on Ryanair scheduled services, regardless of fare. Ryanair sets fares on the basis of the demand for particular flights and by reference to the period remaining to the date of departure of the flight, with higher fares charged on flights with higher levels of demand and for bookings made nearer to the date of departure. In this way, the load factor is very high, revenues are profitable, and the number of daily flights of each aircraft is higher than the airline-sector average. During January 2000, Ryanair converted its host reservation system to a new one; as a result, internet bookings have grown rapidly, accounting for in excess of 96 % of all reservations on a daily basis as of September 2004.

Ryanair revenue is also generated through additional services offered to travellers. Also passengers paying only a nominal fare generate revenue through these services:

- joint ventures with hotels, car rentals, airport parking and downtown transport
- on-board sales of snacks and gift items
- on-board lottery

Ryanair can offer such competitive fares because it has managed to reduce costs by

using secondary, uncongested airports that offer competitive cost terms
eliminating intermediaries in the sale of tickets
cutting booking costs
offering point-to-point routes
offering separate booking for several stretches
eliminating connections
eliminating the assignment of seats
eliminating luggage transfer
reducing delays
using only one type of aircraft

8.9.6 Michelin

Michelin is a French-based multinational, world leader in manufacturing and distribution of tyres, and related products and services [Michelin 2007-1, Michelin 2007-2, Etienvre 2005]. Michelin's key figures are summarized in Table 8.6:

Michelin operates in a global oligopolistic market in which the three main actors (Bridgestone, Michelin, and Goodyear) share over 50 % of the world market. Michelin has a strong brand portfolio with two world-class brands (Michelin and BFGoodrich), strong regional brands (Uniroyal, Kebler, and Warrior), and well positioned private brands covering all market segments. Its headquarters are in Clemont-Ferrand, France; 62 % of its workforce is in Europe, 25 % in North- and South-America, 11 % in Asia,

Table 8.6 Michelin key figures

	2002	2003	2004	2005	2006
Net sales (€ billion)	15.7	15.4	15.0	15.6	16.4
Operating margin (%)	7.8	7.4	8.7	8.8	8.2
Net income to shareholders (€ bln)	.58	.32	.65	.89	.57
Number of shares (million)	133	141	143	143	143
Market capitalization (€ billion)	4.7	5.2	6.8	6.8	10.4
Last quotation (€)	32.9	36.4	47.2	47.5	72.5
Number of employees (thousands)					116
Production sites					69
Countries with production sites					19
Countries with commercial presence					170

Source: [Michelin 2007-1].

and 2 % in Africa. In China, Michelin increased its production in 2006 by 41 %. Michelin products and services are:

1. tyres
2. distribution and services
3. mobility enabling services
4. fleet management
5. suspension and pressure monitoring systems
6. maps and guides
7. lifestyle products

According to the world car industry forecast report quoted in [Michelin 2007-1], cars per inhabitant are projected to grow, between 2006 and 2011, by 8 % in Europe, and by 24 % in India, Brazil, and China; the number of cars in the world is expected to reach two billion by 2050, almost three times the number of cars in 2005. Michelin is actively building its position in this high-growth emerging market, focusing in satisfying local demand by expanding its production capacity, distribution networks, and assistance services. Michelin capacity expands in these regions faster than worldwide with a view to manufacturing tyres close to where they are sold and where labour costs are lower.

In a global economy characterized by increase in mobility, Michelin is taking full responsibility in environmental preservation, investing on road safety, tyre longevity, fuel efficiency, emission reduction, noise reduction, and recycling.

Michelin increased their sales in 2006 by 5.1 % in value, increasing prices on average by 4.7 % despite of a major increase in costs of energy, transport and raw materials (+30 %). Michelin products' prices are 15–25 % higher than those of their competitors because customers are prepared to pay the extra cost of Michelin's technological edge, superior quality, high service performance, and capacity for innovation.

Michelin achieved 20 % productivity gain by implementing a best practice program (Michelin Manufacturing Way, MMW) that transfers the best processes of the leading twelve plants in all remaining 57 plants. MMW has achieved

- reduction of work accident frequency and severity
- improved product quality

- reduction of inventory
- reduction of raw materials waste
- full control of pending orders and constant monitoring of productivity

Michelin invests in personnel training, with an average of 8 full-days-equivalent per year per person and in R&D, with a € 591 million budget, 3.6 % of net sales. Its 4,000 research engineers (3.5 % of the workforce) developed breakthrough solutions in many fields, including: fuel efficiency of tyres, longevity, and increased grip. Michelin introduced the radial technology that reduced rolling resistance by 30 %, a factor contributing by 20–30 % to fuel consumption. New "Michelin green tires" reduce rolling resistance by a further 20 %, and new solutions are coming down the pipe of innovation. New winter high-performance tyres are now on the market adding to innovations that ensure the technological leadership over Michelin's competitors. Michelin managed to garner a considerable store of World Championship titles in all major competitions, from Formula 1 to 24 hours of Le Mans, WRC-Rally world championship, and Paris-Dakar race.

Michelin is committed to sustainability and develops products that help reduce greenhouse emissions and noise, while increasing road safety. A major effort is in recycling used tyres: in Europe, since 2006, old tyres are no longer sent to landfills, but rather to end-of-life recycling industries.

Services are an increasing share of Michelin's business: they include fleet management and assistance in the event of a puncture, two added value services that contribute to the development of business and the captivation of large customers.

Michelin features in this survey of companies whose strategy faces the challenges of the knowledge-based society for its excellence in coping with globalization:

1. to reduce costs and protect employment, Michelin is increasing its capacity in emerging countries where the market is growing and labour is cheaper; conversely, headcount reduction in Europe and North America is mainly achieved with natural attrition
2. by means of MMW, the acquisition of a new manufacturing capacity in a remote location far from the French headquarters—a strategic move to keep manufacturing near to the source of raw materials and to developing markets—is soon brought to the level of excellence of the leading establishments and quality is guaranteed worldwide

Michelin features in this survey also because it is a very good example of the thesis of this book: investments in human resources and in R&D enable to offer quality products that justify business success with higher-than-market prices.

8.10 The Bottom Line

To close this survey of leading enterprises aiming at competitiveness in the knowledge-based society, let us summarize how results are achieved:

- operations are built around the corporate knowledge, capitalizing on the accumulated experience from markets, customers, products, and processes

- operations are concentrated in sectors where the enterprise is a world leader
- operations take full advantage of the use of ICT and of the Internet
- the enterprise values the human resource highly and invests in education and research
- some tasks that used to be performed by company staff are now performed by customers
- basic sales transactions are enhanced by the offer of new revenue-generating services
- transaction costs are reduced
- competition between suppliers increases
- product/service is improved/extended
- relationships with customers improves
- the market is widened

The Internet represents an industrial revolution as significant as the introduction of electricity, the car, or the telephone. This transformation has deeply changed production costs, working procedures, and trade practices.

It may not be essential to be among the first to adopt the Internet in order to be successful; however, it is not sufficient to have e-mail address and a web site to be part of the e-economy. ICTs are nothing but a tool to support a deep transformation of company organization; relationships with and between personnel; knowledge management; relationships with customers, suppliers, banks, and public administration.

At the end, a new business model is achieved, which may be very profitable and from which both the company and the customers will benefit.

This is today's promise of the Internet and it is certainly an objective worth pursuing.

8.11 Recommended Reading

Recommended reading for this chapter are listed in Appendix C and are available on the web site companion to this book, at the URL http://stajano.deis.unibo.it/RQC.htm

Chapter 9
The Position of the EU in the Quest for Competitiveness and Beyond

The European Union is a region with high labour costs. The costs for salaries, social services, education, training, health, and environmental protection can be only partly compensated by increasing productivity of the labour force, improving efficiency of public administration and taking full advantage of the internal market and the monetary union. The possibility of competing in world markets depends on the capacity to demonstrate the superiority of European products and services with reference to their quality, design, innovativeness, and ability to satisfy the requirements of an ever-changing market demand. The greater quality of European products and services might give them the possibility of being competitive despite their cost, originated by high salaries and high standard of living within the EU.

We found in Part One that if we compare EU and U.S. economies, the main differences include: the standard of living, the provision for health care, unemployment benefits, and pensions; the level of participation in productive activities, the level of female employment, the labour productivity, the mobility in the labour market; the level of education, computer literacy, vocational training, and lifelong learning; the diffusion of information and communication technologies; the technological infrastructures and their relative costs; the level of private and public investments in research and development; the companies' access to financial markets and their capacity to reorganize, innovate, and face the challenges of globalization and the knowledge-based society; and – last but not least – a regulatory and institutional framework promoting innovation, mobility, investments, and entrepreneurship.

The comparison also showed that, even though European industry is behind U.S. in key sectors linked to the new economy, it is successful in other sectors, including some high-tech sectors. It is primarily in mature sectors that the European industry beats the U.S. competition. In most cases, the competitive advantage derives from quality rather than price. However, the challenge for the European economy does not come only from the U.S. Analysing the world economy, we noted that emerging countries, and particularly the BRICKs, are growing in importance and are overthrowing the economic and trade relationships. In addition, a new phenomenon is changing the market demand worldwide: an ever growing number of price-conscious customers value *good enough* products better than *best* products, if sufficient quality is offered at budget price.

That large part of the European industry that has acquired a leading position in the international markets in those sectors where the technological challenge has been less aggressive will have to face a new form of competition, in which the maintenance of the competitive advantage will depend on the capacity to integrate innovative content into processes, products, and services. In these sectors – that include those producing household appliances and cars – a strong competitive offer from the emerging countries is present. The EU industries will be able to maintain their competitive position only by increasing high-tech content and innovation.

On the other hand, those companies that have acquired a leading position in the international markets for complex technological products (aerospace, transport, chemical sectors) but that are highly dependent on the acquisition of patents and licenses will have to face new challenges linked to the acceleration of the innovation process and the consequent difficulty in offering products with the most advanced technology content without having an endogenous innovation capability.

In the EU member states where enterprises have low innovation propensity, signs of decline have already been apparent, and negative effects were visible on low-tech activities, too. Because of the ubiquity of technology, a competitive advantage is linked to the introduction of advanced technology also in products and services traditionally considered low-tech (e.g. ceramics, textile industry, shoes, furniture, and tourism). Even low-tech products benefit from high-tech processes used to manufacture or to market them.

We argued that the EU, in order to maintain its share in the global market, ensure sustainable growth and employment, and face the knowledge-based society, should invest in education, training, lifelong learning, research, and innovation; and improve ITC infrastructures, diffusion, and literacy. We argued that lifelong learning is fundamental, not only for the competitiveness, and economic prosperity of the EU, but also for social inclusion, employability, active citizenship, and the personal fulfilment of people. Individuals must be able to update and complement their knowledge, competences, and skills throughout life. We proved that higher investments in R&D are correlated with best trade performance for high-tech products, the fastest-growing segment of international trade.

In the year 2000, the European Union realized that her prosperity was dependent on the ability to face the challenge of the knowledge-based society and formulated the Lisbon Strategy. In the following years the strategy was revamped correcting its flaws, as we described in Chapter 8, "Competitiveness in the Knowledge-based Society". The EU has now placed centre stage the building of a competitive economy in the knowledge-based society and has adopted for the Lisbon's objectives new, strong, and effective instruments supported by major financial investments. The seventh Framework Program – one of the main instruments to implement the strategy, described in Chapter 6, "Framework Programs" – is not only the largest ever financial support to Community R&D, but an initiative designed with an innovative approach and a stronger connection to other Community actions than previous FPs. New features in FP7 include the institution of the European Research Council, a funding agency for frontier research and the European Institute of Technology, a flagship for excellence in innovation, research, and higher education. Community funded R&D can do more that enhancing European competitiveness, it can contribute to EU integration. In fact, although the main objective of EU research policy is to strengthen the scientific and technological basis of EU industrial activities in order to make them more competitive, R&D also contributes to the qualification of the workforce, the realization of the internal market, and the cohesion and integration of the member states.

EU research programs have contributed to EU institutional transformation and have created a very widespread, top-quality European industrial community, proven, and reinforced by hundreds of thousands of person-years of transboundary research collaboration, which has allowed scientists, industrialists, researchers, and Ph.D. students of the different EU countries to know and respect each other and to exchange new ideas and experiences, thus creating the conditions for partnerships and mergers between different companies.

The industrial research community has innovated not only the technological base, which is essential for competitive growth, but also the world of ICT users, thus preparing the Union for the information society and expediting the convergence processes in communication, information processing, the content industry, and the media. Research has achieved important technological results that have been integrated into systems and applications and have created the conditions for the development, transformation, and success of the companies involved. The process initiated by the Framework Programs is a catalyst for innovation inside companies and is particularly important not only for its visible, successful results but also for the transformation that it triggers in individuals, companies, and countries involved.

EU research takes industrialists and researchers outside their own environment and brings them into contact with their most expert European colleagues in a climate of intellectual and technological competition and in a context far from their daily routine, in which new ideas can be conceived and implemented. The EU Framework Program allows a company to take risks that it otherwise could not take by using its own skills and resources and leads industrialists and researchers to think outside the box in an open confrontation far from the direct control of their daily working environment.

So far the member states' response to the Lisbon Strategy is far from homogeneous. Some countries prepare their long term prosperity with large investments in education and R&D, other countries lag far behind. The accession countries demonstrated their ability to benefit from full membership and posted higher GDP growth than the EU average. But this might be a flash in the pan, if the long-term future is not secured by overcoming the labour skills mismatch and their unpreparedness to face the knowledge-based society, two recurrent remarks in the survey of the accession member states three years after the enlargement presented in Chapter 11, "Enlargement of the European Union". These countries, and – to a major extent – the candidate countries are undergoing structural transformations that increase the productivity of the agricultural sector and of other productive activities not tuned with the new economy. Such transformations create redundancies and mismatches in the labour market with the risk to exclude a non negligible part of the population from active productive employment. Only very focused educational and lifelong learning programs can transform the above mentioned redundancies in valuable resources contributing to the personal, economic, and social development of the persons involved and avoiding exclusion of the young generations and of the elder workers not fit to the changing structure of the productive society.

The evolution and revamping of the Lisbon Strategy is happening in a decade when the EU is undergoing the major institutional changes of enlargement and reform of the Treaty. The *political* Union, an idea waiting to happen for 50 years might eventually come about by tying the loose ends of the Nice Treaty with the ratification of the reform Lisbon Treaty on EU, a process analysed in detail in Chapter 10, "From the Treaty of Rome to the Reform Treaty of Lisbon". We saw that both on implementing the Lisbon objectives and in pursuing the institutional changes, member states are not harmonized, synchronized, and unanimous. The vision of multi-speed Union is taking ground and manifests itself in prioritizing national interests to shared values and in sharpening institutional instruments such as the enhanced cooperations to enable overcoming the stumbling blocks introduced by reluctant members.

A multi-faceted Union was visible during the 2007 negotiations for the Lisbon Treaty; in the constitution of groups of three (or – respectively – four) countries in dealing with the crisis in Iran; in the year-end 2007 economic crisis; in member states' attitude towards enlargement to Turkey; and – ex post – in the member states' economic performance and in their educational, research, and other Lisbon-strategy-related attainments. In facts the multi-faceted nature of the EU is in her genes and, by the way, in her motto: "united in diversity." European diversity is not a problem, rather one of her strengths that makes EU fit to cope with the complexity and variety of the globalized world.

The leading position of European northern countries that posted their economic and technological performance over and above the U.S. is a testimonial of Europe's capability to compete worldwide; however, there is no place for complacency, because, in a fair comparison, individual American states should be considered separately, in which case only one European country would rank among the first 10 most competitive economies of the knowledge-based society. The Asian competition is aggravating the long-term European firms' market position.

Can the whole of the EU find ways to successfully position herself in front of her competitors? Is there a way to prepare a bright future for the next generations of European citizens? Or is the time come to give way to new countries which can take the economic lead that was in the hands of the European people for centuries? Answers will become gloomy if the ongoing institutional transformation does not find a successful solution with the ratification of the Treaty on EU before the 2009 EP elections. A strong and united Union would on the contrary be able to focus her action on the building a better future for her citizens and strengthening her external action with the reinforced instruments of the full time standing president and the high representative for the Union foreign affairs and security policy. R&D can help overcoming the fragmentation of a multi-speed EU: R&D policy can help creating a bottom-up process of cohesion among the components of a "variable geometry" EU. Research programs have created a community of scientists, industrialists, and researchers united in temporary and voluntary associations based on common development objectives. These associations have often defined the rules for integration on an industrial scale rather than just at the level of research. Working together and with the help of EU funding, some partnerships have been created from the bottom up and have contributed to the implementation of an EU-wide process of innovation, and cohesion that has proven to be much more effective than any top-down action. On a broader scale, R&D policy has been implemented in the early 1990s anticipating institutional instruments, such as majority voting, before their enshrinement in the Treaty on EU.

It does not sound realistic to expect that the proposed remedy based on education, lifelong learning, research, and innovation will solve all problems for all EU member states. However, it might reduce the aggravation of a trend that has seen several EU countries, mainly those on the "garlic belt" lose positions in the WEF rankings and decline under the combined pressure of the most advanced countries in the Triad and of the emerging economies.

But even if the EU could at most succeed in maintaining her current position, Europe shall continue to have a word to say in proposing to the world a model of development and a system of values. The author is convinced that the European citizens, under the EU flag, can ensure that Europe continues to be one of the reference points in the word and inspire mankind with her values, her vision, her dream of a better future funded on solidarity, multicultural tolerance, peace, and prosperity. It is encouraging that a landmark book on "the European dream" was written by an American author [Rifkin 2004].

Some heartening evidence of the contribution that the people of Europe can give to the construction of a better world when they anticipate the use of the new instruments of foreign policy that the institutional transformation of the EU is offering are epitomized by important achievements of the recent years: the United Nations interim peace-keeping force in Lebanon (2006); the United Nations General Assembly resolution – following a EU proposal – calling for a global moratorium on executions with a view to eventually abolishing the death penalty entirely (2007); the EU leadership in securing success of the United Nations Climate Change Conference in Bali, Indonesia (2007); the message from the EU to the rest of the world with the ongoing EU enlargement negotiations with Turkey, telling that EU supports the way to democracy with solidarity and assistance and that the "Clash of Civilizations" is not the ineluctable destiny of mankind. The EU offers an alternative model to the exclusive, sectarian, and closed society propagated by religious radicalism and to the pipe-dream to export democracy with the instruments of war.

9.1 Recommended Reading

Recommended reading for this chapter are listed in Appendix C and are available on the web site companion to this book, at the URL http://stajano.deis.unibo.it/RQC.htm

Part III
Overview of the Institutional Context of the European Union

Chapter 10
From the Treaty of Rome to the Reform Treaty of Lisbon

The European Union has undergone profound changes over the last 15 years. Today, after her enlargement to 27 members and the reform of her treaty, the Union is preparing to face the challenges of the 21st century with renewed instruments. The situation at the time of writing (Year End 2007) is in a state of transition because the new *Reform Lisbon Treaty,* finalized by an Intergovernmental Conference during 2007 and signed by the heads of state and prime ministers on 13 December 2007, has not yet been ratified by the member states. The Constitutional Treaty, agreed to and signed in 2004, has not been enforced, as its ratification by the member states came to a halt in 2005, following the negative outcome of referendums in France and in The Netherlands. For the time being, the preexisting institutional structure is in force. In this chapter, we will describe the institutional structure as will result from the Reform Lisbon Treaty, but we will also occasionally mention the current situation as it follows from the Treaty of Nice now in force.

10.1 Evolution of the Treaties

At the end of the Second World War, a number of European leaders became convinced that the only way to secure a lasting peace between their countries was to unite them economically and politically. So, in 1950, the French foreign affairs minister Robert Schumann proposed integrating the coal and steel industries of western Europe. As a result, in 1951, the European Coal and Steel Community (ECSC) was set up, with six members: Belgium, France, West Germany, Italy, Luxembourg, and The Netherlands. The power to take decisions about the coal and steel industry in these countries was placed in the hands of an independent, supranational body called the *High Authority*. Jean Monnet was its first president [Fontaine 2003].

10.1.1 The Treaty of Rome

The ECSC was such a success that, within a few years, these same six countries decided to go further and integrate other sectors of their economies. On 25 March 1957,

A. Stajano, *Research, Quality, Competitiveness,*
© Springer Science+Business Media, LLC 2009

they signed the Treaties of Rome, creating the European Atomic Energy Community (Euratom) and the European Economic Community (EEC). The member states set about removing trade barriers between them and forming a *common market*.

The Treaty of Rome founding the European Economic Community (EEC) came into force on 1 January 1958. It set the task for the European Community:

> by establishing a common market and progressively approximating the economic policies of member states, to promote throughout the Community a harmonious development of economic activities, a continuous and balanced expansion, an increase in stability, an accelerated raising of the standard of living and closer relations between the states belonging to it.

In 1967, the institutions of the three European communities (ECSC, EEC, and Euratom) were merged. From this point on, there was a single Commission and a single Council of Ministers as well as the European Parliament.

The Treaty of Rome dealt mainly with economic and trade matters. Customs duties between the six countries were completely removed on 1 July 1968 and common policies – notably on trade and agriculture – were also set up during the 1960s.

10.1.2 The Treaty of Maastricht

The Treaty of Rome was changed several times: by the Single European Act (1986) establishing, by 1992, freedom of movement of persons, goods, services, and capital; and on the occasion of the entry of new member states. The most important institutional reforms were signed in Maastricht on 7 February 1992 with the adoption of the Maastricht Treaty, which became effective on 1 November 1993. This treaty extended the range of action of the European Community, introducing common foreign and security policy and monetary union with a single currency; it strengthened the Community institutions and gave them broader responsibilities. With a view to underlining the move beyond economic and trade matters, the treaty changed the name of the European Economic Community to the European Union. The Treaty of Maastricht enshrined, for the first time, the principle of *subsidiarity*, which is essential to the way the Union works. It means that the EU and her institutions act only if action is more effective at EU level than at the national or local level. This principle ensures that the EU does not interfere unnecessarily in her citizens' daily lives. The identity of European countries is a valuable asset to be preserved: the integration of member states into the European Union must never be confused with a process towards uniformity – which is something Europeans definitely reject [Fontaine 2003]. The motto of the EU is *united in diversity*. While the new members of the United States of America that joined the 13-states nation established in Philadelphia in 1786 were assimilated into the culture, language and values of the founding fathers, the European countries joining the EU are integrated into the European Union preserving their identity, their language and their cultural heritage. Diversity is to be considered an asset rather than a problem; it does increase complexity and introduces some intricacies, it helps, however, the European society to face the challenges of globalization.

The Maastricht Treaty and its subsequent amendments (Amsterdam 1997 and Nice 2001) are referred to by the name *Treaty on European Union*. The Treaty on European Union regulated the relations between the 15 member states (EU15) constituting the EU at the time of its enforcement: Austria, Belgium, Denmark, Finland, France, Germany, Greece, Ireland, Italy, Luxembourg, The Netherlands, Portugal, Spain, Sweden, and the United Kingdom. In the transition period until the enforcement of the Reform Lisbon Treaty, the Treaty on European Union regulates the relations among all the 27 current member states, namely, those just mentioned and the ones that accessed the EU in 2004 (the Czech Republic, Estonia, Cyprus, Latvia, Lithuania, Hungary, Malta, Poland, Slovenia, and Slovakia) and in 2007 (Bulgaria and Romania). The Treaty on European Union introduced a three-pillar structure that characterized the institutional organization until 2004. The three pillars are the Community pillar, the Common Foreign and Security Policy pillar, and the Justice and Home Affairs pillar.

The first pillar deals with the internal market common agricultural policies, cohesion, trade, social affairs, economic and monetary union, and single currency; economic and social cohesion, environment, research, and education; and cooperation and development. The second pillar regulates political union and common foreign, defence, and security policy. The third pillar regulates cooperation in law enforcement, criminal justice, civil judicial matters, and internal affairs; and common policies on asylum and immigration.

The member states of the EU have established cooperation treaties with other European countries, with the aim of further extending the European common market and of paving the way to subsequent Union enlargements. This has led to the European Economic Area Treaty and the Agreements for accession.

The European Free Trade Association (EFTA) Convention (1960) and the European Economic Area (EEA) Agreement (1993) regulate relations between the states of the Union and Norway, Iceland, the Swiss Confederation, and Liechtenstein, guaranteeing the free movement of persons, goods, capital, and services.

The Agreements for accession to the Union regulate relations between the European Union and countries requesting access to the EU. They deal with trade, environment, culture, research, provisions concerning Union enlargement, and political consultations (e.g. thirteen countries candidate for accession were part of the European Convention).

10.1.3 The Treaty of Amsterdam

The Treaty of Amsterdam (1997) [EC 2002–8] extended the Treaty of Maastricht and laid the foundations for the radical change to be completed with the Lisbon Treaty. The Treaty of Amsterdam set the rules for governments and defined citizens' rights, paved the way to a more democratic and more social Europe, and introduced considerable changes in common foreign policy and the free movement of goods. It provided new guarantees for the protection of fundamental rights in the

European Union, particularly with respect to gender equality, non-discrimination, and treatment of personal data.

The Amsterdam Treaty had five major objectives:

1. putting the Union's focus on employment and citizens' rights
2. eliminating the last remaining obstacles to free movement
3. strengthening security
4. enabling Europe to exert a stronger influence at the world level
5. improving the effectiveness of Union institutions with a view to the forthcoming enlargement

By signing the Amsterdam Treaty, governments agreed to undertake concrete measures to coordinate their economic policies and bring their national employment policies into line with the Community's economic policy. They also committed themselves to ensuring a qualified workforce, capable of adapting to changes, as well as labour markets capable of rapidly reacting to transformations in the economy. Every year, the member states' governments must submit, according to the Stability and Growth Pact, a three-year forecast about their public finances and Community institutions carry out controls to harmonize economic policies among member states. Concerning this point, see Chapter 14, "Economic and Monetary Union".

The Amsterdam Treaty was envisaged in order to eliminate the last remaining obstacles to free movement. The Amsterdam Treaty and its subsequent modifications enforce the Schengen Convention (for all EU15 member states with the exception of the United Kingdom and Ireland and including also Iceland, Norway, and the Swiss Federation) [Council 2000–1] regarding the abolition of border checks on roads and at airports, and strengthen security measures. A large number of common regulations on visas, asylum rights, checks at external borders, and cooperation between police forces and customs authorities have been laid down to ensure that free movement of persons across the Schengen area does not adversely affect law and order. The Schengen Convention was extended on 21 December 2007, with what can be seen as the final lifting of the Iron Curtain between the former Soviet bloc and the West. The Schengen zone, within which people can travel without showing passports, extended to nine new countries, namely the eight eastern and central European countries of the 2004 enlargement plus Malta. The Schengen zone includes now 25 countries: the move will boost business and tourism, while measures are in place to strengthen the control at the external borders. The move is also an important step towards the full integration of the European citizens of the enlarged Union.

Cooperation between police forces and the other authorities dealing with criminal matters, once based on intergovernmental conventions is now being incorporated in the EU legislation by means of the Prüm Treaty, an international police cooperation agreement signed (on 27 May 2005) by Belgium, Germany, Spain, France, Luxembourg, the Netherlands, and Austria. It regulates the stepping up of cross-border co-operation, particularly in combating terrorism and cross-border crime, with information exchange on DNA-profiles, fingerprints, and vehicle number-plates. The same goes for judicial cooperation on criminal matters and cooperation in the fight against terrorism, organized crime, crimes against people and children, drug and arms traffic, international fraud, and corruption.

The Amsterdam Treaty was devised to resolve the contradictions emerging from the first disappointing experiences at the international level that took place after the Maastricht Treaty had given the Union responsibilities in matters of foreign and security policy. Heads of state and government agree on common strategies, which are to guide the Union's actions. The European Council decides, usually by unanimity, which actions should be taken. The president of the Council represents the Union and is assisted in this task by the Secretary General of the Council of the European Union, who helps enforce political decisions and has the role and title of High Representative for Common Foreign and Security Policy. There was still much to be achieved before the Union managed to speak with one voice, as demonstrated in 2002 and 2003 by the split of EU member states on the issue of war in Iraq. A major milestone was achieved in 2007 with the Lisbon Treaty, instituting the High Representative of the Union for Foreign Affairs and Security.

The Amsterdam Treaty considerably increased Parliament's responsibilities, and the codecision procedure is now applied in almost every case. The *assent* of the EP is required for special cases such as: sanctions that could be imposed by the Council on a member state in case of serious and persistent violation of basic rights; applications for membership; major international agreements; and the adoption of a common procedure in the election of members of Parliament.

For about a decade, there has been a progressive increase in the number of areas in which decisions can be made by qualified majority voting. With a view to the forthcoming enlargement, the Treaty of Amsterdam (and later the Treaty of Nice) extended the areas in which decisions can be made by a qualified majority. However, unanimity remains the rule in decision-making regarding constitutional matters and particularly sensitive problem areas such as tax issues.

The Amsterdam Treaty assigned more powers to the president of the Commission. He is appointed by the heads of state and government, but his election becomes effective only after the European Parliament has given its consent. Once he has been confirmed, the new president will designate, in agreement with the governments, the other members of the Commission, who are jointly appointed as a group after Parliament approval.

The Treaty of Amsterdam created the formal possibility of a certain number of member states establishing enhanced cooperation between themselves on matters covered by the Treaties, using the institutions and procedures of the European Union.

10.1.4 The Treaty of Nice

The European Council agreed on the Treaty of Nice [EC 2001-6] in December 2000; the treaty was signed on 26 February 2001 and became effective on 1 February 2003. It implemented further changes relating to the functioning of the institutions and replaced unanimity voting with qualified majority voting in many decision-making areas. The Treaty of Nice also brought changes in areas such as common foreign and security policy and the composition, functioning, and role of Community institutions.

The institutional reform approved in Nice has been described as technical and limited. The treaty, in fact, does not change the institutional balance but rather makes some adjustments, mainly on the functioning and composition of the institutions and regarding enhanced cooperation.

One of the main issues concerns the Council of the European Union and the new weighting of votes in a 27-country Council. The provisions of the treaty are more advantageous to medium-sized countries, such as Spain and Poland. In 2002 and 2003, when the reform of the treaties was discussed at the Intergovernmental Conference, these countries did not want to give up the favourable conditions under the Treaty of Nice. They prevented the approval in 2003 of the European Convention proposal regarding – inter alia – new criteria for a qualified majority decision-making system.

The Treaty of Nice also increased the number of areas in which a group of member states (at least half of them) can establish *enhanced cooperation*. As a result, closer cooperation is favoured among those countries in the Union that want to extend their relations beyond the form of integration envisaged by the treaties and outside the Union institutional context, thus setting their own pace and objectives. An example of enhanced cooperation is the Schengen Convention. We will see that the Treaty of Lisbon revises this possibility with a view to adapting it the context of the enlargement of the Union to 27 member states.

10.1.5 The European Convention

The Treaty of Nice laid the ground for the preparation of a reform of the Treaties, which had become necessary given the Union's impending enlargement. To this end, the Council of the European Union set up the European Convention [Council 2002]. The Convention started its work in March 2002 and presented its conclusions in July 2003.

Ambassador Renato Ruggiero, minister of foreign affairs of the Italian Republic until the end of 2001, made the following statement at Montecitorio Palace (the lower Italian Chamber) on 31 October 2001 regarding the issue of reform of the treaties:

> We must work for the reorganization and constitutionalization of the Treaties, which are a source of transparency and democratic legitimacy. The European Constitution will have to incorporate the Chart of Fundamental Rights, as well as some provisions on the European Union's institutions, objectives and competences, and will be founded on the basic principles of solidarity and subsidiarity.
>
> The European Constitution will provide present and future member states with an ethical code and will strengthen the Union's civil and material identity, founded on the respect of national identities, essential both for the states and for Europe. In this context, the model of a federation of national states is an advanced compromise between the different views and perspectives within the Union.

The present institutions should not be radically changed, but rather improved and strengthened so that the entire system can work in a more democratic way.

I think that we will also have to work in order to:

- respect and extend the community approach (I'm thinking, in particular, about the future of our common foreign, defense and security policy) and extend the use of qualified majority voting in the Council, as suggested also by the Italian president Ciampi;
- think about the possibility of having a president of the Commission elected by the European citizens as a starting point to strengthen the role and legitimacy of the Commission, which is the engine of integration and guarantees compliance with the treaties. Or else, the Commission president should be elected by the members of the European Parliament;
- rationalize the work of the Council and strengthen the role of coordination of the Council of General Affairs;
- strengthen the role of the European Parliament, which ensures the democratic legitimacy of all the decisions made by the Union, by extending codecision to all legislative matters;
- improve relations between the European Parliament and the national parliaments.

10.1.5.1 Composition of the Convention

The Convention was joined by the main players involved in the debate on the future of the Union. The European Council appointed Mr. Valéry Giscard d'Estaing (a past president of the French Republic) as president of the Convention and Mr. Giuliano Amato (a past prime minister of the Italian Republic) and Mr. Jean-Luc Dehaene (a past prime minister of the Kingdom of Belgium) as vice-presidents. Besides the president and the two vice-presidents, the Convention comprised:

- 15 representatives of the heads of state or government of the member states (1 for each member state),
- 13 representatives of the heads of state or government of the countries applying for accession[1] (1 for each country),
- 30 representatives of the national Parliaments of the member states (2 for each member state),
- 26 representatives of the national Parliaments of the countries applying for accession (2 for each candidate country),
- 16 representatives of the European Parliament, and
- 2 representatives of the European Commission.

The Convention met several times between March 2002 and June 2003 and put forward its recommendations in a new draft of the constitutional treaty, which will replace and amend the content of the current treaties. The draft was presented on 19 June 2003 by Giscard d'Estaing at the European Council of Thessaloniki, Greece, which welcomed it as a good starting point for future amendments to the treaties. The final text drawn up by the Convention (see recommended reading at the end of

---

[1] In the year 2002, the countries whose application for accession had been accepted were: Bulgaria, Cyprus, Czech Republic, Estonia, Hungary, Latvia, Lithuania, Malta, Poland, Romania, Slovakia, Slovenia, and Turkey.

this chapter) was given to Silvio Berlusconi, then president of the Council of the European Union, in Rome, Italy, in mid-July 2003. The final announcement made by the Presidency of the European Council of Thessaloniki, Greece (19 June 2003), reads:

> The European Council welcomes the Draft Constitutional Treaty presented by the president of the Convention, Valéry Giscard d'Estaing. This presentation marks a historic step in the direction of furthering the objectives of European integration:
>
> - bringing our Union closer to her citizens,
> - strengthening our Union's democratic character,
> - facilitating our Union's capacity to make decisions, especially after her enlargement,
> - enhancing our Union's ability to act as a coherent and unified force in the international system and effectively deal with the challenges globalization and interdependence create.

10.1.5.2 The 2003–2004 Intergovernmental Conference

The amendments to the Treaty on European Union had to be agreed upon by the governments of all member states. The process of debating and amending the treaty is known as Intergovernmental Conference (or IGC), a conference of the government representatives of member states. An IGC is called by the president of the Council of the EU, on the recommendations of the Council and after consulting the European Parliament, the Commission, and, if required, the European Central Bank.

On 29 and 30 September 2003, the European Council approved the convening of an IGC, which opened in Rome on 4 October 2003 with a meeting of the heads of state and government. The sessions of the IGC are attended by the ministers for foreign affairs, and chaired by the minister for foreign affairs of the country holding the rotating presidency of the Council.

The Italian Presidency (second semester of 2003) had hoped for a conclusion of the work of the Intergovernmental Conference by December 2003. However, it was not possible to reach an agreement on all the controversial issues. The diversified constituency of the Convention had expressed the position of actors not represented in the IGC. It is no surprise that the governments of the member states strenuously defended the privileges accorded to their countries by the treaties in force. The most heated debate was about the weight of votes of each member state in the European Council. Spain and Poland defended the Treaty of Nice and the decisions made on that occasion regarding weights of countries in majority voting, and did not accept the proposal put forward by the European Convention.

The failure of the European Council summit of December 2003 cannot be attributed only to the ineffective negotiations led by the Italian Presidency. There was (and still is) a profound split between member states concerning the war in Iraq, in particular between the countries in favour of keeping within the law and respecting the role of the United Nations (the group led by France, Germany, and Belgium) and the pro-Americans and interventionists (the group of countries led by Great Britain and Spain, and supported by Italy, Poland, and almost all other enlargement countries). The weight of France and Germany in the IGC, both in favour of the Convention's draft of treaty, was weakened by the role played by these two countries in the Economic and Financial Affairs (Ecofin) Council of November 2003, which led to a weakening of the EU through the watering down of the Pact for Growth and Stability, the same pact that France and Germany had vigorously defended when other (smaller) countries were going to be sanctioned. On 13 December 2003, after the failure of the European Council summit, President Prodi said [EC 2003-4]:

> Today, no agreement has been reached. It would have been a setback, and nobody wanted it. Now, we need a pause for reflection in order to prepare future ideas. The spirit of the Convention is still alive. We agreed on almost every item on the agenda. Not on all items.

Everybody agreed in 2003 that it was becoming ever more urgent to reach agreement on the constitutional treaty, especially since a new European Parliament had to be elected and enlargement to 25 countries had officially started. This urgency was emphasized in January 2004 by the then president of the Commission, Romano Prodi [Prodi 2004]; by the then president of the Council, the Irish prime minister, Bertie Ahern [Ahern 2004]; by the commissioner responsible for institutional reforms, Michel Barnier [Barnier 2004]; and on different occasions by the president of the Italian Republic, Carlo Azeglio Ciampi [Ciampi 2003-2, Ciampi 2004].

Nevertheless, the possibility that a constitutional treaty could enter into force before the installation of the new Parliament, in the summer of 2004, was soon ruled out. As a matter of fact, it was apparent that even if participant countries could manage to reach a unanimous agreement on the text of a treaty during the Irish Presidency before the EP elections (which they did), such a treaty would only become effective after the ratification by the 25 member states.

The negotiations of the Irish Presidency, during the first semester of 2004, were facilitated by the change in attitude of the new government in office in Spain after the political elections in March 2004: the new prime minister, Mr. José Luis Rodríguez Zapatero, adopted a different approach towards Europe and the U.S. from that of the outgoing prime minister, Mr. José Aznar. This led the Polish prime minister in office at the time, Mr. Leszek Miller, to change his cabinet's attitude in the European negotiations in order not to be isolated in a negative position. The Irish Presidency was able to successfully issue an agreed-upon Constitutional Treaty at the June 2004 European Council. However, the process towards establishing a new Treaty on EU had not come close to finishing, yet. The Treaty, carefully reviewed and translated into the 20 languages of the enlarged European Union, was eventually signed with great pomp and ceremony in Rome on 29 October 2004 in the context of what was referred to as Berlusconi's *catering diplomacy* [Peel 2004], to depict the unrivaled ability of Berlusconi to organize wining and dining while other leaders write down the political agreements.

10.1.6 The Constitutional Treaty

A major transformation was intended to be introduced on 29 October 2004 in Rome with the signing of the Constitutional Treaty[2]. The Constitution was conceived as an

---

[2] Following an established practice, the Constitutional Treaty will be also called in this book – for short – *the Constitution*.

evolution of the Treaty on European Union that would introduce significant changes, with a view to creating an efficient institutional structure, capable of coping with the challenges of an enlarged Union. The Constitution put forward a single text to replace all existing Treaties, achieving clarity and readability, in addition to other major substantial improvements described later.

After being signed, the text of the Constitution should have been ratified by each of the signatory countries in compliance with their constitutional procedures. This process was interrupted following the negative outcome of referendums in France and in The Netherlands. On the basis of the legal and historical traditions of EU countries, the Constitution ratification would have required either one or the other or even both of the following two types of procedures:

- the parliamentary way: the text is approved after one or two houses of Parliament in the country have voted on a text regarding the ratification of an international treaty;
- referendum: a referendum is called and the citizens themselves are asked to vote either for or against the ratification of the Treaty.

There might have been a need to modify or combine the two above-mentioned procedures if so demanded by the constitutional law of a given country, or it might have been necessary in a given country to undergo beforehand a modification of the national Constitution if the latter was incompatible with the Union's Constitutional Treaty. This was, for example, the case for the French Republic.

The Treaty would have entered into force if it had been ratified by all member states and once this ratification had been officially notified by all signatory countries, it would have become effective. In 2004, it was assumed that this process would be completed by 1 November 2006.

The possibility that some member states might not have ratified by that time was considered in 2004 by some political scientists as a challenging problem, since the Union would face a crisis for which no legal instruments appeared to be available. Mario Monti, at the end of his term in office as competition Commissioner, proposed [Bonanni 2004] that non-ratifying countries should consider leaving the Union. However, although the Constitution foresees, for the first time, a *voluntary withdrawal* clause, this clause cannot be applied by countries refusing to ratify, because it would not be applicable without the Constitution entering into force. Lucia Serena Rossi, a political scientist of the European Policy Center in Brussels, suggested [Rossi 2004] several ways out of this *less than theoretical* problem, one of which would be to sneak a clause of voluntary withdrawal into the Accession Treaty for Romania or Bulgaria.

The Constitutional Treaty was made of four parts and a number of annexes and protocols. The full text is in the recommended reading web site companion to this book:

- Part One: Fundamental provisions: definition and objectives of the EU; citizenship, the democratic life; Union competences, Union institutions, Union finances, Union neighbors, Union membership

- Part Two: The Charter of fundamental rights
- Part Three: The policies and functioning of the Union
- Part Four: General and final provisions

In April 2005, the European Commission reported that Lithuania, Hungary, Slovenia, and Italy had ratified and that the progress in the process of ratification was: four countries had ratified, and a fifth (Spain) was in the process of doing so. A consultative referendum had been very successfully held in Spain in February and would be followed by parliamentary ratification in May. Later in the first semester of 2005, also Greece, Bulgaria, Austria, Germany, Slovakia, Romania, Belgium and Cyprus ratified the Constitutional Treaty. However, a binding referendum due in France on 29 May 2005 was creating some anxiety because of unfavourable opinion surveys. According to the opinion of unnamed Commission officials, reported by the *Financial Times*, "a rejection of the Treaty by a large founding member would kill it" [Parker et al. 2005–2].

On 29 May 2005, the referendum in France said *no* to the Constitutional Treaty causing damage beyond repair to the ratification process. On 1 June 2005 also, the people of the Netherlands choose to say "no". In the light of these results, the European Council (17 June 2005) with a view to finding ways around this stumbling block, proposed a time of reflection, explanation, and discussion. They declared that "we do not feel that the date initially planned for a report on ratification of the Treaty, 1 November 2006, is still tenable, since those countries which have not yet ratified the Treaty will be unable to furnish a clear reply before mid-2007." The Union would continue to be ruled according to the Treaty of Nice, which most constitutionalists as well as several politicians consider would not be effective with a Union enlarged to 25 or 27 members.

On 6 July 2005, Malta ratified the Constitutional Treaty with a parliamentary procedure. On 10 July Luxembourg voters voted *yes* to ratification. On 9 May 2006, Estonia ratified the Constitutional Treaty. In December 2006, the Finnish Parliament ratified the Constitutional Treaty, bringing to 18 the number of states which have ratified it: 16 member states and two candidate countries that would access in 2007. Seven member states suspended the ratification process: the Czech Republic, Denmark, Ireland, Poland, Portugal, Sweden, and the United Kingdom (see Table 10.1).

10.1.7 The Reform Treaty of Lisbon

The date of mid-2007 was proposed by the Council as a time when – after the general elections in France – the member states, under the influential presidency of Germany, might come to a new start in the ratification process. On the occasion of the celebrations of the 50th anniversary of the signing of the Treaty of Rome, Angela Merkel, the German Chancellor and President of the European Council, presented, on 24 February 2007, a declaration agreed among the heads of state and

Table 10.1 Ratification of the European Constitution

Ratified	Procedure	Date
Austria	Parliamentary	25 May 2005
Belgium	Parliamentary	13 June 2005
Bulgaria	Parliamentary	11 May 2005
Cyprus	Parliamentary	30 June 2005
Estonia	Parliamentary	9 May 2006
Finland	Parliamentary	5 December 2006
Germany	Parliamentary	May 2005
Greece	Parliamentary	19 April 2005
Hungary	Parliamentary	20 December 2004
Italy	Parliamentary	6 April 2005
Latvia	Parliamentary	2 June: 2005
Lithuania	Parliamentary	11 November 2004
Luxembourg	Referendum + Parliamentary	10 July 2005
		25 October 2005
Malta	Parliamentary	6 July 2005
Romania	Parliamentary	17 May 2005
Slovakia	Parliamentary	11 May 2005
Slovenia	Parliamentary	1 February 2005
Spain	Consultative	
	Referendum + Parliamentary	20 February 2005
		May 2005
Rejected	Procedure	Date
France	Referendum	29 May 2005
The Netherlands	Referendum	1 June 2005
Member states that suspended the process of ratification		
Czech Republic	Poland	United Kingdom
Denmark	Portugal	
Ireland	Sweden	

Source: BBC, 25 March 2007 and European Union, at the URL: http://europa.eu.int/constitution/ratification_en.htm

government stating: "we are united in our aim of placing the European Union on a renewed common basis before the European Parliament elections in 2009."

On 25 March 2007, Angela Merkel, declared in front of the Council at the official ceremony to celebrate the 50th anniversary of the signing of the Treaties of Rome: "we are united in our aim of placing the European Union on a renewed common basis before the European Parliament elections in 2009. I am working to ensure that a roadmap for this can be adopted at the close of Germany's EU Presidency, and I am counting on your support." The support of the Council was reached at the June 2007 Summit closing the German presidency, after a marathon session, "thanks partly to the determined chairmanship of Germany's Angela Merkel, partly to the pizzazz of France's hyperactive new president, Nicolas Sarkozy" [Economist 2007-2]. In the June 2007 European Council, 22 countries worked to resurrect the defunct Constitution, while two other camps were for a lower level compromise. On one side Poland, to protect the privilege assured by the Nice Treaty on the weight of their votes in the Council; on another camp France and The Netherlands in consideration of their voters who had said *no*; and the UK and the Czech Republic in consideration of their voters who would have said *no* if asked [Economist 2007-3]. The June 2007 summit ended with a compromise where the United Kingdom secured opt-outs from the Charter of Fundamental Rights and majority voting in justice and home affairs, as well as "state-like trappings, such as an anthem [, a motto,] and a flag" [Economist 2007-2]; France imposed dropping the reference to free and undistorted competition and Poland obtained a delay to 1 November 2014 in the date when the new voting algorithm in Council would take effect. "If the [Reform] Treaty is ratified without popular votes, the politicians will be able to congratulate themselves on having smuggled through the back door what they had failed to bring in through the front" [Economist 2007-2].

The agreement reached in the European Council of 21-22 June 2007 was on a *Reform Treaty* to be finalized by year end 2007 by an Intergovernmental Conference (IGC). The Intergovernmental Conference was opened under the Portuguese Presidency on 23 July 2007 by the Portuguese Minister of Foreign Affairs, Luís Amado, with a mandate to reformulate the Constitutional Treaty by making it more concise, while keeping its main features. The constitutional concept, which consisted in repealing all existing Treaties and replacing them by a single text called "Constitution", is abandoned. The Reform Treaty would introduce into the existing Treaties, which remain in force, the innovations resulting from the 2004 IGC.

The Portuguese Presidency tabled a draft Reform Treaty, for discussion in the IGC with the aim to decide, by December 2007, on the unanimous adoption of the amendments to be made to the Treaties. A final text was eventually adopted by the European Council on 20 October 2007 and officially signed on 13 December 2007 in Lisbon by the EU27 heads of state and prime ministers. The Treaty will come into force when ratified by each and every of the 27 member states, which is expected to happen before the 2009 EP elections, and possibly by year-end 2008. One week after the signature in Lisbon, Hungary was the first country to ratify the Treaty, on 17 December 2007.

The agreement was a compromise reached after a horse-trading negotiation which enabled to converge to a "package deal" that avoided yet another cycle of discussions. The main features of the defunct Constitution – a full-time standing post of European Commission president, a high representative of the Union for foreign affairs and security policy, a new, simple algorithm to weight the votes in the Council, and an extension of qualified majority voting – were preserved, despite in a watered-down formulation that made them acceptable to the less Euro-enthusiast countries. The result is a matter of consequence for the future of the EU, but to reach the deal, concessions were made to countries that had expressed reserves: Italy won one extra seat in the European Parliament; Poland won a guarantee that a provision would be enshrined in the treaty allowing small groups of states to delay EU decisions for a "reasonable time" – in practice, several months; other minor concessions were granted to Austria and Bulgaria.

The newly appointed British prime minister, Gordon Brown, missed the signing ceremony. "Amid growing claims the UK premier is becoming detached from the rest of the European Union," *The Financial Times'* correspondent from Brussels wrote. Brown declared to be satisfied with the final deal that "ensures that British sovereignty is not undermined", as he said in a news conference where he mentioned 19 times the phrase "British national interests."

Stephen Castle, a Brussels correspondent of *The New York Times* wrote on NYT of 14 December 2007 [Castle 2007]:

> Gathering in a 16th-century monastery in Lisbon, European Union leaders signed a new treaty on Thursday that changes the way the 27-member bloc is run. Those changes include the creation of a permanent post of European Union president to represent Europe on the world stage. The treaty replaces the failed attempt to create a European constitution, which foundered after French and Dutch voters rejected it in referendums two years ago.
>
> The post of president, with a term of two and a half years, will replace the unwieldy system by which European Union leaders and nations rotate holding the presidency every six months. The new treaty also alters the way decisions are made in the union. More decisions will be made by majority vote instead of by unanimous agreement, a threshold that critics said had hindered the union's reaction to some major issues. Foreign policy duties will be concentrated in a single new representative. Likely contenders for the new presidency include Tony Blair, the former British prime minister; Prime Minister Anders Fogh Rasmussen of Denmark; Prime Minister Jean-Claude Juncker of Luxembourg; and a former Polish president, Aleksander Kwasniewski.
>
> The European leaders hailed the treaty, saying it would help the union overcome the political drift that has troubled it since the humbling defeats in the referendums in France and the Netherlands in 2005. Its main achievements so far have included a common currency, the euro, in 13 member states; open borders among Continental members; and the extension of membership to several former Communist countries in 2004.
>
> Blaming a scheduling clash, Mr. Brown—alone among the leaders—arrived late and signed the treaty alone. The European Union is deeply unpopular among some sections of the British public, as it is in several other European countries, and Britain's popular press is skeptical about the benefits of European integration and the surrender of national sovereignty. Mr. Brown's absence from the ceremony enraged supporters of the European Union. "It's a complete and absolute scandal," said Andrew Duff, a pro-European Liberal Democrat member of the European Parliament.

The main features of the Lisbon Treaty with respect to the Nice Treaty now in force are: a higher profile for the EU, simplification, effectiveness, transparency, democracy, and legitimacy.

Higher profile: The full-time standing president of the European Council and the high representative of the Union for foreign affairs and security policy for foreign affairs are two major new roles introduced by the Lisbon Treaty that ensure continuity and impetus in the Union's initiatives, as well as the external representation of the Union with a single authoritative voice in international bodies (UN Security Council, World Trade Organization (WTO), etc.). The high representative of the Union for foreign affairs and security policy is, however, a watered-down version of Union minister for foreign affairs envisaged by the Constitutional Treaty. The change was imposed by the UK to avoid that the British status of permanent member in the UN Security Council might be dwarfed by the EU representative.

Simplification means not only the relevant formal process, deriving from a single treaty encompassing all European legislation, but also the reduction of the number of legal instruments from 36 to 6, the functioning of the institutions, and their adaptability to a dynamic Union, which is likely to change in the years to come, without the need for a change in the Treaty on EU at each step of the institutional transformation. The size of the College of Commissioners of the European Commission is reduced from one per member state to two-thirds thereof, on a rotating basis.

Effectiveness is improved by limiting the power of veto, by clarifying the areas of exclusive vs. shared responsibility of the Union and of the member states, and by introducing a procedure for voluntary withdrawal from the Union.

Transparency is improved by clarification of competences of the individual institutions and of the Union vs. the member states, and by two additional features: the Council, when acting as a legislator, meets in public sessions; and citizens have the right to access documents produced by the Union's institutions.

Democracy is increased by the modified legislating rules, the augmented powers of the European Parliament, the representation of minorities, and the involvement of national Parliaments in the law-making process, allowing them to scrap draft measures.

Legitimacy is improved through democratic participation as a cornerstone of the Union; citizens can initiate legislative proposals; and new obligations are imposed with regard to consultation.

The complexity of harmonizing the activities of the Union at 27+ members induced the legislator to further develop in the reformed Treaty on European Union the institution of *enhanced cooperation* between member states on matters covered by the Treaties, introduced with the Treaty of Amsterdam. The number of member states required for launching the procedure is now eight, and other members can join at any time; the scope of cooperation has been extended to the common foreign and security policy; the right of veto which the member states enjoyed over the establishment of enhanced cooperation has disappeared.

Enhanced cooperation can be undertaken only as a last resort, when it has been established within the Council that the objectives of such cooperation cannot be attained by applying the relevant provisions of the Treaties. The cooperation must contribute to enhancing the process of integration within the Union and must not undermine the single market or the Union's economic and social cohesion. Furthermore, it must not create a barrier to or discrimination in trade between the member states and must not distort competition between them.

Enhanced cooperation may result to be an effective instrument to keep the Union going in areas where groups of members evolve at different speed and a way to prevent the EU to proceed at the pace of her slowest or most Euroskeptic member. Such approach (with different denomination) is not new in the European Union and is one way to implement a *multi-speed EU* or *variable-geometry EU*. It proved effective and efficient in the implementation of major achievements such as the Schengen Convention, allowing traveling in the EU without showing passports; the Prüm Treaty, an international police cooperation agreement combating terrorism and cross-border crime; or the Eurosystem, the authority of the Euro-zone,

Table 10.2 Overlapping circles

Country	EU	Candidate	NATO	Schengen	Euro	Prüm
AT	*			*	*	*
BE	*		*	*	*	*
BG	*		*			
HR		*				
CY	*				*	
CZ	*		*	*		
DK	*		*	*		
EE	*		*	*	*	
FI	*			*	*	
FR	*		*	*	*	*
DE	*		*	*	*	*
EL	*		*	*	*	*
HU	*		*	*		
IS			*	*		
IE	*					
IT	*		*	*	*	
LV	*		*	*		
LI						
LT	*		*	*		
LU	*		*	*	*	*
MK		*				
MT	*			*	*	
NO			*	*		
PL	*		*	*		
PT	*		*	*	*	
RO	*		*			
SK	*		*	*		
SI	*		*	*	*	
ES	*		*	*	*	
SE	*			*		
CH				*		
NL	*		*	*	*	*
TR		*	*			
UK	*					

Source: European Commission, 2007 and [Economist 2007-25].

regulating within the Economic and Monetary Union the monetary policy of countries that adopted the euro. Table 10.2 gives a picture of which European states have ratified different treaties and conventions.

An example of enhanced cooperation in an area near to the subject of this book is the "European enhanced cooperation on vocational education and training," answering to the need for a European dimension to education and training in the context of the transition towards a knowledge based economy capable of sustainable economic growth with more and better jobs [EC 2002-10].

In the present chapter, we describe the institutions, the legislative acts, and the activities of the European Union as they are regulated by the Lisbon Treaty on European Union as signed in December 2007, assuming it will come into force after ratification before the 2009 elections of the European Parliament. We will however, where appropriate, mention the clauses currently in force before ratification, as resulting from the Nice Treaty. The level of detail in our description is that required for the subsequent study of the community policies for sustainable growth. For an in-depth study, we refer to the recommended reading at the end of this chapter, which include also the full text of the Lisbon Treaty.

10.1.7.1 Founding Principles

The Charter of Fundamental Rights of the European Union was formally proclaimed in Nice at the European Council in December 2000 [Council 2000–2]. The Charter had been incorporated in the Constitutional Treaty and constitutes its Part Two. According to the European Council agreement of July 2007, an Article in the Lisbon Treaty addresses fundamental rights and contain a cross reference to the Charter on fundamental rights (see recommended reading), as agreed in the 2004 IGC, giving it legally binding value and setting out the scope of its application. The Charter was born from the need to define an essential nucleus of fundamental values to be the basis of the government system, which had become more complex after the entry in force of the Economic and Monetary Union, and in view of the forthcoming enlargement [Levi 2003]. Shared sovereignty [Holm 2001] between member states on matters of vital importance such as economic policy, foreign policy, security policy, and judicial cooperation requires a constitutional framework in which to settle disputes in a way that respects the rights of the citizens and of the single states. The preamble of the Charter of Fundamental Rights of the European Union (2000) states:

> The peoples of Europe, in creating an ever closer union among them, are resolved to share a peaceful future based on common values. Conscious of its spiritual and moral heritage, the Union is founded on the indivisible, universal values of human dignity, freedom, equality and solidarity; it is based on the principles of democracy and the rule of law. It places the individual at the heart of its activities, by establishing the citizenship of the Union and by creating an area of freedom, security and justice. The Union contributes to the preservation and to the development of these common values while respecting the diversity of the cultures and traditions of the peoples of Europe as well as the national identities of the member states and the organization of their public authorities at national, regional and local levels; it seeks to promote balanced and sustainable development and ensures free movement of persons, goods, services and capital, and the freedom of establishment.

In the second line of the preamble, the word *peace* (peaceful) is mentioned. This is the only time this word appears in the Charter of Fundamental Rights; in fact there is no explicit and strong repudiation of war. Another significant absence is the failure to recall the Christian roots[3] of the values that characterize the civil society of the member states. This second subject was discussed at length in the proceedings of the European Convention and of the Inter-Governmental Conference. On the other hand, the first matter was not sufficiently discussed. As regards the repudiation of war, the author feels that the inspirations of the founding fathers have been betrayed, in particular those of Altiero Spinelli.

The Charter of fundamental rights is made up of 54 articles structured in seven chapters:

Chapter I: Dignity

Human dignity; right to life; right to the integrity of the person; prohibition of torture and inhuman or degrading treatment or punishment; prohibition of slavery and forced labour

Chapter II: Freedoms

Right to liberty and security; respect for private and family life; protection of personal data; right to marry and right to found a family; freedom of thought, conscience, and religion; freedom of expression and information; freedom of assembly and of association; freedom of the arts and sciences; right to education; freedom to choose an occupation and right to engage in work; freedom to conduct a business; right to property; right to asylum; protection in the event of removal; expulsion or extradition

Chapter III: Equality

Equality before the law; nondiscrimination; cultural, religious, and linguistic diversity; equality between men and women; the rights of the child; the rights of the elderly; integration of persons with disabilities

Chapter IV: Solidarity

Workers' right to information and consultation within the undertaking; right of collective bargaining and action; right of access to placement services; protection in the event of unjustified dismissal; fair and just working conditions; prohibition of child labour and protection of young people at work; family and professional life; social security and social assistance; health care; access to services of general economic interest; environmental protection; consumer protection

Chapter V: Citizens' Rights

Right to vote and to stand as a candidate at elections to the European Parliament; right to vote and to stand as a candidate at municipal elections; right to good administration; right of access to documents; Ombudsman; right to petition; freedom of movement and of residence; diplomatic and consular protection

---

[3] The French psychoanalyst Marie Balmary analysed [Balmary 1999] religious references in the texts of the declarations of rights, starting from the Declaration of Independence (Philadelphia, 4 July 1776) to the Declaration of the Rights of Man and of the Citizen, (France 1789), the 1946 French Constitution, and the Universal Declaration of Human Rights of the United Nations, 1948. God, the creator of mankind, is well visible in the Philadelphia declaration. Then his presence becomes philosophical in the French 1789 declaration to disappear altogether in the declarations of 1946 and 1948.

Chapter VI: Justice

Right to an effective remedy and to a fair trial; presumption of innocence and right of defence; principles of legality and proportionality of criminal offences and penalties; right not to be tried or punished twice in criminal proceedings for the same criminal offence

Chapter VII: General Provisions

Scope of guaranteed rights; level of protection; prohibition of abuse of rights

At the end of the year 2000, when the Charter was formally proclaimed at the European Council summit, the representatives of the Institutions of the European Community made the following statements [EP 2000].

Romano Prodi, president of the European Commission, stated:

In the eyes of the European Commission, by proclaiming the Charter of Fundamental Rights, the European Union institutions have committed themselves to respecting the Charter in everything they do and in every policy they promote [...]. The citizens of Europe can rely on the Commission to ensure that the Charter will be respected [...].

Nicole Fontaine, president of the European Parliament, stated:

A signature represents a commitment [...]. I trust that all the citizens of the Union will understand that from now on [...] the Charter will be the law guiding the actions of the Assembly [...]. From now on it will be the point of reference for all the Parliament acts which have a direct or indirect bearing on the lives of citizens throughout the Union.

10.2 Institutions of the European Union

10.2.1 European Parliament

The European Parliament (EP) [Bomberg et al. (ed.) 2003] is the organ of democratic expression and of political control of the European Community. The EP is made up of members elected every five years by universal suffrage. They were 732 seats in the EP's sixth term before the 2007 enlargement (2004–2006) and 785 seats from 2007 to 2009. From 2009 to 2014, the seats will be 751. The constituency of the EP is shown in Table 10.3. The President of the European Parliament is included in the count.

The Parliament's origins go back to the 1950s and the founding treaties. Since 1979, the Members of the European Parliament (MEPs) have been directly elected by the citizens they represent. Previously, MEPs were elected from and by the national Parliaments.

Parliamentary elections are held every five years, and every EU citizen who is registered as a voter in his or her country is entitled to vote. Thus the EP expresses the democratic will of the Union's 493 million citizens and represents their interests in discussions with the other EU institutions. In 2002, Pat Cox was elected EP president for the period 2002–2004, replacing Nicole , who had served as president during the first half of the fifth term (1999–2002). Josep Borrell Fontelles, Member of the Socialist Group, was elected president of the new European Parliament, after

Table 10.3 Number of seats in the EP

	1999 2004	2004 2006	2007 2009	2009 2014
Austria	21	18	18	19
Belgium	25	24	24	22
Bulgaria	–	–	18	18
Cyprus	–	6	6	6
Czech Republic	–	24	24	22
Denmark	16	14	14	13
Estonia	–	6	6	6
Finland	16	14	14	13
France	87	78	78	74
Germany	99	99	99	96
Greece	25	24	24	22
Hungary	–	24	24	22
Ireland	15	13	13	12
Italy	87	78	78	73
Latvia	–	9	9	9
Lithuania	–	13	13	12
Luxembourg	6	6	6	6
Malta	–	5	5	6
The Netherlands	31	27	27	26
Poland	–	54	54	51
Portugal	25	24	24	22
Romania	–	–	35	33
Slovakia	–	14	14	13
Slovenia	–	7	7	8
Spain	64	54	54	54
Sweden	22	19	19	20
United Kingdom	87	78	78	73
TOTAL	**626**	**732**	**785**	**751**

Source: European Parliament, 2007.

the elections in 2004 for the first half of the term; Hans-Gert Poettering was elected president in 2007 for the second half of the term. The present sixth term of the EP is in office until 2009. After the accession of Romania and Bulgaria, new MEPs sit in the Parliament (35 and 18, respectively), and the total number of seats adds up to 785. The number of seats from 1999 to 2009 is presented in Table 10.3.

The Lisbon Treaty lays down that the maximum number of seats is 751, thus increasing the current number laid down in the Treaty of Nice. The Lisbon Treaty breaks with the tradition of enshrining in the treaties the detailed breakdown of seats between the member states. Instead, the Lisbon Treaty establishes an allocation rule, which states that representation of citizens is digressively proportional. The minimum number of seats per member state is to be six, in order to make sure that, even in the least populous member states, all the major shades of political opinion will have a chance to be represented in the European Parliament. The maximum number of seats per member state, which is 96, is also laid down in the Lisbon Treaty for the first time.

The European Parliament works in France, Belgium, and Luxembourg. The monthly plenary sessions, which all MEPs attend, are held in Strasbourg (France) – the Parliament's *seat*. Parliamentary committee meetings and any additional plenary sessions are held in Brussels (Belgium), while Luxembourg is home to the administrative offices.

Parliament has three main roles:

1. it shares with the Council the power to legislate. The fact that it is a directly-elected body helps guarantee the democratic legitimacy of European law;
2. it exercises democratic control over all EU institutions and in particular over the Commission. It has the power to approve or reject the nomination of the College of Commissioners, and it has the right to censure the Commission as a whole;
3. it shares with the Council the authority over the EU budget and can therefore influence EU spending. At the end of the procedure, it adopts or rejects the budget in its entirety.

1. The power to legislate
   The most common procedure for adopting (i.e. passing) EU legislation is *codecision*. This places the European Parliament and the Council on an equal footing, and the laws passed using this procedure are joint acts of the Council and Parliament. It applies to legislation in a wide range of fields. On a range of other proposals, Parliament must be consulted, and its approval is required for certain important political or institutional decisions. Parliament also provides an impetus for new legislation by examining the Commission's annual work program, considering what new laws would be appropriate, and asking the Commission to put forward proposals.
2. Democratic control
   Parliament exercises democratic control over the other European institutions. It does so in several ways. The Lisbon Treaty lays down that from 2009 onward, the president of the Commission will be elected by the European Parliament by a majority of its members, acting on a proposal from the European Council. This proposal will have to take into account the results of the European elections. Once the Commission president is elected, EP interviews individually and approves jointly all the designated members of the Commission (nominated by the member states). The Commissioners and their president are appointed en bloc by Parliament's approval. Parliament can dismiss the Commission by passing a *motion of censure* calling for mass resignation. Further details are given in the European Commission section later in this chapter. The members of the Commission attend plenary sessions of Parliament and meetings of the parliamentary committees, maintaining a continual dialogue between the two institutions. Parliament also monitors the work of the Council: MEPs regularly ask the Council written and oral questions, and the president of the Council attends the plenary sessions and takes part in important debates. Parliament works closely with the

Council in certain areas, such as common foreign and security policy and judicial cooperation, as well as on some issues of common interest such as asylum and immigration policy and measures to combat drug abuse, fraud, and international crime. The Council Presidency keeps Parliament informed on all these subjects.

3. Budget authority

The EU's annual budget is decided jointly by Parliament and the Council of the European Union. Parliament debates it in two successive readings, and it does not come into force until it has been signed by the president of Parliament.

Parliament's Committee on Budgetary Control (Cocobu) monitors how the budget is spent, and each year Parliament decides whether to approve the Commission's handling of the budget for the previous financial year. This approval process is technically known as *granting a discharge*.

10.2.2 European Council

The Lisbon Treaty establishes the European Council as an institution of the Union and gives it a clearly defined role, distinguishing its work from that of the Council of Ministers.

The European Council consists of the heads of state and government of the member states. It includes its president and the president of the Commission. The high representative of the Union for foreign affairs and security policy will also take part in its work.

The European Council should not be confused with the Council of Ministers, which consists of the representatives of the member states at ministerial level and will be discussed in a later section in this chapter.

The European Council is the initiator of the main political initiatives of the Union and is the arbitration body for problems that are not solved by the Council of Ministers. The European Council plays a key role as regards nominations: for example, it proposes the president of the Commission to the European Parliament and appoints, in agreement with the president of the Commission, the high representative of the Union for foreign affairs and security policy. The European Council meets quarterly, bringing together the heads of government of the member states: these are usually prime ministers or premiers, with the exception of those countries, like France or Finland, which are represented by the president of the Republic *and* the prime minister because – according to the national constitution – the institutional responsibilities are divided between the two functions. For this reason, the European Council is described as a meeting of heads of state and government of the member states of the European Union.

The Lisbon Treaty establishes a permanent president of the European Council, who will take on the work currently assigned to rotating Presidencies. He/she will be elected by qualified majority, for a term of two and a half years, renewable once. The previous rotation of the Council Presidency every six months has been considered to be a source of inefficiency and discontinuity. At times, it has prevented the undertaking of major actions that would have required a long period of political continuity and stability.

The president of the European Council may not hold a national mandate at the same time. He or she will chair the European Council, drive forward its work, and ensure its proper preparation and continuity in cooperation with the president of the Commission, on the basis of the General Affairs Council's work. The president also shall endeavour to facilitate cohesion and consensus within the European Council and present a report to the European Parliament after each of its meetings. We reported earlier an anticipation appearing on the international press in 2007 concerning likely contenders for the EU Council presidency. According to Stephen Castle, writing from Brussels on the New York Times on 14 December 2007 [Castle 2007] they would be: Tony Blair, the former British prime minister (supported by the French president Nicholas Sarkozy); Prime Minister Anders Fogh Rasmussen of Denmark; Prime Minister Jean-Claude Juncker of Luxembourg; and a former Polish president, Aleksander Kwasniewski.

The European Council does not exercise legislative functions. All European directives must be adopted by the Council of Ministers, in most cases jointly with the European Parliament. The Lisbon Treaty stipulates that certain decisions, of a more constitutional nature, will be taken by the European Council, such as those relating to the composition of the European Parliament, the arrangements for the rotating Presidency of the Council, the system of equal rotation for the composition of the Commission, the suspension of the rights of a member state in the event of a serious and persistent breach of the values of the Union, and the changeover from a legal basis of unanimous voting to qualified majority voting.

Finally, the Lisbon Treaty stipulates that decisions of the European Council are to be taken by consensus, except where the Treaty provides otherwise.

10.2.3 High Representative of the Union for Foreign Affairs and Security Policy

The creation of the post of Union minister for foreign affairs was one of the major innovations of the Constitution. The Lisbon Treaty proposes the creation of this function under the name of *High Representative of the Union for Foreign Affairs and Security Policy*. The change has a profound meaning and derives from a compromise with the UK not to give to the EU any "state-like trapping" that might imply that one day the EU sits in the Security Council of the UN as an institution representing the whole Union, including UK with her permanent member status. The purpose of introducing the role of high representative of the Union for foreign affairs and security policy is to make the Union's external action more effective and coherent: the high representative of the Union for foreign affairs and security policy becomes the unique voice of the Union's foreign and security policy. This institutional innovation is the result of merging the functions of the High Representative for the common foreign and security policy and the external relations Commissioner. Once the

Lisbon Treaty is ratified, the high representative of the Union for foreign affairs
and security policy will be appointed by the European Council acting by qualified
majority, with the agreement of the president of the Commission. The high representative of the Union for foreign affairs and security policy will also be one of the
vice-presidents of the Commission. In this capacity, he or she will be part of the
Commission.

The high representative of the Union for foreign affairs and security policy will
be both the Council's representative for the common foreign and security policy and
one of the Commission's vice-presidents. He or she will conduct the Union's common foreign and security policy and, for this purpose, will have a right of initiative
in foreign policy matters. He or she will implement that policy under mandate from
the Council of Ministers, conducting political dialogue on the Union's behalf and
expressing the Union's position in international organizations and at international
conferences. The High Representative will perform a similar role in the area of
common security and defence policy, bound under this mandate, by the collegiate
principle governing the Commission.

He or she will also be responsible for coordinating member states' action in international forums. In this capacity, he or she may, where the Union has defined a position on a subject that is on the United Nations Security Council agenda, be called
upon by the member states sitting on the Security Council to present the Union's
position. This role will be played on a casa-by-case basis following agreements
taken by the European Council on the subject in question.

Finally, the high representative of the Union for foreign affairs and security policy will be in charge of a diplomatic service with delegations in almost 125 countries. The Lisbon Treaty provides for a European External Action Service to be set
up to assist the High Representative in his or her functions.

10.2.4 Council of Ministers

The Council of Ministers [Bomberg et al. (ed.) 2003] is the EU's main decision-
making institution. Council meetings are attended by foreign affairs ministers of
member states or ministers of member states assigned to other functions, depending
on the issues on the agenda.

The Council of Ministers, which represents the member states, adopts the community legal acts (see Section 10.3, "Acts of the Institutions") and has legislative
power jointly with the European parliament. Jointly with the latter it exercises legislative and budgetary functions. It also has policymaking and coordinating functions. To summarize, Council has six key responsibilities:

1. To pass European legislation; in many fields, it legislates jointly with the European Parliament
2. To coordinate the broad economic policies of the member states
3. To conclude international agreements between the EU and one or more states or
   international organizations

4. To approve the EU's budget, jointly with the European Parliament
5. To develop the EU's Common Foreign and Security Policy (CFSP), based on guidelines set by the European Council
6. To coordinate cooperation between the national courts and police forces in criminal matters

The Lisbon Treaty changes the voting system within the Council of Ministers to qualified majority voting. It states that the Council of Ministers shall meet in different configurations. Altogether there are nine different configurations for the Council of Ministers:

- General affairs and external relations
- Economic and financial affairs (Ecofin)
- Justice and home affairs
- Employment, social policy, health, and consumer affairs
- Competitiveness (internal market, industry, and research)
- Transport, telecommunications, and energy
- Agriculture and fisheries
- Environment
- Education, youth, and culture

The Lisbon Treaty stipulates that the Presidency of all configurations of the Council of Ministers, other than foreign affairs, be held by member state representatives on the basis of an equal rotation system, adopted by the European Council. The Foreign Affairs Council is to be chaired by the high representative of the Union for foreign affairs and security policy.

The Lisbon Treaty stipulates that the Presidency of all Council configurations, other than Foreign Affairs, is to be held by member state representatives on the basis of equal rotation. The Foreign Affairs Council is chaired by the high representative of the Union for foreign affairs and security policy. The Lisbon Treaty does not specify how this rotation system is to work, but states that rules will be established in a European decision adopted by the European Council acting by a qualified majority. The decision taken in the year 2006 is presented in Table 10.4. This solution has the advantage of flexibility. The system may be changed, if appropriate, without having to amend the Treaty on EU.

The IGC agreed on the details of such a decision. These are included in a declaration appended to the Final Act of the IGC, stating that:

- The Presidency of the Council shall be held by pre-established groups of three member states for a period of 18 months. The groups shall be made up on a basis of equal rotation among the member states, taking into account their diversity and geographical balance within the Union.
- Each member of the group shall in turn chair for a six-month period all configurations of the Council. The other members of the group shall assist the Chair in all its responsibilities on the basis of their common program. Members of the team may decide alternative arrangements among themselves.

Table 10.4 Rotation of the Presidency of the Council of Ministers from 2003 to 2020

	First semester	Second semester
2003	Greece	Italy
2004	Ireland	The Netherlands
2005	Luxembourg	United Kingdom
2006	Austria	Finland
2007	Germany	Portugal
2008	Slovenia	France
2009	Czech Republic	Sweden
2010	Spain	Belgium
2011	Hungary	Poland
2012	Denmark	Cyprus
2013	Ireland	Lithuania
2014	Greece	Italy
2015	Latvia	Luxembourg
2016	The Netherlands	Slovakia
2017	Malta	United Kingdom
2018	Estonia	Bulgaria
2019	Austria	Romania
2020	Finland	

Source: European Council 2006

- The European Council should start to prepare this decision as soon as the Lisbon Treaty is signed and should reach political agreement within six months.

10.2.4.1 Unanimity and Qualified Majority

Decisions in the Council of Ministers are taken by vote. By default, the voting procedure in the Council of Ministers is qualified majority voting. This means that, for a proposal to be adopted, it needs the support of a minimum number of votes as specified below. However, in some particularly sensitive areas indicated below, Council decisions have to be unanimous. In other words, each member state has the power of veto in those areas.

The extension of qualified majority voting is a central part of the institutional reform of the Union associated with enlargement. Unanimous agreement was the basis of decisions in the late 1950s between the six EEC member states and remained so until the 1990s. Unanimity was often hard to achieve between 12 or 15 states: in an enlarged Union of 27 or more countries, it would be virtually impossible. If the EU kept trying to operate under her initial rules, it would be paralysed – unable to act in many important fields. So the Treaties of Amsterdam and of Nice changed the rules, paving the way to the present major reform, and allowing the Council to take decisions by qualified majority voting in quite a number of areas that used to require unanimity.

The acceptance of the principle of qualified majority voting is more than a procedural change to improve the efficiency of decision taking. It has to do with the very nature of the EU, and with her values. Accepting the position of the majority is an approach to sharing sovereignty with peers, with a view to achieving the Union's objectives, in a spirit of solidarity that makes common objectives prevail over national interests.

Following the ratification of the Lisbon Treaty, the following fields will remain subject to unanimity:

- taxation
- harmonization in the field of social security and social protection
- certain provisions in the field of justice and home affairs (the European prosecutor, family law, operational police cooperation, etc.)
- flexibility clause (Article I-18), allowing the Union to act to achieve one of her objectives in the absence of a specific legal basis in the Treaty
- common foreign and security policy, with the exception of certain clearly defined cases
- common security and defence policy, with the exception of the establishment of permanent structured cooperation
- finances of the Union (own resources, the multiannual financial framework)
- membership of the Union (opening of accession negotiations, association, serious violations of the Union's values, etc.)
- citizenship (granting of new rights to European citizens, anti-discrimination measures)
- certain institutional issues including electoral system and composition of the Parliament; certain appointments; the composition of the Committee of the Regions and the European Economic and Social Committee; the seats of the institutions; the language regime; the revision of the Treaty on EU, including the bridging clauses

The Reform Lisbon Treaty changes both the field of application of qualified majority voting and the method, which is made much simpler and is now defined also in case of subsequent enlargements. Prior to the enforcement of the Lisbon Treaty, the vote of each country has a weight that depends on the number of its inhabitants, in a non-proportional manner adjusted in favour of the less populous countries. Until 1 May 2004, the number of votes each country could cast was as in Table 10.5.

Until 1 May 2004, the minimum number of votes required to reach a qualified majority was 62 out of the total of 87 (i.e. 71.3 %). The power of veto could be exercised by a group of at least five member states that made up more than 25 votes. For a six-month period from 1 May 2004, when new member states joined the EU, transitional arrangements for the weighting of the votes applied.

From November 2004 and until the enforcement of the Lisbon Treaty, the number of votes each country (including the new member states) can cast is as in Table 10.6.

From November 2004 and until the enforcement of the Lisbon Treaty, a qualified majority is reached if two conditions are fulfilled:

1. a majority of member states (in some cases a two-thirds majority) approve, and
2. a minimum of votes is cast in favour, that is to say 72.3 % of the total (roughly the same share as under the previous system).

Table 10.5 Number of votes
in the European Council per
country from 1995 to May
2004

Countries	Votes
Germany, France, Italy, United Kingdom	10
Spain	8
Belgium, Greece, The Netherlands, Portugal	5
Austria, Sweden	4
Denmark, Ireland, Finland	3
Luxembourg	2
Total:	**87**

Source: European Commission, 2004.

Table 10.6 Number of votes in the European Council per country from November 2004 to 2006
and from 2007 to 2014

Countries	Votes 2004–2006	Votes 2007–2014
Germany, France, Italy, and the United Kingdom	29	29
Spain and Poland	27	27
Romania	–	14
The Netherlands	13	13
Belgium, the Czech Republic, Greece, Hungary, and Portugal	12	12
Bulgaria	–	12
Austria and Sweden	10	10
Denmark, Ireland, Lithuania, Slovakia, and Finland	7	7
Cyprus, Estonia, Latvia, Luxembourg, and Slovenia	4	4
Malta	3	3
Total	**321**	**345**

Source: European Commission, 2004, 2007.

In addition, a member state may ask for confirmation that the votes in favour represent at least 62 % of the total population of the Union. If this is not the case, the decision will not be adopted.

A completely new (and much simpler) system of qualified majority voting is introduced by the Lisbon Treaty. It is known as *double majority*. This system will become effective with the enforcement of the Lisbon Treaty. The weighting of votes, once subject to long and difficult negotiations between member states on the occasion of new enlargements, will be replaced. Instead of the three criteria required until now for a qualified majority (threshold of weighted votes, majority of member states and 62 % of the population of the Union), only two criteria will apply from 1st November 2014: a majority of the member states must be in favour (55 % of the members of the Council, comprising at least 15 of them) and, in addition, the favourable states must represent at least 65 % of the population of the Union.

The Lisbon Treaty therefore breaks with the weighting of votes in the Council and replaces it with a simple, effective, and flexible system that takes into account the twofold nature of the Union – a Union of states and of peoples. The equality of member states is respected, as each one has one vote, while their different population sizes are also taken into account.

10.2.4.2 Coreper

In Brussels, each EU member state has a permanent team (*Representation*) that represents it and defends its national interest at the EU level. The head of each representation is, in effect, his or her country's ambassador to the EU. These ambassadors (known as *permanent representatives*) meet weekly within the Permanent Representatives Committee (Coreper). The role of this committee is to prepare the work of the Council of Ministers.

10.2.4.3 The General Secretariat

The Presidency of the Council is assisted by the General Secretariat, which prepares and ensures the smooth functioning of the Council's work at all levels. In 1999, Javier Solana was appointed Secretary-General of the Council of Ministers, a role that he will continue to cover in the period 2004–2009. He is also High Representative for the Common Foreign and Security Policy (CFSP), a prefiguration of the role that the Constitution had envisaged for the Union minister for foreign affairs. In the capacity of High Representative for CFSP, he helps the Council to draft and implement political decisions. He also engages in political dialogue, on the Council's behalf, with non-EU countries.

10.2.5 European Commission

The European Commission is often referred to as the executive organ of the Community. However, this definition is restrictive as the functions of the Commission are numerous.

In fact, the European Commission [Bomberg et al. (ed.) 2003, Majone 1996] has four main roles:

1. To propose legislation to Parliament and the Council
2. To manage and implement EU policies and the budget
3. To enforce the Treaty on European Union and European law (jointly with the Court of Justice)
4. To represent the Union on the international stage, for example, by negotiating agreements between the EU and other countries

The Lisbon Treaty confirms the functions of the Commission and supplements the existing rules concerning the number and origin of its members. The Commission takes its decisions collectively, by simple majority.

The European Commission is currently (2007) made up of 27 members proposed by the member states and appointed by Parliament for five years. The Commission until 2004 had two members from each of the most heavily populated member states (France, Germany, Great Britain, Italy, and Spain) and one from each of the other EU countries. When 10 more countries joined the EU on 1 May 2004, 10 more commissioners were temporarily added and the number of commissioners rose to 30. If this system had remained in force, the Commission would have had too many members and would have been unworkable. It was decided that from the date when the 2004–2009 Commission took office (22 November 2004), there would be only one commissioner per country.

A new Commission is appointed every five years, within six months of the elections to the European Parliament. The text of the Lisbon Treaty stipulates that the first Commission appointed under the provisions of the Lisbon Treaty shall consist of one national of each member state, including its president and the high representative of the Union for foreign affairs and security policy. As of 2014, the Commission will be reduced in size and will consist of a number of members corresponding to two-thirds of the number of member states. The nationality of the Commissioners will be determined by a system of rotation absolutely fair to all countries. The European Council, acting unanimously, may nevertheless decide to alter this number.

The procedure to appoint the Commission is as follows:

- the member state governments agree together in the European Council on who to designate as the new Commission president;
- the European Parliament elects the Commission president designated by the Council;
- the Commission president-designate, in discussion with the member state governments, chooses the other Commission members;
- the new Parliament then interviews one by one all the designated members and gives its opinion on the entire *College*. EP approval appoints the new Commission, which can then officially start work.

In the appointment of the Commission for the 2004–2009 term, the EP exercised its institutional power during the hearings of the designated commissioners and expressed reservations about some of them. The Commission president-designate, José Manuel Durão Barroso, did not present his College to the EP on the due date of 27 October 2004, and withdrew his team, declaring [Browne 2004]:

> I have come to the conclusion that if a vote is taken today, the outcome will not be positive for European Institutions or for the European project. In these circumstances, I have decided not to submit a New Commission for your approval today.

Mr. Barroso asked for a delay of up to one month, during which the outgoing Commission would stay in office. The composition of the College was changed, and the new Commissioners-designate were interviewed by the EP, which eventually approved the whole team in the plenary session of 17 November 2004. The new Commission took office on the following 21 November 2004. The names and responsibilities of the Commissioners are among the recommended reading at the end of this chapter, available on the recommended reading web site companion to this book.

Table 10.7 The presidents of the European Commission

	From	To
Walter Hallstein	10 January 1958	5 July 1967
Jean Rey	6 July 1967	30 June 1970
Franco Maria Malfatti	1 July 1970	21 March 1972
Sicco Mansholt	22 March 1972	5 January 1973
François Xavier Ortoli	6 January 1973	5 January 1977
Roy Jenkins	6 January 1977	5 January 1981
Gaston Thorn	6 January 1981	5 January 1985
Jacques Delors	6 January 1985	6 January 1995
Jacques Santer	23 January 1995	15 September 1999
Romano Prodi	16 September 1999	20 November 2004
José Manuel Durão Barroso	21 November 2004	–

The Commission remains politically answerable to Parliament, which has the power to dismiss it by adopting a motion of censure. The Commission attends all the sessions of the European Parliament, where it must clarify and justify its policies. It also replies regularly to written and oral questions posed by MEPs.

The presidents of the European Commission from 1958 to 2009 are listed in Table 10.7.

10.2.6 Court of Justice of the European Union

The Court of Justice was set up in 1952 with the name of Court of Justice of the European Communities under the Treaty of Paris establishing the European Coal and Steel Community (ECSC).

The role of the Court is to ensure the application of the treaties, to enforce community law, and to hold the right of interpretation of the treaties and of the legal acts of the European institutions. The Court of Justice may verify the failure by a member state to fulfil one of its obligations specified in the treaties, control the legality of institutional acts that are contested before the court, and finally verify the failure by the European Parliament, Council, or Commission to fulfil the obligation to deliberate, thereby violating the treaties. The Court of Justice is the only competent decision-making body that upon the request of national courts may pronounce a decision on the interpretation of the treaties or on the validity and interpretation of the acts adopted by the institutions.

The Court is composed of one judge per member state so that all the EU's national legal systems are represented. After the 2004 enlargement, there is still one judge per member state, but for the sake of efficiency the Court can sit as a *Grand Chamber* of only 13 judges instead of always having to meet in a plenary session.

The Court is assisted by eight *advocates-general*. Their role is to present reasoned opinions on the cases brought before the Court. They must do so publicly and impartially. The judges and advocates-general are either former members of the highest national courts or highly competent lawyers who can be relied on to show impartiality. They are appointed by joint agreement of the governments of the member states. Each is appointed for a term of six years, after which they may be reappointed up to twice for a further period of three years.

To help the Court of Justice cope with the thousands of cases brought before it, and to offer citizens better legal protection, a *Court of First Instance* was created in 1989. This Court is responsible for giving rulings on certain kinds of cases, particularly actions brought by private individuals and cases relating to unfair competition between businesses.

10.2.7 European Central Bank

The European Central Bank (ECB), together with the national central banks, constitutes the European System of Central Banks (ESCB). The European Central Bank conducts the monetary policy of the Union, together with Eurosystem, the subset of ESCB composed of the national central banks of the member states whose currency is the euro. The primary objective of the European System of Central Banks is to maintain price stability.

The European Central Bank is the only institution that may authorize the issue of euro. It is independent from Union institutions, bodies, offices, and agencies and from the governments of the member states. Some of the former responsibilities of the national Central Banks of the EU member states have been transferred to the ECB, as we will see in detail in Chapter 14, "Economic and Monetary Union". This change shows an important aspect of the building of the EU: the member states, by transferring some of their sovereignty to collegial bodies of the Union, while giving away prerogatives of independence and rights of action, gain by participating in decision making at a higher level, contribute to the achievement of the Union's objectives, and become active in processes that might have been otherwise imposed on them by other states with a greater political weight.

10.2.8 Court of Auditors

The Court of Auditors is the institution tasked to examine the accounts of all Union revenue and expenditure, and to ensure sound financial management. It checks that all the Union's revenue has been received and all her expenditure is incurred in a lawful and regular manner. The Court's main role is to check that the EU budget is correctly implemented and to guarantee that the EU financial system operates efficiently and openly.

The Court of Auditors has 27 members, one national from each member state. The members of the Court are appointed by the Council for a renewable term of six years subject to consultation with the European parliament. Even after further enlargements, the number of members will correspond to the number of EU states, but, for the sake of efficiency, the Court can set up *chambers* (with only a few members each) to adopt certain types of reports or opinions.

10.2.9 Economic and Social Committee

The Council and the Commission are supported by the Economic and Social Committee, the advisory body that allows the active participation of representatives of organizations of employers, of the employed, and of other parties representative of civil society, notably in socioeconomic, civic, professional, and cultural areas. The Economic and Social Committee had 222 members until May 2004. After the enlargement, 340 members were appointed. In the years to come, according to the Lisbon Treaty, the Council will decide on its composition, which will not exceed 350 members appointed by the Council following the proposal of the member states.

The Committee represents the different categories of the economic and social world. It must be consulted before a large variety of decisions may be taken (on employment, social, vocational training matters, etc.), and on its own initiative it may also give opinions on other matters.

10.2.10 Committee of the Regions

The Committee of the Regions is an advisory body consisting of representatives of regional and local bodies who hold either a regional- or local-authority electoral mandate or are politically accountable to an elected assembly. The Committee of the Regions had, until May 2004, 222 members appointed for four years by the Council, following the proposal of the member states. According to the Lisbon Treaty, the Council will decide on its composition, which will not exceed 350 members appointed for a five-year renewable term. The Council and the Commission consult the Committee in the cases foreseen by the treaty. The Committee can also adopt opinions on its own initiative.

10.2.11 Ombudsman

The European Ombudsman, elected by the European Parliament for the duration of the legislative term, receives, examines, and reports on complaints about maladministration in the activities of Union institutions, bodies, offices, or agencies. He or she is entitled to receive complaints from citizens or businesses and institutions of the Union.

10.3 Acts of the Institutions

The Lisbon Treaty involves the national Parliaments in the legislative process:

> The Commission shall forward its draft European legislative acts and its amended drafts to national Parliaments at the same time as to the Union legislator. Draft European legislative acts shall be justified with regard to the principles of subsidiarity [...]. Any national Parliament or any chamber of a national Parliament may, within eight weeks from the date of transmission of a draft European legislative act, send to the presidents of the European

Parliament, Council and Commission a reasoned opinion stating why it considers that the draft in question does not comply with the principle of subsidiarity.

The characteristics of the legislative procedure regulated by the Lisbon Treaty are as follows:

1. The Commission consults and proposes
2. The Council of Ministers and the European Parliament legislate
3. The Court of Justice ensures that the law is observed
4. National Parliaments can have their say and
5. The European Economic and Social Committee and the Committee of the Regions become involved.

To exercise the Union's competences, the Lisbon Treaty envisages fewer legal instruments than those considered in the previous legislation. This is a significant step towards clarity and simplification. As many as 36 previously existing instruments will be reduced to the following five:

Legislative acts:

European *directive* is a legislative act binding as regards the result to be achieved; the national authorities of each member state choose the form and methods of enforcement.

Nonlegislative acts:

European *regulation* is a nonlegislative binding act of general application.

European *decision* is a nonlegislative binding act that may be of nongeneral application. A decision that specifies those to whom it is addressed shall be binding on them alone.

Acts of no binding force:

Recommendation is a nonbinding act of an institution that makes suggestions to a legal body to operate in a given context according to specific terms.

Opinion is a nonbinding deliberation of an institution on a specific matter.

European Directives shall be adopted, on the basis of proposals from the Commission, jointly by the European Parliament and the Council. European Directives may delegate to the Commission the power to adopt delegated European regulations to supplement or amend certain nonessential elements of the Directive. Member states shall adopt all measures of national law necessary to implement legally binding Union acts.

Examples of acts of the institutions of the European Union follow:

1. European Regulation 2002/2195 on the harmonization of public tenders open to competition in the Union
2. European Regulation 2003/1829 on the obligation to label food packages that contain genetically modified food

3. Directive 2002/58 on the obligation to make laws at national level regarding protection of data concerning people and to ensure the confidentiality of electronic communication
4. Directive 1998/71 on the obligation to make laws at the national level regarding protection of rights in the industrial design context
5. Directive 2000/53 on end-of-life vehicles. This Directive assigns to the manufacturers the task of taking in the cars and demolishing them. The Directive establishes that within 2005 the member states must make laws to regulate the demolition of vehicles as follows: 85 % of the weight of the vehicle must be recovered and recycled (at present the percentage of recycled material is around 75 %, and concerns the metal content). The recovery/recycling objectives will increase respectively up to 95 % and 85 % by the year 2015. This means that the quantity of material eliminated as waste will drop from the current 25 % to less than 5 %.
6. European Decision C 2003/1082, dated 2/4/2003, on the compatibility of the merger of two companies in the media industry, Newscorp and Telepiù, with the laws on competition in the internal market
7. European Recommendation of the European Parliament on the search made of the Ankara headquarters of the Human Rights Association in Turkey (adopted by the European Parliament on 15 May 2003):

> The European Parliament notes that Turkey has amended most of its legislation regarding freedom of speech and association, but regrets that these amendments still leave ample scope for repressive actions by the police and that little has changed on the ground. Urges the Turkish government to take concrete measures to prove its commitment to respecting human rights and to review its legislation on this matter; calls on the Turkish government to implement the judicial reforms it has announced and to abolish the State Security Courts, which represent an obstacle to the development of the rule of law in Turkey. Parliament reminds the Turkish government that reforms represent an element which will be duly taken into account by the Commission in the drafting of its December 2004 report to the Council on Turkey's fulfillment of the Copenhagen political criteria. Stresses that political will on the part of Turkey to make radical changes to guarantee the respect of the rule of law is essential for progress towards EU membership.

8. European Recommendation 1128/2001/IJH, dated 27 June 2002, made by the Ombudsman about the complaint filed by a Dutch NGO concerning the failure to have access to information held by the European Commission concerning trade agreements with the U.S. (Transatlantic Business Dialogue [TABD])
9. European Decision (binding only for the Italian administration) of the European Commission, dated 9 October 1998, on the plan to transfer flights from the Linate airport to the Malpensa airport (two airports near Milan, Italy)

10.4 Activities of the European Union

The tasks of the European Community, introduced by the Treaty of Rome and the treaties that followed, gave rise to the many different activities listed in Table 10.8. Part Two of this book dealt with activities related to research, technology, and the

Table 10.8 Areas of activity of the European Union

Areas of activity		
Agriculture	Energy	Information Society
Audiovisual	Enlargement	Institutional Affairs
Budget	Enterprise	Internal Market
Competition	Environment	Justice and Home Affairs
Consumer	External Relations	Public Health
Culture	External Trade	Regional Policy
Customs	Fisheries	Research and Innovation
Development	Food Safety	Taxation
Economic and Monetary Affairs	Foreign and Security Policy Fraud	Transport
Education, Training, Youth	Humanitarian Aid	
Employment and Social Affairs	Human Rights	

information society. Some aspects of other areas of intervention of the Union that concern policies for sustainable development are studied in some depth in the following chapters. In particular, due to their importance as regards industrial policy, we will address: the internal market, competition policy, budget, and economic and monetary policy.

For a description of other areas of activity see [Sbragia 2003] and the recommended reading for this chapter, where descriptions of the agricultural and regional policies have been included in order to satisfy the legitimate question of readers who asks themselves why such a significant percentage of the community budget has been allocated to the Common Agricultural Policy.

10.5 Implementation of Community Policies

EU directives are one of the main instruments for the harmonization of legislation that, in the states of the Union, governs activities for which there is shared sovereignty between member states. There can be a great delay between the moment of formulation of a community policy by means of a directive and its subsequent enforcement in each and every member state. The European Commission supervises timely promulgation and enactment of implementation laws regarding the directives in the national corpus iuris of each state. Any delay can lead to interinstitutional interventions and even to an appeal at the Court of Justice.

Figure 10.1 shows the reduction in delay of the transposition of directives regarding the internal market into national legislation. Delays and complex harmonization problems can develop in cases where, in the member states, the regions and not the central administration are responsible for legislation concerning the community directive. In such cases, a further level of transposition of the directive is introduced. An example of this type of situation is the delays experienced by some member states regarding environmental protection laws [Sbragia 2000].

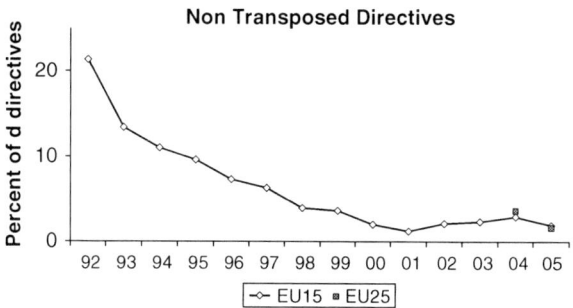

Fig. 10.1 In the period 1992–2005, the percentage of nontransposed directives on the implementation of the internal market decreased. In 2005, the AC10 performed better than the old EU countries: the EU15 value is 1.9 % and the EU25 value is 1.6 %. (European Commission. Internal Market Scoreboard (2002))

Another cause of delay can relate to the method of adoption of the directive. If the directive concerns an area of activity for which decisions were taken by the Council by qualified majority, then in those states where the government did not favour the directive, situations may occur at the parliamentary level that will delay the promulgation of the implementation laws. Finally, some national laws require implementation regulations issued by administrative bodies and this process may lead to further delays.

Legislation has an actual impact on the behaviour of persons and companies only if control systems have been set up to ensure observance of the rules and to generate repressive measures and sanctions to guard against infringement of those rules.

The description above explains how in practice there may be a discrepancy between community policies and their implementation in the territory, or a possible lack of homogeneity in implementation by the various member states and by the different regions of a given state.

The European Commission is responsible for supervising the implementation of the directives. The Commission exercises its role as supervisor by evaluating the violations of the community law in the member states and opening cases, which may even lead to an appeal at the Court of Justice. Figure 10.2 shows the trend of the number of cases of infringements of the law concerning the internal market and documents a significant increase in the number of cases in the last 10 years.

Recently the situation has got worse as regards both transposition and infringements. Almost all member states showed an increase in the transposition deficit between November 2002 and May 2003 (see Fig. 10.3). Cases of infringement increased by 6 % in the same period. Figure 10.4 shows the balance of infringements of the community law concerning the implementation of the internal market as of February 2003.

In the 50 years of life of the Union, a large community corpus iuris has been created (called *acquis communautaire*) that must be acquired by the states requesting accession and must then be transposed into their national legislation. The

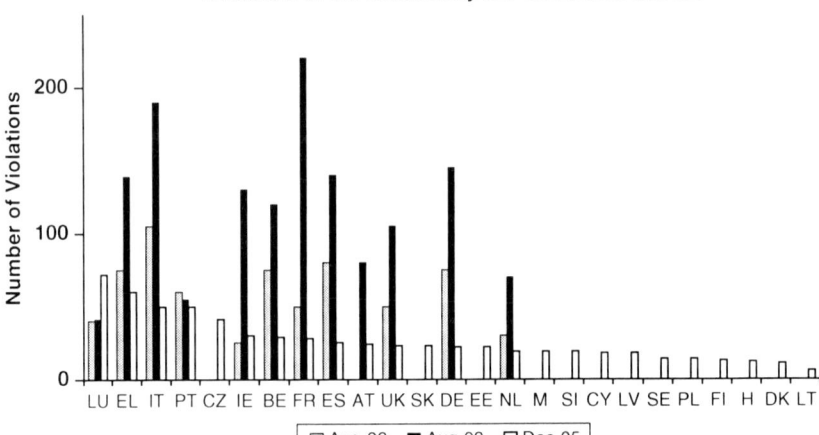

Fig. 10.2 Increase in violations of the community law on the implementation of the internal market in the period 1992–2005. (European Commission. Internal Market Scoreboard (2002))

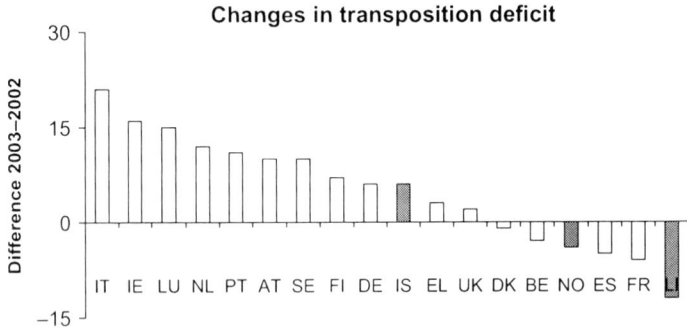

Fig. 10.3 Changes in transposition deficit of directives on the implementation of the internal market from November 2002 to May 2003. EFTA countries are reported in *light grey*. Negative values represent an improvement. Within EU15, only DK, BE, ES, and FR reduced the transposition deficit. (European Commission. Internal Market Scoreboard (2003))

complexity of this operation must not be underestimated and the energy and time necessary to implement completely the acquisition and harmonization process must not be underrated.

The less than satisfactory situation of the increase in the transposition deficit in 2003 gave rise to remedial actions, with a view to improving the process in consideration of the expected increased complexity of the transposition process with the upcoming enlargements. The remedies were very successful, and the AC10 countries performed better than the EU15 countries, lowering the EU25 average percentage of non transposed internal market directives to 1.6 in 2005, while the EU15 average was 1.9 (see Fig. 10.5).

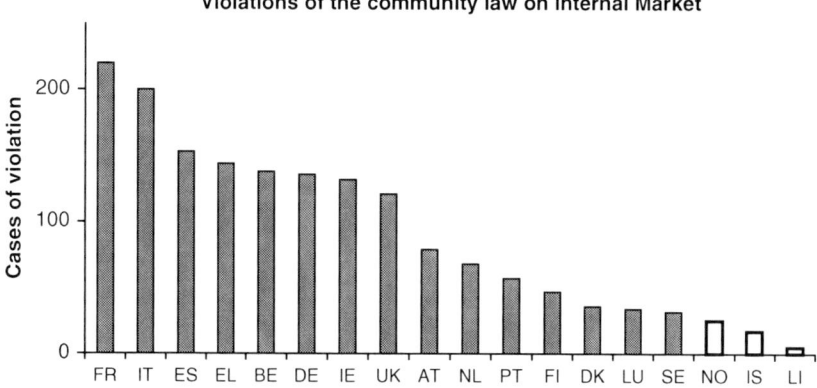

Fig. 10.4 Number of cases of violation of the community law on the implementation of the internal market in February 2003. EU15 data are in *grey* columns; EFTA data are in *white* columns. (European Commission. Internal Market Scoreboard (2003))

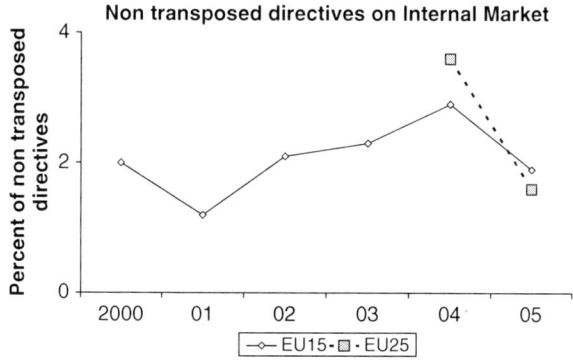

Fig. 10.5 Percentage of non transposed directives in the community law on the implementation of the internal market from 2000 to 2005 in EU15 and EU25. (European Commission. Internal Market Scoreboard (2003, 2005))

10.6 Recommended Reading

Recommended reading for this chapter are listed in Appendix C and are available on the web site companion to this book, at the URL http://stajano.deis.unibo.it/RQC.htm

Chapter 11
Enlargement of the European Union

The 13th of December 2002 is an historic date for the European Union. Only 13 years after the fall of the Berlin Wall (1989) and the unravelling of the Soviet Union (1991), the European Union decided to let 12 other countries join in order to create an area of peace, development, and solidarity in Europe. Ten of the 12 countries, namely, the Czech Republic, Cyprus, Estonia, Hungary, Latvia, Lithuania, Malta, Poland, Slovakia, and Slovenia, accessed the EU in 2004; two other countries, Bulgaria and Romania, accessed in 2007. This model will be extended to the entire Balkan Peninsula over the next decades, as heralded by the start of negotiations with other countries in the Balkans, namely, Croatia and the Former Yugoslav Republic of Macedonia, in 2005. The European Council of December 2004 decided that 3 October 2005 would be the starting date for the negotiations on Turkey's accession to the European Union.

Mr. Romano Prodi, European Commission president, said in an interview for the BBC reported by George Parker [Parker 2004-2]:

> The EU would probably be complete once it had taken in the three outstanding applicants – Bulgaria, Romania and Turkey – and the countries of the western Balkans, including Croatia and Serbia.

After the fall of the Roman Empire, several unsuccessful attempts were made to unite Europe. They failed because the approach was based on domination and conquest. The EEC and now the EU are proposing a new model: history has proved that democracy cannot be exported through the force of arms [Prodi 2004-1]. German Chancellor Gerhard Schröder, attending a ceremony on the border of Poland and the Czech Republic on 1 May 2004, said, "Those who lived through the Second World War and its aftermath would not have thought this possible" [Parker 2004-2].

Ambassador Renato Ruggiero, minister of foreign affairs of the Italian Republic until the end of 2001, made the following statement on enlargement at Montecitorio Palace (the lower Italian Chamber) on 31 October 2001:

> *Enlargement:* it is an irreversible process of great moral and political value. The fall of the Berlin Wall would be meaningless if we didn't try to bring all the countries of this continent back together in one Europe. As suggested by the Madrid European Council in 1995, enlargement is a categorical imperative, a political necessity, and a great social and economic opportunity. The present members of the Union have realized that enlargement

is necessary in order to overcome residual nationalisms and open up a new chapter in the history of Europe. In due time we must also let Russia join in this process, as it is rightly demanding to be a natural member of the Union. The creation of a single economic space including Russia is the key to the building up of this new European dimension.

In the following sections of this chapter, the accession countries and the candidate countries are described. A further extended analysis of the geographical, demographic, and economic data of these countries in comparison with EU15 is presented in Chapter 2, "Overview of Member States" and in Chapter 3, 'Competitiveness of the European Economy". Among the descriptions of the candidate countries, a special focus is on Turkey because the discussion on enlargement to Turkey is a test of different visions about identity and values of the European Union. Section 11.6, "The 2004 Enlargement Three Years On," analyses the situation three years after the 2004 accession. Recommended reading at the end of this chapter and at the end of Chapter 2, "Overview of Member States" (available on the recommended reading web site) include more data on the accession countries.

11.1 What is There to Gain?

What is there to gain? And who is gaining from it? There is only one answer to these two basic questions that the man in the street is asking: both the older 15 members of the Union and the accession countries gain from enlargement. The first effect of enlargement is the creation of a space of peace, stability, and prosperity in Europe, characterized by economic growth and new jobs both in the new and old member states; the quest for peace in Europe, waiting to happen for half a century, is now coming about in the whole of Europe, including the Balkans, and the EU may become a paradigm for the rest of the world. The second effect is an improvement in the quality of life (environment, fight against crime, and fight against illegal immigration). The third effect is the opportunity to strengthen the role of the Union in the world.

Specific benefits for the new countries derive from: prospects of political stability and sustainable development; balancing of public accounts; new jobs; improvements in competitiveness; EU financial support for investments in infrastructures; foreign investments; and more trade in the enlarged internal market.

Specific benefits for the states of EU15 include: possibility of moving manufacturing activities to regions where a qualified workforce is available at lower costs; in increased competitiveness; increase in trade in the enlarged internal market; improved environmental control; improved control on crime, and illegal immigration.

Norman Davies reported in the *Financial Times* [Davies 2004]:

> The enlargement will extend the common market by an area that contains over 70 million people: the population of the incoming members is similar to that of Germany. What is more, their infrastructure is in urgent need of improvement and their consumers are in the early stages of raising their buying power. This means that the more developed countries of the EU, which can export goods, services and capital, are being handed a golden opportunity.

Many companies see the enlargement of the EU as highly beneficial, with lower labour costs, simplified tax procedures, and fewer cross-border checks. "Europe needs to expand because it has to evolve," said Ernie Reading, managing director of Autoliv Spring Dynamics, a UK-based manufacturer of spring cassettes for car seat belts that has established manufacturing facilities in Estonia and Hungary [Moules 2004].

These benefits had already started before 2004. Indeed, in central and eastern Europe, new stable democracies have emerged featuring high economic growth rates and offering new employment opportunities. Trade between the 10 countries that became members of the EU in 2004 (AC10) and the previous 15 members of the European Union (EU15) is increasing (favourable balance of trade EU15 to AC10 of €17 billion in the year 2000), as are investments in the new countries, which are constantly improving the quality of their production processes and products. Environmental protection is also improving. The accession countries might turn out to be the economic locomotive that pulls the whole EU15 out of her recession and triggers the growth of the European economy.

11.2 Criteria for Accession to the European Union

Countries applying for accession to the EU must comply with a set of criteria decided by the European Council in 1993. These criteria, often referred to as the Copenhagen criteria, are as follows:

- stability of institutions guaranteeing democracy, the rule of law, human rights, and respect for and protection of minorities
- the existence of a functioning market economy as well as the capacity to cope with competitive pressure and market forces within the Union
- the ability to take on the obligations of membership including adherence to the aims of political, economic, and monetary union
- implementation of the *acquis communautaire*, namely, the transposition in the national legislation of the common body of Community legislation developed over 50 years

The conditions for success are created by actions on both sides: EU15 and the accession countries:

- the new countries must make some adjustments to comply with accession criteria.
- the Union must prepare herself to welcoming them through:

   - an institutional reform through the European Convention and the reform of the Treaty on EU
   - budget measures in support of the transition phase
   - verification of the process of adjustment

Table 11.1 For each candidate country, 33 negotiation tables are set up by the Commission

Accession negotiation tables	
Free movement of goods	Statistics
Free movement of workers	Employment and social policy
Free supply of services	Enterprise and industrial policy
Free Movement of Capitals	Trans European networks
Public procurement	Regional policy
Company legislation	Judiciary and fundamental rights
Intellectual property legislation	Justice, freedom, and security
Competition policy	Science and research
Financial services	Education and culture
Information society and media	Environmental policy
Agriculture	Consumer and health protection
Food safety	Customs Union
Fisheries	External relations
Transport policy	Foreign, security, and defence policy
Energy	Financial control
Taxation	Financial and budgetary provisions
Economic and Monetary Union	

This last point, namely, the verification of the process through which the new countries are bringing their institutions, their legislations, and their economy into line with those of the Union, has been going on for over 10 years. Negotiations for accession took place for each candidate country individually. Negotiations with Turkey and other candidate countries will go on with the same procedures. Negotiations are held on 33 different chapters at as many negotiating tables, dealing with each one of the topics listed in Table 11.1.

As regards budget measures in support of the transition phase, the European Union gave €3 billion a year as financial aid to the 12 candidate countries agreed to in 2002 (corresponding to about 0.5% of their GDP), while the rest of the costs incurred were paid for by the governments and the private sectors of the candidate countries. The GDP of all the 2004 accession countries is in the order of 4% of the GDP of EU15. President Prodi declared on 28 April 2004 [Prodi 2004-2]:

> Even if the EU intervention were as high as 5% of the GDP of the accession countries [namely 10 times the amount granted], this would be only 0.2% of EU15's GDP. This would not be too expensive a way to buy peace after centuries of bloodshed.

The positive effect on the Union's economy is linked to the enlargement of the internal market to include a further 100 million citizens. This will lead to economic growth in the EU countries and, particularly, in the enlargement countries (as experienced after enlargement to Ireland, Portugal, and Spain).

11.3 Accession Countries and the Euro

After joining the Union, all the accession countries will – in due time – adopt the euro, provided they fulfil all the requirements prescribed by the treaty and summarized below. Further details on this point follow in Chapter 14, "Economic and Monetary Union".

- Economic stability inside an open market economy via membership of the "Exchange Rate Mechanism (ERM)"
- Stability of the national currency exchange rate
- Participation in the internal market
- Budget deficit below 3% of GDP
- Public debt below 60% of GDP
- Inflation in line with the EU best-performing countries

On 20 October 2004, the European Commission and the European Central Bank published their respective Convergence Reports on the progress made in the fulfil-ment of the obligations regarding the achievement of EMU by the member states, including the new member states. The ECB's Convergence Report concluded that none of the accession countries assessed fulfilled all the conditions for adopting the euro at that time. The situation evolved so that Slovenia could adopt the euro in 2007. Two other countries qualified in 2007 and have set dates for euro adop-tion: Cyprus and Malta will introduce the euro on 1 January 2008. Six accession countries are in the Exchange Rate Mechanism (ERM) in 2007: Cyprus, Estonia, Latvia, Lithuania, Malta, and the Slovak Republic. Poland, the Czech Republic, and Hungary are taking a slower track, because they are battling to curb budget deficits [Parker 2004-3]; Romania and Bulgaria do not meet yet the requirements to enter ERM with a view to adopting the euro.

11.4 First Wave of Enlargement

A group of 10 countries accessed the EU on 1 May 2004. They are: the Czech Republic, Cyprus, Estonia, Hungary, Latvia, Lithuania, Malta, Poland, Slovakia, and Slovenia. The Copenhagen European Council decided in December 2002 that the above 10 countries could draw up an Accession Treaty with the European Union. This treaty was actually signed in Athens on 16 April 2003.

Each enlargement country (except Cyprus) held a referendum (see Table 11.2) to have accession to the European Union approved by her citizens, and each of the EU15 states ratified the accession Treaty with an act of the national Parliament. Accession countries participated in the election of the European Parliament in 2004, and the European Commission was enlarged in May 2004 to include Commissioners from the new countries. Basic data on the accession countries[1] are presented in Tables 11.3 and 11.4.

---

[1] In the description of the various countries, we will use some economic indicators whose definition can be found in the Glossary.

Table 11.2 Results of
referendum on accession in
candidate countries

Country	Percentage of YES	Date
Czech Republic	77.33	14 June 2003
Estonia	66.83	14 Sep 2003
Hungary	83.76	12 Apr 2003
Latvia	67.44	20 Sep 2003
Lithuania	91.07	11 May 2003
Malta	53.60	08 Mar 2003
Poland	76.87	08 June 2003
Slovakia	92.46	17 May 2003
Slovenia	89.61	23 Mar 2003

Source: European Commission, 2005.

Table 11.3 Accession
countries, first wave (2004).
Geographical and
demographical data

Country	Capital	Surface ($km^2 \times 000$)	Population (million)
Cyprus	Nicosia	9	0.7
Czech Republic	Prague	79	10
Estonia	Tallinn	45	2
Hungary	Budapest	93	10
Latvia	Riga	64	2
Lithuania	Vilnius	65	4
Malta	Valletta	0.3	0.4
Poland	Warsaw	313	39
Slovakia	Bratislava	49	5
Slovenia	Ljubljana	20	2
	Total AC10	**737**	**75**
	EU15	**3,193**	**375**

Source: OECD and European Commission, 2003.

11.5 An Overview of the 2004 Enlargement

We present in the following pages some information about each of the accession
countries. This overview is further expanded in the recommended reading at the
end of the chapter, which also shows images of the accession countries and of the
people living in them. The accessing countries joined the EU with their cultural
heritage and the legacy of ancient traditions. The level of literacy of their population
is higher than that of EU15. Their universities have long and great traditions. Their
contribution to art and science is a testimony of a very high cultural standard. Thirty
one citizens from five of the 10 accession countries are included among the Nobel
Prize laureates up to 2004 (see Fig. 11.2). On 1 May 2004 Norman Davies, an
academic based in Oxford, wrote on the Financial Times [Davies 2004]:

> Now, when enlargement is at last going ahead, the European Union is in a shaky state and
> new questions are being asked. People are no longer just listing the benefits that the EU
> can bring to countries such as Poland or Hungary. They are also wondering what Poland or
> Hungary or Estonia or Slovenia might bring to the EU: to put it crudely, what is in it for us?
> Fortunately, the answer contains a long list of potential positives.

Table 11.4 Accession countries, first wave (2004). Economic data

Country	pc GDP ppp[a] (€ x000) 2006	pc GDP change[d] (%) 2000–05	Average hourly wage (€)[b] 2003	Inflation[c] (%) 2003	Unemployment (%)[c] 2003
Cyprus[e]	21.7	39[f]	18.7	4.0	3.4
Czech Rep	18.7	94	6.0	0.1	9.9
Estonia	16.0	123	5.6	1.3	10.1
Hungary	15.6	119	5.4	4.7	5.9
Latvia	13.1	109	3.5	2.9	8.6
Lithuania	13.5	127	4.3	−1.2	10.3
Malta	17.6	42	9.9	0.4	7.0
Poland	12.6	57	6.7	0.7	20.0
Slovakia	14.8	105	4.7	8.6	15.2
Slovenia	20.5	62	11.2	5.6	11.2
EU15	26.5	44[g]	19.6	2.7	9.3
EU25	24.6				

Source [a]Eurostat 2007; [b]OECD, European Commission, and [Agentur 2003]; [c]World Bank (2005), [WorldFactbook 2005]; [d]World Bank (2007); [e]Data relative to the Greek-Cypriot sector; [f]2000–2004; [g]EU12: the countries in the Euro area.

The entrant states have enlarged the internal market and can also provide a size-able pool of both skilled and unskilled labour. It does not follow, however, that western Europe may be faced with an uncontrollable horde of immigrants. In some instances, western companies may choose to relocate to the east, to make use of cheaper labour on the spot. In other cases, they may import labour. The fact is that many countries with rapidly ageing demographics are short of workers. In this respect, workforce and skills are an important consideration [Davies 2004].

11.5.1 Cyprus

Cyprus, the third largest island in the Mediterranean, is situated at the crossroads of the European, Asian, and African continents (see Fig. 11.1). The size of Cyprus is $9,000 \, \text{km}^2$ (of which one-third is the Turk-Cypriot sector), about one-tenth of the size of Portugal. Because of her strategic position, Cyprus has played a major part in the history of the eastern Mediterranean from ancient times. The population of the island is 700,000, of which 18% are Turkish-Cypriots.

During the Middle Ages, the island became one of the world's richest countries. In the Modern Age, the Ottoman Empire conquered Cyprus to the Venetians, and the local Greek Orthodox Christian community was faced with the immigration of Ottoman Muslims. Both communities lived in all parts of the island with peaceful coexistence and little intermarriage [Wolleh 2002].

With the decline of the Ottoman Empire in the 19th century, the Great Powers of Europe became concerned about the situation in Cyprus. Britain saw her trade routes towards India threatened. In 1878, London concluded a military treaty with Turkey called the Convention of Defensive Alliance. It assigned the island of Cyprus

Fig. 11.1 The enlargement of the European Union includes 10 countries -the Czech Republic (CZ), Cyprus (CY), Estonia (EE), Hungary (HU), Latvia (LV), Lithuania (LT), Malta (MT), Poland (PL), Slovakia (SK), and Slovenia (SI) – that accessed the EU in 2004 and four candidate countries of which two – Bulgaria (BG) and Romania (RO) – will access the EU in 2007. Three more countries – Croatia (HR), the Former Yugoslav Republic of Macedonia (MK), and Turkey (TR) – are candidates for accession

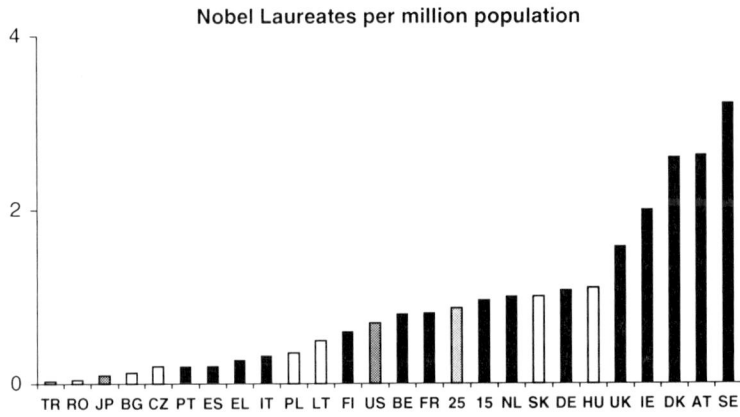

Fig. 11.2 Nobel laureates per million population in the Triad countries. The accession countries are in white. The numbers 15 and 25 stand for EU15 and EU25. Japanese, U.S., and EU25 average are in gray, EU15 average and EU15 countries are in black. Luxembourg has two Nobel laureates, but is not presented in the chart. (The Nobel Foundation 2005)

to be occupied and administered by Great Britain and ended 300 years of Ottoman rule. Cyprus became once again strategically important as a permanent base for the British fleet. The British rule secured the sea route to India via the Suez Canal, which had been opened in 1869; however it gave rise to nationalistic movements and upset the equilibrium between the two communities in the island that had peacefully coexisted since the 16th century. In 1914, Great Britain annexed the island after Turkey had entered the war on Germany's side. In 1925, Cyprus was declared a British Crown Colony. Taxation and the British refusal to consider "$Ένωσης$" – Union with Greece – led to a rebellion of Greek-Cypriots in 1931. Within the Turkish-Cypriot community, a nationalistic movement considered dividing the island in two ethnic communities. In 1955 the Greek-Cypriot National Organization of Cypriot Fighters (EOKA) went in arms for $Ένωσης$ against the British rule; the British used the Turk-Cypriots against EOKA and the relations between the two communities started to worsen [Wolleh 2002] in a process that seems today hopeless, despite the dedicated efforts of senior diplomats and negotiators and the personal intervention, over the years, of five UN Secretaries General [Ker-Lindsay 2005].

Cyprus received independence in 1960. Tensions between the Greek-Cypriot majority and Turkish-Cypriot minority came to a head in December 1963, when violence broke out in the capital of Nicosia. Despite the deployment of UN peacekeepers in 1964, sporadic violence continued, forcing most Turkish-Cypriots into enclaves throughout the island. In 1974, a Greek-sponsored attempt to seize the government was met by military intervention from Turkey, which soon controlled more than a third of the island. The two ethnic communities that had been present over the whole island were separated in the northern Turkish-Cypriot sector and the southern Greek-Cypriot sector that were (and still are) divided, since 1974, by a thin buffer zone occupied by the United Nations Forces in Cyprus (Unficyp).

In 1983, the Turkish-held area declared herself the *Turkish Republic of Northern Cyprus (TRNC)*, but it is recognized only by Turkey. The Republic of Cyprus applied for EU membership in July 1990 and her application was endorsed by the EU in June 1993. Turkey feared that the EU accession of the Republic of Cyprus might delay or impede her own candidacy. On 4 November 2004 the Turkish prime minister, Bulent Ecevit, announced that Turkey would consider annexing Northern Cyprus if the Republic of Cyprus would join the Union [Ker-Lindsay 2005]. Greece pushed for accession of the Republic of Cyprus as part of the AC10 2004 enlargement, reserving to veto the accession treaty, should Cyprus not be included in the deal. Turkey did not recognize the legitimacy of the Republic of Cyprus and supported the intransigent position of the TRNC leader, Rauf R. Denktash.

The Cyprus dispute affected the relations between Turkey and the EU and between Turkey and Greece. The EU put a condition to the start of the accession negotiations with Turkey, asking that Ankara should recognize the Republic of Cyprus. Relations between Turkey and Greece have improved greatly over the past few years and the Greek government (that in the 1990s had offered the intervention of rescue teams on the occasion of a major earthquake in Izmir) supported the Turkish request for accession.

In April 2004, the two-year round of UN-brokered direct talks – between the leaders of the Greek-Cypriot and Turkish-Cypriot communities to reach an agreement to reunite the divided island with a federal solution by constituting the United Cyprus Republic – ended when the Greek-Cypriots rejected the UN unity plan in a referendum. George Lacovou, then Cypriot minister for foreign affairs, wrote [Lacovou 2004]:

> Even though it has not been possible for the Greek-Cypriots to accept the proposed UN plan for the settlement of the Cyprus problem, the Greek-Cypriots continue to desire and strive for the reunification of Cyprus in conditions of security, rule of law, and respect of human rights and fundamental freedoms. The vision of the government of Cyprus is to upgrade the role of the country as a regional trade and service center and a base from which European and international trade can expand in the wider area of the southeastern Mediterranean and the Middle East.

Hopes for a solution were suggested in 2007, when the dismantling of the wall dividing Nicosia was reported by the Financial Times and other international media [Hope 2007, Euractiv 2007]; this event did not, however change the status of deadlock in the Cyprus dispute.

The policy of the EU with regard to the Turkish Cypriot community was set out by the General Affairs Council on 26 April 2004, just before Cyprus joined the EU [EC 2004-11]:

> The Turkish Cypriot community has expressed their clear desire for a future within the European Union. The Council is determined to put an end to the isolation of the Turkish Cypriot community and to facilitate the reunification of Cyprus by encouraging the economic development of the Turkish Cypriot community. The Council invited the Commission to bring forward comprehensive proposals to this end with particular emphasis on the economic integration of the island and on improving contact between the two communities and with the EU.

Although only the internationally recognized Greek-Cypriot controlled *Republic of Cyprus* joined the EU on 1 May 2004, every Cypriot carrying a Cyprus passport has the status of a European citizen. In other words: the door is open for the integration of the Turkish-Cypriots into the EU. A specific protocol on Cyprus is attached to the Accession Treaty, which foresees that in the absence of a settlement, the application of the *acquis* shall be suspended to the northern part of the island until the Council unanimously decides otherwise, on the basis of a proposal by the Commission.

In December 2004 [EC 2004-14], the agreement to start accession negotiations with Turkey was made conditional to the extension of the Turkey-EU Customs Union to all members of the EU, as a first step towards Turkey's recognition of the Republic of Cyprus. However the adaptation of the Ankara Agreement, taking account of the accession of the 10 new member states should be interpreted only as the as an obligation to opening Turkish ports and airports to Greek-Cypriot ships and planes, as it had been the case already years earlier; recognition would come gradually, according to Turkey, along with the progress of accession negotiations.

11.5.1.1 Economy

Mineral resources include copper, pyrites, chrome, asbestos, and gypsum. Timber is also important. Agricultural products include cereal grains, olives, citrus, potatoes, and cotton; in addition, the Greek sector grows deciduous fruits and wine grapes, and the Turkish side grows tobacco, vegetables, and table grapes. Sheep, goats, poultry, hogs, and some cattle are raised. Fishing is an important industry in the Turkish sector and the Greek side has a strong manufacturing economy (processed foods and beverages, paper, chemicals, textiles, metal products, and refined petroleum). Many European elderly citizens choose Cyprus, as well as Malta, as their second home, or the country in which to spend their retirement years in a sunny environment, much cheaper than many of the cold northern member states, at least for the time being. Other foreigners, notably Russian citizens, buy real estate on the island as an investment. Money laundering might be one of the factors affecting major real estate investments made recently in both sectors.

Basic economic data[2] published by various quoted sources[3] are presented in Table 11.5 for the southern Greek-Cypriot sector and for the northern Turkish-Cypriot sector. More information is available in the recommended reading at the end of the chapter.

The Greek sector is considerably more prosperous than the Turkish side, which is heavily dependent on aid from Turkey: Ker-Lindsay, the Director of Civilitas Research (an independent policy-orientated think tank in Cyprus), reports that Turkey spends 250 $ million/year to support the Turkish-Cypriots [Ker-Lindsay 2005].

The Republic of Cyprus during the accession negotiations proved to have one of the soundest economic records among the AC10 and in 2007 qualified for the adoption of euro on 1 January 2008. The economy is, however very strongly dependent in both sectors upon international tourism (15–20% of GDP in the south), which makes it vulnerable and dependent on political instability and general economic trends and political atmosphere in Europe and Asia.

The legal tender is the Cypriot Pound (CYP) in the Greek-Cypriot sector (1 € = 0.60 CYP on 14 May 2003 and 1 € = 0.58 CYP on 6 May 2007) and the Turkish Lira (TRL) in the Turkish-Cypriot sector (1 € = 1.76 TRL on 14 May 2003 and 1 € = 1.84 TRL on 6 May 2007). The Republic of Cyprus will adopt the euro on 1 January 2008 at the rate of 1 € = 0.585274 CYP. The Cypriot government has prepared a National Changeover Plan and setup a Strategic Communication Campaign for the adoption of the euro aiming at informing every single citizen in Cyprus about the transition from the Cyprus pound to the euro. Moreover it aims at clarifying that this change is very simple and that it is a process which involves all citizens personally.

---

[2] For definitions of the economic indicators presented, see Glossary.

[3] In this table as well as in other tables in this Chapter, in some case the value of an indicator in a given year published by the different sources is not quite the same, and different values are reported one after the other. This discrepancy reflects the difficult process of convergence of statistical procedures used by different national or supranational statistical services or research bodies.

Table 11.5 Basic economic data for Cyprus

Greek-Cypriot sector		2000	2001	2002	2003	2004	2005	2006
GDP[c]	($bln)		10.5	13.3	15.8	16.9	18.2	
GDP[i]	($bln)	9.2	9.5	10.5	13.2	15.4	17.1	18.3
GDP[h]	($bln)							16.4
GDP[f]	(€bln)			11.2	11.8	12.7	13.6	14.5
pcGDP ppp[h]	($x000)				19.2			22.7
pcGDP ppp[c]	($x000)							22.2
GDP Growth[c]	(%)			2.0	1.8	4.2	3.9	3.8
GDP Growth[h]	(%)				1.9			3.7
Current Account Balance[h]	(%GDP)				−4.0			−5.9
Current Account Balance[c]	(%GDP)			−3.6	−2.2	−5.2	−5.56	−6.1
Current Account Balance[f]	(%GDP)			−4.5	−6.3	−4.1	−2.4	
Government Balance[h]	(%GDP)				−5.7			−1.8
Total Debt[h]	(%GDP)				62.3			64.8
Total Debt[f]	(%GDP)			65.2	69.7	71.7	70.3	
Inflation (CPI)[c]	(%)			2.8	4.1	2.3	2.6	2.5
Inflation (CPI)[h]	(%)				4.1		2.8	
Unemployment[h]	(%)				3.4		5.5	
Exports[h]	(%GDP)				7.7			8.2
Imports[h]	(%GDP)				34.0			35.4
School enrollment, tertiary[a]	(%)	20	22	25	32	3		
Internet Users[f]	(/1000)		247	300				
FDI Inflows[c]	(%GDP)							7.9[d]
Stock market capitalization[a]	(%GDP)	48	65	48	36	32		
€Exchange Rate[f]	(CYP)			0.576	0.575	0.584	0.582	0.577
GDP % in agriculture[h]					4.1		3.7	
GDP % in industry[h]					20.3		19.6	
GDP % in services[h]					75.6		76.8	

Turkish-Cypriot sector		2000	2001	2002	2003	2004	2005	2006
GDP[c]	($bln)		0.91	0.94	1.28	1.77	2.34	2.67
GDP[h]	($bln)				1.2[g]			4.54[g]
pcGDP ppp[h]	($x000)				5.6			7.1
GDP Growth[h]	(%)				2.6			10.6
Current Account Balance[c]	(%GDP)		−1.9	1.5	1.5	−0.8	−11.8	−10.6
Total Debt[h]	(%GDP)				62.3			
Inflation (CPI)[c]	(%)		76.8	24.5	12.6	11.6	2.7	11.4
Inflation (CPI)[h]	(%)				12.6	9.1		
Unemployment[h]	(%)				5.6	5.6		
Exports[h]	(%GDP)				3.8			1.5
Imports[h]	(%GDP)				25.1			26.4
GDP % in agriculture[h]					10.6			
GDP % in industry[h]					20.5			
GDP % in services[h]					68.9			

Source: [a]The World Bank (2005, 2007); [h]http://www.cia.gov/cia/publications/factbook; [d]Average Average 2002–2006; [c]The Economist Intelligence Unit, 2007; [f]Eurostat, 2007; [g]Purchasing Power Parity (ppp); [i] International Monetary Fund.

Tax rates are low. The corporate tax rate was harmonized at 10% for all companies from 2005. Value-added tax is charged at the EU minimum rate of 15% on most goods (some goods are exempt or subject to reduced rates). The employers' social insurance contribution rate is 6.3% of gross wages – the same as for employees – while the self-employed pay 11.6%. The government pays 9.3% for public-sector employees, while their contribution is only 3.3%.

The European Commission analysis and recommendations in the 2004 Broad Guidelines of the Economic Policies include [EC 2004-10]:

> The small economy, with tourism as a mainstay, has shown some resilience in the face of a difficult international environment. The relatively good growth performance was supported by a market economy dominated by the private sector and with, *inter alia*, a highly developed financial and legal system, generally prudent economic policies, and a flexible labor market. Policies in Cyprus should aim at achieving a high degree of sustainable convergence, in particular: as regards the consolidation of public finances, ensure a reduction of the general government deficit on a sustainable basis; to increase the diversification of the economy towards higher value added activities, step up efforts to increase the adequacy of skilled human capital, promote R&D and innovation, in particular in the business sector, and improve conditions to facilitate ICT diffusion.

The fall 2007 European Commission Economic forecast for Cyprus [EC 2007-27] is presented as recommended reading at the end of Chapter 14, "Economic and Monetary Union,". It is summarized as "healthy growth fuelled by strong domestic demand" and it includes the following:

> Economic activity in Cyprus remained strong in the first half of 2007, recording a growth rate of $3^{1}/_{3}$%. Economic activity has being mainly driven by robust domestic demand.
> GDP is projected to continue growing firmly at just below 4%. Private consumption will continue to be robust. It will be mainly driven by construction, underpinned not only by strong demand for dwellings by non-residents, but also by large infrastructure and other projects. Moreover, as in January 2008 Cyprus joins the euro area, confidence effects should also sustain total investment.
> Following the developments in the oil and commodity markets, HICP inflation accelerated in the first nine months of 2007. As a result, HICP is expected to average at about 2% for 2007. HICP inflation is projected to increase slightly to 2.3% in 2008.
> For 2007, the government deficit is projected at 1.0% of GDP. Debt is projected to keep on a decreasing path attaining about $49^{1}/_{2}$% of GDP by 2009 (it was 65.2% in 2006).

11.5.2 Czech Republic

This landlocked country is situated in the geographic centre of Europe (see Fig. 11.1). The size of the Czech Republic is 79 thousand km^2, or about as large as Austria. The population of the Czech Republic is 10 million. Almost 95% of the people are Czech, with small minorities of Slovaks, Germans, Poles, Gypsies, and Hungarians. Coal and lignite are in abundant supply. There are also deposits of mercury, antimony, tin, lead, zinc, and iron ore, as well as a number of major European uranium deposits.

The territory of the Czech Republic was part of the Austro-Hungarian Empire, and has been until 1938 one of the most economically developed and industrialized parts of Europe. As the only country in Central Europe to remain a democracy before the Second World War, the then Czechoslovakia was among the 10 most developed industrial states of the world.

Following the First World War, the closely related Czechs and Slovaks of the former Austro-Hungarian Empire merged to form Czechoslovakia. After Second World War, Czechoslovakia fell within the Soviet sphere of influence. In 1968, an invasion by Warsaw Pact troops ended the efforts of the country's leaders to liberalize Communist party rule and create *socialism with a human face*. Anti-Soviet demonstrations the following year ushered in a period of harsh repression. With the collapse of Soviet authority in 1989, Czechoslovakia regained her freedom through a peaceful *velvet revolution*. On 1 January 1993, the country underwent a *velvet divorce* into her two national components, the Czech Republic and Slovakia. Now a member of NATO (1999), the Czech Republic has moved towards integration in world markets, a development that poses both opportunities and risks.

The attractiveness of the Czech Republic and especially of her capital city, Prague, lies in her remarkable historical and architectural heritage stretching back over 1,000 years, and brings over 10 million visitors a year to the Czech Republic. Throughout the centuries, Prague has preserved her unrivaled richness of historical monuments of different styles. Castles and chateaux built in the past centuries still dominate the Czech landscape.

Czechs are dependable and skilled workers, and punctual and businesslike people. "The Bohemians are not bohemians," wrote Veronika Fantová, an expert in the Czech National Resource Centre for Vocational Guidance [Fantová 2004].

The Czech Republic, a developed economy with a GDP per capita of around 80% of the European Union average, has been recovering from recession since mid-1999. Growth in 2000–2001 was led by exports to the European Union, especially Germany, and foreign investment, while domestic demand is reviving. The rate of corruption remains one of the highest among OECD countries. The government is currently looking over several reform plans to cut the deficit to 3.0%, but pension and healthcare reforms are continuing without clear prospects for agreement and implementation. The prime minister, Mirek Topolanek of the Civic Democratic Party, won a confidence vote in parliament in January 2007 and runs a coalition without absolute majority in Parliament. A government's public finance reform bill has been given first-reading approval by parliament. However, winning approval in the two subsequent readings will be difficult, owing to dissent and the reform package is therefore likely to be watered down considerably [Economist 2007-16].

11.5.2.1 Economy

Agriculture plays a comparatively small role alongside traditional engineering and other industries. The chief crops are corn, sugar beets, potatoes, wheat, barley, and rye. Among the country's livestock are hogs, cattle, sheep, and poultry. In state hands during the Communist era, most of the Czech Republic's agricultural and industrial sectors were privatized relatively quickly and showed appreciable growth in the early 1990s. Foreign investment was widely sought. An economic slowdown beginning in 1997, however, revealed problems in the transition from government control to a privatized economy, as many large industrial conglomerates with thousands of employees lost money and sought government aid instead of revamping. In 1999–2000, most of the state-owned banks were privatized, with the government assuming responsibility for bad loans.

Manufacturing is the chief economic activity, especially the production of cars (Skoda is a Czech company), machine tools, and machinery. Processing industries (iron and steel, chemicals, and agrifood) are highly developed. Other industries include electronics and glass. After the fall of communism, trade was reoriented to the EU, reaching 84% of total trade in the year 2006 (Germany is first with 34% of total trade, followed by and Slovakia with 9%). Recent efforts have increased trade with many other countries including Russia and China. The largest supplier is EU25 (70%). Accession to the EU gave further impetus and direction to structural reform. In early 2004, the government passed increases in the Value Added Tax (VAT) and tightened eligibility for social benefits with the intention to bring the public finance gap down. However, significant increases in social spending in the run-up to June 2006 elections slowed the move towards meeting this goal. Negotiations on pension and healthcare reforms are continuing without clear prospects for agreement and implementation. Privatization of the state-owned telecommunications firm Cesky Telecom took place in 2005. Intensified restructuring among large enterprises, improvements in the financial sector, and effective use of available EU funds should strengthen output growth [Factbook 2007].

The current account deficit has declined as demand for Czech products in the European Union has increased. It widened substantially, to $ 6.1 billion, in 2006, as a solid trade performance was offset by increased repatriation of profits by foreign-owned companies [Economist 2007-4]. Inflation is under control: at or below EU15 average in the 2000s.

The national currency is the Czech Koruna (CZK); 1 € = 31.44 CZK on 14 May 2003 and 1 € = 28.14 CZK on 6 May 2007. The government is considering a target date of 2012 for euro adoption, implying that the Czech Republic should enter the EU's exchange-rate mechanism (ERM) in 2009 at the latest.

In September 2003 parliament approved a gradual reduction in the corporate tax rate to 24% by January 2006, from 31%. The two lowest personal income-tax rates were reduced to 12% and 19% in January 2006, from 15% and 20%, respectively, with the top rate remaining at 40%. Indirect taxes play an increasingly important role within the tax regime. Since joining the EU, most goods and services are subject to VAT at 19%.

The European Commission analysis and recommendations in the Broad Guidelines of the Economic Policies [EC 2004-10] are presented as recommended reading at the end of the Chapter 14, "Economic and Monetary Union,". Those relative to the Czech Republic point to: labour market, the business environment, the deficiencies of the education system, the skills mismatch between offer and demand, and the slow transition to a knowledge-based economy. They include:

After the economic slowdown of 1997–1998, the Czech Republic's annual real GDP growth averaged about 3 per cent. Inflation reached very low levels (1.4 per cent in 2002 and –0.1 per cent in 2003) and the monetary authorities were undershooting the inflation target. The growth performance has been underpinned by restructuring efforts, but output has not yet fully reached its estimated potential.

Policies in the Czech Republic should aim at achieving a high degree of sustainable convergence, in particular as regards the consolidation of public finances. The economy has been characterized by relatively low growth in comparison with the other new member states. Therefore, the further consolidation of public finances needs to be supported by policies which assist in accelerating the growth performance into the medium and long term.

This requires, in particular, an effort to tackle the remaining structural shortcomings, notably in the labour market and in the business environment, and to speed up the transition towards a knowledge-based economy. Against this background, the Czech Republic faces four major challenges:

1. urgently ensure a further reduction of the general government deficit on a sustainable basis and the long-term sustainability of public finances,
2. continue to address the structural problems in the labour market,
3. improve conditions for an accelerated productivity growth,
4. promote entrepreneurship and SMEs.

Factors contributing to low productivity growth could be the limited flexibility of the education system, the low effectiveness of R&D and innovation, and the low use of ICT. The limited flexibility of the education and training system in responding to the changing skills requirements manifests itself through the qualification mismatches in the labour market. Educational attainment (the percentage of the population aged 20–24 with at least upper secondary education) is very high (92 per cent), but the share of tertiary graduates is one of the lowest in the EU. Despite a relatively high level of expenditure on R&D in comparison with the new members, the resulting innovation activity, as measured by the number of patents, is very low, pointing to low effectiveness of R&D. There seems to be only limited cooperation between the research institutions (e.g. universities and the Czech Academy of Sciences) and the private sector.

Recognizing these problems, the government has adopted the National R&D Policy for 2004–2008 to promote R&D but results will largely depend on its full implementation.

Despite the progress in establishing a well-functioning competition framework and continuous developments in the legal environment there seem to remain significant barriers to entrepreneurship in the Czech Republic. This is reflected *inter alia* in the low share of GDP accounted for by SMEs (less than 40%). Factors contributing to this situation appear to be the weaknesses in the business environment which have a disproportionately heavier impact on SMEs, e.g. regarding regulations and legal enforcement, excessive administrative burden on companies and limited access to finances.

Meeting the challenges outlined above requires broad structural reforms: reduce the general government deficit; reform the health care and pension systems to ensure their financial sustainability; strengthen labor supply by reforming tax benefit systems to eliminate disincentives to work and enhance occupational and regional mobility by reducing skill mismatches; improve the efficiency and quality of the education and training system and its responsiveness to changing skills requirements.

Table 11.6 presents basic economic data published by various quoted sources. More information is available in the recommended reading at the end of the chapter.

The fall 2007 European Commission Economic forecast for the Czech Republic [EC 2007-27] is presented as recommended reading at the end of the chapter. It is

Table 11.6 Basic economic data for the Czech Republic

Czech Republic		2000	2001	2002	2003	2004	2005	2006
GDP[a]	($bln)	56.7	61.8	75.2	91.2	108.2	124.4	
GDP[i]	($bln)	55.7	60.9	73.8	90.4	107.1	125.7	33.2
GDP[b]	($bln)			69.5				118.9
GDP[e]	($bln)		75.3	91.4	108.2	124.0	141.8	
GDP[f]	(€bln)			80.0	80.9	87.2	99.7	113.1
pcGDP ppp[b]	($x000)			15.7				21.6
GDP Growth[a,e]	(%)	3.6					6.1	4.6[e,h]
GDP Growth[e]	(%)			1.9	3.6	4.2	6.1	6.1
GDP Growth[b]	(%)					2.9		6.2
Current Account Balance[a]	(%GDP)			−6.5				
Government Balance[b]	(%GDP)			−6.3	−6.0			−2.0
Government Balance[f]	(%GDP)			−6.8	−6.6	−2.9	−2.6	
Total Debt[b]	(%GDP)			38.1				29.1
Total Debt[f]	(%GDP)			28.8	30.0	30.6	30.5	
Inflation (GDP deflator)[a]	(%)		5	3	1	4	1	
Inflation (CPI)[b]	(%)			1.4	−0.1			2.7
Inflation (CPI)[e]	(%)			1.8	0.1	2.8	1.9	2.5
Unemployment[b]	(%)			7.3	9.9			8.4
Exports[a]	(%GDP)		65	60	62	71	72	
Imports[a]	(%GDP)		68	62	64	72	70	
High-tech Exports[a]	(%Manuf. Exp.)		10	13	13	13	13	
School enrollment, tertiary[a]	(%)	29	31	35	37	43		
Internet Users[a]	(/1000)		147	255	235	252	269	
ITC Expenditure[f]	(%GDP)				2.7	2.8	2.9	
€ Exchange Rate[f]	(CZK)			34.068	30.804	31.846	31.891	29.782
FDI Inflows[a]	(%GDP)		9.1	11.3	2.2	4.1		6.3[e]
Stock market capitalization	(%GDP)	19	15	21	19	29	31	
GDP % in agriculture[b]					3.1		2.8	
GDP % in industry[b]					35.5		37.8	
GDP % in services[b]					61.4		59.4	

Source: [a]The World Bank (2005, 2007); [b]http://www.cia.gov/cia/publications/factbook; [e]The Economist Intelligence Unit 2007; [f]Eurostat, 2007; [h]Average 2002–2006; [i]International Monetary Fund.

summarized as "moderating growth buoyed by consumer demand" and it includes the following:

> Growth has been mainly propelled by strong consumer demand. External demand has remained robust, with exports slightly outpacing imports. The strong rise in consumer spending has been underpinned by a number of factors including climbing wage levels, falling unemployment, expanding credit, moderate inflation and an increase in social benefits. On the supply side, the largest contribution has come from manufacturing, where growth in output has been spread over most sectors, in particular, electronics, transport equipment, and the automotive sector.

The government deficit dropped to below 3% of GDP in 2006, and is expected to worsen to about $3^1/_2$% of GDP in 2007. The government debt ratio is likely to be constrained to about $30^1/_2$% of GDP by 2009, partially benefiting from privatization revenue.

11.5.3 *Estonia*

Estonia is a northeastern country (see Fig. 11.1), with a climate of icy, snowy winters, and long-light summers. Tallinn, Estonia's capital city is about 80 kilometres south of Helsinki, across the Gulf of Finland. Estonia's surface area is 45 thousand km^2, about the same size as The Netherlands. Estonia is mostly flat, with many lakes and islands; much of her land is farmed or forested, with industrial production concentrated around Tallinn and in the northeast. Estonia is sparsely populated with around 1.4 million people. About demography, Brian Groom, a journalist of the Financial Times, reports [Groom 2004]:

> Europe has got used to thinking of herself as an irredeemably ageing continent, doomed to a futile struggle to compete with younger rivals. Now, from the far northeast corner of the enlarged EU, comes a tiny message of hope. Estonia is having a baby boom.

The capital, Tallinn, is an important port and one of the best-preserved medieval cities in Europe. It is a city of grey towers topped with red tiles, of stone stairs beneath arching gateways, and of narrow winding streets, cobbled pavements and towering ramparts.

The independent Republic of Estonia was born in the aftermath of the First World War in 1918. It was subsequently occupied by the Soviet Union (1940–1941, 1944–1991) and Nazi Germany (1941–1944). A resurgence of Estonian national identity began in the late 1980s. The most evident (but peaceful) protests occurred in 1988, when large numbers of Estonians came together to sing national songs in the so-called *singing revolution*, and in 1989, when people across all three Baltic countries joined hands to form a massive human chain.

In 1991, Estonia declared the restoration of her independence, which was quickly recognized by other countries. Since then, Estonian governments have pursued a liberal free-trade policy and have embraced new technologies, resulting in a rapid transformation to a market economy. In 1993, Estonia signed a free-trade agreement with her fellow Baltic states, Latvia and Lithuania. Reforms carried out after 1992 were comprehensive and systematic.

Estonia, and to a lesser extent the other countries washed by the Baltic Sea, are experiencing an unprecedented boom whereby Nordic rigor, the creativity of the Slav people, western capitals, and an unpolluted environment are attracting the giants of European economy: Nokia, Volvo, ABB, to name a few. Estonia's magic moment is visible in her success in attracting foreign investment. Her GDP has been growing between 7% and 11% in the period 2000–2006; in 2005 it was 2.4 times its value in the year 2000, with the largest increase among the accession countries. The Economist wrote in May 2007: "[Estonia is] the closest that eastern Europe has to a real economic and political success story" [Economist 2007-5].

This move is not without problems for EU15. Christopher Brown-Humes wrote in the *Financial Times* of 15 May 2004 [Brown-Humes 2004]:

Estonia's entry into the EU, while officially welcomed, has caused widespread unease in Finland. Wages and corporate tax rates are far lower in the Baltic state, which lies just across the Gulf of Finland and whose capital Tallinn is only 18 minutes by helicopter from Helsinki. Some Finnish companies, such as Elcoteq, a sub-contractor of Nokia, have a substantial amount of production in Estonia.

The country began small-scale privatization in 1991 and during the 1990s auctioned off several larger industries; it has also actively sought foreign investment. The nation exports light industrial products, machinery, food, wood products, textiles, and electric power. A recent development is the Estonians' love of information technology. Estonia is the leading country for Internet connections per capita among the accession countries. It even ranks ahead of EU member states such as Belgium or France. In August 2000, the government of Estonia changed her cabinet meeting to paperless sessions, using a web-based document system. Around the country, there were, in 2005, 700 public Internet access points (PIAP), which guarantee free Internet access to all. The location of PIAPs is advertised on the roads of the country with special traffic signs that have not been seen so far in other EU states. Internet access is a constitutional right, granted by law to all Estonian citizens in February 2000. All Estonian schools are connected, with one computer every 20 students on average.

It would not be realistic, however, to ignore the problems that Estonia is faced with in her transition and growth: conditions in the labour market have deteriorated gradually following the country's independence. The skill mismatch resulting from the restructuring of the economy brought to a rise in unemployment to $14\frac{1}{2}$ per cent by mid-2000, while the employment rate declined from 65 per cent in 1997 to 60.6 per cent in 2000; the level of productivity is very low compared to the EU average (42% of the EU15 level in 2002). While labour productivity growth remained strong (7% per annum) over 1997–2000, it has slowed down since 2000 due to the lack of qualifications in the workforce and the low level of R&D and innovation [EC 2004-10].

11.5.3.1 Economy

As a result of the transition to a new economic system, Estonia's gross domestic product (GDP) decreased sharply in the years 1991–1994. By 1995, the recession phase was over, and economic growth was the fastest in 1997. Because of a crisis in the financial sector, foreign demand began to decline in 1998. The same year saw a crisis in the Russian market, and as a result, Estonia's GDP decreased by 1.1% in 1999. In 2000, the growth rate of Estonia's economy increased rapidly to 6%, driven by economic integration with EU member states. This high rate of growth continued in 2001. Important exports include machinery and electrical equipment, wood, and textiles products. Tourism and transit trade also make an important contribution to the economy. Finland and Sweden are amongst Estonia's biggest partners in business, investment, and tourism. Estonia's major trade partners include the Russian Federation but relations with Russia where compromised in spring 2007 with Russia over-reacting to the removal of a Soviet-era war memorial from Tallinn.

The pace of growth of the Estonian economy outpaced the European average several fold in recent years, wrote Juhan Parts, the young Estonian prime minister [Parts 2004]. Economic growth is the prerequisite for catching up with the *old* Europe. Estonia enjoys a low level of public debt (5% of GDP), continues to follow the principles of a balanced budget, with a policy aimed at sustainable economic growth, and further development of the internal market, while complying with the Stability and Growth Pact. Estonia supports the implementation of the Lisbon strategy in improving EU competitiveness.

In the years when it was part of the Soviet Union, Estonia provided the USSR with gas and oil produced from her large supply of oil shale. It is still the world's second largest producer of oil shale. The majority of her workforce is involved in industry, which also includes shipbuilding, phosphate mining, and the manufacture of electronics and telecommunications equipment, electric motors, excavators, cement, furniture, textiles, and clothing. Her efficient agricultural sector employs some 20% of the labour force and produces meat (largely pork), dairy products, potatoes, flax, and sugar beets. Fishing is also important. Peat, limestone, dolomite, marl, clays (for cement and earthenware), sand (for the glass industry), phosphorite (for fertilizers), and timber are important natural resources.

The national currency is the Estonian Kroon (EEK); 1 € = 15.65 EEK on 14 May 2003 and 1 € = 15.6465 on 6 May 2007. Estonia entered the EU's Exchange Rate Mechanism in preparation of euro adoption at a date not yet fixed at the moment. Table 11.7 presents basic economic data published by various quoted sources. More information is available in the recommended reading at the end of the chapter.

Estonia has a "flat" system of personal income tax, with the single tax rate set at 22% in 2007. The government is committed to reducing this rate by 1% point per year until it reaches 18% in 2011. However, social security contributions are set at 33%. Estonia currently only levies corporate income tax on distributed profits, which has led to tensions with some other EU members. Value-added tax (VAT) is levied at 18%, although medicines, books and periodicals have a VAT rate of 5%.

The European Commission analysis and recommendations in the Broad Guidelines of the Economic Policies [EC 2004-10] are presented as recommended reading at the end of Chapter 14, "Economic and Monetary Union,". Those relative to Estonia point, among other things, to the lack of qualifications in the workforce and the low level of R&D and innovation. They include:

> Against the background of weak external demand, macroeconomic performance in Estonia remained solid in 2003, but the current account deficit widened to 13.7 per cent of GDP. GDP growth of 4.8 per cent was underpinned both by buoyant investment and private consumption growth. Inflation, which had accelerated to almost 7 per cent in mid-2001, receded to 1.3 per cent in 2003. The general government surplus widened to 1.8 per cent of GDP in 2002 and further to 2.6 per cent in 2003, on account of strong activity and improved tax collection, and despite additional spending approved by parliament.
>
> Policies in Estonia should aim at achieving a high degree of sustainable convergence, and a narrowing of the current account deficit. Estonia faces four major challenges:

Table 11.7 Basic economic data for Estonia

Estonia		2000	2001	2002	2003	2004	2005	2006
GDP[a]	($bln)		6.0	7.0	9.2	11.2	13.1	
GDP[i]	($bln)	5.5	6.0	7.0	9.1	11.2	12.3	13.5
GDP[b]	($bln)				8.4			13.6
GDP[e]	($bln)			7.3	9.6	11.6	13.8	16.4
GDP[f]	(€bln)			7.8	8.5	9.4	11.1	13.1
pcGDP ppp[b]	($x000)				12.3			19.6
GDP Growth[a]	(%)		6.0	7.0	7.0	8.0	10.0	
GDP Growth[b]	(%)				4.7			9.8
GDP Growth[e]	(%)			8.0	7.1	8.1	10.5	11.4
GDP Growth[e]	(%)							9.0[h]
Current Account Balance[b]	(%GDP)				−12.1			−14.8
Government Balance[b]	(%GDP)				2.4			2.7
Government Balance[f]	(%GDP)			1.02	4.0	1.5	1.6	
Total Debt[b]	(%GDP)				7.4			3.6
Total Debt[f]	(%GDP)			5.5	6.1	4.0	4.8	
Inflation (GDP deflator)[a]	(%)		6	4	2	3	6	
Inflation (CPI)[b]	(%)				1.3			
Inflation (CPI)[e]	(%)			3.6	1.3	3.0	4.1	4.4
Unemployment[b]	(%)	14.5[c]			10.1			4.5
Exports[a]	(%GDP)		84	74	74	78	84	
Imports[a]	(%GDP)		87	81	82	86	90	
High-tech Exports[a]	(%Manuf. Exp.)		19	12	13	14	18	
School enrollment, tertiary[a]	(%)	56	60	63	64	65		
Internet Users[a]	(/1000)		315	327	443	497	513	
ITC Expenditure[f]	(%GDP)				2.3	2.7	2.9	
FDI Inflows[a,e]	(%GDP)		9.1	4.0	10.0	8.6	22.9	10.7[e,h]
€ Exchange Rate[f]	(EEK)			15.647	15.647	15.647	15.647	15.647
Stock market capitalization[a]	(%GDP)	34	25	35	41	55	27	
GDP % in agriculture[a]				5	5	4	4	4
GDP % in industry[a]				27	28	28	29	29
GDP % in services[a]				68	68	67	67	67

Source: [a]The World Bank (2005, 2007); [b]http://www.cia.gov/cia/publications/factbook; [c]European Commission, 2004; [e]The Economist Intelligence Unit, 2007; [f]Eurostat, 2007; [h]Average 2002–2006; [i]International Monetary Fund.

1. address the sizeable current account deficit, to which an appropriate fiscal policy should contribute,
2. address the structural problems in the labor market,
3. improve conditions for increasing productivity,
4. develop effective competition in network industries.

Two main factors could constitute a handicap for future increases in productivity growth: the lack of qualifications in the workforce and the low level of R&D and innovation. These two factors also explain that FDI are mainly concentrated in areas that do not require important R&D and qualifications. The lack of qualifications in the workforce results from inefficiencies in the education system, as public expenditures on education are relatively high (6.7 per cent of GDP in 2000). Particularly worrying is the fact that the education system is not sufficiently close to the needs of the labor market, creating a mismatch.

A well-designed R&D policy can also have an important role in increasing productivity growth. Currently Estonia has a relatively low level of R&D expenditures, below that of other European countries (0.8 per cent of GDP in 2001), and the lowest rate of business R&D among the new members. This is partly due to the weak links between the academic sector and the business community. In this area, a new strategy has been approved which foresees a significant increase in total R&D expenditures over the period 2002–2006.

Estonia is recommended to:

1. implement a fiscal policy aimed at narrowing of the current account deficit
2. adopt policies with particular emphasis on re-integrating the long-term unemployed, by promoting vocational training and life-long learning, and by setting up an effective institutional framework that is supporting job creation
3. encourage social partners to ensure that wage developments – including changes to the minimum wage legislation – do not hinder employment growth, and hold up the recovery in the labor market
4. improve the efficiency and quality of the education system and vocational training in order to reduce the mismatch on the labor market
5. implement the R&D strategy approved in 2001 and, in particular, promote the stronger involvement of the business sector in R&D spending
6. proceed with the liberalization of the electricity market, strengthen the independence of the regulator and ensure effective competition in telecommunications

The fall 2007 European Commission Economic forecast for Estonia [EC 2007-27] is presented as recommended reading at the end of Chapter 14, "Economic and Monetary Union,". It is summarized as "slower growth contributes to reducing imbalances" and it includes the following:

After having grown at a rate above 10% for a year and a half, the Estonian economy started to decelerate at the beginning of 2007 with the rate of economic growth reaching 8.8% year-on-year during the first half.

In 2007 the population of working age started to shrink due to a natural decrease, with little net impact assumed from net migration. Some of those who left the country in the wave of opening labor markets in the EU will return. However, limited scope for further increases in the labor force will be a significant constraint on increasing capacity going forward.

Shortages of skilled labor in the context of strong growth triggered a leap in wages in the first half of 2007. Robust wage growth is expected to continue in 2008 and 2009, driving up labor costs. Inflation has worsened in 2007 more than foreseen, reflecting domestic demand pressures, developments in some world market prices, and administrative price increases. This has resulted in an upward revision of the HICP inflation forecast for 2007 from the spring forecast to $6^1/4\%$.

In 2007 tax revenues are expected to again considerably exceed those initially planned, resulting in a general government surplus around 3% of GDP. The level of public debt, already the lowest in the EU (4.8% in 2005), will continue to decrease in 2009.

11.5.4 Hungary

Hungary is a landlocked country in central Europe (see Fig. 11.1). Her size is 93 thousand km^2, about the same as Austria. The Danube River forms part of Hungary's northwestern border with Slovakia, and then flows south through Budapest, dividing Hungary into two regions. Since the inauguration of the canal linking the Danube to the Main in 1992, goods can be carried from the Black Sea to the North Sea.

Budapest, located along the two embankments of the Danube, is the largest city and the capital. It is also the cultural, economic, and industrial centre of Hungary. The population of Hungary is about 10 million, of which approximately 1.8 million live in Budapest.

Hungary is a highly musical country; her violinists and pianists are celebrated virtuosi worldwide. Hungary was the homeland of Franz Liszt, Béla Bartók, and Zoltan Kodaly, whose music was inspired by the rich national folk traditions. Hungary has more than 5,000 public libraries and more than 100 public museums are maintained throughout the country. The Hungarian education system produces very capable programmers, engineers, and mathematicians. It is no surprise that the Hungarians may pride themselves on the large number (11) of Nobel Prize winners the country has produced over the years [Albert 2004]. Hungary has, among the EU countries, the highest number of books published yearly per capita. The University of Szeged is the best-performing university in the accession countries, according to a study [Shanghai Univ. 2004] ranking world universities, reported in Chapter 3, "Competitiveness of the European Economy".

Hungary was part of the polyglot Austro-Hungarian Empire, which collapsed during the First World War. The country fell under Communist rule following Second World War . In 1956, a revolt and the announced withdrawal from the Warsaw Pact were met with a massive military intervention by Moscow. During the 1956 revolt, Cardinal Joseph Mindszenty, who had been imprisoned by the regime in 1949, was liberated and received asylum in the U.S. embassy, where he lived until 1971. Mindszenty has been presented by the U.S. and Vatican propaganda as the symbol of that part of the Hungarian nation that opposed the Communist rule; through him the silent Hungarian Catholic Church would have played the role of preserving the national identity during the period of Communist regime. The view of Hungarian citizens on the role of Mindszenty is controversial and not comparable with the perception expressed by Lithuanian or Polish citizens on the role played by the Catholic Church in their country during the cold war.

Under the leadership of Janos Kadar in 1968, Hungary began liberalizing her economy, moving the first steps towards market economy, and introducing what in the west was called *goulash Communism*. Hungary held her first post-communist multiparty elections in 1990 and "by the second half of the decade, over 80% of the country's GDP already came from the private sector," wrote Dénes Albert [Albert 2004], a Reuters correspondent from Budapest.

Hungarian foreign policy has been marked by continuity since 1990, even though four different governments have ruled the country since then. Membership in the EU has been the overriding strategic objective. Hungary joined NATO on 12 March 1999; it provided airfields and logistical support for the NATO air campaign against the Federal Republic of Yugoslavia (FRY)[4] in March/April 1999.

---

[4] FRY, the Federal Republic of Yugoslavia, consisted, in 2007, of two Republics (Serbia and Montenegro) and two autonomous provinces (Kosovo and Vojvodina). Kosovo unilaterally declared independence from Serbia proclaiming the Republic of Kosovo on 17 February 2008.

The present government is a centre-left majority coalition, led by Ferenc Gyurcsany and comprising the Hungarian Socialist Party and the Hungarian Liberal Party. The coalition took office in June 2006 and currently has 210 of the 386 seats in parliament. A tape leaked in September 2006 revealed him admitting "that his government had systematically lied about the country's parlous public finances to win the parliamentary elections in April" [Economist 2006-4]. The government's support is weak, owing to the introduction of fiscal austerity measures and the audiotape scandal. There is a risk that Mr. Gyurcsany could be ousted before the end of term (2010), should the government's popularity remain low.

One week after the signing of the Lisbon Treaty by the members of the European Council, Hungary became the first of the 27 European Union countries to ratify the reform Treaty on EU in a parliamentary vote on 17 December 2007. "There is a fundamental agreement on the question whether Hungary's place is in the cooperation for European integration," Prime Minister Ferenc Gyurcsany said before the vote.

11.5.4.1 Economy

Hungary has long been an agricultural country. Slightly over 50% of Hungary's land is arable. With highly diversified crop and livestock production, the country is self-sufficient in food and in the mid-1990s was making about 15% of her export earnings from agriculture. Corn, wheat, barley, sugar beets, potatoes, sunflower seeds, and grapes are the major crops. Pigs, cattle, and sheep are raised. Through the 1980s, two-thirds of agricultural output came from collective and state farms.

Hungary has been an important producer of bauxite and deposits of copper, natural gas, coal, oil, and uranium have also been exploited. Mining was drastically curtailed in the 1990s as the country moved to a market economy and found that it was not cost-effective to exploit the country's minerals at world prices. Gas and oil production gradually declined owing to the exhaustion of reserves. Hungary since Second World War has become heavily industrialized. Industry is well diversified; major products include steel, chemicals, pharmaceuticals, cement, processed food, textiles, and motor vehicles. About one third of Hungarian industry is located in or near Budapest. Other industrial centres are Gyr, Miskolc, and Pécs. Through the 1980s, industry was largely nationally owned. Hungary's economy underwent difficult readjustment in the 1990s making the transition from a centrally planned to a market economy, with a per capita income nearly two-thirds that of the EU-25 average. It moved from producing goods chiefly for export to the USSR to developing a market-based economy and finding new trading partners. By the end of 1995, almost all retail trade had been privatized, and less than half of all economic output originated from state-owned enterprises. Economic reforms also brought high unemployment and inflation. Unemployment has persisted above 6%. Hungary's labour force participation rate of 57% is one of the lowest in the Organization for Economic Cooperation and Development (OECD). Inflation that reached 14% in 1998 had remained high around the 10% mark until mid-2001 but finally started to decline, to a year-on-year rate of 4.6% in September 2002 and to 3.7% in 2006.

This followed the adoption of an inflation-targeting monetary policy by the National Bank, with the aim of achieving *price stability* in 2005.

Today Hungary continues to demonstrate strong economic growth and her economy is one of the most prosperous in eastern Europe. Her GDP is made up of 65% services and 4% farming. Hungary is the stronghold of foreign high-tech investments from Germany and Japan, attracted by competitive total labour costs, a liberal foreign trade regime, and a predictable and business-friendly policy framework. The factory in Gyor, northwestern Hungary, of the German car manufacturer Audi is the main engine production site for Volkswagen's (VW) Audi subsidiary. Audi is Hungary's largest exporter, with sales of €3.72 billion in 2004.

Since 1997, economic growth has been impressive, with the Hungarian economy recording growth rates around 4% a year. GDP growth at 5.2% in 2000 was the highest since transition, mainly pushed by export growth rates of over 20%. Inflation had remained high around the 10% mark until mid-2001, but finally started to decline, to a year-on-year rate of 4.6% in September 2002. This followed the adoption of an inflation-targeting monetary policy by the National Bank, with the aim of achieving price stability. The current account deficit has been on a declining trend towards low levels of around 3% of GDP, which reflects primarily a slowdown of import-led investment. Public debt in 2001 was 58.5% of GDP. The national currency is the Forint (HUF); 1 € = 244.50 HUF on 14 May 2003 and 1 € = 245.94 HUF on 6 May 2007. The adoption of the euro is not expected until 2013 or 2014.

The corporate tax rate on reinvested profit fell to 16% in 2004. The two-tier system of VAT levies a basic rate of 20%. Pharmaceuticals and a few other items are taxed at 5%. At 40.8% in 2004, the implicit tax on labour in Hungary (all personal income and payroll taxes, plus compulsory social security contributions, expressed as a share of the total compensation of employees) was among the highest in Europe.

The economies of Hungary and the EU are increasingly integrated. In 2001, Hungary's share in the EU's trade with the rest of the world reached 2.4% of EU exports and 2.4% of EU imports. Hungary's overall trade has grown with extraordinary dynamism: while export volumes have increased by more than 50% since 1989, imports have tripled. Hungary's export and import figures with the EU have quadrupled since 1989. The trade deficit (€1.62 billion in 2000 and €1.25 billion at the end of 2001) reflects Hungary's efforts to reequip the economy with capital goods. Her main trading partners are other countries in the EU (especially Germany, Austria, and Italy). The trade deficit remains high, at 4.7% of GDP, covered by the inflow of foreign direct investment.

The large majority (84%) of Hungarian citizens had said *yes* in the 2003 referendum about accession (Table 11.2). In 2006 the Eurobarometer reports a more lukewarm attitude towards the EU [EC 2006-2].

This was the first time in the history of Eurobarometer surveys in Hungary when the figure for those who thought membership offered no advantages was higher than that of people who thought membership was advantageous. Over the past six months, there has been a dramatic change in Hungary in the way people expect their own lives and the situation of the country to develop during the next year. At a poll in the spring of 2006, two-thirds of those surveyed thought their personal situation would improve during the following 12 months, but this percentage had dropped by more than half by the fall. An even bigger decrease was evident in expectations on the development of the country's economic situation. A quarter of those interviewed expected an improvement in the spring of 2006, but this figure had dropped to 8% by September. The Eurobarometer survey also sought to find out what people thought were the most important challenges their country was facing. In the list compiled by Hungarians, the development of the economic situation was at the top, followed by unemployment. During the past six months, there has been a significant increase in the number of people who thought increasing prices were a major problem, so this issue replaced the healthcare system in third position.

These findings confirm the personal impression gained by visiting Hungary: while young people see prospects of a prosperous future coming from EU membership, many citizens in their late forties and older experience difficulty in finding their way in a liberalized open-market economy where entrepreneurship is demanded of them, after a lifetime of dependency from a centrally-controlled economy that did not induce the development of personal initiative.

Table 11.8 presents basic economic data published by various quoted sources. More information is available in the recommended reading at the end of the chapter.

The European Commission analysis and recommendations in the Broad Guidelines of the Economic Policies [EC 2004-10] are presented as recommended reading at the end of Chapter 14, "Economic and Monetary Union,". Those relative to Hungary cover the whole spectrum of the economy and include a quest for educational and research investments. They include:

Following a relatively high real GDP growth of 3.5 per cent in 2002, real GDP growth in 2003 slowed to below 3 per cent. While unemployment is among the lowest of the acceding countries, the participation rate is very low compared to the EU average. Serious structural shortcomings in the labor market render further increase in employment difficult: notably a lack of regional mobility, skill mismatches and disincentives from the benefit systems. After the high fiscal deficit of 9.3 per cent of GDP in 2002, fiscal policy became restrictive again in 2003.

Policies in Hungary should aim at achieving a high degree of sustainable convergence, in particular as regards the consolidation of public finances. Hungary implemented a number of structural reforms and has achieved important catch-up with the EU. However there are areas which pose further challenges to the country. Hungary should aim to adopt measures in order to make the education system more effective and R&D has to be encouraged in order to achieve higher productivity. Competition in the network industries needs to be further improved and the independence of the regulators has to be strengthened.

While showing a slight improvement since the beginning of 2003, Hungary's cost competitiveness has shown a significant deterioration in the years 2000–2002. The decline in cost competitiveness was partly due to the fast increase in wages. Real wage growth exceeded productivity growth in the last three years, led by a rise in minimum wages and by the public sector, but also by a delayed adaptation of the corporate sector to the new low-inflation environment.

Productivity growth has been decreasing since 2000. Factors contributing to the slowdown in productivity growth have been the relatively low level of R&D (0.95 per cent of GDP in 2001), inefficiencies in the education system, and instability in the business environment.

The number of total tertiary graduates in science and technology per 1000 of population (aged 20–29) has been decreasing during the last years: in 2001 it was 3.7 which is one of the lowest in the EU. In the field of research and development, applied research is lagging behind, partly due to the still low level of business R&D expenditures, despite fiscal

Table 11.8 Basic economic data for Hungary

Hungary		2000	2001	2002	2003	2004	2005	2006
GDP[a]	($bln)		52.3	65.6	83.1	100.8	109.2	
GDP[i]	($bln)	46.4	51.8	64.9	82.8	99.4	107.1	115.4
GDP[b]	($bln)		51.8					113.1
GDP[e]	($bln)			66.7	84.4	102.2	11.6	112.0
GDP[f]	(€bln)			70.1	74.7	82.3	88.8	89.2
pcGDP ppp[b]	($x000)				13.9			17.3
GDP Growth[a]	(%)		3.8	3.3	3	5	4	
GDP Growth[b]	(%)					2.9		3.8
GDP Growth[e]	(%)			4.3	4.1	4.9	4.2	3.9
Current Account Balance[b]	(%GDP)		−2.1					−7.4
Government Balance[b]	(%GDP)					−9.4		−9.6
Government Balance[f]	(%GDP)			−8.4	−6.4	−5.4	−6.1	
Total Debt[b]	(%GDP)		58.5	53.1	57			68.6
Total Debt[f]	(%GDP)			55.1		57.1	58.4	
Inflation (GDP deflator)[a]	(%)		8	9	7	4	3	
Inflation (CPI)[b]	(%)				4.7			3.7
Inflation (CPI)[e]	(%)			5.4	4.9	6.8	3.7	4.1
Unemployment[b]	(%)				5.9			7.4
Exports[a]	(%GDP)		73	64	62	64	66	
Imports[a]	(%GDP)		74	66	66	68	69	
Exports Growth[a]	(%)	14.1[i]	9.1					
High-tech Exports[a]	(%Manuf. Exp.)		24	25	26	29	25	
School enrollment, tertiary[a]	(%)	37	40	45	52	60		
Internet Users[a]	(/1000)		145	158	237	267	297	
ITC Expenditure[f]	(%GDP)					2.1	2.4	2.4
FDI Inflows[a]	(%GDP)			7.5	4.6	2.6	4.6	5.9
€Exchange Rate[f]	(HUF)			256.59	242.96	253.62	251.66	248.05
Stock market capitalization[a]	(%GDP)	26	20	20	20	28	30	
GDP % in agriculture[a]				4	4	3	4	
GDP % in industry[a]				32	31	30	31	
GDP % in services[a]				64	66	66	65	

Source: [a]The World Bank (2005, 2007); [b]http://www.cia.gov/cia/publications/factbook; [e]The Economist Intelligence Unit, 2007; [f]Eurostat, 2007; [i]International Monetary Fund.

incentives, and limited cooperation between companies and research institutes. Public R&D expenditures are also relatively low, not only as a share of GDP but also as a share of total government expenditures (R&D expenditures by the government represented 0.9 per cent of total government expenditures in 2001).

Meeting the challenges outlined above requires to:

1. reduce the general government deficit in a credible and sustainable way

2. strengthen labor supply by removing obstacles to regional mobility

3. ensure that the tax and benefit systems support employment and provide incentives to enter or remain in the labor market
4. encourage a reform of wage setting which allows wages to better reflect productivity
5. promote the stronger involvement of the private sector in R&D and innovation, strengthen the link between business and research institutions, ensure sufficient resources to improve the quality of research and support the transfer of knowledge through FDI
6. improve the efficiency of the education system, increase its flexibility in order to better adapt to the skill needs of the market and ensure adequate resources for vocational training and education
7. ensure stability of legislation and government policies to create a business environment, more supportive to entrepreneurship
8. proceed with the liberalization of the network industries, increase effectiveness of competition and the independence of the network regulators

The fall 2007 European Commission Economic forecast for Hungary [EC 2007-27] is presented as recommended reading at the end of Chapter 14, "Economic and Monetary Union,". It is summarized as "private-sector-based gradual economic recovery after slowdown" and it includes the following:

In 2007, economic growth has slowed down considerably, reflecting to a large extent the consolidating efforts of the government which started in mid-2006. Although GDP growth was surprisingly weak in the second quarter this year (0.1 % quarter-on-quarter) it is expected to pace up and reach 2% in 2007. Economic performance is characterized by a pronounced divergence in its main components, with a marked contraction in domestic demand counterbalanced by a strong export performance.

The economy is expected to gradually recover in both 2008 and 2009, as a result of improving domestic demand and a supportive external environment. Throughout the forecast horizon, the private sector is anticipated to increase its contribution to growth, in parallel with continued public sector retrenchment. In reaction to the substantial decrease in real disposable income, households reduced their consumption only moderately while further increasing their indebtedness. Private and public investment may further benefit from the increasing inflow of EU structural funds (around 2.5% of GDP).

In 2007, private sector wages rose by more than 10%, though part of the increase may be due to the government's efforts to legalize side payments to workers. Public sector employment has dropped by more than 4%.

After the very high budgetary outturn recorded in 2006 (a deficit of 9.2% of GDP), the deficit is expected to considerably decrease to 6.4% of GDP in 2007 as a result of the fiscal adjustment measures. The general government deficit is expected to decline to 4.2% of GDP in 2008 and to improve slightly to 3.8% of GDP in 2009, due to continuing expenditure moderation. After the strong increase in the debt-to-GDP ratio from 61.6% in 2005 to 65.5% in 2006, a more moderate increase to 66.1% and 66.3% is expected in 2007 and 2008 respectively, chiefly as a result of the declining budget deficit. The debt ratio is projected to fall somewhat to around 66% in 2009.

11.5.5 Latvia

Latvia is one of the three Baltic countries (see Fig. 11.1). Her size is $64,500\,\mathrm{km}^2$, about the same as Ireland. The climate is temperate and wet with moderate winters and long, sunny, warm days in summer. The terrain is flat; 30% is arable, with the land often too wet and in need of drainage. Approximately 85% of agricultural land has been improved by drainage. About one third of the 2 million population (7,47,000) lives in the capital city Riga and its surroundings.

Poland conquered the territory in 1562 and occupied it until Sweden took over the country in 1629, ruling over it until 1721. Then the country passed to Russia. From 1721 until 1918, the Latvians remained Russian subjects, although they preserved their language, customs, and folklore. The Russian Revolution in 1917 gave them their opportunity for freedom, and the Latvian republic was proclaimed on 18 November 1918. The republic lasted little more than 20 years. Plagued by political instability, Latvia essentially became a dictatorship under president Karlis Ulmanis. Latvia was occupied by Russian troops in 1939 and incorporated into the Soviet Union in 1940. German armies occupied the nation from 1941 to 1944. Of the 70,000 Jews living in Latvia during Second World War, 95% were massacred. In 1944, Russia once again took control of Latvia. After the war, Latvia became fully integrated in the Soviet Union and was subject to collectivization and central planning. As the country had a solid agricultural and industrial tradition, much of its production from collectivized farms and from new factories was destined for other parts of the Soviet Union. A significant migration took place, and Russian became the predominant language. As a result, by the late 1980s, native Latvians formed only over half of the population. With the new policies and openness introduced by Mikhaïl Gorbachev, the Soviet Republics, including Latvia, obtained some economic autonomy, and Latvian identity gradually manifested itself again.

Latvia was one of the most economically well-off and industrialized parts of the Soviet Union. When a coup against Soviet President Mikhaïl Gorbachev failed in 1991, the Baltic nations saw an opportunity to free themselves from Soviet domination, and, following the actions of Lithuania and Estonia, Latvia declared her independence on 21 August 1991. Although the last Russian troops left in 1994, the status of the Russian minority (some 30% of the population) remains of concern to Moscow. European countries and most other nations quickly recognized Latvia's independence, and on 2 September 1991, President George H.W. Bush announced full diplomatic recognition of Latvia, Estonia, and Lithuania. The Soviet Union recognized Latvia's independence on 6 September, and UN membership followed on 17 September 1991. Latvia joined the World Trade Organization in February 1999.

Nature and culture – especially music and songs – are very important in the life of Latvians. Literature has also been flourishing. Works of painters and sculptors, including contemporary artists, can be admired and bought in the many galleries of Riga. In the early 1890s, many Art-Nouveau buildings were erected in Riga. Today most of them can still be admired, some of them beautifully restored.

Ms. Vaira Vikke-Freiberga, the president of Latvia, a returned émigré, wrote [Vikke-Freiberga 2004]:

> Latvia has become the outer border of the European Union on the eastern side. Beyond that border we have nations which have suffered the same system of oppression and tyranny but that have not, as yet, managed to go as far as we have in terms of reforms, in terms of structural change in their societies, nor indeed in terms of commitment to values that the rest of Europe represents. We in Latvia like to think of ourselves as a success story of change. We have come a long way in a very short time. We have striven to adopt not only the

acquis communautaire of the EU, but we have striven to adjust to the values and principles that Europe has developed. We think of Europe as a lighthouse that spreads its light to the world in a dialogue of civilizations. The smaller countries have experience of and a need for dialogue. We are aware that we are different, without a sense of inferiority. We come to the EU with our experience of understanding and reconciliation of differences. We small nations are aware of the details. There is a role that small countries may take.

The four-party centre-right coalition under the prime minister, Aigars Kalvitis, commands a majority in parliament, but a serious corruption scandal makes it unlikely that it can survive until the end of his term (2010). The government has announced a set of measures to reduce inflation and plans to balance the budget in 2007–08. A western-style living standard is building up in Latvia at an unexpected speed and inflation might come out of control [Economist 2006-7].

11.5.5.1 Economy

Table 11.9 presents basic economic data published by various quoted sources. More information is available in the recommended reading at the end of the chapter. Dairying and stock raising remain integral to the agricultural sector, which employs more than 15% of the labour force and contributes to 4% of the GDP. Grain, sugar beets, potatoes, and vegetables are also important.

The nation has valuable timber resources. Latvia has been engaged in transforming the state-run economy, inherited from her years as a Soviet republic, into a market economy; reforms include the abolition of price controls and the initiation of privatization. GDP is growing at rates between 3% and over 10% from 2003. It is expected that GDP growth will remain at potential growth of 5–6 per cent per year in the medium term. High international oil prices, increases in food prices and rising wage costs are exerting inflationary pressures; average annual consumer price inflation remained high in 2006, and might rise in 2007, owing to rises in regulated utility prices, and further wage increases. Construction is booming and wages are spiralling to a point that inflation might come out of control and delay the prospects of an early euro adoption. The majority of companies, banks, and real estate have been privatized, although the state still holds sizable stakes in a few large enterprises. The country has encouraged foreign investment. The current account deficit – more than 15% of GDP in 2006 – and inflation remain major concerns.

The national currency is the Latvian Lat (LVL); 1 € = 0.57 LVL on 14 May 2003 and 1 € = 0.70 LVL on 6 May 2007. Adopting the euro is not considered in the short term, inflation being far too high: Latvia entered the EU's Exchange Rate Mechanism in preparation of euro adoption but the Latvian government said in June 2007 that the initial target of 2008 is not realistic and that euro might be adopted by 2012.

Latvia has a "flat" system of personal income tax, with the single tax rate at 25%. The corporate profit tax is 15%, and employers also pay social security contributions at 24% of salary (the employee pays a further 9%) and a real estate tax of 1.5%. Value-added tax (VAT) is levied at 18%, with a lower rate of 5% for medicines, communications and certain utilities.

Table 11.9 Basic economic data for Latvia

Latvia		2000	2001	2002	2003	2004	2005	2006
GDP[a]	($bln)		8.3	9.3	11.2	13.7	15.8	
GDP[i]	($bln)	7.7	8.2	9.2	11.1	13.7	15.4	16.8
GDP[b]	($bln)				9.7			16.1
GDP[c]	($bln)			9.3	11.2	13.8	16.0	20.1
GDP[f]	(€bln)			9.9	10.0	13.0	16.2	
pcGDP ppp[b]	($x000)				10.2			15.4
GDP Growth[a]	(%)		8	6	7	9	10	
GDP Growth[b]	(%)					8.2		10.2
GDP Growth[c]	(%)			6.5	7.2	8.7	10.6	11.9
GDP Growth[e]	(%)							9.0[h]
Current Account Balance[b]	(%GDP)				−8.0			15[c]
Government Balance[b,c]	(%GDP)		−1.6[c]	−3.0[c]	−1.8[b,c]			−1.7[b]
Government Balance[f]	(%GDP)			−2.3	−1.2	−0.9	0.2	
Total Debt[b]	(%GDP)					14.4		11
Total Debt[f]	(%GDP)			13.5	14.4	14.6	11.9	
Inflation (GDP deflator)[a]	(%)		2	4	4	7	9	
Inflation (CPI)[b]	(%)					2.9		6.8
Inflation (CPI)[e]	(%)			1.8	2.9	6.2	6.7	6.5
Inflation (CPI)[e]	(%)							4.8[h]
Unemployment[b]	(%)					8.6		6.5
Exports[a]	(%GDP)		42	41	42	44	48	
Imports[a]	(%GDP)		51	51	55	60	62	
Exports Growth[a]						8.2		7.9[d]
High-tech Exports[a]	(%Manuf. Exp.)		4	4	4	5	5	
School enrollment, tertiary[a]	(%)	56	63	67	71	74		
Internet Users[a]	(/1000)		72	133		350	448	
ITC Expenditure[f]	(%GDP)					1.9	2.1	2.2
FDI Inflows[a,c]	(%GDP)		1.6	2.7	2.7	4.6	4.6	4.6[c,h]
€Exchange Rate[f]	(LVL)			0.56	0.581	0.641	0.665	0.696
Stock market capitalization[a]	(%GDP)	7	8	8	10	12	16	
GDP % in agriculture[a]			5	5	4	4	4	
GDP % in industry[a]			23	23	22	22	22	
GDP % in services[a]			72	72	74	73	74	

Source: [a]The World Bank (2005, 2007); [b]http://www.cia.gov/cia/publications/factbook; [c]European Commission, 2004; [d]Average Average 2002–2006; [e]The Economist Intelligence Unit, 2007; [f]Eurostat, 2007; [h]Average 2002–2006; [i]International Monetary Fund.

Latvia is an important industrial centre, and this sector employs about 40% of the workforce. Her industries are extremely diversified and include the manufacture of motor vehicles, street and railroad cars, synthetic fibres, agricultural machinery, pharmaceuticals, electrical equipment, electronics, information technologies, and textiles. Food and dairy processing, distilling, and shipbuilding are also significant, and tourism has developed as a source of foreign income. The largest trading partner is the European Union (mainly Germany, Sweden, and the United Kingdom), followed by the Russian Federation and other former Soviet republics.

The European Commission analysis and recommendations in the Broad Guidelines of the Economic Policies [EC 2004-10] are presented as recommended reading at the end of Chapter 14, "Economic and Monetary Union,". Those relative to Latvia point to: inadequate education system, workforce mismatch with the market demand, low mobility, low productivity, and low business participation in R&D activities. They include:

Robust economic performance was registered in Latvia in recent years despite a weak external environment. Private consumption and gross fixed capital formation have been particularly strong and are the driving factors for growth.

Latvia made remarkable progress implementing structural reforms in recent years, but still faces some challenges, including the functioning of the public administration, that could impair Latvia's absorption of EU structural funds. These challenges if not fully met, may weaken Latvia's ability to stay on a path of strong growth.

The unemployment rate declined slowly and varies considerably across regions. The smooth functioning of the labor market is hampered by a number of structural problems, including inadequate education and training systems, skills mismatches and low geographical mobility. Competition on Latvian product markets has been boosted by the privatization of most State-owned companies, and the level of State aid remains lower than the EU average. The few remaining competition problems are mainly concentrated in network industries. Although Latvia benefits from low labor costs and taxes, the low level of labor productivity remains a major concern.

Policies in Latvia should aim at achieving a high degree of sustainable convergence, in particular as regards the consolidation of public finances. Both the further consolidation of public finances and solutions to address the underutilization of Latvia's human resources need to be supported by policies which assist in perpetuating the strong growth performance into the medium and long term. This requires in particular a strengthening and diversification of the growth and employment base by fostering the entrepreneurial climate. Furthermore, the productivity level needs to be raised and Latvia should prepare for the eventual transition to a knowledge-based economy.

Labor productivity in Latvia is the lowest in the EU (under 40 per cent of the EU15 average). Over the period 1995–2003, it has increased relatively to the EU average but its growth rate slowed down in recent years. Several factors bear on Latvia's productivity performance. First, the educational system still suffers from efficiency, contents, and external partnerships problems, and the links between higher education and industries are still underdeveloped. Second, R&D – in particular the share of business R&D – and innovation activities appear to be limited, both because of a lack of public funding and the absence of critical mass for most Latvian companies to carry out research activities. Productivity growth is also constrained by the low level of entrepreneurial activity as companies face heavy local regulations, difficulties in accessing finance, and a nascent entrepreneurial culture. Despite high levels of business investments, productivity levels are still held back by low capital deepening and a still relatively low level of physical infrastructure. All these factors also contribute to explain why Latvia remains specialized in relatively low-tech sectors and in transit activities without much value-added.

Meeting the challenges outlined above requires broad structural reforms and Latvia is recommended to:

1. reduce the general government deficit in a credible and sustainable way
2. increase the efficiency, quality and accessibility of the education and training systems, and their responsiveness to the labor markets' needs
3. encourage R&D and innovation, in particular in the business sector

4. encourage an entrepreneurial culture
5. revise the tax and benefit system in order to make work pay more, in particular by increasing the efficiency of social spending
6. strengthen labor supply by pursuing efforts to better adapt the qualifications of the workforce to the requirements of the labor market and by facilitating labor mobility, especially through improvements in transport infrastructure

The fall 2007 European Commission Economic forecast for Latvia [EC 2007-27] is presented as recommended reading at the end of Chapter 14, "Economic and Monetary Union,". It is summarized as "consequences of overheating will linger" and it includes the following:

After growth of 11.9% in 2006, GDP data for the first two quarters of 2007 give very little sign of a slowdown, with 11.2% annual growth in the first quarter and 11.0% in the second. From the expenditure side, growth was still driven by private consumption and investment. There are strong indications that a slowdown of domestic consumption and a correction to the overheated real estate sector have already begun. The government's anti-inflation plan has helped to end the price boom in the residential real estate market and in slowing down mortgage lending. The annual growth of monthly retail sales and the registration of new cars also suggest that a slowdown of the economy has been ongoing in recent months. From the production side, growth is still strongest in those sectors which are satisfying domestic demand. The largest segment, real estate, renting and business activities, has continued its robust growth, but by the second quarter the year-on-year growth rate of its gross value-added fell below 10%, representing a marked slowdown from 18% in 2006. The second largest segment of the economy, wholesale and retail trade has also been very dynamic with a growth rate of around 15% in the first half. In a sharp contrast to these segments, manufacturing has stagnated, largely due to the poor performance of its main component, wood processing. Producers of tradable goods are facing difficulties due to tight labor market conditions, which have led to rapidly increasing unit labor costs and constraints on output.

Continuing rapid real wage growth will provide stimulus for households, but negative wealth effects from falls in nominal house prices, increasing debt service payments and less optimistic expectations will limit private consumption growth. Investment in residential buildings is expected to slow, helping to ease the overheating situation in the construction sector and in the whole economy. However, public investment projects will continue and even increase, which will provide a cushion against a too-sharp adjustment. On the external side, the product composition of Latvian exports does not favor a very rapid expansion of output, but exports are expected to be boosted by the wood industry redirecting sales from domestic to export markets. Overall, GDP growth is expected to remain driven by domestic demand, but the negative contribution of net exports is expected to decrease substantially.

Domestic overheating, unfavorable international price developments (increasing food and energy prices) and regulated price increases have resulted in annual HICP inflation and core inflation reaching double-digit levels (11.5% and 11.1% respectively in October 2007). Inflation is poised to remain high in 2007 and early 2008 as a consequence of food, heating and other regulated price increases and accession-related excise harmonization. Inflation is expected to decline below double-digit levels only from mid-2008 onwards, but the moderation will be slow, as the consequence of heightened inflationary expectations, high wage growth and a sharp increase in the price of imported gas are likely to be felt even in the second half of the year. The rate of unemployment is projected to stay below 6% during the forecast period. However, it is expected that the structure of employment will change, with the construction and retail sectors losing employees in favour of other sectors.

The fiscal outturn for 2006 is reported to be a deficit of $1/2$% of GDP. For 2007, a surplus of about 1% of GDP is projected. The debt ratio, already one of the lowest in the EU (11.9% in 2005), should decrease somewhat further.

11.5.6 *Lithuania*

Lithuania is biggest of the three Baltic states, located at the western end of the east European plain, on the shores of the Baltic Sea (see Fig. 11.1). The size of Lithuania is 65, 301 km^2, almost twice the size of The Netherlands. Lithuania consists predominantly of gently rolling plains and extensive forests.

About 56% of inhabitants (4 million in total) live in the three biggest cities: Vilnius, Kaunas, and port Klaipeda. Vilnius was established as the capital of the Grand Duchy of Lithuania in mid-14th century. Her Renaissance and Baroque old town centre houses Lithuania's oldest university, established by the Jesuits in 1579.

The name *Lithuania* first appeared in written sources in 1009 AD. The modern Lithuanian state was established in 1918. Independent between the two World Wars, Lithuania was annexed by the USSR in 1940 and regained her independence in 1990 after 50 years of foreign rule. Lithuanians are primarily Roman Catholic; the Catholic worship, opposed by the Soviet regime, has contributed to preserving the national identity and its underground practice was for many not only a spiritual experience but also a way to witness and preserve ideals and values different from those imposed by the regime. A strong role was played by the Catholic Church also in Poland and – to a lesser extent – in Hungary.

Mikhaïl Gorbachev's *perestroika* in 1985 had a crucial impact on Lithuania and her liberalization process, namely, the *Singing Revolution*. In June 1988, the first organized opposition to the Communist Party, the Sajudis, was founded and participated in elections to the Congress of Peoples' Deputies, the highest body of the Soviet administration. A mass protest, gathering people from the three Baltic states, was held on the 50th anniversary of the Molotov–Ribbentrop pact on 23 August 1989: about two million people linked hands in a human *Baltic chain* stretching 650 kilometres from Vilnius to Tallinn.

On 24 February 1990, Sajudis won 106 seats out of a total of 114 in the local Supreme Council elections: the Council restored Lithuania's independence on 11 March 1990. Moscow refused to accept this vote and attempted, on 13 January 1991 – while the world's attention was focused on the Iraqi war – to overthrow Lithuania's legitimate government. The crackdown, carried out by armed forces against unarmed citizens, resulted in 14 deaths at the now historical TV tower. Lithuania, which was the first Baltic state to restore independence, paved the way for a peaceful and bloodless restoration process in Latvia and Estonia.

Lithuania joined the United Nations on 17 September 1991, and Russian troops were finally withdrawn from Lithuanian territory on 31 August 1993. Lithuania joined NATO in 2004. Lithuania has gained membership in the World Trade Organization and moved ahead in due time with the transformations required to join the EU.

A four-party coalition government led by Gediminas Kirkilas of the Social Democratic Party is in office from July 2006 until next elections, due in October 2008. The coalition does not have a majority in Parliament but the opposition party Homeland Union agreed to support it. In 2003, Lithuania had been given the name of *Baltic Tiger*, being at the time the fastest-growing economy in Europe, with a two-digit GDP growth [Economist 2003-5]:

look to Lithuania, the southernmost of the three small Baltic states, and for most of the 1990s the quietest and sleepiest of them. Sleepy no more, Lithuania romped home last year as Europe's top-performing economy.

Later however, Lithuania has been mostly a story of scandal and missed opportunity. That may be why as many as 450,000 people – a tenth of the population – have left to work abroad. Today public tolerance of corruption seems to be lessening, and the law-enforcement authorities are showing impressive new vigour in prosecuting it [Economist 2006-6].

Mr. Algirdas Brazauskas, the prime minister of Lithuania at accession – whose left-leaning government collapsed in 2006 due to a corruption scandal – wrote [Brazauskas 2004]:

> The new member states have to become the connecting link between the EU and its eastern neighbors. Tight contacts with the northern countries, the strategic Lithuanian and Polish partnership, and trilateral cooperation between the three Baltic states may become cooperation models for expanding the circle of stable and friendly neighbors.

11.5.6.1 Economy

Primarily agricultural before 1940, Lithuania has since developed considerable industry, including food processing, shipbuilding, and the manufacture of machinery and machine tools, metal products, major appliances, electronic components, motors, textiles, and electrical equipment. In the agricultural sector, dairy farming and stock raising are carried out extensively, and grains, flax, sugar beets, potatoes, and vegetables are grown.

Lithuania has a modern highway system, several international airports, and the major ice-free seaport of Klaipeda. The country is relatively poor of natural resources; however, the United Nations rates it 52nd among 174 world nations according to the human social development index. Lithuania has restructured her economy and is integrating it into the European Union. Lithuania is rapidly modernizing her economy in all productive sectors: services, industry, and agriculture. The countries of the European Union are major investors in Lithuania. The Russian Federation, Germany, Latvia, and Poland were the main trading partners in the 1990s, while now trade has been increasingly oriented towards the west, with main partners being the UK, Germany, Italy, France, and Sweden. Lithuania, the Baltic state that conducted the most trade with Russia, has been slowly rebounding from the 1998 Russian financial crisis. Relations with Russia have worsened considerably following Russia's closure of the pipeline to the Mazeikiai refinery in July 2006. Bilateral tensions are unlikely to improve quickly.

Since 1997, development has been generally positive against a challenging international economic backdrop. On average, real GDP has grown by 3.6% annually and exports by 7.4% annually. The development of private consumption has been relatively stable (4.6% annually). Up to 2003, low and stable inflation has been one of the main achievements of economic policy (from 8.8% in 1997 to -1.4% in the third quarter of 2002). High unemployment, at 12.5% in 2001 up to 13.1% in 2002 and at 10.7 in 2003, and weak consumption have held back recovery. Unemployment fell to 3.7% in 2006, while wages grew 17.6%, contributing to rising inflation to 3.8% in 2006, with negative effects on productivity and FDI.

The current account deficit has fallen significantly (from 12.1% of GDP in 1998 to 4.8% in 2001 and 5.6% in 2003). The current account balance has improved, and at the same time the inflow of foreign direct investment has been relatively stable.

In the 1990s, Lithuania benefited from her adherence to strict fiscal and monetary policies as it followed a program of privatization and increased foreign investment. Important steps have been taken in the area of structural reforms, and privatization of the large, state-owned utilities, particularly in the energy sector, is under way. The level of state aid has declined significantly (from 1.36% of GDP in 1997 to 0.23% in 2000). The stable monetary framework of the Currency Board Arrangement and sound fiscal policy have both contributed to the achievement of internal and external balance. Lithuania has a "flat" personal income tax: the rate fell from 33% to 27% in July 2006, and the government plans to reduce to 24% in 2008. The corporate profit tax of 15% was temporarily increased to 19% in 2006 (the so-called social tax), but cut to 18% in 2007. Employers pay social security contributions at 31% of salary. Value-added tax (VAT) is levied at 18%, with rates of 9% for state-financed construction and renovation, and 5% for hotel accommodation and certain cultural and medical goods.

Table 11.10 presents basic economic data published by various quoted sources. More information is available in the recommended reading at the end of the chapter.

The repegging of the Litas (LTL) from the U.S. dollar to the euro took place smoothly on 2 February 2002 (1 € = 3.45 LTL in 2003; 1 € = 3.45 LTL on 6 May 2007). Lithuania entered the EU's Exchange Rate Mechanism in preparation of euro adoption but her high inflation might delay joining the euro zone until 2011.

According to the Lithuanian Statistics Department, between January and November 2002, 48.9% of Lithuanian exports (main partners: the UK, Germany, Sweden) and 45.3% of imports (main partners: Germany, Italy, France) were made with the European Union. Looking at trade by sectors, machinery and vehicles represent the largest sector for EU exports to Lithuania. Textile and mineral products are the largest sectors for EU imports from Lithuania.

The European Commission analysis and recommendations in the Broad Guidelines of the Economic Policies [EC 2004-10] are presented as recommended reading at the end of Chapter 14, "Economic and Monetary Union,". Those relative to Lithuania point to: unemployment, productivity, deficiencies of the education system, low level of R&D, and workers' skill mismatch with market needs. They include:

> In spite of a weak external environment, Lithuania's macroeconomic performance remained particularly strong in the last two years. Real GDP growth accelerated rapidly to 8.9 per cent in 2003, primarily supported by strong investment and private consumption, although export growth remained robust. A large nominal effective appreciation of the litas, together with strong productivity growth that attenuated wage inflation, led to a decline in price levels.
>
> Policies in Lithuania should aim at achieving a high degree of sustainable convergence. The labor market is a source of concern. Decreasing the high unemployment rate and a further consolidation of public finances will be crucial to enhance macroeconomic stability in

Table 11.10 Basic economic data for Lithuania

Lithuania		2000	2001	2002	2003	2004	2005	2006
GDP[a]	($bln)		12.1	14.1	18.6	22.5	25.6	
GDP[i]	($bln)	11.4	12.1	14.1	18.4	22.2	23.7	25.4
GDP[b]	($bln)				18.2			30.2
GDP[e]	($bln)			14.2	18.6	22.5	25.7	29.8
GDP[f]	(€bln)			15.0	16.5	18.1	20.6	23.7
pcGDP ppp[b]	($x000)					11.4		15.1
GDP Growth[a]	(%)		6	7	11	7	7	
GDP Growth[b]	(%)				6.1			7.4
GDP Growth[c]	(%)				8.9			
GDP Growth[e]	(%)			6.9	10.3	7.3	7.6	7.5
GDP Growth[e]	(%)							7.9[h]
Current Account Balance[b]	(%GDP)		-4.8		-5.6			
Government Balance[c,b]	(%GDP)			-1.4[c]	-1.7[b]			-1.1[b]
Government Balance[f]	(%GDP)			-1.4	-1.2	-1.5	-0.5	
Total Debt[b]	(%GDP)				23.6			18
Total Debt[f]	(%GDP)			22.3	21.2	19.5	18.7	
Inflation (GDP deflator)[a]	(%)		0	0	-1	3	6	
Inflation (CPI)[b]	(%)				-1.2			3.8
Inflation (CPI)[c]	(%)			0.3	-1.2	1.2	2.7	3.8
Inflation (CPI)[e]	(%)							1.3[h]
Unemployment[b]	(%)		12.5	13.1	10.7			3.7
Exports[a]	(%GDP)		50	53	51	52	52	58
Imports[a]	(%GDP)		55	59	57	59	65	
Exports Growth[a]	(%)				23.1			9.4[d]
High-tech Exports[a]	(%Manuf. Exp.)		5	4	5	5	6	
School enrollment, tertiary[a]	(%)	51	57	62	69	73		
Internet Users[a](/1000)			72	144	201	282	358	
ITC Expenditure[f]	(%GDP)					1.3	1.4	1.6
FDI Inflows[a]	(%GDP)			3.7	5.0	1.0	3.4	4.0
FDI Inflows[e]	(%GDP)							3.9[h]
Stock market capitalization[a]	(%GDP)	14	10	10	19	19	32	
€Exchange Rate[f]	(LTL)			3.582	3.459	3.453	3.453	3.453
GDP % in agriculture[a]				7	7	6	6	6
GDP % in industry[a]				31	30	32	33	33
GDP % in services[a]				62	63	62	61	61

Source: [a]The World Bank (2005, 2007); [b]http://www.cia.gov/cia/publications/factbook; [c]European Commission, 2004; [d]Average Average 2002–2006; [e]The Economist Intelligence Unit, 2007; [f]Eurostat, 2007; [h]Average 2002–2006; [i]International Monetary Fund.

the medium term. Preserving Lithuania's competitiveness will be of paramount importance for a rapid and sustained convergence with the EU economies. Further structural changes will be necessary to maintain the current productivity growth trend, which is required to close the substantial productivity gap between Lithuania and the EU average.

Lithuania has had high productivity growth after the Russian crisis in 1999, but the productivity level is still very low at 42 per cent of the EU15 average. Also the recent high productivity growth seems to have been partly caused by one-off effects due to better capacity utilization of existing resources. With a GDP per capita level at 39 per cent of the EU15 average in 2002 a high and sustainable productivity growth is necessary to reduce the income gap with the EU. In this respect, deficiencies of the education system and the low level of R&D and innovation are a handicap.

Firstly, despite high public spending on education and a high number of tertiary graduates there is a mismatch between the skills acquired in the education and training systems and the needs in the business sector.

The government is preparing a program for the implementation of an education strategy stretching to 2012, but further efforts may be necessary to adapt the education and training systems to the future needs as the economic structure develops. Secondly, a large share of the current economic structure is based on low technology activities. A change of the structure will require higher R&D and more innovation, which currently are among the lowest in the new member states. Maintaining high levels of FDI could act as a catalyst through knowledge transfer and thereby contribute to faster structural changes in the economy towards higher value-added sectors and improved productivity. Further development of the physical infrastructure could also contribute to maintaining the high productivity growth. IT expenditure has increased slightly from a very low level and low IT penetration could be a hindering factor for improving productivity.

Despite deregulation in most network industries, effective competition remains weak in all of them, except mobile telephony and road transport. The fixed telephony market was fully liberalized in January 2003, but the incumbent is still the only player in the market. The telephone market regulator does not yet seem to have adequate resources to effectively promote competition. The implementation of the EU legislation on railway market opening has not yet been completed and the infrastructure is poorly developed, especially the interconnections with Poland. The opening-up of the energy markets have led to few visible benefits for consumers and high concentration ratios persist, both in supply and distribution. The electricity market was opened-up for large customers in January 2002, corresponding to a fourth of the electricity consumption. Further deregulation is foreseen in steps and a privatization of electricity distributors is currently underway. However, about 80 percent of all electricity is produced by a nuclear power plant and a lack of interconnection capacity with other acceding countries prevents integration with the EU electricity market. Also the gas market has been liberalized for large customers, corresponding to 80 per cent of consumption. However, there are few independent players in the market and there is no interconnection with the western European gas network.

Meeting the challenges outlined above requires broad structural reforms. Lithuania is recommended to:

1. Enhance regional mobility and reduce skill mismatches, whilst ensuring the efficiency of education, retraining measures and other active labor market policies
2. Improve the combined incentive effects of taxes and benefits
3. Pursue low budget deficits in a credible and sustainable way
4. Avoid pro-cyclical fiscal policies that prevent a further reduction of the general government deficit
5. Improve the efficiency and quality of the education and training systems and their responsiveness to labor market needs
6. Promote R&D and innovation, strengthen the links between research institutes and the business sector and support knowledge transfer through FDI and higher IT penetration
7. Pursue liberalization and enforce effective competition in energy, telecommunication and railway markets
8. Create and improve interconnection capacities with neighboring EU member states

The fall 2007 European Commission Economic forecast for Lithuania [EC 2007-27] is presented as recommended reading at the end of Chapter 14, "Economic and Monetary Union,". It is summarized as "strong growth but high external imbalances and inflationary pressures" and it includes the following:

The expected slowdown has not yet clearly set in, and data for the first half of 2007 suggest that economic growth may accelerate further to around $8^1/_2\%$, thereby increasing risks of overheating. Domestic demand remains the main driver of growth. A large part of investment is directed into the currently booming construction and real estate sectors. Consumer spending is also forecast to remain strong, as suggested by high growth of credit and retail sales. Household disposable income has been buoyed by the 6 percent points cut in the personal income tax rate in July 2006, cash payments in March 2007 as compensation for earlier savings losses during the shift from the Russian Ruble to the new national currency, high wage growth and rising employment. The rapid deterioration of the trade balance seen in the first eight months of 2007 (affected by capacity restrictions in oil refining) looks set to be confirmed as the trend for the whole year.

Real GDP growth is projected to gradually weaken but should remain above 7% in 2008 and 6% in 2009, with domestic demand continuing to be the engine of growth. Investment is expected to continue to lead growth in 2008–2009. However, supply-side constraints have emerged as a consequence of the tight labor market. Private consumption growth is forecast to remain high, buoyed by high wage growth and a further 3 percent points cut in personal income tax in January 2008, public sector wage increases and higher social transfers. However, consumption growth will be lower than in previous years because of the upturn in interest rates and an expected moderation in credit growth.

Employment growth picked up in the first half of 2007, while the unemployment rate declined further to just above 4% and earnings accelerated rapidly. Several sectors are experiencing increasing labor shortages, reinforced by emigration, which has put upward pressures on labor costs. Employment growth is expected to peak in 2007, due to increased participation rates, before moderating in 2008–2009. Nominal wage growth is expected to remain high and far exceed productivity growth. Unemployment is expected to be stable at a rate of some 4%. Against this background, unit labor costs will increase at a relatively fast rate and competitiveness will deteriorate, especially in labor intensive sectors.

Both headline and core inflation in 2007 have been increasing, fuelled by overheating, higher energy tariffs and accession-related excise harmonization. Food price increases have also strongly fuelled inflation. Inflation for 2007 as a whole is projected to average around $5^1/_2\%$, up from below 4% in 2006. In 2008, inflation is expected to edge up further to $6^1/_2\%$, prompted by a new round of increases in excise duties (for tobacco and fuel) and higher electricity prices. In 2009, inflation is expected to ease slightly but to stay at a high level. Short-term risks are on the upside, notably through possible higher food prices, wage increases and higher inflation expectations. A possible further gas price increase at the beginning of 2008 (by 30–50%) by Lithuania's sole gas supplier, possibly reaching West European price levels – not assumed in the forecast – also poses a substantial risk.

In 2007, the general government deficit is likely to widen to around 0.9% of GDP (2006, 0.6%), in line with the government target. The general government debt-to-GDP ratio is expected to decrease and remain broadly stable at around 17%.

11.5.7 Malta

Malta, a crossroads between Europe and Africa, at the southern tip of the European continent (see Fig. 11.1), is a melting pot of civilizations in the heart of the Mediterranean Sea. The size of Malta is $300\,\mathrm{km}^2$, about one thousandth the size of

Italy. Malta is the smallest of the EU27 states. Her population is four hundred thousand people. The Maltese are descendants of ancient Carthaginians and Phoenicians with an admixture of Arab and Italian. The ethnic aliens are largely retired British nationals.

Malta boasts a rich legacy from her centuries-old history, from megalithic temples – unique in the world – to her capital, Valletta, a jewel of baroque architecture, and her massive fortifications, which witnessed the bravery of the Maltese people over the centuries.

With the division of the Roman Empire in 395 AD, Malta was assigned to the eastern portion dominated by Constantinople. Between 870 and 1090, it came under Arab rule. In 1091, the Norman noble Roger I, then ruler of Sicily, came to Malta with a small retinue and defeated the Arabs. The Knights of St. John (Malta), who obtained the three habitable Maltese islands of Malta, Gozo, and Comino from Charles V in 1530, reached their highest fame when they withstood an attack by superior Turkish forces in 1565. Napoleon seized Malta in 1798, but the French forces were ousted by British troops the next year, and British rule was confirmed by the Treaty of Paris in 1814. The island staunchly supported the UK through both World Wars and remained in the Commonwealth when it became independent in 1964. The courage and endurance of the Maltese people was recognized when the United Kingdom awarded to Malta the George Cross in 1942, which is now an integral part of the national flag. Malta is not a member of the North Atlantic Treaty Organization (NATO). However, NATO established its Mediterranean military headquarters in Malta in 1953. Over the last 15 years, the island has become a freight transhipment point, financial centre, and tourist destination. As we mentioned earlier for Cyprus, these two Mediterranean islands that have recently accessed the EU are chosen by many European elderly citizens as their second home, or as the country in which to spend their retirement years, in a sunny environment much cheaper than many of the cold northern member states. In Malta, tourism contributes to GDP by as much as 25%.

Malta has no rivers or lakes, no natural resources, and very few trees. It is, however, of great strategic value and was an important British military base until 1979. Following the withdrawal of British forces, the country faced severe unemployment; it has since made progress in diversifying its economic base. Manufacturing and tourism are now the main industries. Electronics, textiles, processed food, clothing, tobacco products, and construction materials are manufactured. Ship building and repair, performed in state-owned dry docks, are also important. Although the soil is poor, there is some agriculture, producing potatoes, cauliflowers, grapes, wheat, barley, and cut flowers. The beekeeping industry, already renowned from ancient times, is still flourishing. Hogs and chickens are raised. International banking and financial services are growing, and the island is developing as an offshore tax haven. The shortage of water has stimulated the building of desalination plants, which now provide more than half the country's freshwater needs. The main imports are food, petroleum, machinery, and manufactured goods; exports include textiles, clothing, and ships. Most trade is with Italy, the United Kingdom, and Germany.

Still, Malta is not only an island in the sun and an open-air museum in the Mediterranean; it is also an island looking towards the future. Apart from the tourism and manufacturing industries now firmly established, Malta is currently developing her service economy, and it also aims to become a hub for communications in the Mediterranean. For this purpose, Malta has a winning card – her human resources: a flexible labor force easily adaptable to new circumstances and with the great advantage of being multilingual. Malta has a very high population density. English and Maltese, a Semitic dialect, are the official languages, although Italian is also widely spoken.

Malta is a major maritime hub in the middle of the Mediterranean Sea. Her mercantile fleet has no less than 1,300 vessels for a total of 28 million tons. As a comparison, Italy has 450 vessels for 8.5 million tons. The accession of Malta will hopefully result in an important step towards enforcing European maritime safety regulations and standards, reducing the risk of shipwreck of oil tankers that in the past years have polluted the European coasts.

In November 2000, following the disastrous Erika incident off the coast of France, when a Maltese-registered ship sank off the coast of Brittany with extensive environmental repercussions, the EU proposed new laws to accelerate the phasing-in of double-hull tankers instead of single hulls. This will ensure that the transport of oil and oil products is carried out under safe and acceptable conditions, with the least risk of oil pollution, as well as with less risk of loss of human life [Malta-EUIC 2005].

The Commission also launched a ship safety database, known as Equasis, in May 2000. Poor management should be targeted at the source, that is, when a ship is brought to dry dock or out of the repair yard. Ships that have a history of being a danger to the marine environment will, in future, be refused admittance at EU ports [Malta-EUIC 2005].

During the accession negotiations, Malta's enforcement of maritime safety was under particular scrutiny. Malta presented the Action Plan prepared by the Malta Maritime Authority, which is intended to ensure the proper implementation of the maritime transport legislation.

The accession of Malta, in addition to contributing to maritime safety, will hopefully also help to increase control over illegal immigration into Europe from clandestine traffic by sea.

The island had been divided politically over the question of joining the EU. However, in a closely fought contest, with 91% of the electorate going to the polls on 8 March 2003, 53.6% of the population voted positively in answer to the question, "Do you agree that Malta should become a member of the European Union in the enlargement that is to take place on 1 May 2004?" This was the first of the referendums held in the accession countries (see Table 11.2).

Prime Minister Lawrence Gonzi was appointed in March 2004 for a five years term. Within days from his appointment, he delivered a speech to the Malta Federation of Industry in which he committed his government to promote education, lifelong learning, and competitiveness [Malta Info 2004]:

Maltese SMEs have proved to be a major engine of growth and made a major contribution to the development of our economy. Around 99 per cent of enterprises are small and these employ around 60 per cent of private sector employees. Last year, wage increases given by small enterprises were higher than those granted by their large counterparts. The importance of small firms is increasing.

SMEs are the backbone to our commercial and social infrastructure. We know that it is our responsibility in Government to help them succeed. To ensure that our SMEs find their place in the global market we must create the right economic and administrative framework within which they can not just thrive, but prosper. Government also has an important role to play. In particular, macroeconomic stability is an important requisite and to this end, my Government is committed to bring back public finances in order over the medium term.

For the future, we intend to proceed on this course with a view to address those elements that threaten our competitiveness. An important aspect is the way the domestic labor market performs. In this respect, our Government places a great deal of importance on the skills structure of the Maltese workforce. In doing so, we have to ensure that our formal education system is preparing our young people with the knowledge, skills and qualifications demanded by emerging labor markets. Along the years we have increased the investment going to education. We have doubled investment in the University to ensure that technological and managerial competences are firmly rooted as a sure way to connect SMEs to global markets. Formal education needs to lead to further and continuing education and development across life.

It has many times been argued that our geographical isolation may represent a significant barrier to those firms wishing to expand in foreign markets. With the advent of information technology, SMEs located in Malta are now able to overcome these disadvantages allowing them to exploit network advantages. During the past years we have introduced an advanced legislative and regulatory framework within which electronic commerce can be carried out. We have invested heavily in information technology in our schools and the public service. Importantly, we have introduced and advanced the idea of e-government whereby public services can be easily accessed through a computer. This reflects the high percentage of Maltese enterprises that have access to the internet and represents an asset with huge potential which we need to harness to our advantage.

Government is committed to put in place the right framework and to provide the necessary support to enable enterprises to thrive in a global setting. We are also eager to work with the social partners in order to undertake the necessary reforms. But to accomplish this goal, businesses, unions and workers must work constructively and creatively together. Industry needs to be ambitious in responding to the challenge ahead of us, whilst workers, led by their representatives, need to have the confidence and skills to seize these opportunities. It is only if all stakeholders contribute towards this goal that our SMEs will succeed in the global market.

11.5.7.1 Economy

Major resources are limestone, a favourable geographic location, and a productive labour force. Malta produces only about 20% of her food needs, has limited freshwater supplies, and has no domestic energy sources. The economy is dependent on foreign trade, manufacturing (especially electronics and textiles), and tourism. Malta privatized state-controlled firms and liberalized markets while preparing for membership in the Union. The budget deficit rose from 4% of GDP in 1995 to over 11% in 1998 because of structural imbalances. The trend has now reversed: 6.7% in 1999 and 6.6% in 2000. The public debt is increasing. It reached 60.6% of GDP in 2000 and 73.6% in 2003.

Table 11.11 Basic economic data for Malta

Malta		2000	2001	2002	2003	2004	2005	2006
GDP[i]	($bln)	3.8	3.8	4.0	4.8	5.4	5.5	5.7
GDP[b]	($bln)				4.5			5.4
GDP[f]	(€bln)			4.5	4.4	4.4	4.6	4.9
pcGDP ppp[b]	($x000)				17.7			20.3
GDP Growth[b]	(%)				0.8			1.3
Current Account Balance[b]	(%GDP)				−5.9			−17.9
Government Balance[b]	(%GDP)				−6.2			−3.7
Government Balance[c]	(%GDP)				−9.7			
Government Balance[f]	(%GDP)			−5.6	−10.2	−5.1	−3.3	
Total Debt[f]	(%GDP)			61.2	71.3	76.2	74.7	
Total Debt[b]	(%GDP)				73.6			
Inflation (CPI)[b](%)					0.4			2.6
Unemployment[b](%)					7.0			6.8
Exports[b]	(%GDP)				48.3			45.0
Imports[b]	(%GDP)				61.4			75.6
School enrollment, tertiary[a]	(%)	21	25	24	30	26		
Internet Users[a](/1000)			230	253				
€ Exchange Rate[f]	(MTL)		0.403	0.409	0.426	0.428	0.43	
Stock market capitalization[a]	(%GDP)	53	36	33	38	53	74	
GDP % in agriculture[b]					3			3
GDP % in industry[b]					23			23
GDP % in service[b]					74			74

Source: [a]The World Bank (2005, 2007); [b]http://www.cia.gov/cia/publications/factbook; [c]European Commission, 2004; [f]Eurostat, 2007; [i]International Monetary Fund.

Table 11.11 presents basic economic data for Malta published by various quoted sources. More information is available in the recommended reading at the end of the chapter.

The share of imports and exports in GDP are increasing significantly. The export base of the economy is concentrated in a few sectors, mainly in electronics, machinery, and transport equipment (which generated about 75% of total exports in the first half of 2001). Malta is well integrated in terms of trade with the European Union. The latter accounted for around 33% of Malta's exports and 60% of her imports in 2000.

The national currency is the Maltese Lira (MTL); 1 € = 0.43 MTL on 14 May 2003 and still the same rate on 6 May 2007. Malta will adopt the euro on 1 January 2008 at the fixed rate of 1 € = 0.429300 MTL. Measures are taken to avoid any impact on prices. They include price-stability agreements negotiated with the retail sector and closely monitored before and after the changeover, with the active involvement of consumer protection organizations and professional organizations of the retail sector at the national and European level.

The European Commission analysis and recommendations in the Broad Guidelines of the Economic Policies [EC 2004-10] are presented as recommended reading at the end of Chapter 14, "Economic and Monetary Union,". Those relative to

Malta point to: the too low gap between the minimum wage and unemployment benefit, affecting employment participation, and the educational system.
They include:

The increasingly open nature of the Maltese economy, large dependence on tourism revenues and its small size make the economy increasingly vulnerable to external economic
and geopolitical shocks. The difficult international economic environment in the last two
years and restructuring of the public sector brought about modest economic growth, mostly
fuelled by large public consumption. Real GDP growth is currently far below its estimated
potential.

The population ageing implies a significant fiscal risk to the long-term sustainability of
public finances, in particular due to a relatively high level of government debt. The expected
rapid increase in the old-age dependency ratio between the years 2000 and 2050 (from 18
per cent to 38.6 per cent) is to exert significant fiscal pressure on public finances in the
future. While a reform of the 1st pillar of the pension scheme is being planned, it is unclear
when it is going to be implemented.

In Malta, the labor market seems to be sufficiently flexible to adjust to economic shocks
without generating long periods of high unemployment. However, in September 2003 the
employment rate (53.7 per cent) was low compared to the EU average. The relatively low
employment rate was attributable to a low female employment rate (33.1 per cent) as the
equivalent male figure (74.2 per cent) was higher than the EU average. The increase in
total and female employment rates is of paramount importance to widen the base for social
security contributions, in the light of the ageing population.

The gap between the minimum wage and unemployment benefit, especially for the larger
families, remains very low and reduces incentives to work. Early retirement schemes used
as a means to restructure public sector entities should be also limited, efforts being directed
towards re-training.

A significant number of reforms have been introduced to increase competition in the Maltese economy. In order to meeting the challenges outlined above, Malta is recommended to:

1. reduce the government deficit in a credible and sustainable way
2. streamline the tax-benefit system interaction to strengthen the incentives to work and
    reduce taxation on labor to improve its competitiveness
3. improve the quality of secondary education and vocational training
4. strengthen retraining of labor to make it more adaptable in the case of labor-shedding
    and facilitate the return to work of middle-age women
5. pursue the efforts to increase competition in certain sectors such as network industries,
    food industries, and shipbuilding

The fall 2007 European Commission Economic forecast for Malta [EC 2007-27]
is presented as recommended reading at the end of Chapter 14, "Economic and
Monetary Union,". It is summarized as "continuing strong growth in 2007" and it
includes the following:

Economic growth accelerated in the first half of 2007 underpinned by higher domestic
demand, in particular private consumption. GDP is projected to grow by 3.1% in 2007
in real terms. Private consumption is foreseen to grow by 2.7%. This is attributed to the
improvement in disposable income as a result of the increase in employment and earnings
as well as the revised lower income tax rates and the fall in the rate of inflation. An additional
contributory factor to higher private consumption is the cash de-hoarding in anticipation of
the euro changeover.

Real GDP growth is expected to decelerate to 2.8% in 2008, before increasing to 2.9% in
2009. Economic activity is anticipated to be driven primarily by domestic demand in both
years, while the contribution of the external sector should also be positive, albeit lower.

Exports of services are also projected to grow, reflecting a better performance of the tourist industry, IT and remote gaming sectors. After falling to 6.7% of GDP, the current account deficit is anticipated to narrow further to 3.8% of GDP in 2007, on the back of lower imports and a higher value of exports.

Employment growth in the first half of 2007 increased at a brisk pace. For the whole year, employment is expected to rise by some 1.4%. The pace of employment creation is projected to decelerate to around 1.2% in 2008 – mainly owing to the completion of labor-intensive public construction projects – and remain almost unchanged in 2009. Job creation during the forecast period is expected to be generated mainly by the services sector. As a result, the unemployment rate is anticipated to progressively decline to 6.5% of the labor force by 2009. Despite high oil prices, HICP inflation is expected to slow down significantly to 0.8% in 2007. This mainly reflects declines in tourist accommodation prices as well as the authorities' decision to keep electricity and water charges to consumers almost unchanged. In 2008, headline inflation is anticipated to rise to around 2.5%, primarily reflecting higher food prices. For 2009, HICP inflation is projected to decline to 2.2%.

The general government deficit is expected to continue its downward path reaching 1.8% of GDP in 2007, mainly as a result of a lower current expenditure ratio. General government debt in 2006 stood at around $64^3/_3$% of GDP and is expected to fall to around 63% of GDP in 2007. Under the no-policy-change scenario, the debt ratio is projected to decline further to around $59^1/_4$% of GDP by 2009.

11.5.8 Poland

Poland's total surface area is 313 thousand km^2, slightly larger than Italy's. This makes her the sixth largest country in the Union, after France, Spain, Sweden, Germany, and Finland (see Fig. 11.1). Her population is 39 million: Poland is the largest and most populated of the countries that accessed the EU in 2004.

The capital, Warsaw (1.6 million inhabitants), is the country's economic and political centre. Cracow – the country's third largest city – has been her cultural centre since the Middle Ages and was Cultural City of Europe in the year 2000. Other Polish cities like Gdansk, Poznan, Lódz, and Wroclaw are of European importance, as underlined by the candidature of the latter for EXPO 2010.

For centuries, Polish culture has been an integral part of European culture. Among the greatest Polish contributors to European culture are the astronomer Copernicus, the great composer and pianist Fryderyk Chopin and the outstanding scientist Maria Curie-Skodowska. Films by Andrzej Wajda, Krzysztof Kielowski, and Roman Polanski have contributed significantly to European and world cinema. Like the Hungarians, Poles may pride themselves on the large number (14) of Nobel Prize winners the country has produced over the years, of whom the most known abroad is Lech Walesa, the Peace Nobel Prize winner in 1983. Today however, the Polish educational system is not adapted to the requirements of the changing economy and the education levels of the adult population are rather low [EC 2004-19, Economist 2006-1].

The Polish state is over 1,000 years old. In the 16th century, under the Jagiellonian dynasty, Poland was one of the richest and most powerful states on the continent. On 3 May 1791, the Commonwealth of Poland – Lithuania ratified a constitution – the first written constitution of Europe. Soon afterwards, Poland ceased to exist for 123 years, after being partitioned by her neighbours Russia, Austria, and Prussia. The country regained independence in 1918 only to be overrun by Germany and the Soviet Union in Second World War. It became a Soviet satellite country following the war, but one that was comparatively tolerant and progressive. Labour turmoil in 1980 led to the formation of the independent trade union Solidarność (Solidarity) that over time became a political force and by 1990 had swept parliamentary elections and the presidency. Solidarno had the support of the Catholic Church, a strong and influential presence in the life of the Polish people whose religious practice had been associated – since the 17th century – with national resistance to foreign rule or occupation. In 1989, the first partially free elections in Poland's post-war history concluded the Solidarno movement's 10-year struggle for freedom and resulted in the defeat of Poland's communist rulers.

In 1999, Poland joined NATO and began negotiating her full membership in the European Union. Poland, indefinitely faithful to the U.S., is the embodiment of the new Europe, according to the U.S. Defense Secretary Donald Rumsfeld. Paul Wolfowitz, the Deputy U.S. Secretary of Defense, thanked the Poles for joining in a war that much of Europe had repudiated and continues to oppose. His message was clear: "History, especially Europe's in the last century, has proved that it is smarter to side with the U.S. than against it. We will not forget Poland's commitment," he promised [Boyer 2004].

Poland, after joining the EU, proved to be an assertive defender of her national interests, in particular in the negotiations for the Lisbon Treaty. The unstable coalition government lead by Jaroslaw Kaczynski has antagonized Germany and other key EU member states, making it difficult for Poland to make alliances within the EU to pursue its interests [Economist 2007-6]. "Mr. Kaczynski won a delay on new voting rules that will reduce Poland's weight. But the victory was marred by his graceless approach, using Poland's wartime suffering at German hands as a bargaining chip" [Economist 2007-7]. Soon after the June 2007 European Council summit, *The Economist* wrote:

If you want to join the family you will have to fit in. And you had better be humble and grateful. That, roughly, is the source of the misunderstanding between Poland and the old democracies of Europe. Lech and Jaroslaw Kaczynski, the twins who are the country's president and prime minister, are outraged that the European Union is treating Poland as a new member of a club. Of course Poland won't meet all the EU's standards (and, the Kaczynskis note, many old members break them too). But does not western Europe owe it a huge moral debt? The twins almost brought a recent EU summit to a standstill by insisting that a new treaty give smaller countries greater voting rights, rather than have such rights determined directly by population size.

The Kaczynskis have won a battle. But they risk making Poland as Greece used to be: unpopular, expensive and, most dangerously, marginal. Perhaps only America can persuade the avowedly Atlanticist twins that a strong Poland in a strong EU is hugely preferable to a marginalized Poland in a weak one [Economist 2007-8].

11.5.8.1 Economy

The country has a variety of natural resources, including coal, copper, zinc, iron, gypsum, lignite, and some oil and natural gas reserves; the chief minerals produced are coal, sulphur, copper, lead, and zinc. About 27% of all Poles are employed in agriculture, which contributes to 4% of the GDP (2003). About 60% of the country's land is used for agriculture. Almost one fourth of the population still lives on small, inefficient farms. Poland's main agricultural products include grains, potatoes, sugar beets, fodder, and livestock. Agriculture is mostly privately run and was so even during the Communist years. Poland is generally self-sufficient in food; the main crops are rye, potatoes, beets, wheat, and dairy products. Pigs and sheep are the main livestock. The country's leading manufactures include machinery, iron and steel products, chemicals, ships, food processing, and textiles. Trade, high technology, and the service sector are increasingly playing an important role in employment and restructuring of the national economy.

A *shock therapy* program during the early 1990s enabled the country to transform its economy into one of the most robust in Central Europe, boosting hopes for acceptance to the EU. Poland has steadfastly pursued a policy of liberalizing the economy and today stands out as one of the most successful and open transition economies. The dynamic development of the private sectors is based on the continuous inflow of Foreign Direct Investment and the high level of entrepreneurial activity of the Polish population. GDP growth had been strong and steady in 1993–2000; it fell back in 2001 and 2002 with slowdowns in domestic investment and consumption and the weakening in the global economy; however it increased again from 2003, reaching 6.1% in 2006. The privatization of small and medium-size state-owned companies and a liberal law on establishing new firms have allowed, for the rapid development of a vibrant private sector. In contrast, Poland's large agricultural sector remains handicapped by structural problems, surplus labour, inefficient small farms, and lack of investment. Restructuring and privatization of *sensitive sectors* (e.g. coal, steel, railroads, and energy) has begun. Structural reforms in health care, education, the pension system, and state administration have resulted in larger-than-expected fiscal pressures. Further progress in public finance depends mainly on privatization of Poland's remaining state sector. The government's determination to enter the EU in the shortest possible time affected most aspects of its economic policies. Improving Poland's budget deficit and reining in inflation are priorities.

Krzysztof Bobinski, a prominent journalist based in Warsaw, wrote [Bobinski 2004]:

> The greatest challenge for Polish politicians is the need to get the right balance between offering adequate social services and cutting the budget deficit; however foreign policy is becoming part of the political debate in Poland. Foreign policy priorities tend to concentrate on ensuring that the United States continues to maintain a presence in Europe, with the Polish government happy to demonstrate its reliability as Washington's ally. In Europe Poles will press for an active foreign policy which will ensure good relations with the Russian Federation while supporting the democratization of not only Ukraine, but also Moldova and Belarus.

Like other accession countries, Poland has attracted foreign investments. The Italian automotive group Fiat increased its investments in its plant in Tychy, now strong of a workforce of 2,700 qualified, reliable, and dedicated workers, to produce its award-winning Panda and other entry-level vehicles.

Industry, which had been state controlled, began to be privatized in the early 1990s, although restructuring and privatization of the country's large coal, steel, and chemical industries has moved forward slowly, when it has progressed at all. Prices were freed, subsidies were reduced, and Poland's currency, the Zloty (PLN), was made convertible as the country began the difficult transition to a free-market economy. Reforms initially resulted in high unemployment, hyperinflation, shortages of consumer goods, a large external debt, and a general drop in the standard of living. The situation later stabilized, however, and at the end of the 1990s Poland's economy was among the fastest growing in eastern Europe.

Poland's currency, Zloty, due to political uncertainty has recently depreciated in relation to the euro, while currencies of the other euro-zone aspirants have been appreciating: 1 € = 4.32 PLN on 14 May 2003; 1 € = 3.75 PLN on 6 May 2007. The high budget deficit implies that entry into the euro zone is unlikely before 2012, while the zloty's entry into the EU's exchange-rate mechanism may be expected around 2009 or 2010.

The labour market is characterized by high unemployment and vast migrations of (mainly unskilled) Polish workers to western EU countries. From 2004 to 2006, two million have gone abroad, and a recent survey found another three million planning to do the same [Economist 2007-7]. The *Polish plumber* has become a metaphor for the euroskeptics that fear social dumping from the enlargement countries and oppose the free provision of services across Europe. The migration of the most enterprising workers creates a temporary shortage and drains the market of some of its best talents, but a number of them are likely to come back after one or a few years, enriched with know-how and language skills. One reason for migration of young Polish citizens is the poor Polish universities, where islands of excellence sit in an area of mediocrity [Economist 2006-1].

Since 2002, even though the Zloty appreciated 30%, Poland's exports more than doubled the main export commodities being machinery, transport equipment, and other manufactured goods; Germany, the Russian Federation, Italy, France, and The Netherlands are the most important trading partners.

Basic economic data for Poland from various quoted sources are presented in Table 11.12. More information is available in the reading at the end of the chapter.

The European Commission analysis and recommendations in the Broad Guidelines of the Economic Policies [EC 2004-10] are presented as recommended reading at the end of Chapter 14, "Economic and Monetary Union,". Those relative to Poland point to: unemployment, market imbalance, skill mismatches between labour supply and demand, and low investments in ICT and R&D. They include:

> After a record of strong growth during most of the 1990s, Poland experienced a sharp economic slowdown in 2001–2002. Since the end of 2002, the recovery has gradually gained strength, and real GDP growth accelerated to 3.7 per cent in 2003.
>
> Despite remarkable progress in recent years, Poland still faces serious structural problems that may impair its capacity to stay on a path of strong growth. The unemployment rate has increased rapidly since 1999 up to about 20% in 2003, the highest among the acceding countries. The labor tax wedge is high in Poland and creates a disincentive to work in the official economy. In addition, the combined effects of tax and benefit systems make working or returning to work a costly decision and therefore discourage labor market

Table 11.12 Basic economic data for Poland

Poland		2000	2001	2002	2003	2004	2005	2006
GDP[a]	($bln)		190.3	198.0	216.6	252.7	303.2	
GDP[i]	($bln)	166.6	185.8	191.5	209.5	241.8	312.3	332.6
GDP[b]	($bln)					427.1[g]		542.6[g]
GDP[b]	($bln)							337.0
GDP[f]	(€bln)			209.4	191.7	204.2	244.2	271.5
pcGDP ppp[b]	($x000)					11.1		14.1
GDP Growth[a]	(%)		1	1	4	5	3	
GDP Growth[b]	(%)					3.7		5.3
GDP Growth[c]	(%)			1.4	3.9	5.3	3.6	6.1
Current Account Balance[b]	(%GDP)					−0.1		
Current Account Balance[c]	(%GDP)			−2.54	−2.1	−4.2	−1.7	−2.3
Government Balance[b]	(%GDP)					−2.2	−2.7	
Government Balance[c,f]	(%GDP)	−1.8[c]			−3.2	−4.7	−3.9	−2.5
Total Debt[b]	(%GDP)					34.0		49
Total Debt[f]	(%GDP)				39.8	43.9	41.9	42.5
Inflation, (GDP deflator)[a]	(%)		3	2	0	4	3	
Inflation, (CPI)[b]	(%)					0.7		1.3
Inflation, (CPI)[c]	(%)			1.9	0.8	3.5	2.1	1.0
Unemployment[b,c]	(%)			19.8[c]	20.0			14.9
Exports[a]	(%GDP)		27	29	33	38	37	
Imports[a]	(%GDP)		31	32	36	40	37	
Exports Growth[a]	(%)				13.6			7.8[h]
High-tech Exports[a]	(%Manuf. Exp.)		3	3	3	3	4	
School enrollment, tertiary[a]	(%)	49	54	58	59	61		
Internet Users[a]	(/1000)		99	232	235	236	262	
ITC Expenditure[f]	(%GDP)					1.7	1.9	2.2
FDI Inflows[a]	(%GDP)		3.0	2.1	2.1	5.1	3.2	3.3[d,e]
€ Exchange Rate (PLN)		3.672	3.857	4.4		4.527	4.023	
Stock market capitalization[a]	(%GDP)	18	14	15	17	28	31	
GDP % in agriculture[a]			5	5	4	5	5	
GDP % in industry[a]		29	29	30	31	31		
GDP % in services[a]		65	67	66	64	64		

Source: [a]The World Bank (2005, 2007); [b]http://www.cia.gov/cia/publications/factbook; [c]European Commission, 2004; [d]Average Average 2002–2006; [e]The Economist Intelligence Unit, 2007; [f]Eurostat, 2007; [g]Purchasing Power Parity (ppp); [h]Average 2002–2006; [i]International Monetary Fund.

participation. In parallel, the employment rate has fallen markedly and almost half of the working age population is without work. The smooth functioning of the labor market is hampered by a number of structural problems, including limited responsiveness of wages to labor market conditions, disincentives from the tax and benefit system, skills mismatches and low geographical mobility.

Poland's fiscal position has deteriorated significantly since 2000, as a result of both cyclical factors and the relaxation of fiscal policy. The general government deficit increased from 1.8 per cent of GDP in 2000 to 3.6 per cent in 2002. Until recently, the authorities have shown some reluctance to tackle the fiscal problems, including the rapidly rising debt ratio.

Policies in Poland should aim at achieving a high degree of sustainable convergence, in particular as regards the consolidation of public finances. Moreover, stability-oriented macroeconomic policies need to be complemented by structural reforms aiming at improving Poland's growth performance. In addition to the under-utilization of human resources, the relatively low productivity level limits the capacity of the Polish economy to increase both actual and potential output growth. To address this issue, continued efforts are needed to improve the education and training system and also to create favorable conditions for R&D and technology transfer. In addition, the economy, in particular the agriculture sector, calls for more restructuring and there is room to improve the business environment.

Despite reasonable productivity gains over the period 1995–2002, labor productivity per person employed (in ppp) in Poland was less than half that of the EU15 in 2002 and below the average of the new member states. Poland remains below the EU15 average in terms of investment in ICT and R&D. The low level of business investment in R&D is particularly problematic (only 30 per cent of the total R&D expenditure is financed by firms).

Poland has made substantial progress in developing a large SME sector thanks to recent reforms. In particular, a new enterprise register is being set up, which creates a "one-stop shop" for firms from January 1, 2004. Since January 2004 a single 19 per cent corporate tax is also in application. Finally, a new solvency law was adopted in 2003 bringing the bankruptcy framework in line with the requirements of a modern market economy.

In order to meet the challenges, Poland is recommended to:

1. increase the flexibility of the wage-setting process to ensure that wages better reflect differences in productivity across skills, firms and regions
2. lower the tax burden on labor, together with efforts to widen the tax base on labor and to improve the efficiency of the tax collection and enforcement system
3. strengthen labor supply by pursuing efforts to better adapt the qualifications of the workforce to the requirements of the labor market and by removing obstacles to regional mobility
4. reduce the general government deficit in a credible and sustainable way
5. monitor the reform of the pension system to counter the expected increase in the old-age dependency ratio
6. pursue and reinforce efforts to improve the efficiency and quality of the education and training system and its responsiveness to changing skills requirements.

Despite the record-high unemployment, wages are up, due also to emigration of skilled and unskilled workers attracted by the higher wages in western EU countries. There are no more any *plumbers* in Poland, in facts so many Polish workers migrated in the UK and in other EU western countries that "the Polish plumber" became a metaphor in Europe meaning dumping in the construction and refurbishing business but also in the service sector. The fall 2007 European Commission Economic forecast for Poland [EC 2007-27] is presented as recommended reading at the end of Chapter 14, "Economic and Monetary Union,". It is summarized as "strong fundamentals, but public finances remain a weak spot" and it includes the following:

Economic activity continued to be robust in the first half of 2007. Driven by domestic demand, real GDP growth reached 7.1% year-on-year. Growth was driven by private consumption (6.0%) and investment (25.3%).

Domestic demand will continue to be the main driving force of GDP growth, which is expected to ease to 5.6% in 2008 and 5.2% in 2009. The better situation in the labor market and accelerating real wage growth because of the tightening labor market, a further cut in the tax wedge in 2008 and lower personal income taxes in 2009 are the main factors supporting private consumption.

In the course of 2007 an impressive improvement was seen in the labor market, continuing the trend from 2006. Until August 2007 the unemployment rate fell by more than $4^1/_2\%$ from 13.8% in 2006, which corresponds to a decrease in the number of unemployed by nearly 700,000. Although total employment increased ($1^1/_4\%$ in the first half of 2007), part of the drop in the unemployment rate may be attributed to a falling labor force due to increased early retirement, emigration and a growing number of students.

Annual HICP inflation increased from 2.0% in the first quarter of 2007 to 2.4% in the second. The pressure on inflation from rising wages is contained owing to increased competition among the companies. Unit labor costs are projected to increase by about $4^1/_2\%$ in 2007, as a result of strong wage growth combined with modest productivity increases.

The general government deficit is now expected to improve from 3.8% of GDP in 2006 to 2.7% of GDP in 2007. However, the general government deficit is expected to deteriorate to 3.2% of GDP in 2008 because of a number of deficit increasing measures adopted recently by government and parliament. Gross debt is projected to decrease slightly from 47.6% of GDP in 2006 to 46.8% in 2007 and rebound to 47.1% in 2008 and 2009. Privatization has been stalled since the beginning of 2006. If it is re-activated by the new government formed after the early elections, the debt ratio may decline faster.

11.5.9 Slovakia

Slovakia is a landlocked country located in central Europe (see Fig. 11.1). Her size is 48.8 km^2, about 20% larger than The Netherlands. Her population is 5 million. Present-day Slovakia was settled by Slavic ancestors of modern Slovaks in about the sixth century. They were politically united in the Moravian Empire in the ninth century. In 907, the Germans and the Magyars conquered the Moravian state, and the Slovaks fell under Hungarian control from the 10th century up until 1918. When the Habsburg-ruled Empire collapsed in 1918 following the First World War, the Slovaks joined the Czech lands of Bohemia, Moravia, and part of Silesia to form the new joint state of Czechoslovakia. In March 1939, Germany occupied Czechoslovakia, established a German "protectorate," and created a puppet state out of Slovakia. The country was liberated from the Germans by the Soviet army in the spring of 1945, and Slovakia was restored to her prewar status and rejoined to a new Czechoslovakian state, a Communist nation within Soviet-ruled eastern Europe. Soviet influence collapsed in 1989 and Czechoslovakia once again became free. The Slovaks and the Czechs agreed to separate peacefully on 1 January 1993. Historical, political, and geographic factors have caused Slovakia to experience more difficulty in developing a modern market economy than some of her central European neighbours.

Slovakia has mastered much of the difficult transition from a centrally planned economy to a modern market economy. The Dzurinda government (1998–2006) made excellent progress in 2001 as regards macroeconomic stabilization and structural reform. Major privatizations are nearly complete, the banking sector is almost completely in foreign hands, and foreign investment has picked up. Slovakia's economy exceeded expectations in 2001, despite a recession in key export markets. Revival of domestic demand, partly because of a rise in real wages, offset slowing export growth to help drive the economy to its strongest expansion since 1998 and boosted economic growth up to over 4% in 2003 and over 8% in 2006. Unemployment, rising to 19.8% at the end of 2001 dropped to 10.2 in 2006, but remains the economy's Achilles' heel. The government faces other strong challenges, especially the maintenance of the fiscal balance, cutting budget and current account deficits, and privatization of the Slovak energy and power monopolies. Slovakia joined NATO in 2004: "Slovakia has no alternative to the Euro-Atlantic course," wrote the then prime minister of Slovakia, Mr. Mikuláš Dzurinda [Dzurinda 2004]. After the 2006 elections, Mr. Robert Fico replaced Dzurinda as prime minister and leads a centre-left government that has pledged to soften or reverse some of the economic reforms enacted by the previous, more market-friendly administration. However, the bulk of these reforms might be upheld with the aim of adopting the euro in 2009, although fiscal profligacy, high inflation, and exchange-rate instability pose a risk of delay [Economist 2007-9].

Slovakia had the highest percentage of *yes* (92.46%) among the accession countries in the referendum held in 2003. A survey conducted by Eurobarometer in the fall of 2006 shows a more positive and pro-EU attitude than in the Union on average [EC 2006-2]:

> As far as the expectations for the next 12 months related to the economic situation in the country are concerned, Slovakia belongs to the more optimistic countries in the EU. In the EU25, only 20% of citizens expect that the economic situation in their country will be better in the next 12 months, while in Slovakia 28% of citizens expect improvement. Worsening of the economic situation is expected by 30% of Slovak citizens, that is 5 points less then in the EU25.
>
> Citizens of the Slovak Republic are, compared with the EU25 citizens, tuned more optimistically about their expectations for the next 12 months related to the employment situation in their country. However, as far as expectations for the next 12 months in regard to the personal job situation are concerned, Slovaks are more pessimistic than EU25 citizens. Only 17% of Slovak respondents expect an improvement, i.e., 6 points below the European average.
>
> In fall 2006, 61% of Slovaks considered their country's EU membership to be a good thing (in the EU, on average, 53%), which is 6 points more than in spring 2006 and 11 points more than in fall 2005.
>
> Slovaks also tend to be more positive concerning the assessment of the benefits of EU membership. As many as 71% of Slovak citizens, that is 17 points more than citizens in the EU25 as a whole, believe that their country has benefited from EU membership. Citizens of the Slovak Republic consider unemployment and the economic situation to be the two most important issues in their country. Unemployment is considered to be one of the two most important issues by 44% of Slovaks, which is 4 points more than in the EU25.

11.5.9.1 Economy

Farms, vineyards, orchards, and pastures for stock form the basis of Slovakia's economy. Main crops are wheat, barley, potatoes, sugar beets, hops, and fruit. The mountainous part of Slovakia has vast forests and pastures, used for intensive sheep grazing, and is rich in mineral resources, including high-grade iron ore, copper, magnetite, lead, and zinc. There are also numerous mineral springs, notably at Piešt'any, and many popular resorts. Slovakia has undergone considerable industrialization and urbanization since Second World War. Her industries produce metals and metal products, food, oil and gas, coke, chemicals, synthetic fibres and textiles, machinery, ceramics, and motor vehicles. Her main trading partners are Germany, Austria, Italy, the Czech Republic, and other eastern European countries.

Slovakia's economic growth exceeded expectations in 2001–06 and grew beyond 10% in 2007; low wages, a qualified and committed workforce, and a 19% low flat tax rate have created the conditions for significant foreign investments in the automotive area: Kia, VW, and Peugeot have major factories here. As in Poland, the assembly lines in Slovakia deliver quality products: the VW car factory in Bratislava assembles top-of-range models and the Passat was rated as the best quality car among VW products. Slovakia is becoming the leading country in the world for number of manufactured cars per capita. However Angel Guría, the OECD secretary-general declared in Bratislava on 5 April 2007 at the presentation of the OECD report on Slovakia's performance: "We can't live on the next car factory and think that we're going to have a new car factory every year. Slovakia will also need to look for other ways to keep moving the economy forward" [MacLellan 2007]. Goría acknowledged that Slovakia made good use of the growth opportunities offered by accession, however, he expressed concern on the dangers of an overheating economy and on the high number of unemployed among the young, the women, and the elderly citizens. His suggestions included: flexibility in labour market; higher mobility of workers; raise of mandatory retirement age; reduction of the duration of the three years maternity leave; and improved education across the country.

The national currency is the Slovak Koruna (SKK); 1 € = 42.35 SKK on 14 May 2003 and the exchange rate decreased year after year to 1 € = 33.61 SKK on 6 May 2007 (see Table 11.13). Slovakia is in the Exchange Rate Mechanism with a view to adopting the euro in 2009, a date that at this point is uncertain and should be confirmed by the European Council during 2008.

Value-added tax (VAT) is levied at a rate of 19%, although some healthcare products are taxed at 10%. Corporate and individual income taxes have been levied at a flat rate of 19% since January 2004, with no tax on dividends. Payroll taxes add 48.6% to labor costs. The government relies on indirect taxes for the bulk of its revenue.

Table 11.13 presents basic economic data for Slovakia published by various quoted sources. More information is available in the recommended reading at the end of the chapter.

The European Commission analysis and recommendations in the Broad Guidelines of the Economic Policies [EC 2004-10] are presented as recommended reading at the end of Chapter 14, "Economic and Monetary Union,". Those relative to Slovakia point to: public finances, unemployment, regional disparities, the rigid education system, and the low R&D expenditure. They include:

> After a stabilization-induced slowdown in 1998, Slovakia's real GDP growth has steadily accelerated and exceeded 4% in 2003, for the second year in a row. Nevertheless, output has not yet fully reached its estimated potential. The growth performance has been

Table 11.13 Basic economic data for Slovakia

Slovakia		2000	2001	2002	2003	2004	2005	2006
GDP[a]	($bln)		20.9	24.2	32.7	41.1	46.4	
GDP[i]	($bln)	20.2	20.9	24.2	32.7	41.1	50.3	55.9
GDP[b]	($bln)					32.5		46.9
GDP[c]	($bln)			24.5	33.0	42.0	47.4	55.3
GDP[f]	(€bln)			26.0	29.4	34.0	38.0	43.9
pcGDP ppp[b,c]	($x000)					13.3		17.7[c]
GDP Growth[a]	(%)		4	5	4	5	6	
GDP Growth[b]	(%)					4.2		
GDP Growth[c]	(%)			4.1	4.2	5.4	6.0	8.3
GDP Growth[c,h]	(%)							5.6[h]
Current Account Balance[a,b,c]	(%GDP)				−8.0[a]	−0.9[a,b]		−8.3[c]
Government Balance[c,b]	(%GDP)				−5.7[c]	−5.1[b]	−3.6[c]	−3.3[b]
Government Balance[f,c]	(%GDP)			−7.7	−3.7	−3.0	−2.9	−2.5[c]
Total Debt[b]	(%GDP)					56.3		
Total Debt[f]	(%GDP)				43.3	42.7	41.6	34.5
Inflation (GDP deflator)[a]	(%)		4	4	5	5	2	
Inflation (CPI)[b]	(%)					8.6		4.4
Inflation (CPI)[c]	(%)			3.3	8.6	7.5	2.7	4.5
Unemployment[b,c]	(%)		19[c]			15.2[b]	17[c]	10.2[b]
Exports[a]	(%GDP)		73	72	78	77	79	
Imports[a]	(%GDP)		82	79	79	79	83	
Exports Growth[a]	(%)				5.5	22.6		10.4[d]
High-tech Exports[a]	(%Manuf. Exp.)		4	3	4	5	7	
Internet Users[a]	(/1000)		125	160	256	423	464	
School enrollment, tertiary[a]	(%)	29	30	32	34	36		
ITC Expenditure[f]	(%GDP)					2.0	2.2	2.3
FDI Inflows[a,c]	(%GDP)		7.6	16.9	1.7	3.1	4.1	6.6[c,h]
€ Exchange Rate[f]	(SKK)			43.3	42.70	41.49	40.02	38.60
Stock market capitalization[a]	(%GDP)	6	7	8	9	11	9	
GDP % in agriculture[a]		4	4	4	4	3		
GDP % in industry[a]		30	28	29	30	29		
GDP % in services[a]		66	68	67	67	67		

Source: [a]The World Bank (2005, 2007); [b]http://www.cia.gov/cia/publications/factbook; [c]European Commission, 2004; [d]Average Average 2002–2006; [e]The Economist Intelligence Unit, 2007; [f]Eurostat, 2007; [h]Average 2002–2006; [i]International Monetary Fund.

underpinned by intensified structural reforms. Prime examples were the restructuring and privatization in the banking and non-financial sectors. However, this did not come without cost: the unemployment rate ratcheted up to more than 19 per cent in 2001 and still amounts to around 17 per cent. A broad array of structural shortcomings in the labor market – notably a lack of regional mobility, disincentives from the benefit systems, wage inflexibilities, and skill mismatches – have been hampering its re-absorption capacity and are only being tackled now.

The general government deficit decreased from 5.7 per cent of GDP in 2002 to 3.6 per cent of GDP in 2003.

Slovakia has one of the lowest employment rates (around 57 per cent) and the second highest unemployment rate (currently around 17 per cent) in the acceding countries. Employment is particularly low in the age group over 55. Unemployment is concentrated among the young (below 24) and the low skilled. Regional disparities are high. The underlying structural deficiencies of the labor market are multi-faceted and are now being addressed more decisively. Incentives to work and to leave the informal sector are being strengthened by social benefit and pension reforms, including an increase of the retirement age to a still relatively low level of 62. Regional mobility is being enhanced by financial support for commuters and housing benefits but continues to be limited as the transport infrastructure and the functioning of the housing market improve only gradually. Skill mismatches hinder in particular the reintegration of the long-term unemployed, although retraining measures are being intensified. A still low alignment of the education system with the requirements of a market economy contributes to persistent youth unemployment. Job creation has been fostered by recent amendments to the labor code, which allow more flexible work relationships. However, the wage setting mechanism is not yet flexible enough and does not sufficiently cater for enterprise-specific conditions.

Over the recent years, the government has introduced a number of measures aimed at improving the business environment (e.g. for setting up a new business or concerning bankruptcy legislation). Despite these improvements, there still seem to be important barriers to entrepreneurship in Slovakia.

While retaining relatively high labor productivity growth, the level of productivity is still low (around 58 per cent of the EU average in 2003). Factors contributing to this situation are the lack of flexibility of the education system, coupled with low education expenditures and the weak R&D and innovation activity. The education system does not seem to respond appropriately to the labor market needs. This is particularly the case for secondary schooling which often produces graduates with obsolete skills. Moreover, the share of tertiary graduates is very low. The high share of long-term unemployed poses an additional challenge for the vocational and training programs. Furthermore, expenditure on education has dropped to one of the lowest levels among the new member states in 2001. Responding to these problems, the government has taken first steps towards rationalizing the system, improving its efficiency and increasing the sources of financing. In 2002, the expenditures on R&D accounted for only 0.59 per cent of GDP, less than in most of the acceding countries. Innovation activity, as measured by the number of patent applications, is very low. The government has undertaken some measures to improve the situation regarding R&D (e.g. by improving the legislative framework for R&D), and is committed to increasing the public resources available for R&D support.

To meet the challenges outlined above, Slovakia is recommended to:

1. reduce the general government deficit in a credible and sustainable way
2. strengthen labor supply by removing obstacles to regional mobility and by reducing skill mismatches
3. generate additional labor demand by allowing for more flexibility in the wage setting mechanism
4. lower the very high combined health and social contribution rates, while observing the overall budgetary consolidation constraints, in particular by additional health system reform
5. strengthen the legislative framework supportive to entrepreneurship and improve its enforceability by, in particular, adopting the new bankruptcy legislation and increasing the capacity and transparency of the judicial system
6. improve the efficiency and quality of the education and training system and its responsiveness to changing skills requirements
7. encourage R&D and innovation and support the transfer of knowledge

The fall 2007 European Commission Economic forecast for Slovakia [EC 2007-27] is presented as recommended reading at the end of Chapter 14, "Economic and Monetary Union,". It is summarized as "rapid growth contributes to lower unemployment" and it includes the following:

In 2006, real GDP growth increased to 8 1/4%. It was primarily driven by domestic demand but the external contribution also entered positive territory. Private consumption growth is expected to increase to some 7% in 2007, benefiting from strong real wage and employment growth. Continued big investment projects in the corporate sector should keep gross fixed capital formation growth above 7%. New export capacities in the manufacturing sector are likely to keep export growth at around 20%, outpacing imports by some 4%. A rapidly falling trade deficit should lead to a decrease in net borrowing from the rest of the world to around 4 1/4% of GDP in 2007.

Economic expansion is expected to gradually decelerate to around 7% and 6 1/4% in 2008 and 2009 respectively. Domestic demand is likely to remain the main driving force of growth. A steadily improving labor market situation combined with strong credit growth is expected to continue to support private consumption growth, which, despite slowing down, is likely to remain at above 5% over the forecast period. Gross fixed capital formation should decelerate only marginally as Slovakia continues to attract new FDI projects.

Total employment is expected to increase by some 2% in 2007 while the unemployment rate is likely to decrease to close to 11%. Nominal unit labor costs growth is likely to accelerate from 2007 onwards, due to decelerating productivity growth. Thanks to lower increases in regulated prices in the energy sector at the beginning of 2007 and the positive impact of exchange rate appreciation, HICP inflation is anticipated to drop to around 1 3/3% in 2007. An increase in excise taxes on cigarettes and accelerating electricity, water and food prices are expected to push the annual average HICP inflation up to some 2 1/2% in 2008.

The general government deficit (including the pension reform costs) increased to 3 3/3% of GDP in 2006, but was still some 0.5% lower than foreseen in the 2006 budget, mainly thanks to lower-than-budgeted cofinancing needs for EU-sponsored projects. In 2007, stronger-than-expected economic growth is likely to result in higher tax and dividend revenue which together with lower-than-expected interest expenditure should enable the government to bring the deficit below its target of 2.9% of GDP despite higher-than-foreseen transfers to the second pension pillar. Gross public debt (34.5% in 2005) is expected to remain broadly stable in 2008–2009, thanks mainly to strong GDP growth.

11.5.10 Slovenia

Slovenia is located in southeastern Europe (see Fig. 11.1). Her size is 20 thousand km², two thirds of Belgium. Her population is 2 million. The Slovene lands were part of the Holy Roman Empire and were governed by the Habsburgs of the Austro-Hungarian Empire from the end of the Middle Ages until 1918, when the Slovenes joined the Serbs and Croats in forming a new nation, renamed Yugoslavia in 1929. After Second World War, Slovenia became a republic of the renewed Yugoslavia, which, though Communist, distanced herself from Moscow's rule.

Slovenia held the first multiparty elections in Yugoslavia since Second World War in April 1990. The winning coalition called for independence, and nearly 90% of Slovenia's population voted for independence in a referendum in December 1990. Slovenia declared her independence from Yugoslavia in June 1991, and after defeating the Serb-dominated Yugoslav People's Army in a 10-day war, it quickly won international recognition. It was the most liberal republic within the former Yugoslav federation, and always the most prosperous region, and it has made a smooth transition towards a pluralist democracy and a market economy.

Slovenia is a small country with a wide exposure to the outside world. Slovenia combines the Slavic soul, Germanic straightforwardness, and Latin flair in an incomparable mix of cultures. Slovenia is known, among other things, for her recent Olympic medals in downhill skiing, rowing, shooting, kayaking, and athletics. Slovenia's younger generation of winegrowers on the very borders with Italy have developed a seriously lucrative export business for their high-quality, high-tech, and high-priced white wines. Slovenia hosts one of Europe's leading early music festivals. Endowed with highly skilled human capital and strategically located as a crossroads between the East and West of Europe it is widely perceived as an important contributor to the stabilization and cohesion of the Balkan region.

On Sunday, 23 March 2003, about 90% of Slovenes voted *yes* to the EU accession referendum, seeing the EU membership as a step forward, as Janez Potonik, Slovene Commissioner explains [Potonik 2004]: "Accession is not the end of a journey, but a step, a major one toward achieving the same opportunities the other EU citizens have already enjoyed for a long time."

The only accession country from former Yugoslavia, Slovenia's population of 2 million owe their heritage to their uninterrupted links from the early Middle Ages with the Italian- and German-speaking part of the Holy Roman and Austro-Hungarian Empires. Slovenia is the most prosperous EU accession countries, with a GDP per capita (in terms of purchasing power parity) higher than Greece's and close to Portugal's, and an unemployment rate of 5.1% in 2007, lower than in Germany (6.4%) or France (8.6) [Eurostat 2007-1]. Her main concern about her terms of EU accession is that her relative prosperity will mean it could pay in more than it gets out.

Mr. Janez Jansa, the Slovenian centre-right prime minister, whose government replaced after the 2004 elections a centre-left administration, is faced with a big challenge: after being the fist accession country to adopt the euro on 1 January 2007, Slovenia will be the first accession country to hold the rotating presidency of the European Council on 1 January 2008:

> Running the European Union for six months is a big job, even for a large country. For a small one, it means a rare limelight – and a huge headache. In barely a year's time, and with fewer civil servants than in a big country's foreign ministry alone, Slovenia will be running the 4,000-odd meetings and eight ministerial get-togethers of the EU presidency. With a population of 2m it will be the smallest country to lead the EU, and the first ex-communist state to do so [Economist 2006-2].

The past and present special relations of Slovenia with the other countries in the Balkans might ease the way during the Slovenian EU presidency towards a solution of the proposed independence of Kosovo from Serbia.

11.5.10.1 Economy

Coal is the most abundant natural resource in Slovenia; other resources include lead, zinc, mercury, uranium, and silver, as well as natural gas and petroleum. A nuclear power station at Krško, on Slovenia's side of the border but half-owned by Croatia, produces 39% of the electricity used in Slovenia. It benefits from advanced safety applications developed under the Euratom Framework Program funded by the European Union, and has not been the focus of any safety concerns.

Slovenia enjoys a substantially higher GDP per capita than those of the other transitioning economies of central Europe; her neighbours in western Balkans are doing less well. Average gross wages were $ 1,000 on average in Slovenia during 2000–2001, just less than $ 700 in Croatia, about $ 300 in Bosnia and Herzegovina, and less than $ 200 in Serbia and Montenegro [Economist 2003-4]. Slovenia needs to speed up the privatization process and the dismantling of restrictions on foreign investment. After attracting relatively little foreign direct investment at a level of about 1% of GDP, which is the lowest in the region, Slovenia has recently been winning several large foreign investments; the overall level of FDI increased to over 2% of GDP in 2002, however, it stays well below the level of most accession countries. The government is intensifying economic co-operation with Russia, particularly in energy and steel, and will attempt to attract greenfield investment, despite fierce regional competition. Nevertheless, the uncertainty surrounding the role of foreign companies in the privatization of some key assets signals that the overall attitude towards foreign investment will remain lukewarm [Economist 2007-10]. About 45% of the economy remains in state hands. Despite the global slowdown in 2001, the economy turned in an excellent record on exports, which grew 5%.

Farming and livestock raising are contributing 3% to GDP, the main crops being cereals such as corn and wheat, potatoes, sugar beets, and fruits (particularly grapes). Industry constituted 38% of GDP in Slovenia in 2000. Following the breakup of Yugoslavia, Slovenia's economy has grown unimpeded by the warfare that has devastated other regions, so that today Slovenia is the most industrialized and urbanized of all the former Yugoslav republics. Major industries produce electrical equipment, processed food, textiles, paper and paper products, chemicals, and wood products. Tourism has increased markedly and is now a major industry, with its attractions ranging from spas to the stalactites and stalagmites in the 20 kilometres of underground passages in the Postojna Caves, and from the Bled and Bohinj lakes in the shadows of the Alps to the four-century-old stud farms at Lipica, where the famous Lipizzaner horses come from. Most visitors are from Italy, Germany, and Austria. Iron, steel, aluminium, machine tools, motor vehicles, cement, chemicals, textiles, and leather, as well as light engineering and some electronics, are the main industries. The country's chief trading partners are Germany, Italy, and Croatia.

The Slovene economy has achieved solid growth – averaging 4.3% over the past eight years – while avoiding the major macroeconomic imbalances that have characterized most other transition economies in the region. Slovenia, the richest ex-communist state in Europe, wants to catch up Estonia, the fastest-growing post-communist economy [Economist 2005-3]. GDP per capita at purchasing power parity reached the level of €16,990 in 2002 and €19,000 in 2003, representing over 80% of the EU15 average. Slovenia is among the countries with the smallest public deficit, which despite having increased in 2000, stands at only 29.1% of GDP in

2005, caused for the first time by foreign rather than domestic debt. Unemployment has been declining and stood at 6.4% in 2002.

Inflation, persistently high in the early 2000s, has been an issue, however it was curbed from 2005 (7.5% in 2002; 5.6% in 2003, 2.5% in 2005 and 2006). The national currency was the Tolar (SIT); 1 € = 233.05 SIT on 14 May 2003. Slovenia was the first of AC10 to adopt the euro. Euro replaced Tolar on 1 January 2007 at the rate of 1 € = 239.64 SIT. Corporate income is taxed at a flat rate of 23%, with a scheduled progressive reduction to 20% by 2010. Personal income is taxed progressively, with tax brackets ranging from 16% to 41%.

Some basic economic data for Slovenia published by various quoted sources are presented in Table 11.14. More information is available in the recommended reading at the end of the chapter.

Slovenia has a high level of trade integration with the EU. About 59% of Slovenia's exports go to the EU and 68% of imports come from the EU. The main exports include machinery and transport equipment, chemicals, footwear, furniture, and other household goods.

The European Commission analysis and recommendations in the Broad Guidelines of the Economic Policies [EC 2004-10] are presented as recommended reading at the end of Chapter 14, "Economic and Monetary Union,". Those relative to Slovenia point to: the need to control inflation and to unemployment of the elderly and of low-skilled people; the report acknowledges the higher level of R&D investment, compared to other accession countries; however it identifies problems in transferring results of research to the business community The analysis and recommendations include:

Over the last decade, Slovenia has exhibited stable growth; real GDP has grown steadily at 3–5 per cent since 1993. After a recent slow-down, GDP growth is expected to pick up and gradually draw closer to the potential output growth, estimated at around 4%.

Despite a broadly sound economy and commendable achievements in many policy areas, certain weaknesses remain. While having declined steeply in the last two years, relatively high inflation (5.7 per cent in 2003) is still a cause of some concern and has been recognized as a key policy challenge. The government designed an apt policy to lower inflation in a sustainable way, aiming to create conditions to be able to fully reap the benefits of EU accession. It has pursued structural reforms with the objective to facilitate price liberalization. However, given ineffective competition in the utilities and insufficient flexibility in both in the financial sector and the labor market, a sustainable reduction of inflation still needs to be confirmed.

The restructuring process of the economy has negatively affected the situation on the labor market; employment in manufacturing industry fell markedly over the period 1996–2000. Although the overall picture concerning the labor market in Slovenia is positive, structural unemployment problems persist. The proportion of long-term unemployment is high, in particular amongst older, lows-killed persons. Furthermore, the low employment rate of workers over 55 years of age is a cause of concern, especially with regard to the challenges stemming from an ageing population.

The level of productivity, while the second highest amongst the new member states, remains well below the EU15 average (69.5 per cent in 2003). Labor productivity growth in Slovenia was relatively rapid over the period 1995–1999, with an average annual rise of 4.8 per cent. However, it decelerated thereafter (2 per cent on average in 1999–2002), placing Slovenia amongst the least performing acceding countries in this period of time.

Table 11.14 Basic economic data for Slovenia

Slovenia		2000	2001	2002	2003	2004	2005	2006
GDP[c]	($bln)			22.3	28.1	32.6	34.4	37.3
GDP[i]	($bln)	19.0	19.6	22.1	27.8	32.8	35.1	37.5
GDP[h]	($bln)				36.8[g]			37.64
GDP[f]	(€bln)			22.3	24.3	26.2	27.6	29.7
pcGDP ppp[c,f]	(€x000)			17.0[c]	19.0[c]		20.5[f]	
pcGDP ppp[h]	($x000)				19.0			23.4
GDP Growth[c]	(%)			3.5	2.7	4.4	4.0	5.2
GDP Growth[h]	(%)				2.3			
Current Account Balance[h]	(%GDP)				−1.0			
Current Account Balance[c]	(%GDP)			1.1	−0.8	−2.7	−2.0	−2.5
Government Balance[c,h]	(%GDP)				−0.9			−1.2
Government Balance[f]	(%GDP)			−2.7	−2.8	−2.3	−1.8	
Total Debt[h]	(%GDP)				31.9			29.0
Total Debt[f]	(%GDP)			29.7	29.1	29.5	29.1	
Inflation (CPI)[h]	(%)				5.6			2.4
Inflation (CPI)[c]	(%)			7.5	5.6	3.6	2.5	2.5
Unemployment[b,c]	(%)			6.4[c]	11.2			9.6
Unemployment[f]	(%)							5.3[f]
Exports[a]	(%GDP)				42.6			58.0
Imports[a]	(%GDP)				44.9			62.7
Exports[d]	(%GDP)			49.2	46.4	48.8	52.2	
Imports[d]	(%GDP)			51.9	50.4	54.0	57.3	
Exports Growth[a,c]	(%)		5.0[c]		5.0			
School enrollment, tertiary[a]	(%)	56	61	67	70	74		
Internet Users[d]	(/1000)			420	430	430	560	600
Internet Users[j]	(/1000)					375	440	
ITC Expenditure[f]	(%GDP)				1.8	1.9	2.0	
FDI Inflows[c]	(%GDP)			2.0				
€ Exchange Rate[f]	(SIT)	217.98	225.98	233.85	239.09	239.57		
Stock market capitalization[a]	(%GDP)	13	14	21	25	30	23	
GDP % in agriculture[a,b]		3[a]			3.0			2.3
GDP % in industry[a,b]		38[a]			39.7			34.1
GDP % in services[a,b]		59[a]			57.3			63.6

Source: [a]The World Bank (2005, 2007); [b]http://www.cia.gov/cia/publications/factbook; [c]European Commission, 2004; [d]Statistical Office of Slovenia; [e]The Economist Intelligence Unit, 2007; [f]Eurostat, 2007; [g]Purchasing Power Parity (ppp); [i]International Monetary Fund; [j]University of Ljubljana, RIS Project, 2007.

There is a lack of effectiveness of R&D expenditure to build on relatively good achievements in basic research in terms of transfer of know-how to the business sector, of patenting and of product or process innovation. With R&D expenditures reaching 1.6 per cent of GDP in 2001, Slovenia ranks first amongst the new member states but still remains below the EU average, despite strong growth in direct public funding of R&D and the introduction of fiscal incentives for business R&D. In addition, the share of researchers employed in the business sector is low as compared to the public sector (one-third and two third respectively) and innovative activity appears to be weak in the high tech sector.

To meet the challenges outlined above, Slovakia is recommended to:

1. step-up structural reforms aimed at liberalizing administered prices and advance further with de-indexation, in particular of the wage setting mechanism
2. review the tax and benefit systems, with a focus on labor market participation of older workers, and reassess the measures promoting active ageing by means of lifelong learning activities, as well as address the imbalance between temporary and permanent work conditions
3. further reduce the time and costs necessary to set up a new company and simplify the administrative procedures affecting businesses
4. promote R&D and innovation in the business sector and improve the quality of the tertiary level education system
5. strengthen the administrative capacity of the Competition Protection Office; ease the entry of new competitors in network

The fall 2007 European Commission Economic forecast for Slovenia [EC 2007-27] is presented as recommended reading at the end of Chapter 14, "Economic and Monetary Union,". It is summarized as "strong but decelerating growth with a pick-up in inflationary pressures" and it includes the following:

> Economic growth in Slovenia remained buoyant in 2007. Non-housing construction was particularly buoyant due to, inter alia, motorway construction and other infrastructure investment. Exports of goods and services grew at very high rates during the first half of the year, supported by the adoption of the euro in 2007.
>
> Consumer price inflation increased markedly in 2007 and is projected to reach an average level of 3.5 % for the year as a whole, compared to 2.5 % in 2006. Although euro changeover effects appeared limited initially, some abnormal price increases were reported following the end of the dual price display in June. In 2008, inflation is projected to rise to 3.7 %. Slovenia is especially vulnerable to changes in oil prices due to the high share of energy in HICP. Also, food prices might experience further increases. Wage pressures arise mainly from the government's efforts to reduce income disparities within the public sector, which are expected to lead to a substantial increase of the public sector wage bill in 2008. Due to demonstration effects, significant increases might also be expected for private sector wages. However, inflation is expected to revert to lower levels in 2009.
>
> The general government deficit is expected to narrow to 0.7 % of GDP in 2007, well below the 1.5 % projected in spring. The debt-to-GDP ratio is expected to decline further, falling below 24 % in 2009.

11.6 The 2004 Enlargement Three Years On

The accession of 10 countries in 2004 is one of the great achievements of the European Union and a major milestone towards peace, stability, and prosperity in Europe. The first and foremost benefit for the European citizens is securing a peaceful development for a continent that experiences centuries of bloody wars. Another important accomplishment is that the accession countries achieved a profound democratic and economic transformation and brought 75 more million citizens to share the values and principles that unite the peoples in the European Union. Other major benefits include the expansion of the internal market, the increased trade, the creation of new jobs, easier travel, better chances to study abroad, better protection of the environment and of health, control of illegal migration and effectiveness in fighting organized crime. A further step towards integration of the 2004 accession countries with EU15 was the extension in December 2007 of the Schengen Convention, allowing travelling in the EU without showing passports. The Economist wrote in 2005: "stability and prosperity can be exported without importing instability and poverty in exchange," [Economist 2005-4]; the European Commission wrote reports in 2005 and 2007 [EC 2006-6, EC 2007-27] to documents some facts about the enlargement countries that are included as recommended reading at the end of this chapter.

The European Central Bank's Annual Report 2004 had anticipated, concerning enlargement, that – notwithstanding the historical relevance of the event – accession would not change the characteristics of EU economy, due to the modest contribution to EU GDP of the 10 entrant countries (4.5%), about half the increase of GDP at the 1986 and 1995 enlargements; ECB anticipated that the accession would in the long term contribute positively to EU's economic growth [ECB 2004]:

In terms of the number of countries joining, this enlargement was the most important in the history of European integration. However, the economic weight of the new member states is relatively small compared with previous enlargements. As a result, the accession of the new member states has not fundamentally changed the key characteristics of the EU economy. Economic diversity within the EU has increased, however, as the institutional and structural features of the new member states are in many respects different from those of the other member states. The majority of the new member states have been engaged in a transition process from a centrally planned to a market economy, involving fundamental institutional and structural changes in their economies. Over a longer time horizon, enlargement is likely to contribute positively to economic growth in the EU; indeed, a positive impact was already observed during the decade of preparations for enlargement. Economic activity in the countries forming the EU25, as measured by GDP, was €9.8 billion in 2003, to which, on the basis of current exchange rates, the accession of the new member states contributed 4.5%. From a historical perspective, the economic size of this enlargement was relatively limited. For example, the accession of Spain and Portugal to the European Community in 1986 (which at the time consisted of ten countries) raised the Community's total GDP by slightly more than 8%, and the enlargement to Austria, Finland and Sweden in 1995 increased economic output by just over 7%. As the new member states have a relatively large population in relation to their current level of economic activity, their accession implies a decline in the average level of GDP per capita in the EU. This, however, is expected to change gradually in line with progress in the catching-up process. Compared with the United States and Japan, GDP per capita in the EU25 is relatively low, although the difference vis-à-vis Japan is much more limited in PPP terms (on the basis of figures for 2003).

Pro and cons of enlargement are not unanimously assessed [Economist 2004-6]:

On one side stand economists armed with formulae and tables of data, arguing that migration from the poor countries of central Europe to the rich countries of western Europe will be modest and manageable after ten new members join the European Union in May. On the other stand Eurosceptics, trade unions and some governments, worried that enlargement will bring a rush of migrants chasing jobs and social-security benefits. Almost drowned out are voices from the poor countries themselves, demanding the rights and freedoms of EU membership, but fearing a drain of skilled workers.

The 2004 enlargement has lowered the average per capita income in the EU, as had all previous enlargements. The eight central European countries had in 2001 an average income per head of only 23% of the EU average, a figure that would fall to 18% if Bulgaria and Romania were added [Economist 2004-6]. However, GDP in AC10 is growing at a faster rate than in EU15 (see Fig. 11.3), their trades are developing at a higher rate (see Fig. 11.4), and their inflation is converging to EU15 values (see Fig. 11.5).

11.6.1 Benefits and Challenges in the Enlarged Union

Common positive features of the economy of the accession countries at the end of 2007 include the above mentioned higher-than-EU-average growth of GDP and international trade; reduced unemployment, falling to the lowest level recorded since independence; reduced informal economy, strong internal demand, and sustained foreign investments; wealth and jobs creation via the development of motorways and other public works – sustained by EU funding; growing construction industry, in particular real estate; attraction of an increasing number of international tourists; increase of per capita income.

Some shadows are, however, present and could compromise the competitive edge of the accession countries: the economic growth is hampered by emigration of skilled and entrepreneuring workers, by brain drain, still high unemployment, low participation to productive activities, in particular for women and older and unskilled citizens; skills mismatch between demand and offer on the labour market; inadequate education and vocational training system; insufficient investments in R&D and insufficient involvement of private business in research and innovation. Inflationary pressures are present, due to wage increases, raise in cost of housing and of food and energy prices, and speculative unjustified price increases especially in

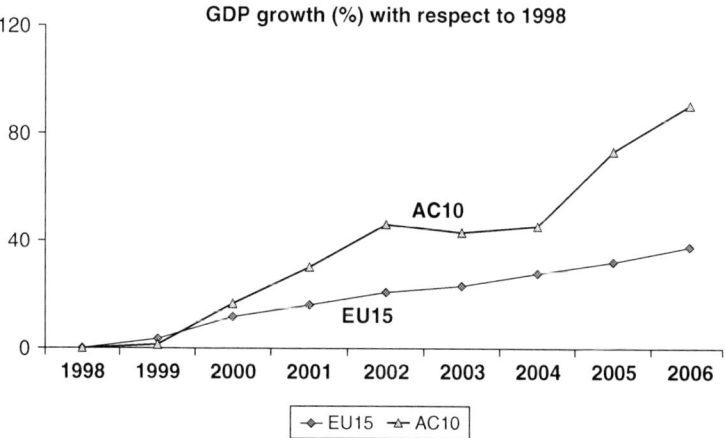

Fig. 11.3 AC10 and EU15 GDP growth with respect to its value in 1998. (Eurostat 2007)

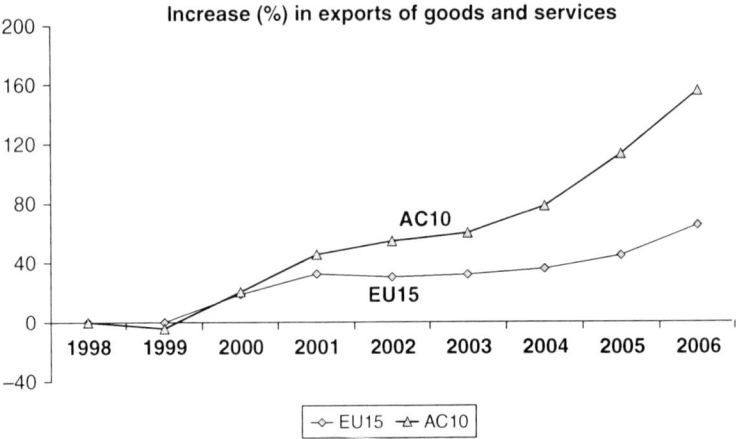

Fig. 11.4 AC10 and EU15 increase in exports with respect to their value in 1998. (Eurostat 2007)

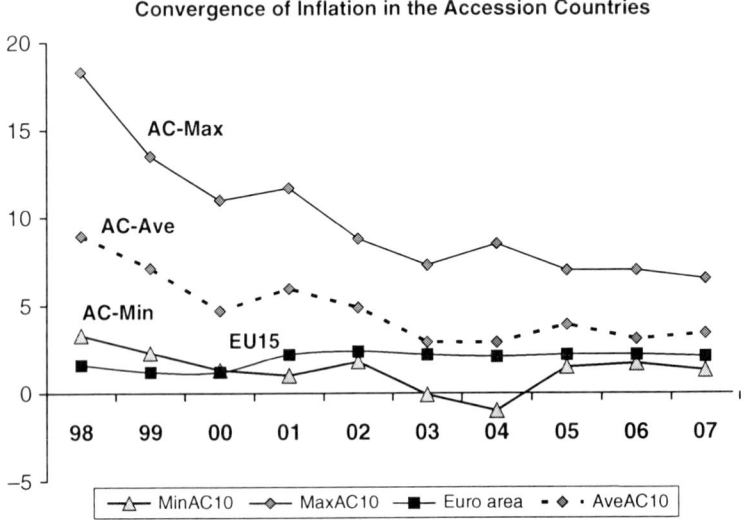

Fig. 11.5 AC10 and EU15 inflation (Consumer Price Index) from 1998 to 2007. For AC10 countries the graph displays the minimum, the average and the maximum values. The best performing AC10 countries have lower inflation than the EU15 average. (Eurostat 2007)

the areas with intense presence of foreign businesspersons and tourists; international speculation in the real estate market, (partly related to money laundering) that – together with uncontrolled increase in prices and wages – might undermine the international attractiveness of foreign investments.

The identification of these problem areas is in support of the thesis of this book: European competitiveness relies on major investments in education, lifelong learning, research, and innovation. These investments, together with macroeconomic policies essential to support a well-balanced economic expansion and the full real-ization of current growth potential, complemented by the development of transport, energy, and telecommunication infrastructures are the conditions for building in the knowledge-based environment of the 21st century the competitive European society that the EU is aiming at. These arguments have been addressed in the second part of this book.

Might the accession countries might become the locomotive that pulls the econ-omy of the European Union out of years of slow growth? Enlargement may con-tribute to the economic growth of Europe but it may well take 50–100 years before the present differences in living standards across Europe are overcome and new dynamics in the society define winners and losers in the 21st century, wiping out historical differences. Already in the second half of the first decade of the new cen-tury, with the economic recovery set to unfold in the European Union, the situation is characterized by significant growth dispersion, with almost all of the recently acceded member states posting with Denmark, Greece, Spain, Ireland, and Luxem-bourg growth rates of more than 3% in 2005, whilst Germany, Italy, and Portugal saw output growth staying below 1%. This might contribute to the development of a multi-speed-EU; however, the participation of the AC10 to the group of countries posting high growth is dependent upon their ability to continue in the transforma-tion of their societal organization, their economies, and their education and research structures, overcoming the weaknesses outlined above.

In the era of globalization, enlargement provides global solutions to problems that could not be approached on a local basis: fight organized crime, money launder-ing, and illegal immigration; efficient transport networks for people, goods, energy, and information; protection of the environment; technological research to increase competitiveness.

A fair picture of the enlargement should not underestimate the great challenges that EU27 is facing. The repeated mention of the same problem areas in the survey of the accession countries in the previous sections identifies the hot areas: participation, mobility and qualification in the labour market, the deficiencies in the education and training system producing lack of skilled workers, inflationary pressures, and low productivity reducing competitiveness. Accession countries must work hard to keep their competitive edge based on lower labour costs and become prepared to face competition in the knowledge-based economy.

Fighting crime was mentioned earlier: enlargement enables the EU to extend its police and justice cooperation to the new member states, thus making the fight against crime and terrorism more effective. The more EU members integrate their crime-fighting efforts with their neighbours, the better they can protect Europe's citizens. Security and justice are areas where the EU can clearly do more to protect her citizens than any one country can alone. The EU has developed police, customs and judicial cooperation to tackle terrorism, organized crime, drugs, trafficking in human beings, and illegal migration that are now part of the common rules that all member states must apply. Problems are still far from being remedied in the accession countries or in (some of) the EU15, but enlargement is for sure a step towards their solution rather than an aggravation.

Another subject has call forth strong emotions in EU15: migration, as reported in the article by Norman Davies [Davies 2004] quoted at the beginning of this chapter. Labour migration from new to old member states has been modest, rarely reaching even 1% of the active working population of the host country. The largest number of migrants came from Poland, and in percentage of the countries' population significant migration is reported from the Baltic republics. Some EU15 countries adopted temporary restrictions but it was proved that they do not work; they rather increase illegal and shadow employment. Three countries (IE, SW, UK) imposed no restrictions and in the outcome of the fall 2006 Eurobarometer [EC 2006-2], citizens from all three countries answered *Agree* to the question "did immigrants contribute to your country?" with a higher percentage than EU25 average. Far from stealing jobs from locals, most incomers have taken jobs that locals shun in agriculture, construction, and care of the elderly and of people with reduced abilities. In addition, most people who move do not want to settle abroad, but to get cash and skills for a better life back at home. In a flexible and deregulated labour market, migration rather than a problem is an opportunity that creates cheaply but legally jobs for the workers who want them.

Outsourcing and relocation is the converse of migration and has also raised much discussion. Studies confirm that relocation of companies from the old to the new member states remains a marginal phenomenon. Moreover, Central and Eastern Europe is not the main destination for relocation, but rather Asia: it is not enlargement that mainly causes outsourcing and relocation, but global competition. Investing in Central and Eastern Europe instead of China or India can help European industry to maintain jobs and create growth throughout Europe. For instance, the expansion of service enterprises instead of being a thread of social dumping is creating jobs in both new and old member states; and the delocalization of west-European car plants in the AC10 countries makes industrial project feasible and successful, while they would not be economical any longer in the EU15 home country due to global competition. Enlargement helps the EU to meet the challenge of globalization by increasing internal and external trade and thus keeping and creating jobs.

11.6.2 The Internal Market in the Enlarged Union

We mentioned earlier the expansion of the internal market as one of the major benefits of enlargement. A question may be raised as to whether the regulations that underlie the internal market have not been watered-down by the accession of 10 new EU members. Distortions of the internal market might result from badly transposed and implemented common rules, whether by old or new members. We saw in Chapter 10, "From the Treaty of Rome to the Reform Treaty of Lisbon", that the new member states are performing substantially better than the old members in transposing and applying EU laws (see Fig. 10.5). So we should be reassured: the internal market can work in an enlarged EU. By opening up a market of 75 million consumers to companies from the old member states, enlargement has strengthened competition in the internal market, which in turn also makes European companies more competitive on world markets. The same rules on internal market, competition, and state aid now apply across all the member states.

The history of the EU proves that there is no contradiction between widening the Union and deepening its integration. The EU has managed to do both: since 1973 the EU has enlarged five times to take in altogether 19 new countries, while at the same time developing the single market, creating the Schengen area of passport-free travel, adopting the euro, and developing a host of other new policies, such as internal security and a stronger foreign policy. The EU has however, increased the difference in wealth between the better off and the poorest regions. This implies the need for rethinking the agricultural and cohesion policies, which may cause some vibrations in a number of EU15 countries that were the main beneficiaries of the EU subsidies.

Enlargement requires some adjustments to the Union's policies and her budget. The first two lines in the Community budget are the Common Agricultural Policy (CAP) and the Cohesion Policy. Both will be deeply affected. CAP is reformed from direct subsidies to rural development and protection of environment, consumers, animal welfare and product quality. Cohesion Policy will continue to privilege regions where pc GDP is lower than 75% of the EU average, but – the average being lowered – some regions in EU15 that qualified for subsidies will no longer get them. However, poorer countries and regions will not be flooded with lots of euro from Brussels, because in no case will aids exceed 4% of the local GDP to ensure safe absorption without harming the national equilibriums.

11.6.3 Citizens' Opinion on Enlargement

The members of EU27 are not unanimous in their judgment about enlargement in the same way as unanimity is not yet reached at the European Council level, on the monetary policy, and on the social policy. In Chapter 9, "The Position of the EU in the Quest for Competitiveness and Beyond" the author will expand on these differences and present his personal vision of a multi-tier EU where groups of members evolve at different speed and not necessarily in the same direction.

The voice of the citizens of AC10 should be heard when assessing enlargement. The Eurobarometer report of the fall 2006 [EC 2006-2] gives interesting indications:

Question QA7: Do you think that your country's membership is a good think?
EU25 answer *a good thing*: 53%
Answer in AC10 with a higher score than EU25: LT, PL, SK, SI, and EE
Answer in AC10 with a lower score than EU25: CK, CY, MT, LV, and HU
Question QA8: Do you think that your country has benefited or not from being a member of the EU?
EU25 answer *benefited*: 54%
Answer in AC10 with a higher score than EU25: LT, PL, EE, SK, SI, CK, LV, and MT
Answer in AC10 with a lower score than EU25: CY and HU
Question QA9: Does the EU conjure up for you a positive or negative image? EU25 answer
positive: 46%
Answer in AC10 with a higher score than EU25: SI, PL, CY, LT, SK, CK, and EE

Answer in AC10 with a lower score than EU25: MT, LV, and HU

Hungary has consistently a lower score than EU average; Cyprus and Latvia have a lower score than EU average on two out of the three questions reported above. Lithuania, Poland, Slovakia, Slovenia, and Estonia have – for all the three questions – a higher score than the EU average. The survey goes deeper in the analysis of the opinion in the different member state. Some findings in two extreme cases of Hungary and Lithuania are reported below, while the whole reports are available in the recommended reading at the end of the chapter.

The pessimism of Hungarian citizens is evident once more in fall 2006. It has now become almost a tradition in the series of Eurobarometer surveys that, of all the Europeans, Hungarians are the least content with their lives. In addition, the ratio of those who expect positive changes in their lives or in the economic situation of the country during the next year has dropped drastically. In Hungary, as well as in the 25 member states on average, there seemed to be a consensus that education and training was the area to which the EU had to pay the most attention in order to develop the economy. This was followed by increasing the efficiency of energy use, making it easier to set up companies and supporting innovation and research. The outstanding proportion of those who rated the situation of the Hungarian economy and employment as bad or rather bad shows that there is strong demand for improving the economic situation. This generally negative picture was unchanged compared to the results six months ago, although there was a slight improvement in the assessment of the employment situation.

Compared with spring 2006, Lithuanians have become even greater optimists when assessing various perspectives of the country and their own personal perspectives. The share of those claiming their life will be better within the next 12 months has increased by 6 % up to 43%. The same increase (up to 37%) is observed among those thinking that the economic situation in Lithuania will improve. The share of respondents thinking that their families' financial situation will improve increased by 5% and the share of those claiming that the employment situation will improve increased by 9%.

In terms of all the above attitudes, Lithuanian residents are considerably greater optimists than the average EU citizen.

Lithuanian levels of optimism are the highest recorded since Eurobarometer surveys started in Lithuania. Comparable figures were observed back in 2004 after Lithuania had joined the European Union, but levels of optimism were even lower then. Undoubtedly, these attitudes are determined both by the country's favorable economic situation, the results of which are probably felt by a great part of society, and by the significant amount of EU support for the country's development, giving optimism for the future. The share of Lithuanian residents claiming they are satisfied with their current life also increased by 4%.

Lithuanians are some of the most favorably disposed towards membership of the EU, i.e. 77% of Lithuanian residents think that EU membership is beneficial for Lithuania, and, in terms of this attitude, Lithuania's figure considerably exceeds those of other EU countries: only slightly more than a half of all EU citizens think that EU membership is beneficial to their countries. The level of favorable disposition towards the EU reached almost the same level in fall 2006 – just after Lithuania had joined the Union.

Likewise, in many recent years, Lithuanians currently attach greatest importance to problems related to crime and economy. Moreover, problems related to the health care system are also included into the top five of the key problems identified. A dominating problem – unemployment – persists in the European Union. On average, 40% of the EU population cited it. In fact, compared with the spring survey, the share of respondents citing this problem has decreased by 9%. The importance of this problem to Lithuanians has also decreased by 8%.

On the other hand, rising prices/inflation is becoming a more urgent problem in Lithuania (7% increase). The discussions relating to the adoption of the Euro and the European Commission's warnings to Lithuania about increasing inflation also had an influence on public opinion on this problem.

Across the whole survey, young people give more positive and optimistic answers both to questions about their confidence in the EU and about their own future, while the elderly show a clearly different and less favourable attitude. This reflects the difficulty for those in their late forties and older to find their way in an open market coming from a lifetime in a centrally planned economy. Earlier in this chapter we wrote that it might take fifty to one hundred years before the difference between the old and the new member states are wiped out. In fact, we should not be too much in a hurry, because it may well take several generations before the new life styles and personal attitudes are adapted to the new societal organization.

11.7 Second Wave: An Overview of the 2007 Enlargement

The Copenhagen European Council decided in December 2002 that Romania and Bulgaria (see Table 11.15) would be allowed to draw up an accession Treaty with the European Union starting from 2007, should the convergence towards the Copenhagen criteria be completed. Accession negotiations were completed in December 2004. The Accession Treaty was signed in April 2005. The Council decision of 2002 was confirmed in 2006 on the basis of a report by the European Commission [EC 2006-3] and the two countries accessed the EU on 1 January 2007.

11.7.1 Bulgaria

Bulgaria is a southeastern European country (see Fig. 11.1) whose size is 111 thousand km^2, or about halfway between the size of Austria and that of Greece. Her population is slightly less than 8 million. Bulgaria is one of the most ancient states on the European continent. It was founded in 681. Her rich historic heritage, coupled with beautiful natural scenery, is most conducive to the development of tourism. The country is famous for its Thracian Gold Treasure. It also boasts nine cultural monuments and natural reserves featuring on the Unesco list.

Table 11.15 Countries that accessed in 2007. Additional data in the following sections addressing individual countries. Surface area is in thousand km^2. GDP is in thousand dollars

Country	Capital	Surface	Population (million)	pc GDP ppp	Inflation (%)	Unemployed (%)
Bulgaria	Sofija	111	8	7.6	2.3	14.3
Romania	Bucharest	238	23	7.0	15.3	7.2

Source: OECD and European Commission, 2003.

Bulgaria struggled with the Byzantine Empire to assert her place in the Balkans, but by the end of the fourteenth century the country was overrun by the Ottoman Turks. Bulgaria regained her independence in 1878, but, having fought on the losing side in both World Wars, it fell within the Soviet sphere of influence and became a People's Republic in 1946. Communist domination ended in 1990, when Bulgaria held her first multiparty election since Second World War. The contentious process of moving towards political democracy and a market economy while combating inflation, unemployment, corruption, and crime began in the 1990s. The country joined NATO in 2004. Reforms and democratization kept Bulgaria on a path towards integration into the EU. Corruption and the lack of determination in prosecuting corruption appear to be one of the major problems in Bulgaria. Mr. Ognian Shentov, of the Centre for the Study of Democracy, a Sofia think-tank, quoted by the Economist, argues that "corruption in Bulgaria is not necessarily the worst in the EU – but it is certainly the most intensively studied." While one optimistic Bulgarian official says [Economist 2006-3]: " 'Italian-style.' Indeed, if the EU can cope with Sicily, why not Bulgaria?"

The country joined NATO in 2004. Reforms and democratization kept Bulgaria on a path towards integration into the EU. Bulgaria is in the honeymoon stage of EU membership. Accession has fuelled a construction boom from the Black Sea to Sofia. Foreigners are pouring money into property; local builders are so scarce that workers are being recruited in Belarus and Ukraine. The economy will grow by over 6% in 2007. So many Bulgarians have taken mortgages that the central bank has introduced loan curbs. The government is under Prime Minister Sergei Stanishev of the Bulgarian Socialist Party, who leads a coalition since August 2005. Next parliamentary elections are due in June 2009 but Stanishev might not be in office by then, due to conflicts within the coalition. After the elections of June 2005, Stanishev has held together his government mainly because of pressure from the EU to maintain the pace of reform in order to qualify for EU membership. The lack of policy cohesion beyond EU issues raises the danger of instability and the prospect of an early parliamentary election. Mr. Stanishev's government is also challenged by the pressure from Parliamentary opposition on opening the files of the state security collaborators during the Communist era, which would reveal secrets that may identify as former agents a number of MPs, ministers and civil servants [Economist 2007-11].

Mr. Stanishev seems unaware that membership brings obligations. But if progress is not made in contrasting organized crime, in a year's time, Bulgaria may face EU sanctions. Contract killings persist in Sofia, as does corruption among prosecutors and judges. Despite efforts to clean up the prosecution service, not a single suspect in a contract killing has been convicted. Worse, the government has stopped trying [Economist 2007-17].

In its 2004 Report on Enlargement [EC 2004-2, EC 2004-3] the European Commission reiterated its recognition of Bulgaria as being a functioning market economy. Furthermore, the 2004 report concluded that Bulgaria should be able to cope with competitive pressure and market forces within the Union. The Commission's Report also mentioned Bulgaria's further good progress in structural reforms over the previous years. This progress holds in particular for the increasing role of the private sector through privatization and the reduction of state aid, the positive development of the banking sector, and some improvements in the regulatory environment. However, further structural reforms were reported to be needed: streamlining the regulatory procedures for the enterprise sector, improving the efficiency of the administrative and judicial system, privatizing remaining public enterprises, restructuring and liberalizing the network industries (particularly in the energy sector), and improving the flexibility of the labour market and the efficiency and quality of the education system.

The 2006 Monitoring Report on the state of preparedness for EU membership of Bulgaria and Romania outlines the Commission's assessment, as of September 2006, of both countries' progress towards accession. It confirms that Bulgaria has made further progress to complete her preparations for membership, demonstrating capacity to apply EU principles and legislation from 1 January 2007. However, the Commission also identifies a number of areas of continuing concern: organized crime, money laundering, corruption, and the judicial system. The report includes the following findings [EC 2006-3]:

> The areas needing immediate action for Bulgaria are the justice system, the fight against corruption, police cooperation and the fight against organized crime, money-laundering, integrated administrative control system for agriculture, transmissible spongiform encephalopathies, and financial control.
>
> Some progress has been made in the reform of the **justice system**. Rules have been introduced establishing objective procedures for the appointment and evaluation of magistrates. Pre-trial proceedings have been improved by the introduction of a fast-track procedure. Further reform of the Supreme Judicial Council is necessary, in particular as regards its accountability and capacity to effectively manage the judiciary, in order to ensure the transparency and efficiency of judicial processes. Difficulties in the implementation of penal procedures persist. The civil procedural code and the judicial system act have not yet been adopted. Amendments to the Constitution need to be adopted.
>
> The legislative framework for the **fight against corruption** has been improved with the adoption of amendments to the laws on political parties and on publicity of property owned by high level officials. All ministers have published their asset declarations on the internet. However, there have been few concrete examples of investigations or prosecution or charges of high level corruption. Corruption remains a problem. The public administration, including tax collecting agencies at the border and local government remain particularly vulnerable.
>
> In the area of **money laundering**, Bulgarian legislation is now largely in line with the *acquis*. However, implementation of legislation is limited to date and so far no successful prosecutions for money laundering can be reported.
>
> The number of cases prosecuted successfully related to **organized crime** is still low. Reliable crime statistics are yet to be established. Law enforcement increased and successful actions have been registered against criminal networks, in some cases in cooperation with EU member states. However, there is still insufficient cooperation between the bodies involved in the fight against organized crime. There is no systematic confiscation of assets of criminals.
>
> Further progress is still needed in a number of other areas, such as social inclusion, social dialogue, anti-discrimination and public health, motor vehicle insurance, nuclear energy and safety, environment, financial management and control of future structural funds, as well as animal diseases.

> Bulgaria has made further progress with macroeconomic stabilization and economic reform. Its current reform path should enable it to cope with competitive pressure and market forces within the Union.
>
> Overall, there has been some progress in the following areas: trafficking in human beings, child protection, and the protection and integration of minorities. Limited progress has been made in the area of detention conditions, treatment of people with disabilities and the mental healthcare and broadcasting. Further action is needed. In addition, Bulgaria needs to ensure the sustainability of public administration reform.

In June 2007 the European Commission's concerns about corruption and maladministration were still very serious and the Commission was considering recommending that some financial aid be withheld from Bulgaria to put pressure on the administration and push government to act. On 27 June 2007, the European Commission issued reports on Romania's and Bulgaria's progress on accompanying measures following accession [EC 2007-2, EC 2007-3]. Romania and Bulgaria escaped legal sanctions but not criticism by the EU over their lack of progress in fighting widespread corruption and organized crime. EU Justice and Home Affairs Commissioner Franco Frattini said the two countries, which joined the EU in January, needed to step up their battle against corruption and work to overcome a backlog of cases waiting to go to court, in order to meet EU rules. "High-level corruption is still one point of weakness, both governments are aware of this," Frattini told reporters after the European Commission adopted report cards on the two EU members. "Too few results are shown concerning practical results, too many indictments still need to be translated into a final decision of a court, that's why we say very frankly progress made in this field is still insufficient" [EC 2007-4]. The report about Bulgaria [EC 2007-2] is presented as recommended reading at the end of the chapter. It includes:

> The Bulgarian Government is committed to judicial reform and cleansing the system of corruption and organized crime. However, the real test can only be met through determined implementation of these actions on the ground every day. There is still a clear weakness in translating these intentions into results.
>
> An implementation program to fight corruption has been adopted. However, the program's implementation lacks clear lines of responsibilities and an efficient coordination mechanism.
>
> Overall, progress achieved in the judicial treatment of high-level corruption cases in Bulgaria is still insufficient. In the first six months of accession, Bulgaria has continued to make progress in remedying weaknesses that could prevent an effective application of EU laws, policies and programs. But, there has not been sufficient time to demonstrate convincing results in key areas.

11.7.1.1 Economy

Until 1989, Bulgaria had a Soviet-style economy in which nearly all agricultural and industrial enterprises were state controlled. A stagnant economy; shortages of food, energy, and consumer goods; an enormous foreign debt; and an obsolete and inefficient industrial complex instigated attempts at market-oriented reform in the 1990s. Traditionally an agricultural country, Bulgaria has been considerably industrialized since Second World War. Leading industries are machine building, metalworking, food processing, engineering, and the production of chemicals, textiles, and electronics. Bulgaria's chief mineral resources include bauxite, copper, lead, zinc, coal, brown coal (lignite), iron ore, oil, and natural gas. There are many mineral springs. Agriculture accounted in 2003 for 12% of the gross domestic product and employed just over 10% of the workforce, over twice the EU25 average. Principal crops are wheat, oilseeds, corn, barley, vegetables, and tobacco. Grapes and other fruit are grown, as are roses, and wine and brandy production is important.

Bulgaria has experienced macroeconomic stability and positive growth rates since a major economic downturn in 1996 led to the fall of the then socialist government. The government in office between 2001 and 2005 has pledged to maintain the fundamental economic policy objectives of its predecessor, i.e. retaining the Currency Board, practicing sound financial policies, accelerating privatization, and pursuing structural reforms. A $300 million standby agreement negotiated with the International Monetary Fund (IMF) at the end of 2001 helped the government maintain economic stability while seeking to overcome high rates of poverty and unemployment.

Despite the difficult global economic situation in 2003, the Bulgarian economy continued to benefit from high growth and stability. Real GDP growth was 4.9% in 2002 and 4.3% in 2003; inflation declined steadily at the beginning of the century, reaching 5.8% in 2002 and 2.3% in 2003 and 2004, but it raised recently, reaching 7.3 in 2006; registered unemployment went down to 13.6% in 2003 (4% lower than the previous year), to reach 9.6% in 2006; the current account deficit increased to 8.4% of GDP in 2003; and public debt continued falling, from above 100% of GDP in 1997 to 46% of GDP at the end of 2003 and slightly less than 30% in 2005. However, the catching-up in terms of GDP per capita was very slow, remaining at only 29% of the EU25 average in 2003, only a 3% increase from 1997.

The national currency is the Lev (BGL) that was redenominated on 5 July 1999 by a factor of one thousand; 1 € = 1.99 BGL in 2003 and 1 € = 1.96 BGL on 6 May 2007. Qualifying for the adoption of euro is not likely before 2011. Personal income tax rates are progressive, ranging from 20% to 24%. The rate of tax on company profits was reduced from 15% in 2006 to 10% in January 2007. Value-added tax (VAT) is levied at a single rate of 20%, although tourism packages sold abroad have a rate of 7%.

Trade relations between the EU and Bulgaria have continued to develop strongly. Bulgaria's trade, once mainly with former Soviet bloc countries is now mainly with Italy, Germany, Greece, Turkey, and France. Turnover in trade with the EC in 2003 was up 7% from 2002 and accounted for 52.4% of Bulgaria's overall trade. In 2003, exports to the EC were up 0.9% from 2002, accounting for 56.5% (€3.77 billion) of Bulgaria's total exports. Rapid import growth is expected to keep the current-account deficit high.

The main industrial exports to the EC were textiles and clothing, and iron and steel. Bulgaria's main agricultural exports to the EC were cereals, oil seeds, oleaginous fruits, and meat. In 2003, imports from the EC were slightly down by 0.6% from 2002, accounting for 49.6% (€4.40 billion) of Bulgaria's total imports. Her main industrial imports from the EU were textiles and clothing. The country's main agricultural imports were meat, fats and oils, and fruits and nuts.

Some basic economic data for Bulgaria published by various quoted sources are presented in Table 11.16. More information is available in the recommended reading at the end of the chapter.

The fall 2007 European Commission Economic forecast for Bulgaria [EC 2007-27] is presented as reading at the end of the chapter. It is summarized as "tight labour market fuelling wage pressures" and it includes the following:

> Real GDP growth accelerated to 6.4% in the first half of 2007, fuelled by continued strong FDI inflows. High wage increases and strong employment growth also boosted private consumption, which grew by close to 7%.
>
> The Bulgarian economy has been showing signs of overheating in 2007 with mounting wage pressures, rising inflation and a widening of external imbalances. Buoyant employment growth and a strong decrease in unemployment have led to a considerable tightening of the labor market, with growing shortages especially for high-skilled labor. Consequently, nominal wage growth has accelerated sharply since the end of 2006 to over 17% in the first half of 2007.
>
> Thanks to very strong revenue growth, the general government budget surplus is expected to reach around 3% of GDP in 2007, which is clearly above initial budget projections. A comparable general government surplus to 2007 is expected in 2008 and 2009. This would imply a further decrease in general government gross debt to below 15% of GDP.

11.7.2 Romania

Romania is a republic in southeast Europe (see Fig. 11.1), whose size, 238 thousand km^2, is slightly less than that of the UK. Romania occupies, roughly, ancient Dacia, which was a Roman province in the second and third centuries AD. The ethnic character of modern Romania seems to have been formed in the Roman period; Christianity was introduced at that time as well. After the Romans left the region, the area was overrun successively by Goths, Huns, Avars, Bulgars, and Magyars. Romania gained full independence in 1881 and was proclaimed a kingdom. The national language, Romanian, has Latin roots, which in the 20th century facilitated Romania's cultural contacts with France and Italy.

In June 1941, Romania joined Germany in her attack on the Soviet Union. Romanian troops helped to take Odessa, but they suffered heavily at Stalingrad in late 1942 and early 1943. In August 1944 two Soviet army groups entered Romania, which surrendered to the USSR and was ordered to fight on the Allied side.

Soviet occupation following Second World War led to the formation of a Communist *people's republic* in 1947 and to the abdication of the king. The decades-long rule of Dictator Nicolae Ceausescu and his Securitate police state became increasingly oppressive and draconian through the 1980s. Ceausescu was overthrown and executed in late 1989. Former communists dominated the government until 1996, when they were swept from power by a fractious coalition of centre-right parties. The country joined NATO in March 2004.

Romania, one of the poorest countries of central and eastern Europe, began the transition from Communism in 1989 with a largely obsolete industrial base and a pattern of output unsuited to the country's needs. Much economic restructuring

Table 11.16 Basic economic data for Bulgaria

Bulgaria		2000	2001	2002	2003	2004	2005	2006
GDP[a]	($bln)		13.6	15.6	19.9	24.3	26.6	
GDP[i]	($bln)	12.6	13.6	15.6	19.9	23.9	26.0	28.3
GDP[b]	($bln)				19.9			27.9
GDP[e]	($bln)			15.6	20.0	24.3	26.7	31.5
GDP[f]	(€bln)			16.6	17.8	19.9	21.9	25.1
pcGDP ppp[b]	($x000)				7.6			10.4
pcGDP ppp[f]	(€x000)							8.6
GDP Growth[a]	(%)		4	4.9	4.3	6	6	5.3[d]
GDP Growth[b]	(%)				4.3			6.5
GDP Growth[c]	(%)			4.9	4.3			
GDP Growth[e]	(%)			4.5	5.0	6.6	6.2	6.1
Current Account Balance[b]	(%GDP)				−8.4			
Current Account Balance[e]	(%GDP)			−2.0	−5.1	−6.9	−12.4	−15.9
Government Balance[c,b]	(%GDP)				0.0			4.0
Government Balance[f]	(%GDP)			0.1	0.3	1.9	3.1	
Total Debt[a,b]	(%GDP)			72.2[a]	66.7[a,b]			25.6[b]
Total Debt[f]	(%GDP)			54.0	46.1	38.6	29.9	
Inflation (GDP deflator)[a]	(%)		7	4	2	5	4	
Inflation (CPI)[b]	(%)			5.8	2.3			6.5
Inflation (CPI)[e]	(%)							7.3
Unemployment[b]	(%)			17.5	14.3			9.6
Exports[a]	(%GDP)		56	53	54	58	61	
Imports[a]	(%GDP)		63	60	63	68	77	
Exports Growth[a]	(%)			7.0	8.0			10.2[d]
High-tech Exports[a]	(%Manuf. Exp.)		3	4	4	4	5	
School enrollment, tertiary[a]	(%)	45	42	40	41	41		
Internet Users[a]	(/1000)			76	80	159	206	
ITC Expenditure[f]	(%GDP)					1.4	1.5	1.8
FDI Inflows[a]	(%GDP)			6.0	5.8	10.5	10.9	9.8
€ Exchange Rate[f]	(BGN)			1.948	1.949	1.949	1.953	1.956
Stock market capitalization[a]	(%GDP)		5	4	5	9	12	19
GDP % in agriculture[a]				14	12	12	11	10
GDP % in industry[a]				30	30	30	31	32
GDP % in services[a]				56	58	58	58	59

Source: [a]The World Bank (2005, 2007); [b]http://www.cia.gov/cia/publications/factbook; [c]European Commission, 2004; [d]Average Average 2002–2006; [e]The Economist Intelligence Unit, 2007; [f]Eurostat, 2007; [i]International Monetary Fund.

remained to be carried out before Romania could achieve her target of joining the EU. Over the past decade, economic restructuring has lagged behind most other countries in the region. Consequently, living standards have continued to fall – real wages are down perhaps 40%. The country emerged in 2000 from a punishing three-year recession thanks to strong demand in EU export markets, and despite the global slowdown in 2001, strong domestic activity in construction, agriculture, and consumption led to a 4.8% growth. A standby agreement with the International Monetary Fund (IMF) covered the period from October 2001 to March 2003, and provided a key opportunity for vigorous privatization, regulatory reform, deficit reduction, and the curbing of inflation.

The requests of the European Commission in the roadmap to accession [EC 2004-4] include completing privatization in the banking sector, continuing the reform of public expenditures and budgetary procedures, and implementing improved regulatory and legal frameworks to support the establishment of a functioning market economy capable of coping with competitive pressure and market forces within the Union.

The 2006 Monitoring Report on the state of preparedness for EU membership of Bulgaria and Romania outlines the Commission's assessment, as of September 2006, of both countries' progress towards accession. It confirms that Romania has made further progress to complete her preparations for membership, demonstrating capacity to apply EU principles and legislation from 1 January 2007. However, the Commission also identifies a number of areas of continuing concern. The report [EC 2006-3] is presented as recommended reading at the end of the chapter; it includes the following findings:

> The areas needing immediate action for Romania are the justice system and the fight against corruption and the integrated administrative control system for agriculture.
>
> Progress has continued in the fight against corruption. However, there needs to be a clear political will to demonstrate the sustainability and irreversibility of the recent positive progress in fighting corruption.
>
> Overall, there has been some progress in the following areas: trafficking in human beings, detention conditions, restitution of property and child protection Limited progress has been made with the treatment of people with disabilities, the mental healthcare and the protection and integration of minorities.Further action is needed. In addition, Romania needs to ensure the sustainability of public administration reform, and to fully align with EU external positions, such as on the International Criminal Court.

Since Romania joined the EU in January 2007, political disagreements inside the coalition of the ruling majority, as well as between the prime minister, Calin Popescu Tariceanu, and the president, Traian Basescu, have plunged the country into turmoil. Outsiders are nervous about the second-biggest new EU member. The EU is pressing for a new ethics watchdog to be set up. The American secretary of state, Condoleezza Rice, refused in April 2007 to meet Mr. Cioroianu, the acting Foreign Affairs minister. A NATO summit planned next year has been shifted, probably to Portugal. Boosted by EU membership, Romania's economy is booming. But its political stock is falling fast. President Basescu has returned to office after parliament's failed attempt to impeach him; he is continuing a campaign to oust the prime minister, so that more political instability is likely; he is continuing his campaign to bring about an early parliamentary election. Romania's political malaise is thus likely to persist, with negative consequences for EU-related reforms and economic policy-making in general [Economist 2007-13, Economist 2007-14]. Mr. Tariceanu's plans for a "great leap" to catch up with western Europe in the fields of infrastructure and welfare carry several dangers for the country, on the current account deficit and on the sustainability of the economic growth in compliance with the criteria of the Treaty on EU, in particular in regard to inflation [Economist 2007-15].

In June 2007, as mentioned in the previous section on Bulgaria, the European Commission issued reports on Romania's and Bulgaria's progress on accompanying measures following accession [EC 2007-2, EC 2007-3]. As mentioned in the previous section, Romania and Bulgaria escaped legal sanctions but not criticism by the EU over their lack of progress in fighting widespread corruption and organized crime. EU Justice and Home Affairs Commissioner Franco Frattini said: "High-level corruption is still one point of weakness, both governments are aware of this, too few results are shown concerning practical results, too many indictments still need to be translated into a final decision of a court, that's why we say very frankly progress made in this field is still insufficient." The summary of the report about Romania [EC 2007-3] is presented as recommended reading at the end of the chapter; it includes:

> The Romanian Government is committed to judicial reform and cleansing the system of corruption. They have prepared the necessary draft laws, action plans and programs. However, the real test can only be met through determined implementation of these actions on the ground every day. There is still a clear weakness in translating these intentions into results.
>
> Deeply rooted problems, notably corruption require the irreversible establishment and effective functioning of sustainable structures at investigative and enforcement level capable of sending strong dissuasive signals. In addition, the structural changes which are needed impact on the society at large and require a step change which goes much beyond the mere fulfillment of the benchmarks. This requires a strong long term commitment by Romania and can only be successful if the strict separation of the executive, legislative and judicial power is respected and if stable political conditions and commitment are in place.

11.7.2.1 Economy

Romania has an inadequate supply of mineral resources and must import raw materials and fuels, although historically it has been an important oil-producing centre. About 25% of the country is forested, and large quantities of timber are cut, especially in Transylvania. Agriculture employs about 27% of the labour force and accounts for 15% of the GDP. The chief crops are corn, sugar beets, potatoes, and various grains.

From 1948 until 1989, Romania had a Soviet-style command economy in which nearly all agricultural and industrial enterprises were state controlled. During those years, it built an economy based largely on heavy industry. Industry contributes over half of the country's gross domestic product (GDP) and accounts for one-third of the labour force. Major manufactures include steel products, machinery, transport vehicles, and chemicals. Romania's tourism industry is growing and Bucharest is known as the "Paris of the East."

Throughout the 1990s and into the next decade, the country's economy lagged as it struggled to make the transition to a market-based economy. Agriculture occupies, in 2005, 32% of the workforce, over six times the EU25 average and generates 10% of the GDP.

Romania is one of the poorest countries in Europe, her GDP per capita at purchasing power parity being in 2005 one-third of the EU25 average. Price increases and food shortages led to civil unrest, and the closing of mines set off large-scale strikes and demonstrations by miners. Inflation was high (15% in 2003) but is coming down rapidly (it was over 45% in 2000). Romania has, however, later than other accession countries curbed inflation to a one-digit percentage. Privatization of state-run industries has proceeded cautiously, with citizens having shares in companies but little knowledge or information about their investments.

Widespread corruption has been and continues to be a problem. In October 2003, the country approved constitutional changes protecting the rights of ethnic minorities and property owners; the amendments were designed to win European Union approval for Romania's accession, but continuing pervasive corruption is a potential stumbling block that delays FDI inflows, despite attractive low wages.

Some basic economic data for Romania published by various quoted sources are presented in Table 11.17. More information is available in the recommended reading at the end of the chapter.

According to data from the year 2005, the most important destination countries of Romanian exports are Italy, at 19.4% (down from 24.3% in 2003); Germany, at 14% (down from 15.7 in 2003); France, at 7.4% (the same as in 2003); Turkey is up by 2% to 7.9%; UK is down by 1.2% to 5.5%. China is an emerging leading supplier and Hungary an emerging leading market. The most important exports from Romania in 2006 are textiles, at 20%; machinery and equipment, at 15%; minerals and fuel, at 11%. Most important imports to Romania in 2006 are machinery and equipment, at 24%; textiles, at 14%; minerals and fuel, at 8%; chemical products, at 8% [Economist 2007-12]. The national currency is the Leu (ROL); 1 € = 38.24 ROL in 2003. A new Leu was introduced in 2005, with a revaluation factor of 10, and 1 € = 3.31 ROL on 6 May 2007. No date has been proposed for euro adoption.

In line with the regional trend towards the introduction of lower tax rates and flat-tax regimes, Romania's government introduced a flat tax of 16% for both personal income and corporate profits, effective from 1 January 2005. Corporate profits had previously been taxed at 25% and personal income taxes ranged from 18% to 40%. Since 2000 there has been a uniform rate of 19% for VAT. The authorities are resisting pressure to raise the VAT rate to compensate for lost revenue following the introduction of the flat-rate 16% tax. Social security contributions are high, despite a series of reductions in recent years and total 49.5%.

The fall 2007 European Commission Economic forecast for Romania [EC 2007-27] is presented as recommended reading at the end of the chapter. It is summarized as "growing fiscal and external imbalances" and it includes the following:

After reaching 7.7% in 2006, real GDP growth slowed to 5.8% in the first half of 2007, mainly due to a deceleration of private consumption and a more negative contribution from net exports. From the supply side, the highest growth rates are observed in the construction

Table 11.17 Basic economic data for Romania

Romania		2000	2001	2002	2003	2004	2005	2006
GDP[a]	($bln)		40.1	45.8	59.5	75.5	98.6	
GDP[i]	($bln)	37.1	40.2	45.8	57.3	71.3	79.9	88.2
GDP[b]	($bln)				155[g]			79.2
GDP[c]	($bln)			45.8	59.5	75.5	97.1	121.9
GDP[f]	(€bln)				52.6	60.8	79.6	97.1
pcGDP ppp[b]	($x000)				7.0			8.8
pcGDP ppp[f]	(€x000)							8.8
GDP Growth[a]	(%)		6	5	4.9	8	4	5.0[d]
GDP Growth[b]	(%)				4.9			6.4
GDP Growth[c]	(%)			6.1[h]				7.7
Current Account Balance[a]	(%GDP)			−3.3	−5.8			
Current Account Balance[c]	(%GDP)							−7.3[h]
Government Balance[c,b]	(%GDP)				−2.3[b]			−2.8[b]
Government Balance[c,f]	(%GDP)			−2.0	−1.7	−1.0.4	−1.7[f]	
Total Debt[a]	(%GDP)				39.8			
Total Debt[b]	(%GDP)				25.5			
Total Debt[f]	(%GDP)			23.8	20.7	18.0	15.2	
Inflation (GDP deflator)[a]	(%)	44	37	23	24	15	12	
Inflation (CPI)[b]	(%)	45.7		22.5	15.3			6.8
Inflation (CPI)[c]	(%)			22.5	15.3	11.9	9.0	12.9[h]
Unemployment[b]	(%)				8.4	7.2		6.1
Exports[a]	(%GDP)		33	35	35	36	33	
Imports[a]	(%GDP)		41	41	42	45	43	
Exports Growth[a]	(%)			16.9	27.0			7.1[d]
High-tech Exports[a]	(%Manuf. Exp.)		6	3	4	3	3	
School enrollment, tertiary[a]	(%)	24	28	32	36	40		
Internet Users[a]	(/1000)		45	101	184	208		
ITC Expenditure[f]	(%GDP)				1.3	1.6	1.9	
FDI Inflows[a,c]	(%GDP)		2.9	2.5	3.1	8.5	6.7	6.0[c,h]
€ Exchange Rate[f]	(ROL)			2.6	3.127	3.755	4.051	3.621
Stock market capitalization[a]	(%GDP)	3	5	10	9	16	21	
GDP % in agriculture[a]				15	13	13	14	10
GDP % in industry[a]				37	38	35	35	35
GDP % in services[a]				48	49	52	51	55

Source: [a]The World Bank (2005, 2007); [b]http://www.cia.gov/cia/publications/factbook; [c]European Commission, 2004; [d]Average Average 2002–2006; [e]The Economist Intelligence Unit, 2007; [f]Eurostat, 2007; [g]Purchasing Power Parity (ppp); [h]Average 2002–2006; [i]International Monetary Fund.

sector. Strong domestic demand dynamics will likely result in a further widening of the trade deficit to 15 1/2% of GDP in 2007.

In spite of the contraction of several labor-intensive industries (especially textiles and clothing), overall employment is expected to grow by 1 1/4% in 2007. Unemployment is projected to decline slightly over 2008–2009.

In combination with strong demand-pull pressures, annual inflation is projected to average 4 3/4% in 2007.

The budget deficit is expected to deteriorate from 1.9% of GDP in 2006 to 2.7% of GDP in 2007, mainly due to an increase in government consumption, in particular public wages, and social transfers. The debt-to-GDP ratio is projected to increase by 1% between 2007 and 2009 to around 13 1/2% mainly due to higher deficits.

11.8 A New Set of Candidate Countries Includes Turkey: A Litmus Test for European Identity

As shown in Table 11.18, the third wave of possible enlargement involves three countries: Croatia, the Former Yugoslav Republic of Macedonia (referred to as Macedonia later in this book), and Turkey. Croatia had tabled a request for accession in 2003, as had Turkey in 1987, following initial steps dating back to 1959. The start of negotiations with Croatia, initially fixed in March 2005, was later postponed to October 2005, when Croatia could prove to be fully cooperating with the UN war-crimes prosecutor; the European Council fixed on 5 October 2005 the starting date of accession negotiations with Turkey; Macedonia is a candidate for accession to the EU since December 2005, however, at the time of writing (Year End 2007), a starting date for negotiations has not yet been agreed. Other countries in the Balkans: Albania, Bosnia and Herzegovina, Montenegro, Serbia, and Kosovo have currently the status of *potentially candidate* countries.

From 1 January 2007, the EU is using, the Instrument of Pre-Accession (IPA), a new financial tool for promoting modernization, reform and alignment with the EU's legal order. Candidate countries as well as potential candidate countries are eligible to funding under the new instrument. In particular, beneficiaries will be able to receive support to increase the administrative capacity and establish the correct management structures necessary to take responsibility of the management of assistance.

IPA assigns to Croatia €138.5 million in 2007 and €146.0 and 151.2 million in the two following years; IPA assigns to the Former Yugoslav Rep. of Macedonia €58.5 million in 2007 and €70.2 million and €81.8 million in the two following years. The western Balkan countries and Turkey will benefit from almost €11.5 billion over the next seven years [EC 2006-7].

Table 11.18 Data relative to Croatia, Macedonia, and Turkey. Additional data are presented in the following sections addressing the individual countries. Surface area is in thousand km². GDP is in thousand dollars

Country	Capital	Surface	Population (million)	pc GDP ppp	Inflation (%)	Unemployed (%)
Croatia	Zagreb	57	4.5	10.6	1.8	19.5
Macedonia	Skopje	26	2.0	8.3	3	36
Turkey	Ankara	781	71.2	6.7	9.3	10.5

Source: OECD and European Commission, 2003 The World Bank (2005), and http://www.cia.gov/cia/publications/factbook

11.8.1 Croatia

The Republic of Croatia is in the northwest corner of the Balkan Peninsula (see Fig. 11.1). Zagreb is the capital. The size of Croatia is 57 thousand km^2, or halfway between the size of The Netherlands and that of Ireland. The total population is 4.6 million. The principal ethnic majority is the Croats, who account for 78% of the population, while 12.7% are Serbs.

Croatia, at one time the Roman province of Pannonia, was settled in the seventh century by the Croats. They converted to Christianity between the seventh and the ninth centuries and adopted the Roman alphabet under the suzerainty of Charlemagne. In 925, the Croats defeated Byzantine and Frankish invaders and established their own independent kingdom, which reached its peak during the 11th century. A civil war ensued in 1089, which later led to the country being conquered by the Hungarians in 1091. Croatia was under Hungarian ruling until the collapse of Austria-Hungary in 1918 following her defeat in the First World War. On 29 October 1918, Croatia proclaimed her independence and joined with Montenegro, Serbia, and Slovenia to form the Kingdom of Serbs, Croats, and Slovenes. The name was changed to Yugoslavia in 1929.

When Germany invaded Yugoslavia in 1941, Croatia became a Nazi puppet state. Croatian fascists, the Ustachi, slaughtered countless Serbs and Jews during the war. After Germany was defeated in 1945, Croatia was made into a republic of the newly reestablished communist nation of Yugoslavia; however, Croatian nationalism persisted. After Yugoslavian leader Marshal Tito's death in 1980, Croatia's demands for independence began multiplying.

In 1990, free elections were held, and the communists were defeated by a nationalist party. Although Croatia declared her independence from Yugoslavia in 1991, it took four years of sporadic, but often bitter, fighting before occupying Serb armies were almost completely cleared from Croatian lands. Under UN supervision, the last Serb-held enclave in eastern Slavonia was returned to Croatia in 1998. In 2003, Croatia formally submitted her application to join the EU. Croatia is one of the beneficiaries of the EU's stabilization and association process for southeastern Europe. EU accession negotiations began on 3 October 2005 and are continuing. The European Commission published on 8 November 2006 a Memo on key findings about the progress towards accession [EC 2006-4], which is presented as recommended reading at the end of the chapter. It includes the following remarks for Croatia:

Croatia continues to meet the Copenhagen political criteria and the political situation has continued to improve.

However, reform is at an early stage and there is considerable scope for improvement in the judiciary, public administration and in the fight against corruption. Further progress is essential in the protection of minorities, on refugee return, and in the conduct of war crimes trials, including witness protection.

As regards economic criteria, Croatia can be regarded as a functioning market economy, provided that it vigorously implements its reform program to remove the significant remaining weaknesses.

Inflation is low, the exchange rate stable, fiscal consolidation has continued and growth has accelerated slightly.

Croatia has improved its ability to take on the obligations of membership. However, in many cases enforcement is weak and administrative capacity remains uneven.

The World Bank [World Bank 2007-3] identifies a number of challenges facing Croatia in her transition towards full EU membership:

Macroeconomic stability has yet to translate into strong growth. The economy continues to rely on public investment and consumption, tourism and other selected but highly subsidized industries. Croatia needs to find new engines of growth in terms of reducing the state burden and shifting towards increased productivity and private sector-driven growth instead of excessive public investments and consumption.

A faster pace of structural reforms is needed. Slow structural reform remains a key constraint to the investment required to promote growth and reduce the high level of unemployment, which stood at 16.7 percent in May 2006, according to official figures.

Downsizing the presence of the state. At 60 percent of the economy, the private sector share in Croatia is below that of any of the so-called EU8 countries (the Central European and Baltic countries which joined the EU in 2004). A large number of public and state-owned enterprises still need to be privatized or liquidated.

Strengthening the rule of law. An inefficient judicial system is hindering the exit of unviable firms. The execution and enforcement of contracts and property rights are weak. The country needs to create stable, clear, effective, and predictable laws and institutions that will promote investments and growth.

Converging to EU agriculture and environment standards and institutions. In Croatia, where there is a long tradition of agriculture, numerous possibilities for diverse modes of production exist. Reforms are essential to raise the competitiveness of domestic producers and improve the quality of living of rural populations.

Improving the efficiency of public spending.

Ensuring equal distribution of growth benefits through labor, education and social sector reforms.

11.8.1.1 Economy

Natural resources include oil, some coal, bauxite, low-grade iron ore, calcium, gypsum, natural asphalt, silica, mica, clays, salt, and hydropower.

Before the dissolution of Yugoslavia, the Republic of Croatia, after Slovenia, was the most prosperous and industrialized area, with a per capita output one-third above the Yugoslav average. The economy emerged from her mild recession in 2000 with tourism as the main factor. The tourist industry has not yet been revamped and maintains the flavour of the centralized planning of the past regime. Structural unemployment remains a key negative element in Croatia's economy. The country is likely to experience only moderate growth without disciplined fiscal and structural reform. Structural reforms are needed in order to complete the transition to a market economy and tap into Croatia's growth potential. These reforms cover fiscal policy and public administration, improved property and creditor rights, a functioning legal and judiciary system, agriculture and environment adjustment, and education reform. Croatia has opportunities for economic growth through European integration. The growth impact will come not only from increased access to EU and integration with markets, but also from the improved investment climate that would result from aligning Croatian policies and institutions with EU best practices.

In 2005, real GDP growth was 4.3%, up from 3.8% in 2004, mainly driven by domestic demand. Net exports added only 0.1 percentage point to real growth. In the first half of 2006, real GDP accelerated further to 4.8% year-on-year, largely due to stronger private investment. In 2006, industrial production rose by 4.1%; average per-capita income further increased to an estimated 47% of the EU-25 average (in purchasing power standards) in 2005. Overall, economic growth continued on the back of stronger private investment.

The current account deficit widened from 5% of GDP in 2004 to 6.4% in 2005 and further to 7.7%5 in the second quarter of 2006. In the same period, net foreign direct investment (FDI) grew to 4.6% of GDP and covered 60% of the current account deficit. FDI were largely driven by capital increases and takeovers rather than by privatization or greenfield investments. The officially registered unemployment rate has continued to decline to 15.7% in July 2006 compared to 16.9% in the same month a year earlier. Still, a relatively high unemployment rate and limited job turnover and job creation remain some of the most pressing economic problems.

Croatia is upgrading her telecommunication infrastructure (mobile: 3.0 million phones in 2005 – up from 2.6 million in 2003; 1.9 million main line in 2005 – up from 1.8 million in 2002) by participating in Adria-1, a joint fibre-optic project with Germany, Albania, and Greece. Major industries are chemicals and plastics, machine tools, fabricated metal, electronics, pig iron and rolled steel products, aluminium, paper, wood products, construction materials, textiles, shipbuilding, petroleum and petroleum refining, food and beverages, and tourism. Major trading partners are Italy, Bosnia and Herzegovina, Germany, Slovenia, Austria, the Russian Federation, and France.

The national currency is the Kuna (HRK); 1 € = 7.72 HRK in 2003 and 1 € = 7.35 HRK on 6 May 2007.

Personal income is taxed at four rates of between 15% and 45%. There are three rates for VAT: zero, 10% and 22%. The corporate tax rate is set at 20%. Some basic economic data published by various quoted sources for Croatia are presented in Table 11.19. More information is available in the recommended reading at the end of the chapter.

The fall 2007 European Commission Economic forecast for Croatia [EC 2007-27] is presented as recommended reading at the end of the chapter. It is summarized as "robust performance continues" and it includes the following:

In the first half of 2007, real GDP growth accelerated to 6.8%, up from 4.8% in 2006, driven by continued efficiency gains as well as accelerated employment growth. Average annual consumer price inflation declined slightly from to 3.3% in 2005 to 3.2% in 2006. Despite a recent increase in agricultural and food prices, average inflation declined further to 2.2% in the first nine months of 2007. Real wage growth remained broadly in line with productivity gains.

The economic performance is expected to remain robust over 2008–2009. Growth will continue to be driven by strong domestic demand.

Due to a stable environment with relatively low inflation, a significant increase in wage pressures is unlikely to occur, despite expected wage hikes in some public companies. Average real wages are forecast to grow more slowly than average productivity gains, leading to a lowering of real unit labor costs. Despite recent increases in agricultural and food prices,

Table 11.19 Basic economic data for Croatia

Croatia		2000	2001	2002	2003	2004	2005	2006
GDP[a]	($bln)				28.7	35.3	38.5	
GDP[b]	($bln)	18.4	19.9	22.8	28.8	33.2	35.1	37.9
GDP[b]	($bln)							37.4
GDP[f]	(€bln)			24.4	26.2	28.4	30.9	
pcGDP ppp[b]	($x000)					10.6		13.2
pcGDP ppp[f]	(€x000)							12.0
GDP Growth[a]	(%)					3.8	4.3	
GDP Growth[b]	(%)				4.3			4.4
GDP Growth[c]	(%)					3.8	4.3	
Current Account Balance[b]	(%GDP)					−6.1		
Government Balance[c,b]	(%GDP)					5[c]	6.4[c]	7.7[c]
Government Balance[a]	(%GDP)					−4.1	−3.6	
Government Balance[f]	(%GDP)			−4.1	−4.5	−5.0	−3.9	
Total Debt[b]	(%GDP)					75.3		56.2
Total Debt[f]	(%GDP)			40.0	40.9	43.7	44.2	
Inflation (GDP deflator)[a]	(%)		4	4	4	3	3	
Inflation (CPI)[a,c]	(%)					2.1	3.3	3.6[c]
Inflation (CPI)[b]	(%)					1.8		3.4
Unemployment[b]	(%)				19.5			17.2
Unemployment[c]	(%)						16.9	15.7
Exports[a]	(%GDP)		48	45	47	47	47	
Imports[a]	(%GDP)		54	56	58	57	56	
Exports Growth[a]	(%)				4.8	5.4	4.6	
High-tech Exports[a]	(%Manuf. Exp.)		10	12	12	13	12	
School enrollment, tertiary[a]	(%)	31	33	36	39	42		
Internet Users[a]	(/1000)		117	178	228	299	327	
ITC Expenditure[f]	(%GDP)							
FDI Inflows[a,c]	(%GDP)	6.7	4.9	6.9	3.5	4.6		4.6[c]
Stock market capitalization[a]	(%GDP)	15	17	17	21	31	34	
GDP % in agriculture[a]			9	9	7	7.2	7.0	
GDP % in industry[a]			30	29	30	30.3	30.8	
GDP % in services[a]			60	62	62	62.5	62.2	

Source: [a]The World Bank (2005, 2007); [b]http://www.cia.gov/cia/publications/factbook; [c]European Commission, 2004; [f]Eurostat, 2007;

annual average consumer price inflation is forecast to further decrease to 2.5% in 2007. It is expected to slightly increase to 3% in 2008 and 2009, driven by further adjustments of administrative prices and catching-up effects.

The general government deficit is projected to remain at around 2.2% in 2007. The public debt ratio is expected to be reduced to slightly above 37% of GDP by 2009. Debt reduction (was 44.2% in 2005) will continue to be supported by revenues from privatization, although they will slightly decline relative to GDP over the forecast horizon.

11.8.2 The Former Yugoslav Republic of Macedonia

Macedonia, the southernmost part of the Socialist Federative Republic of Yugoslavia, gained its independence peacefully in September 1991, but Greece's objection to the new state's use of what it considered a Hellenic name and symbols delayed international recognition, which occurred under the provisional designation of "the Former Yugoslav Republic of Macedonia." In 1995, Greece lifted a 20-month trade embargo and the two countries agreed to normalize relations. The United States began referring to Macedonia by its constitutional name, Republic of Macedonia, in 2004 and negotiations continue between Greece and Macedonia to resolve the name issue [WorldFactbook 2007].

The territory which forms today the Republic of Macedonia previously came under a number of different states and former empires. The first recorded state on the territory was the Thraco-Illyrian kingdom of Paionia in the fourth century BC. Parts of the territory passed to the successive rule of: ancient Macedon (originally centred in today's Greek Macedonia), the Roman Republic, the Roman Empire, the Byzantine Empire, the Bulgarian Empires, the Serbian Empire, and the Ottoman Empire.

Macedonia's current borders were fixed shortly after Second World War when the government of the then People's Federal Republic of Yugoslavia established the People's Republic of Macedonia, recognizing the region as a separate nation within Yugoslavia. The Republic of Macedonia is a landlocked country that is geographically clearly defined by a central valley formed by the Vardar river and framed along its borders by mountain ranges. The surface area of Macedonia is 25.7 km^2, halfway between the size of Slovenia and that of Belgium. The population is 2 million, of which two thirds consist of ethnic Macedonians while there is a sizeable Albanian minority of 25%, and smaller minorities of Turks and Roma. The Republic's terrain is mostly rugged, located between the Šara and Osogovo, which frame the valley of the Vardar river. Three large lakes – Lake Ohrid, Lake Prespa, and Dojran Lake – lie on the southern borders of the Republic, at her border with Albania and Greece. Ohrid is considered to be one of the oldest lakes and biotopes in the world. The region is seismically active and has been the site of destructive earthquakes in the past, most recently in 1963 when Skopje, Macedonia's capital, was heavily damaged by a major earthquake, killing over one thousand people.

Skopje is a synthesis of old and new, of antiquity, modernity, and a future rich with possibility. In its varied architecture, the city attests to a long history of great civilizations and turbulent events. Here one can find sleek modern hotels above cobble-stoned Ottoman streets, stately neoclassical homes right around the corner from grand old Yugoslav-era buildings, chic cafes, shopping malls, and brightly-coloured new offices, as well as the more humble brickwork of Byzantine churches.

Macedonia escaped the ethnic bloodletting in Bosnia and Croatia of 1991–'95. However, Macedonia, once considered the model in the Balkans of a multicultural society, has been in crisis since 2001. The two ethnic groups used to share power in government in the early 1990s, in a functioning pluralistic society but the conflicts that had been restrained by the Communist regime gradually deteriorated in independent Macedonia and were exacerbated at the end of the decade with the influx of 360,000 Kosovar refugees (15% of the population). Unemployment ranged between 30% and 70%: economic factors aggravated the tensions between the mainly urban Macedonian and the largely rural Albanian communities and nurtured a feeling of mutual mistrust [Broughton et al. 2002].

The refugees have long since returned to Kosovo, and the Macedonian government generally espoused tolerance. In March 2001, however, Kosovar-Albanian guerrillas infiltrated Macedonia, in an effort to rouse Albanian nationalism. Some ethnic Albanians, angered by perceived political and economic inequities, launched an insurgency that eventually won the support of the majority of Macedonia's Albanian population and led to the internationally-brokered Framework Agreement, which ended the fighting by establishing a set of new laws enhancing the rights of minorities. The undetermined status of neighboring Kosovo, implementation of the Framework Agreement, and a weak economy continue to be challenges for Macedonia [WorldFactbook 2007]. Fighting escalated alarmingly around Tetovo, Macedonia's largest ethnic-Albanian city. The shooting of eight Slavs in April 2001 worsened tensions, even as a new government took office. NATO deployed some 4,500 troops on a one-month mission to collect rebel guns, with the hesitant support of Macedonia's president. Most troops have since withdrawn, but the peace remains uneasy. Elections in September 2002 seemed to endorse moderation, handing a social-democrat prime minister the challenge of sharing power in coalition with Albanians.

The Former Yugoslav Republic of Macedonia submitted an application for EU membership on 22 March 2004 and the Commission analysed the country's application on the basis of its capacity to meet the criteria set by the Copenhagen European Council of 1993. In line with these criteria it has formulated three main conclusions [EC 2007-5]:

1. The Former Yugoslav Republic of Macedonia is well on its way to satisfy the political criteria for EU membership. It is a functioning democracy, with stable institutions which generally guarantee the rule of law and respect for fundamental rights.

2. On the economic criteria, the Former Yugoslav Republic of Macedonia has taken important steps towards establishing a functioning market economy. However, on the basis of the present situation the Commission considers that the country would not be able to cope with competitive pressure and market forces within the Union in the medium term.

3. On the country's ability to assume the obligations of membership, a detailed analysis is based on the 33 Chapters of the acquis which will eventually serve as the basis for accession negotiations.

Overall, the Former Yugoslav Republic of Macedonia should be in a position to take on most of the obligations of membership in the medium term (five years), but major efforts to ensure the effective implementation and enforcement of legislation will be necessary. Macedonia obtained the status of country candidate for accession in December 2005.

The 2006 Commission Report on progress made by this country in preparing for EU membership states that on Macedonia has achieved a sufficient degree of macroeconomic stability. It has made important progress in reducing barriers to market entry and exit, in strengthening the legal and institutional framework and in improving the transparency and accountability of public procedures. However, the full functioning of the market economy is still impeded by weaknesses in the judiciary, administrative bottlenecks, a low degree of legal certainty, a high number of unsettled property disputes, and considerable labour market imbalances. In 2005, on average per-capita GDP in purchasing power standards was 26% of the EU-25 average. Overall, unemployment has started to decline, although from a very high level. Consumer price inflation has increased to 3.3% during the first nine months of 2006, compared to 0.4% during the same period in 2005. Due to a financing transaction, public debt temporarily increased from 36.6% of GDP at the end of 2004, to 40.1% of GDP at the end of 2005 [EC 2006-4]. The European Commission draws conclusions from the report and published on 8 November 2006 a Memo on key findings about the progress towards accession [EC 2006-5]. It includes the following remarks for Macedonia:

> The Former Yugoslav Republic of Macedonia is well on the way to satisfy the political criteria. However, the country should step up its efforts in a number of areas. The independence and professionalism of the state administration, as well as administrative capacity, need to be strengthened.
>
> The country is well advanced in establishing a functioning market economy. However, institutional weaknesses remain, such as cumbersome administrative procedures, corruption, as well as a low degree of legal certainty, affecting the business climate and a proper functioning of the market economy. Labor and financial markets are functioning badly, and the informal sector distorts the economy.
>
> The country has made some progress concerning EU legal order. However, the country still faces major challenges in implementing and effectively enforcing the legislation.

The Macedonian government hoped to receive a date for the start of negotiation by the end of 2007. However, the European Council made clear that further steps have to be considered in the light of the debate on the enlargement strategy and significant further progress is needed to respond to the issues expressed in the Commission report and to the criteria for membership. The European Union's Enlargement Commissioner, Olli Rehn, has issued an unusually stark warning to Macedonia, on 8 February 2007, worried by the tension between the ruling conservatives and Macedonia's largest ethnic Albanian party. Mr. Rehn described developments in the country over the last year as "alarming and a cause for concern". The 2007 Commission Report on Macedonia and the communication to Council "Enlargement Strategy and Main Challenges 2007–2008" [EC 2007-20, EC 2007-21] state that the conditions for the opening of accession negotiations have still not been met.

While economy is growing, unemployment remains very high and FDIs low. Greater political consensus has to be found to push forward with reforms:

> The Former Yugoslav Republic of Macedonia is not yet sufficiently prepared to implement the *acquis* on justice, freedom and security. In the field of *science and research* there was little progress in alignment and the capacity of the scientific institutions remains weak. The country adopted policy measures which enhanced alignment in the area of *education and culture*. However, the resources to implement the policy reforms are not sufficient. The country should continue its preparations with a view to future participation in the community programs Lifelong Learning and Youth in Action. Progress has been made on developing the legislative framework in the field of *environment*, but implementation of the legislation remains in an early stage, especially in areas which require major investment. Administrative capacity and financial resources are still inadequate. On *consumer and health protection*, some progress has been made, regarding both legislative alignment and building institutional capacity for consumer protection and healthcare. More human and financial resources are needed to allow full implementation of legislation, strategies and action plans. Progress has been substantial in the area of *customs union*. As regards administrative capacity, significant progress has been made in fighting illegal trade and corruption and in collecting revenues. Overall, the Customs Administration needs to continue its modernization efforts. There has been some progress in the areas of *external relations* and of *foreign, security and defense policy*. However, the institutional and administrative capacity is not yet sufficient to enable the country to participate fully in the EU policies in these areas. Progress in the area of *financial control* has been limited to the public internal financial control. The administrative capacity of the responsible institutions remains inadequate to meet the obligations arising from the *acquis*. There has been no particular progress in the area of *financial and budgetary provisions*. Appropriate coordination structures, implementing rules and administrative strengthening will need to be established in due course.

11.8.2.1 Economy

At independence in September 1991, Macedonia was the least developed of the Yugoslav republics, producing a mere 5% of the total federal output of goods and services. The collapse of Yugoslavia ended transfer payments from the central government and eliminated advantages from inclusion in a de facto free trade area. This led to a period of economic decline with high inflation, large fiscal deficits, and almost no foreign investment. An absence of infrastructure, UN sanctions on the downsized Yugoslavia, and a Greek economic embargo over a dispute about the country's constitutional name and flag hindered economic growth until 1996. GDP subsequently rose each year through 2000. The positive trend came to an abrupt halt with the 2001 armed conflict between the government and ethnic Albanian rebels. The fiscal balance and balance of payments deteriorated severely, and reforms were stalled. Despite these developments, inflation has remained modest and the exchange rate stable.

The economy shrank 4.5% because of decreased trade, intermittent border closures, increased deficit spending on security needs, and investor uncertainty. Growth barely recovered in 2002 to 0.9%, then averaged 4% per year during 2003–06. Macedonia has maintained macroeconomic stability with low inflation, but it has lagged the region in attracting foreign investment, and job growth has been anaemic.

Table 11.20 Basic economic data for Macedonia

Macedonia		2000	2001	2002	2003	2004	2005	2006
GDP[a]	($bln)		3.4	3.8	4.6	5.4	5.8	
GDP[i]	($bln)	3.6	3.4	3.8	4.6	4.7	5.0	5.2
GDP[b]	($bln)							6.2
pcGDP ppp[b]	($x000)							8.3
GDP Growth[a]	(%)		−5	1	3	4	4	
GDP Growth[b]	(%)							3.1
GDP Growth[c]	(%)							
Current Account Balance[a]	(%GDP)					−7.7	−1.4	
Total Debt[a]	(%GDP)					38.1		
Total Debt[b]	(%GDP)							41.5
Total Debt[c]	(%GDP)					36.6	40.1	
Inflation (GDP deflator)[a]	(%)		4	3	0	1	3	
Inflation (CPI)[a,b]	(%)	5	5	1.5	1	−0.4	0.5	3[b]
Unemployment[b]	(%)							36
Exports[a]	(%GDP)		43	38	38	40	45	
Imports[a]	(%GDP)		57	58	55	60	62	
Exports Growth[a]	(%)					11.0	9.3	
High-tech Exports[a]	(%Manuf. Exp.)		1	1	1	1	1	
School enrollment, tertiary[a]	(%)	23	24	27	27	28		
Internet Users[a]	(/1000)		35	49	62	78	79	
FDI Inflows[a,c]	(%GDP)		12.8	2.1	2.1	2.9	1.7	6.0[c]
Stock market capitalization[a]	(%GDP)	0	1	5	8	8	11	
Stock market capitalization[a]	(%GDP)					16	18	20
GDP % in agriculture[a]			12	12	13	13	13	
GDP % in industry[a]			32	30	31	29	29	
GDP % in services[a]			56	57	56	58	58	

Source: [a]The World Bank (2005, 2007); [b]http://www.cia.gov/cia/publications/factbook; [c]European Commission, 2004; [i]International Monetary Fund.

The structural reform agenda was reintroduced, and the large fiscal and external imbalances were corrected. With the economic environment improving and peace and stability restored, growth is gradually returning to pre-conflict levels.

The exchange rate of the Denar against the euro was 1 € = 60.027 Denar on 30 June 2007.

Some basic economic data published by various quoted sources for Macedonia are presented in Table 11.20. Macedonia has an extensive shadow market, estimated to be more than 20 percent of GDP that falls outside official statistics [World Bank 2007-1, WorldFactbook 2007]. More information is available in the recommended reading at the end of the chapter.

The fall 2007 European Commission Economic forecast for Macedonia [EC 2007-27] is presented as recommended reading at the end of the chapter. It is summarized as "some acceleration of growth ahead" and it includes the following:

Output expanded by some 5½% during the first half of 2007, compared to 3½% the year before. Industrial output rose only moderately so far, reflecting lower production of electricity and textiles. For the whole year, real GDP growth is expected to be close to 5%. Employment increased by nearly 4% in the first half of 2007, mainly due to a strong rise in self-employment out of the shadow economy.

The authorities have recently adopted additional spending measures, which could bring the deficit close to 1% of GDP. The debt ratio is likely to drop significantly to some 27% of GDP by the end of the year.

Unemployment is likely to decline to some 33% of the labor force. Consumer price inflation is likely to remain low. Increasing wages and higher energy prices will lead to acceleration in inflation. However, the de facto peg to the euro will help to keep import prices close to EU levels and the reduction in the tax burden will dampen the increase in the cost of living. Thus, annual inflation is likely to remain below 3% over 2008–2009.

In 2008, the deficit is expected to reach $1^1/_2$% of GDP. In 2009, higher revenues, benefiting from the stronger role of domestic demand and improved efficiency in tax collection, could reduce the deficit towards $1^1/_4$% of GDP. The debt ratio is expected to continue to decline, and could fall to some 23% by 2009.

11.8.3 Turkey

This section provides a picture of Turkey along a different format than the one used for the AC12 countries and the other countries of the ongoing enlargement negotiations, due to the interest in expanding on the EU-TR negotiation history and progress. Our great interest in these negotiations relies on the relevance of the case in order to better understand the diverse visions of EU identity as unveiled by the attitudes of the actors involved in the process.

Many European leaders are looking at whether Turkey is capable of meeting the requirements set by the Copenhagen criteria for accession [EC 1995] and of implementing the institutional reforms and the economic and social transformations that they imply. Others fear that Turkey might do so and eventually qualify for membership. The author thinks that Turkey belongs in Europe and that the EU will be safer, stronger, and richer when Turkey achieves full membership. He sees the danger that a less-than-friendly attitude of some member states in the European Council and the length of open-ended negotiations might weaken the support that the Turkish people expressed in 2002 in favour of the European project and that Turkey might accept other alliances that would weaken the European Union and extinguish hopes of a larger multicultural Community rallying on shared values. Western studies of EU-TR relations suggest that economic, political, and cultural issues work against Turkey's accession, though security issues work in her favour [Arikan 2003]. Therefore on balance, the European leaders who think that Turkey should not join recognize nevertheless that the European Union needs it on her side and as a consequence they tend to develop a containment policy. Turkey needs changes to qualify for membership and those proposed by the EU should be implemented anyway and could be realized unilaterally by Turkey. However the EU model makes sense to Turkey and the EU is a catalyst of change that offers a series of templates to implement it as well as technical and financial assistance. Turkey's accession and even just the ongoing negotiations illustrate the European way to "export" democracy and values: without instruments of war, rather via solidarity and assistance. Presenting an alternative model to the exclusive, sectarian, and closed society propagated by radical Islamists, an enlarged Europe including Turkey could play an inestimable role in future relations between the "West" and the Islamic world.

This section develops the above arguments with the depth appropriate to this chapter. This section is structured as follows:

1. Geography
2. History
3. Economy
4. The process of EU enlargement to Turkey
5. Critical areas in the process of EU enlargement to Turkey
6. Expected attitude of EU27 member states in the European Council concerning Turkish full membership
7. Effect of Turkish accession on the EU
8. Effect of full membership on Turkey
9. Final considerations on Turkish full membership.

A more elaborate analysis is published in [Stajano 2008] and is presented as recommended reading at the end of this chapter. It addresses also opinions concerning EU enlargement to Turkey: in the EU, in the U.S., and in Turkey; as well as European Union's pre-accession policy for Turkey.

11.8.3.1 Geography

Turkey is at the northeast end of the Mediterranean Sea in southeastern Europe and southwestern Asia. Turkey's size is 781 thousand km^2, about twice the size of Germany. Her neighbours are Greece and Bulgaria on the North–West, Georgia on the North–East, Armenia and Iran on the East, and Iraq and Syria on the South (see Fig. 11.6). Turkey would be by far the largest EU member state. The size of

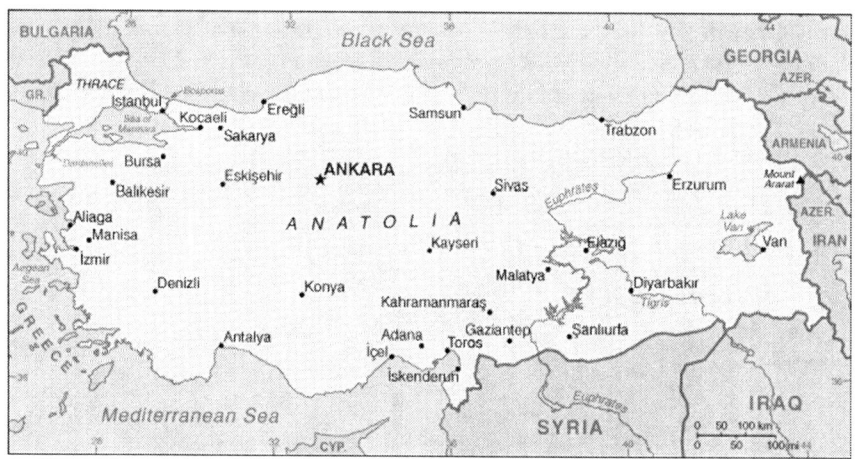

Fig. 11.6 A map of Turkey. (WorldFactBook 2007)

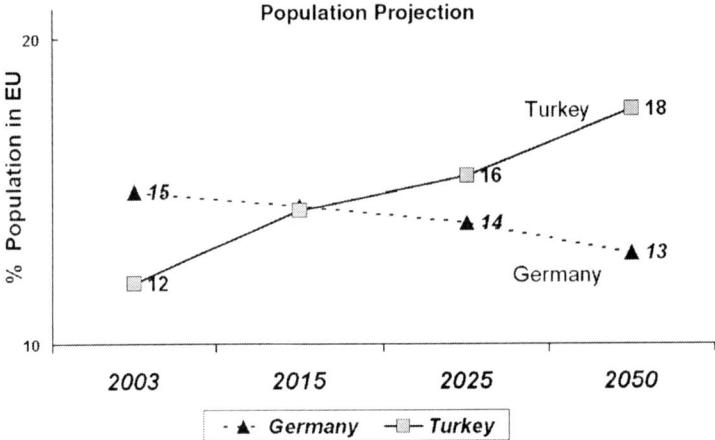

Fig. 11.7 Projected population in Turkey and in Germany, as a percentage of total EU population, in the period 2003–2050. (European Commission 2004)

the geographically European part of Turkey is smaller than that of Belgium, and is about 3% of the total area. Anatolia, the Asian region of Turkey, is about the size of Germany and Finland together. Turkey has one of the Middle East's best road and rail systems, which includes the Baghdad Railway.

The total population of Turkey was 72 million in the year 2005, of which about 10 million lived in the European region. Over 29% of Turks are below 15 years of age and only 8% are above 60. As a comparison these percentages are 14 and 26 for Italy and 14 and 25 for Germany [Economist 2007-1]. The youth of the Turkish population is a great asset for the country and makes a marked difference with ageing western Europe. It demands, however, substantial investments in all levels of education in order to qualify the population to the market requirements in a changing economy. It should be considered that by the time of the eventual accession, Turkey might well be the most populous European member state, overtaking Germany, owing to a higher birth rate in Turkey (see Fig. 11.7). Large numbers of Turks are employed in western Europe, especially in Germany (see Table 11.21).

Table 11.21 Turkish presence in EU member states (Thousands)

Country	Total	TR nationality	EU naturalized
Germany	2642	1912	730
France	370	196	174
The Netherlands	270	96	174
Austria	200	120	80
Belgium	110	67	43
UK	70	37	33
Denmark	53	39	14
Sweden	37	14	23

Source: Report of the Independent Commission on Turkey [British Council 2004].

11.8.3.2 History

Turkey is referred to as *the cradle of civilization*. Evidence of settlements in Anatolia dates back to 1900 BC. The Persian Empire occupied the area in the sixth century BC, giving way to the Roman Empire, and then later to the Byzantine Empire and the Ottoman Empire, which at the peak of its power stretched from the Persian Gulf to western Algeria. Lasting for 600 years up to the beginning of the 20th century, the Ottoman Empire was not only one of the most powerful empires in the history of the Mediterranean region but also generated a great cultural outpouring of Islamic art, architecture, and literature. Contrary to most EU27 countries, Turkey in the past 700 years was never under the ruling of a foreign power.

Turkey's current boundaries were drawn in 1923 at the Conference of Lausanne. After the First World War, Mustafa Kemal Atatürk rehabilitated Turkey under a new democratic system. Turkey was established as a republic, with Kendal Atatürk as the first president. With Ataturk's reforms, Turkey began to develop into a modern secular state and modernization, reform, and industrialization began under his direction. Atatürk created present-day Turkey, looking at Europe as a model, starting a process of westernization that continued in the 20th century, and would be crowned with the eventual EU membership. Ataturk's reforms included the renunciation of Sharia law, the adoption of a new Civil Code modelled on that of Switzerland, the replacement of the Arabic by the Latin alphabet for writing the Turkish language, and the granting of political rights to women (earlier than in most EU countries).

Some Europeans consider Turkey as an Islamic country and associate her with the Arab world. Turks, however, take a distance from Arabs and would value the EU membership as recognition of their belonging in Europe and in the European culture.

After Atatürk's death in 1938, parliamentary government and a multiparty system gradually took root in Turkey, despite periods of instability and brief intervals of military rule. Neutral during most of Second World War, Turkey took no active part in the conflict, but on 23 February 1945, declared war on Germany and Japan, a necessary precondition for participation in the Conference on International Organization, held in San Francisco in April 1945, from which the United Nations (UN) emerged. Turkey thereby became one of the 51 original members of the world organization. The Council of Europe, guardian of European values and principles, admitted Turkey as full member in August 1949. Turkey joined the Organization for European Economic Cooperation (OEEC) in 1950, became a full member of NATO in 1952, and became an associate member of the European Common Market in the year 1963.

After a period of one-party rule, an experiment with multi-party politics led to the 1950 election victory of the opposition Democratic Party and the peaceful transfer of power. Since then, Turkish political parties have multiplied, but democracy has been fractured by periods of instability and intermittent military coups (1960, 1971, 1980), which in each case eventually resulted in a return of political power to civilians.

Since the second half of the 20th century, as a consequence of events described elsewhere [Wolleh 2002] and mentioned in the Section 11.5.1, "Cyprus", the relations between Turkey and Greece and between Turkey and the Republic of Cyprus have been critical on various occasions, due to claims on territorial waters in the

Aegean Sea and the conflicts on the island of Cyprus, divided in two sectors: the Turkish Republic of Northern Cyprus (which is recognized only by Turkey) and the Republic of Cyprus, separated by a UN buffer zone since 1964.

In November 2002, Turkish elections had been won by the recently formed Justice and Development Party (AKP) that had placed Turkish EU membership high on its political agenda. However, its leader, Mr. Recep Tayyip Erdoan, was at first barred from becoming prime minister because of a conviction for "inciting religious hatred" by reciting an Islamic poem at a rally in 1998. Another popular AKP leader, Mr. Abdullah Gül, served as prime minister until Turkish law was amended to permit Mr. Erdoan to run for a seat in parliament again, which he easily won. Mr. Gül resigned as prime minister, making way for Mr. Erdoan, who enjoyed a stronger parliamentary support than any prime minister in the EU, due to the Turkish electoral law, that has a blockage threshold of 10% of voters to achieve party representation in Parliament and assures a high majority premium to a relative majority party: AKP percentage in the 2002–2007 Parliament was of 66%, while its percentage in the polls had been less than 35%.

In an effort to make herself more attractive for potential EU membership, Turkey has begun revamping some of her repressive laws and policies. In 2003, her parliament passed a law reducing the military's role in political life, and offered partial amnesty to PKK members, many of whom have sought refuge in Northern Iraq. In 2004, Turkish state television broadcast the first Kurdish language program, and the government freed four Kurdish activists from prison. Turkey also abolished the death penalty in all but exceptional cases.

Mr. Erdoğan supported energetically the Turkish EU candidacy in the difficult period from 2002 to 2007. His leadership was challenged during the campaign for the July 2007 elections, in a changed political climate where that part of the electorate that is inspired by the Kemalist vision of a secular state opposed to the pro-Islamic AKP's position. The weakening of Mr. Erdoan's position appeared evident in spring 2007 when millions of citizens rallied in protest of the attempt to elect the AKP minister for Foreign Affairs, Mr. Abdullah Gül as President of the Republic. However, Erdoan triumphed in the July 2007 elections.

The results (see Fig. 11.8) give representation to four parties in the 550-seats Parliament, according to an electoral law by which an increase in percentage of votes may imply a reduction of the seats if a higher number of parties go above the blocking threshold of 10% (see Table 11.22). The new Parliament has 50 women members, up from 24 in 2002.

In the electoral campaign the political parties put EU membership on the back burner because candidates feared the risk of emphasizing on issues that implied too high an emotional involvement, due to the frustrations originated by the length and uncertainty of the negotiation process, perceived as humiliating by a great number of citizens. "The EU doesn't sell in Anatolia," said Mr. Murat Mercon, an AKP candidate quoted by the Economist [Economist 2007-18]. In his contacts with Turkish people, both intellectuals and men-in-the-street the author has perceived that Turks often feign other options to the EU membership but in the bottom of their heart they know they belong in Europe and if they do not express it too loudly it is for the

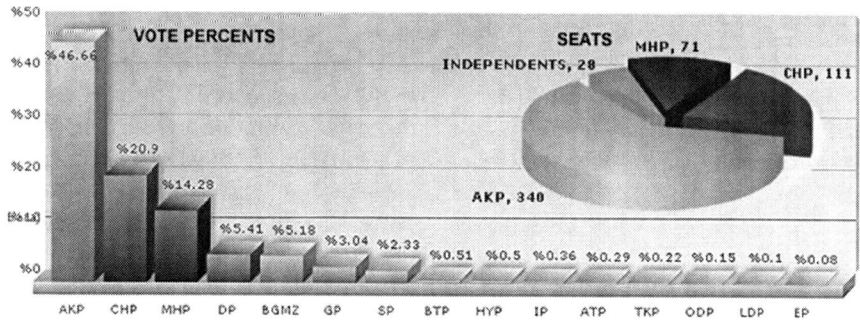

Fig. 11.8 Votes and seats in the 22 July 2007 elections. ([TDN 2007-3] © *Turkish Daily News*)

Table 11.22 Votes and seats in the Turkish general elections of 2002 and 2007

	2002			2007		
	Votes (%)	Seats (#)	Seats (%)	Votes (%)	Seats (#)	Seats (%)
AKP	34.3	363	66	46.7	341	62
CHP	19.4	178	32	20.9	112	20
MHP	8.3			14.3	71	13
Independents	16.2	9	2	5.2	26	5
Other	21.8	0	0	12.9	0	0
Total	100	550	100	100	550	100

Source: [Coppola 2007].

fear of being offended in their dignity and in their pride. Many Turks look at the accession of Cyprus in the EU before the settlement of the controversial division of the island as unfair to the Turkish-Cypriots and the 2007 accession of Romania and Bulgaria as a way to apply double standards in the application of the economic part of the Copenhagen criteria. While the author has no doubt that the European Commission is conducting and has conducted the negotiations in full fairness and impartiality, he can understand the feeling and the frustration of Turkish citizens, injured and deflated by centuries of unjustified denigration [Özben 2006].

The AKP leader, Mr. Recep Tayyip Erdoan, had a strong enough majority to rule, however not strong enough to elect the president of the Republic in the initial two voting requiring a two-thirds majority (367 seats) that he could only achieve by allying with other political forces. From the third voting, absolute majority suffices, but a two-thirds quorum of MPs presence is needed, which gives some kind of veto power to the minority, a power that was exercised in spring 2007, but not on 28 August 2007, when Abdullah Gül was elected president of the Turkish Republic. The participation in the voting of MPs from other parties than AKP is a positive sign that suggests that Mr. Gül can be the president of all Turks.

11.8.3.3 Economy

Turkey's dynamic economy is a complex mix of modern industry and commerce along with a traditional agriculture sector that in 2005 still accounted for 34% of employment and 11% of GDP. Principal minerals extracted are coal, chromium, lignite, copper and iron ores, antimony, mercury, and boron. Some petroleum is produced. The construction of a $ 3 billion, 1,500-km oil pipeline running from Baku, Azerbaijan, to the Mediterranean port city of Ceyhan, and expected to bolster the Turkish economy, began in September 2002, a time that dissident commentators [Rall 2002] have correlated to the war in Afghanistan after the 9/11 terrorist attack in the U.S.

Turkey's chief crops are tobacco, cotton, wheat, barley, corn, rye, oats, rice, olives, figs, raisins, sugar beets, and fruit. Large numbers of sheep, goats (including many mohair-producing Angora goats), and cattle are raised.

Great strides have been made since the 1970s to strengthen and diversify the economy. In recent years, the economic situation has been marked by erratic economic growth and serious imbalances. The most productive farmland is in west Turkey, but in recent years the country has instituted the massive southeast Anatolia Project to use the Tigris and Euphrates rivers for irrigation and hydroelectric power. This project envisions the construction of 22 dams and 19 hydroelectric plants on the two rivers. In the late 1990s the giant Atatürk Dam and Reservoir on the upper Euphrates, as well as two other dams and adjacent power plants, was completed, and six more were under construction. Although plagued by the Kurdish conflict and bitterly opposed by Syria and Iraq (who are concerned that the downstream water flow from the rivers to them will be severely impeded), the project is continuing. The government's goal is to transform arid southeastern Turkey into a prosperous agricultural-industrial region.

Turkey has a strong and rapidly growing private sector, yet the state still plays a role in basic industry, banking, transport, and communication but not in manufacturing, largely privatized. Contrary to the Central and Eastern European Countries, Turkey has been open to market economy for many decades. However, the existence of a functioning market economy as requested by the Copenhagen criteria, is compromised by incomplete liberalization, macroeconomic instability, a banking system partially under state control, and an attitude towards populism that supports rent-seeking lobbies taking advantage of diffused corruption of civil servants, discretionary extra-budget public funds, and non transparent state aid [Uur 2004-1, Eder 2004]. Real GDP growth has exceeded 6% in most years, but this strong expansion was interrupted by erratic GDP growth values and sharp declines in output in 1994, 1999, and 2001. Comparing GDP growth in Turkey and in the 10 countries of the 2004 accession (AC10), Turkey features record high growth on several years, offset by severe contractions. Meanwhile the public sector fiscal deficit regularly exceeded 10% of GDP – due in large part to the huge burden of interest payments, which in 2001 accounted for more than 50% of central government spending – while inflation remained in the double digit range until 2003. Perhaps because of these problems, foreign direct investment in Turkey remained low until 2005 (less than $ 1 billion annually).

The European Commission wrote some encouraging statements about Turkish economy in the 2005 Report about accession negotiations [EC 2005-4] that reflect a more positive view than that expressed by the scholars quoted above:

> As regards the economic criteria, Turkey can be regarded as a functioning market economy, as long as it firmly maintains its recent stabilization and reform achievements. Turkey should also be able to cope with competitive pressure and market forces within the Union in the medium term, provided that it firmly maintains its stabilization policy and takes further decisive steps towards structural reforms.

Mehmet Hadra, the head of the Turkish-Arab Business Association, reports that, in 2007, FDI of $ 3 billion is coming from Arab Gulf countries, with a target of $ 10 billion by 2010 [TDN 2007-2]. According to *The Economist*, FDI globally rose to $ 19 billion in 2007 from $ 800 million in 1997 [Economist 2007-18]. In late 2000 and early 2001, a growing trade deficit and serious weaknesses in the banking sector plunged the economy into crisis, forcing Ankara to float the Turkish Lira and pushing the country into recession. This crisis was a serious setback for Turkey's economy; but it also showed its resilience, dynamism, and flexibility. Prospects since 2002 have been much better, due to a far-reaching reform program backed by strong financial support from the International Monetary Fund, tighter fiscal policy, a major bank restructuring program, and the enactment of numerous other economic reforms.

The distribution of wealth in Turkey is characterized by higher inequality than in the EU27 countries, and by the emergence of high income groups. Disparities of regional income present a serious problem, causing large-scale migration flows inside Turkey. The Marmara (Istanbul) region has a population of 17.3 million and a per capita income standing at 153% of the Turkish average; the 9 million people of the Aegean region earn 130% of the average income, Central Anatolia has 11.6 million people earning 97% of average income, while East Anatolia's 8.1 million people have the lowest income, at 28% of the average [British Council 2004]. The ratio of pc GDP in the richest region to its value in the poorest region is 5.62 (resulting from €4,588 / €815), while this value is less than 3 in any other EU27 country, e.g. it is 2.83 in Germany (resulting from €42,800 / €15,072) [Burrell 2005-2].

The leading industrial centres are Istanbul, Ankara, Karabük, Bursa, Izmir, Adana, Samsun, and Diyarbakir; Kayseri (the home town of Abullah Gül) is increasing in importance as one of the growing centres of the pro-Islamic business community. Turkish industry mainly depends on the private sector activities. The share of public sector in the manufacturing industry has been decreased through privatization activities in recent years. Currently, more than 80% of production and about 95% of gross fixed investment in the manufacturing industry is realized by the private sector. The most important industry – and largest exporter – is textiles and clothing, which is almost entirely in private hands. Besides textiles and clothing, the country's chief manufactures include processed food, iron and steel, petroleum, construction materials (especially cement), forest products, wine, and chemical fertilizers. Turkey is also noted for the manufacture of carpets, Meerschaum pipes and artefacts, and pottery. Some of these products rely on low cost specialized workforce, often in the shadow economy. The much needed change towards higher welfare and reduction of shadow economy might reduce the availability of cheap labour and compromise the competitiveness of traditional carpet manufacturing, unless innovative processes and advanced technology rejuvenates a centuries-old industry.

Tourist trade gives a substantial contribution to the national economy and to trade balance, thanks to the great attractions of a rich cultural heritage, the lively and beautiful Istanbul, the sunny beaches and the national parks. Turkey ranks 12th in the list of tourist destinations in the world, with 17 million visitors in 2004. The annual value of Turkey's imports is usually considerably higher than that of her exports. Exports were 19.7% of GDP and imports 28.8% of GDP in 2003 [Burrell 2005-2]. Trade deficit was over TRL 10 trillion (€6 million) in that year. The EU is the largest TR's trade partner (46% of imports and 52% of exports in 2003; 46.7% of total import and for 54.6% of total export in 2005) [Economist 2007-1]. Leading EU trade partners are Germany, the United Kingdom, and Italy. The next regions in the import ranking were Middle-Eastern countries and U.S. in 1990 and changed into the Confederation of Independent States (CIS) and Middle-Eastern countries in 2003; while the next regions in the export ranking were Middle-Eastern countries and U.S. in 1990 and changed into CIS and U.S. in 2003 [Burrell 2005-2]. Chief imports are machinery, crude oil, metals, pharmaceuticals, plastics, rubber, and chemicals; principal exports are textiles and clothing, iron and steel products, agricultural produce, and minerals.

The national currency is the new Turkish Lira (TRL), which was revalued by a factor of one million on 1 January 2005. 1 € = 1.76 TRL on 27 April 2005 and 1.75 TRL on 11 July 2007. Details on inflation in Turkey are presented below in the next Section 11.8.3.4, "The process of EU enlargement to Turkey". Some basic data for Turkey's economy published by different quoted sources are presented in Table 11.23.

The fall 2007 European Commission Economic forecast for Turkey [EC 2007-27] is presented as recommended reading at the end of the chapter. It is summarized as "sustained growth in a context of reduced fiscal imbalances" and it includes the following:

Real GDP growth fell from 6.1% in 2006 to 5.3% in the first half of 2007. Public consumption and investment increased by respectively 8% and 25% in the first six months of 2007. Imports grew at 7% in 2006 and by 6% in the first half of 2007. Export growth rose from 8 1/2% in 2006 to 13 1/2% in the first six months of 2007. Inflation fell significantly after going up in early 2006, in large part due to faltering price pressures in the service sector. Consumer price inflation fell from around 11% in 2006 to 7% in the third quarter of 2007, still far above the year-end target of 4%. Meanwhile high real interest rates of above 10% per annum are inviting large capital inflows and putting upward pressure on the Turkish lira.

The overall picture for 2008–2009 looks rather favorable. Turkey should be able to increase export growth in particular in tourism while the monetary and fiscal policy mix will support the disinflation process. A gradual decline in inflationary pressures will allow a fall in interest rates and improve the investment climate.

In line with economic expansion, employment is forecast to increase by about 1 1/2% per year. The trend in declining inflationary pressures is likely to continue as from 2007.

The general government balance is expected to change from a surplus of 1/2% of GDP in 2006 to a deficit of 1% of GDP in 2007. In 2008 and 2009, a budgetary surplus of

Table 11.23 Basic data for Turkey

Turkey		2000	2001	2002	2003	2004	2005	2006
Population[b]	(million)					68.9		71.2
Median age[b]	(years)					27.3		28.6
Birth rate[a]	(births/1,000 population)					1.7		1.6
GDP[a]	($bln)		145.2	188.9	240.4	302.8	362.5	
GDP[i]	($bln)	204.9	153.5	184.8	239.8	300.1	340.3	367.3
GDP[b]	($bln)							358.2
GDP[f]	(€bln)			192	212	242	290	
GDP ppp[b,g]	($bln)					458.2		627.2
pcGDP ppp[b]	($x000)					6.7		8.9
pcGDP ppp[f]	(€x000)							6.9
GDP Growth[a]	(%)		−7	8	6	9	7	
GDP Growth[b]	(%)					5.8		5.2
GDP Growth[c]	(%)	7.4	−7.4	7.8	5.4			
Current Account Balance[a]	(%GDP)				−0.8	−2.8		
Government Balance[f]	(%GDP)				−12.9	−11.3	−5.7	−1.2
Total Debt[a]	(%GDP)				71.3	61.2		
Total Debt[b]	(%GDP)					78.7		64.7
Total Debt[f]	(%GDP)				93.1		76.9	69.6
Inflation (GDP deflator)[a]	(%)		55	44	23	10	5	
Inflation (CPI)[b]	(%)					25.3		9.8
Inflation (CPI)[c]	(%)	68.8	39	68.5	29.5	18.4		
Unemployment[b]	(%)					10.5		10.2
Exports[a]	(%GDP)			34	29	27	29	27
Imports[a]	(%GDP)			31	31	31	35	34
Exports Growth[a]	(%)					16.0		5.6[d]
High-tech Exports[a]	(%Manuf. Exp.)			4	2	2	2	2
School enrollment, tertiary[a]	(%)	23	23	24	28	29		
Internet Users[a]	(/1000)			51	62	85	144	222
ICT Expenditure[h]	(%GDP)			2.2	1.7			
FDI Inflows[a]	(%GDP)			2.3	0.6	0.7	1.0	2.7
Stock market capitalization[a]	(%GDP)	35	32	18	28	32	45	
GDP % in agriculture[a]				13	13	13	13	12
GDP % in industry[a]				26	24	22	22	24
GDP % in services[a]				61	63	65	65	64

Source: [a]The World Bank (2005, 2007); [b]http://www.cia.gov/cia/publications/factbook/; [c]European Commission, 2004; [d]Average 2003–2007; [f]Eurostat, 2007; [g]Purchasing Power Parity (ppp); [h]EITO, European IT Observatory, 2003; [i]International Monetary Fund.

respectively 1/4% of GDP and 1 1/4 is projected as a result of gradual fiscal tightening and falling interest rates. General government debt is expected to further decline gradually, albeit at a decelerating pace, to less than 50% in 2009.

11.8.3.4 The Process of EU Enlargement to Turkey

The idea of Turkey accessing the EU has been discussed for decades and has animated hot political debates, with strong reservations form the EU side, although Turkey had achieved early recognition (earlier than the majority of EU27 member states) in international forums, as mentioned above in Section "History".

The enlargement process started as early as 1959, when Turkey tabled an application for associate membership [Arikan 2003], as a reaction to Greece's application, in the fear that TR-EL relations might be biased in favour of Greece, should only the latter be associated with the EEC. If the enlargement process will end with Turkey's accession, this will anyway happen not earlier than the second half of the 2010s. In 1963, the EEC–Turkey association agreement, known as the *Ankara Agreement* (AA) [EC 1964], was signed. It laid down the basic objectives of relations between Turkey and the EEC, such as the continuous and balanced strengthening of trade and economic relations and the establishment of a Customs Union (CU), eventually signed in 1995. Turkey considered the AA as an important confirmation of her European identity, and in 1987, tabled a request for accession. Twelve years after the request, a major breakthrough in Turkish relations with the European Union took place at the Helsinki European Council of 10–11 December 1999, which concluded that "Turkey is a candidate State destined to join the Union on the basis of the same criteria as applied to the other candidate States." With this decision, Turkey was put firmly on the track towards accession; however – contrary to other accession countries – Turkey did not receive a timetable for accession negotiations. The European Council eventually decided on 17 December 2004 that Turkey qualified to start accession negotiations, and fixed the starting date of 3 October 2005. The Turkish diplomats coming to Brussels for the negotiations might not be back home for supper: the process will take 10–20 years.

The historic agreement to start negotiations [EC 2004-14] is based on five points:

1. the start of accession negotiations is on 3 October 2005
2. adaptation before 3 October 2005 of the Ankara Agreement, taking account of the accession of the 10 new member states
3. final agreements may include derogations and transitory disposition settlements
4. negotiations will not be finalized before 2014
5. there is no guarantee regarding the final outcome of the negotiations

The second point meant the extension of the Turkey-EU Customs Union to all members of the EU, resulting in an obligation to opening Turkish ports and airports to Greek-Cypriot ships and planes, which had been open already years earlier; recognition of the Republic of Cyprus would come gradually, according to Turkey, along with the progress of accession negotiations.

The position among the EU25 member states about the start of accession negotiations was not unanimous. *The Economist* [Economist 2004-4] reported:

> Some suspect that Tony Blair, the UK prime minister, or, worse still, George W Bush, the U.S. president, are pushing Turkish membership precisely because they want the EU to revert to being little more than a glorified free-trade area.

In the same article, *The Economist* made it clear (see Fig. 11.9) that the outcome of the negotiations is uncertain. It must be noted that even if the negotiations were to bring Turkey into compliance with all the Copenhagen criteria, a new treaty stating the accession of Turkey would have to be ratified by all the EU member states, possibly no less than 28 of them by 2014+. Some countries might approach ratification via a referendum, the outcome of which is, for the time being, far from certainly positive; on the contrary, it is a riskier challenge than the negotiations themselves.

Negotiations started in a climate of uncertainty due to reasons that in part are under the power of Turkey's authorities to resolve, and in part are not, because they are associated with lukewarm acceptance – by several EU governments – of a large and poor country often depicted as an Islamic state. Turkey should, to resolve also these issues, make a big effort in becoming better known, addressing legitimate concerns but also misconceptions and anxieties, so as to turn public opinion around and become more respected by the other European countries. Professor Dr. Haluk Kabaalioğlu, the newly elected president of the Turkish Economic Development Foundation, a nongovernmental organization representing the interests of the private sector in relations with the European Union, in a press conference in Istanbul on 9 July 2007 announced a project to address decision-makers, civil servants, and experts in Brussels with the view to making them more familiar with Turkey. The

Fig. 11.9 There is no guarantee on the final outcome of accession negotiations for Turkey. ([Economist 2004-4] © *The Economist*. Reproduced with kind permission)

project includes the endowment of chairs in two French universities and a program to make the EU better known in Turkey [TDN 2007-1].

After the start of the negotiations in October 2005, the EU has continued in the containment policy that had characterized EU-Turkey relations for decades: encourage the process towards meeting the accession criteria and keep Turkey within the influence of the EU, while leaving Turk EU membership in a far-away and undefined future [Arikan 2003]. This approach is both hypocritical and short-sighted and creates a feeling of mistrust in Turkey and the not totally unfounded accusation of using double standards. Turkey may be difficult to digest, but her participation in ESDP is too important to be neglected; Turkey's participation in ESDP requires in turn her full involvement in the decision process. This is only possible with full membership [Çayhan 2003].

The EU was helped in legitimating her containment policy by Turkey's slow pace in fully implementing the reforms needed to meet the membership requirements. Turkey did not recognize the Republic of Cyprus and continued to deny vessels and aircraft flying the Cyprus flag, or whose last port of call was in Cyprus, access to her ports. Such restrictions on shipping often preclude the most economical way of transport and therefore result in a barrier to the free movement of goods and to trade. They infringe the Customs Union Decision [EC 2005-4]. This was a major stumbling block in the negotiation process, which lead the European Council to suspend the negotiation of eight out of 35 chapters in December 2006 [LaGro et al. (eds.) 2007]. However, progress was acknowledged in some areas so that, as an example, the negotiation on the science and research chapter was satisfactory closed. The Commission reported [EC 2006-1]:

> Turkey's research policy resulted in significantly increased budgets for research and development: nearly fivefold compared to 2002 levels. New universities have been opened in 15 cities. Improvements were also achieved in Turkey's science and research capacities including its gradually more successful participation in the Framework Program. Turkey's success rate under FP6 improved, mostly in obtaining small projects. However, EU funding is not achieving its potential. Taking into account actions that Turkey has taken with respect to mobility of researchers, science and society and percentage of GDP for Science and Technology Action Plan measures, Turkey is already well integrated into the European Research Area.

In June 2007, after the election of Mr. Nicolas Sarkozy in the French presidential elections, the attitude of France towards Turkey worsened; Jean Christou wrote on a Cyprus paper [Christou 2007]: "If Cyprus is a mere thorn in the side of Turkey's EU accession, then France has suddenly become a millstone around its neck." Sarkozy's government opposed the opening of the negotiation chapter on the Economic and Monetary Union, perceived by the Turkey-unfriendly French administration as an irreversible step towards full membership [Adnkronos 2007]. The attitude of the French government confirms the opinion expressed by Jørgensen and LaGro: "no matter the subject matter being negotiated technically, the closing of all chapters will be a political decision" [LaGro et al. (eds.) 2007].

11.8.3.5 Critical Areas in the Process of EU Enlargement to Turkey

The debate about the accession of Turkey became very intense, in particular between 2002 and 2004 during the works of the European Convention, when the case of Turkey was a test bed for the definition of the European identity. Many European politicians were against the accession of Turkey and some were even against the start of negotiations. Some of them presented technical arguments [Ramonet 2004]; others appealed to the values and principles around which the European people unite, and others expressed the fear that an enlargement to Turkey would weaken the European project. The debate unveiled the lack of respect of and knowledge about Turkey in Europe, the power of prejudices against aliens, and – most important – the anxiety of the western culture towards Islam.

One of the many problems concerning the accession of Turkey is related to the recognition of the Republic of Cyprus. The 2007 Commission report [EC 2007-2] stressed that Ankara must normalize its relations with Cyprus and honour a 2005 pact to open its ports and airports to the island republic. Other problems include her size and population, poverty, a non completely working market economy, a high percentage of informal economy, human rights, torture, gender equality, social inclusion, minority protection, and – last but not least – impact on European security and defence policy by extending the external borders in a politically instable region.

The *size* of Turkey may have an adverse impact on the cost and risk to the EU of Turkish membership [Arikan 2003]. Social cohesion among member states and the Common Agricultural Policy (CAP) generate very high costs with enlargement countries when their GDP pc is low and their agricultural GDP share is high. The membership of Turkey would cause the so called "statistical effect" of lowering the average EU GDP pc and, as a result, 20 regions in eight EU27 countries (CZ, DE, ES, FR, GR, IT, PT, and UK) would no longer qualify for Objective 1 status, namely for cohesion funds contribution, being no longer below the 75% of the reducing EU GDP pc average. This might create a strong pressure in the European Council with eight countries having internal policy reasons for not supporting Turkish membership. As shown in Table 11.23, Turkey's economy is of the order of €300 billion, corresponding to 2.6% of the total EU GDP. Liberalization and other measures will affect economically Turkey more than the rest of the Union.

With reference to the concerns about the Turkish *population*, some countries fear that TR accession might result in possibly uncontrollable mass migration into western (richer) EU states. Western European countries, however, should not expect a flow of migrants from Turkey after the accession, for two reasons: firstly, the whole process is meant to create better living conditions in Turkey, and secondly, clauses in the Accession Treaty might restrict or at least delay the free movement of labour by many years, making it possible when hopefully it would no longer be desired by the Turkish people. In 2005, three million citizens of Turkish origin lived in the EU; Harry Flam, a Swedish economist, estimates [Flam 2004] that, without restrictions on migration, up to 1.8 million Turks might move to other EU countries after accession; however, the destination of migrants is not expected to be evenly distributed among EU member states; those countries with large Turkish communities like Germany (see Table 11.21) may well receive the major share of migration flows. The number of 1.8 million migrants amounts to less than 0.4% of the EU27 population and would not create a problem if positive provisions in housing, education, health care, and social services are offered together with measures to avoid migrants' concentration in regions or neighbourhoods where they might create national enclaves instead of integrating with the local population. Turkey's population would be reflected also in Turkey's weight in the decisions of the Council of Ministers and in her high number of seats in the European Parliament. Turkey would become one of the heavyweight decision makers; however, the impact of a large representation of Turkish MPs would be much reduced by the fact that voting in the European Parliament normally follows party lines rather than the national positions of member states.

About *poverty* in Turkey, we would like to refer to Fig. 7.26. There *The Financial Times* shows that also Greece, Portugal, and Spain were also quite poor with respect to the European average at the time of their accessions in 1981 and 1986, respectively. Nevertheless, the 1980s enlargements were beneficial both to the entering countries and to those already in the EEC. After becoming members of the EEC, the three countries experienced a major increase in their GDP pc [Economist 2004-3] by being integrated in the then buildup of the internal market (see Fig. 11.10). We should also expect a comparable growth of GDP pc for Turkey once its -eventually achieved- full membership would initiate processes that support the expansion of her economy. Some significant transformations are ongoing already: in the 40 years from the Ankara Agreement to 2004, Turkey has undergone major changes, and the most significant ones have occurred since 2002, under the leadership of Erdoan. He has undertaken many reforms of the institutional

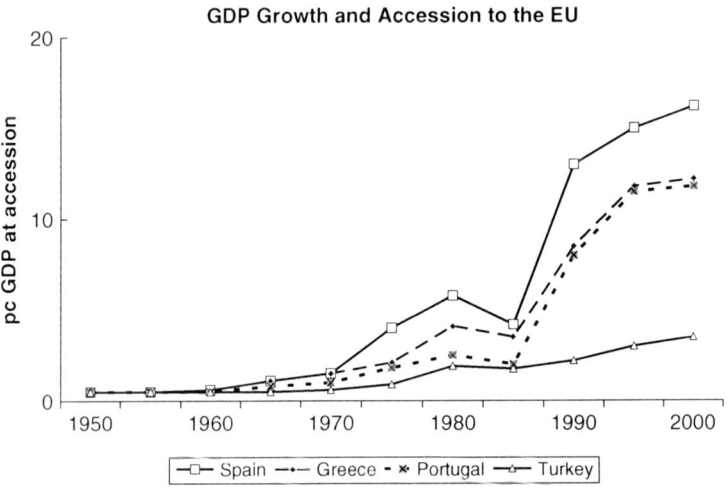

Fig. 11.10 pc GDP in Spain, Greece, Portugal and Turkey in the period 1950–2000. ([Economist 2004-3])

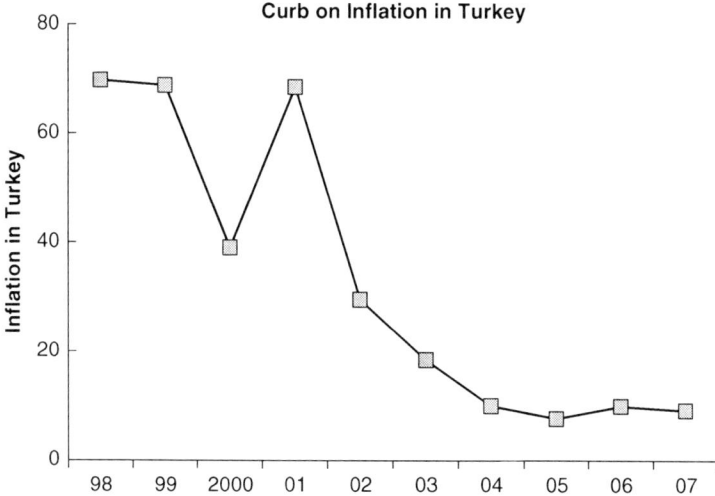

Fig. 11.11 Inflation in Turkey in the period 1998–2007. (European Commission 2005, 2007, [Bryant 2007])

organization of the country (civil control over the powerful army, elimination of the military-controlled State Security Courts, reform of the judiciary system, abolition of the death penalty) and improved human rights and respect towards minorities. He has achieved historic results in the control of the budget, the growth of GDP, that has been in the range of 5% to 8% since 2002, and the reduction of inflation, that had been near 70% until 2001 and dropped into the single digits in 2004 (see Fig. 11.11). Turkey went a long way towards convergence to EU values; however, she needs to curb inflation further and to keep the public debt below 60% of her GDP.

Western Europeans visiting Istanbul do not perceive Turkey as a poor country. They are impressed by the liveliness of the metropolis, her airports, her infrastructures, the public transport system, the cleanliness and the safety of the town. Business visitors are impressed by the low bureaucracy and the business-like attitude of the productive community. Critics may say that not all Turkey is like Istanbul, and that this town is not a complete picture of Turkey. But likewise a foreigner visiting only Milan and Bolzano while missing Naples and Gela would have a very partial and rosy picture of the author's beautiful country.

In comparison to other Central and Eastern European Countries (CEECs), Turkey is more prepared to be integrated in the EU internal market, not having been under state-controlled central planning in the past decades. Turkey is, however – despite the encouraging statement of the European Commission reported earlier – still far from a fully functioning *market economy* and her accession might create a risk to the EU by increasing the economic diversity of her members. This issue was also perceived in connection with the countries of the 2007 enlargement; however the size of Turkey's economy is well ahead of that of Romania and Bulgaria.

Concerns presented in the 2006 Commission Report on accession negotiations [EC 2006-1] on the subject of Turkey's market economy include: too high a percentage of Turkish economy is in the shadow; state aid distorts competition; immunity and anti-corruption measures prove to be inadequate; the level of participation of women in the labour market is too low and women remain vulnerable to discrimination policies; job creation in industry and services does not compensate for the reduction of employment in agricultural sector; agricultural policy must move away from price support to income support and align with the reformed CAP; and – last not least – full implementation of free trade with EU countries is compromised by the Cyprus dispute and the denial of access in Turkey of vessels with the flag of the Republic of Cyprus. The 2006 Commission report is, nevertheless, also pointing at many areas where significant progress was made, often however not yet sufficient to comply with the criteria. Issues associated with religious differences, cultural differences and size are not addressed by the report to maintain the ostensible objectivity of the Copenhagen criteria [McLaren et al. 2003].

The economic instability of recent years is coupled with a two tier labour market, whose formal component is rigid while her flexible shadow-economy component, accounting for over 50% of employment, is precarious and exempt from taxation and workers' protection [Eder 2003]. A large informal sector is identified also by OECD, together with low confidence and weak governance, as one of the major problem areas in Turkey's economic development [OECD 2004].

The accession of Turkey could divert EU funding away from programs in support of sustainable growth. The most developed countries might be against a reduced emphasis on the programs for competitiveness of the kind of those proposed under the Lisbon Agenda. It is estimated that the net EU budget cost of Turkish accession would exceed 20 €billion (16%) at 2004 prices [Oscam 2005].

With regard to *fundamental rights*, the 2006 Commission report can acknowledge only limited progress. A law for the establishment of an Ombudsman is expected to contribute to transparency and accountability of public activities. A law on abolition of death penalty at all times was ratified in February 2006 and the development of a legislative framework against torture has continued. However further efforts are required for the change to percolate to police stations, barracks, and courts of justice, since cases of human rights violations have still been reported. Freedom of thought, conscience and religion is respected; restrictions on freedom of assembly have diminished. Problems are reported on the rights to property for minority groups. Gender equality and human rights receive growing attention; women representation in the parliament to elected in the 2007 general election has changed the spring 2000 picture of a 96% of male MPs representation [McLaren et al. 2003] by more than doubling the women presence in the Parliament; in practice women's rights are not always protected, in particular in the poorest areas. There was no progress in aligning Turkey with international and EU standards on minority rights. There was progress in the fight against corruption, however corruption remains widespread and anti-corruption authorities and policies remain weak.

The proposed legislation that would allow women to cover their heads on university campuses and other public buildings will most likely pass the Turkish Parliament, where Erdoan's party and its allies have more than enough votes to alter the constitution. This is not perceived by the Parliament opposition and by western observers as the freedom for a lady to cover or not to cover her head, rather it would be an alarming move towards a deviation from the Kemalist secular ruling of the Turk Republic.

After the inclusion of the European Security and Defense Policy (ESDP) in the Treaty on EU in 1992, enlargements have included the support of the democratization process and of institutional reforms, with a view to increasing stability in Europe, and a contribution to the *security* and defence policy. Security had been an important consideration during the Cold War in the pre-accession negotiations with Greece, Cyprus and Malta. The end of the Cold War and, later, the 9/11 attack changed the general picture and, in particular, the role of Turkey's geopolitical role [Burrel 2005-3]. Now security threads are feared from the Southeastern crisis areas in Turkey's proximity, so that today defence and security considerations are of prominent importance in the ongoing negotiations with Turkey. This country had been for decades the southern bulwark of western powers against the Union of Soviet Socialist Republic (USSR) and has, after the U.S., the largest army in NATO with an extensive peacekeeping and war-fighting experience. It would constitute a valuable asset for the European Security and Defense Policy [Çayhan 2003] and could now become the European southeastern stronghold in front of the middle-eastern countries whose political instability and anti-western attitude challenge the world's equilibrium. Concerning ESDP, Turkey is more homogeneous with EU15 than the CEECs, having been associated with NATO well ahead other EU members. It is however questioned if the extended Turkish border and the political instability and anti-western attitude of the would-be new EU neighbours (Syria, Iraq, Iran, Armenia, and Georgia) that Turkey's accession would entail would not create a thread to EU security [Arikan 2003]. By the way, borders and neighbours are relevant also to illegal immigration and to human health protection in case of illegal traffic of live animals.

11.8.3.6   Expected Attitude of EU27 Member States in the European Council Concerning Turkish Full Membership

Turkey's main priority should be to enhance a political dialogue with each and every member state, since Turkey's accession will not be an easy process [LaGro et al. (eds.) 2007]. If and when the European Commission will close all negotiation chapters, the decision to propose a date for accession must be taken at unanimity by the European Council. Its outcome is not certain because even if each chapter will be closed satisfactorily, accession remains a political decision, and the attitude of member states is not in all cases in favour.

At the time of writing (Year End 2007) some EU countries have expressed a clear position; for most countries the balance between *pro* and *contra* is still undefined. By the time the Council is faced with the decision of a date for Turkey's accession, changes may have occurred in worldwide international relations as well as in the governments of EU27. Turkey, meanwhile, will have deepened and widened her transformation, for a successful conclusion of the accession negotiations. This may change the picture perceived now.

The main arguments in favour of accession can be summarized as follows:

1. Improved ESDP
2. Prospects of solving the Cyprus dispute
3. Bridge to the East
4. Secured energy transport
5. Improved trade
6. Alignment to U.S. policy
7. Desire to water-down the European project to a glorified internal market
8. Need of workforce
9. Positive experience with Turkish migrants
10. Xenophilia

The main arguments against accession can be summarized as follows:

1. Fear of watering-down the European project
2. Limit the geographical eligibility to EU membership
3. Proximity of Turkey to troubled regions
4. Size of Turkey and her heavyweight in Council decisions and EP representation
5. Fear of Turkey, perceived as an Islamic state
6. Low Turkish GDP pc
7. Xenophobia
8. Religious Xenophobia
9. Loss of national share of cohesion funds
10. Loss in CAP subsidies
11. Fear of uncontrolled migrations

Points 4, 9, and 10 in the *against* list might be addressed by the EU during the negotiation years, so that there be no issue about them at the time of accession: the possibility of derogations and specific arrangements have been anticipated before the start of negotiations in 2005; the Council decision rules could be adapted to the presence of the new large an populous member. Some points in either list might find their converse in the other, e.g. the watering down of the European project could be seen positively by Euroskeptics or negatively by strong believers in the European values.

The anticipation of which EU government will be in favour of Turkey at the eventual European Council meeting when the Commission will table a positive opinion for accession and a date will be proposed is not easy a task at this point in time. The author's perception is that the Greece, Italy, Sweden and the UK can be expected to be favourable; Austria, France, Germany, The Netherlands are countries on which, in the author's view, Turkish diplomatic efforts should be concentrated in the coming years, so that a presently unfavourable position may be reversed. Some other countries appear to offer a lukewarm or not unanimous support. For many countries the author is not able to say on which side they stand at this point in time. A factor that might increase the support to the full Turkish membership is the attitude of the U.S. and the interest of many countries to align to the American foreign policy. But it would not be realistic to pretend to be able to anticipate hard political decisions to be taken perhaps 10 years from now, when the instable world politics may have changed the background against which the political leaders will make up their mind.

11.8.3.7 Effect of Turkish Accession on the EU

If the Lisbon Treaty on the EU will come into effect with the new voting system, then the EU will not be deeply affected by Turkey's accession. Should the Nice Treaty regulate the functioning of the EU institutions, then Turkey's accession would reduce the functioning capability of the EU institutions [Hoekman et al. (eds.) 2005]. The EU budget will be affected: it is estimated that the net EU budget cost of Turkish accession (namely not considering Turkey's contribution to the EU budget) would exceed 20 €billion (16%) at 2004 prices [Oscam 2005]. The financial perspectives at the time of Turkey's accession are not known now, but we can anticipate that accession of Turkey could divert EU funding away from programs in support of sustainable growth. The most developed countries might be against a reduced emphasis on the Lisbon-like programs for competitiveness that might be in operation at the time.

The relative poverty of Turkey will have a "statistical effect" on the eligibility of a number of EU regions to Objective 1 cohesion funds. CAP will also be affected. EU production in agriculture and trade of agricultural products will be affected, with better Turkish products offered on the enlarged EU market. Turkish and western agrifood industries may contribute to the enhancement of Turkish agricultural products and extend the shelf life of fresh products that could be offered throughout the EU.

The European industries would have full access to a large and growing market characterized by a high percentage of young citizens. Increasing FDI and delocalization in a country with available workforce may be an opportunity if delocalization is not only seeking cheap labour but also takes into account the Turkish demand for products, services, and infrastructures as well as the texture of local economy.

Migration in western European countries is not expected to become an issue, because restrictions and delays would be imposed in the accession treaty, authorizing migration only when the migratory pressure would be reduced. The improved conditions in Turkey would in fact reduce the migratory pressure over the years, if the structural reforms will succeed in granting jobs to the workers made redundant by increased productivity in agriculture.

Security and defence policy will be enhanced by Turkey's full membership, and hopefully, once that Greece and Turkey are on equal grounds in the European Union, the territorial waters disputes [Arikan 2003] and the case of Cyprus [Wolleh 2002] will more likely find the solution that today appears difficult to reach.

11.8.3.8 Effect of Full Membership on Turkey

The achievement of full membership, for which Turkey has been longing for half a century, will have deep effects on Turkey and will be a major milestone in her history. Once the EU formally recognizes that Turkey belongs in Europe, the attainment of democracy and the respect of fundamental rights will be irreversible. The long march towards westernization, initiated by Kemal Atatürk will be completed and Turkey will commit to the values around which the European people unite.

The prospect of a subsequent adoption in due time of the euro, will increase the stability of Turkish economy, increase FDIs, and help the containment of the budget and of the national debt within the parameters imposed by the Treaty on EU. The full integration in the internal market will be the natural completion of the Customs Union, and Turkey will at last participate in all decision making and legislating institutions that shape the European policies.

Over several decades, the cohesion policy will help reducing the economic gap between Turkey and the richer western European countries, while EMU will accompany Turkey in an economic sustainable growth with contained inflation. In this way the competitive edge of lower wages will sustain foreign trade and foreign investments.

Turkey is faced with many severe challenges: to transform the society so that the transposition of the *acquis* has effect on the lives of Turkish citizens and improves all aspects of the civil society; to confirm and to strengthen the secular nature of the country against pressure from Islamic fundamentalists; to increase transparency of all aspects of public activity; to dampen inflation further; to curb corruption, populism and undue immunities; to create jobs for the workforce made redundant by increased productivity in agriculture; to educate the young Turkish citizens so that they can find a job and contribute to the economic growth; to create lifelong learning programs to support labour mobility and coping with a changing demand in the labour market; to invest in research and development so that competitiveness of the Turkish economy helps sustain international competition.

11.8.3.9 Final Considerations on Turkish Full Membership

The EU model makes sense to Turkey and the proposed changes (on many topics, including: market discipline, control over state aid, competitiveness, liberalization, corruption, judiciary, respect of minorities, education, competitiveness, environment control, integration of transport networks, and more) should be implemented anyway. All changes to the Turkish society and economy required by the EU could be done by Turkey for her own sake unilaterally or with one-to-one alliances and agreements. However, the EU is a catalyst for the reform process and offers a series of templates to implement the change as well as technical and financial assistance. Turkey has interest in learning from the experience of other large EU countries that are able to align to the Treaty requirements only via the pressure that the EU authorities exercise on the national administration, helping the government to exact heavy tolls that could not be imposed or would not be accepted otherwise [Hoekman et al. (eds.) 2005].

The greatest moral value of the eventual accession of Turkey and of the accession negotiations lies in the approach offered by the EU: the willingness to accompany Turkey in her democratic and economic growth and in her respect of citizens' fundamental rights, in continuity with a process that has been ongoing since the 1960s. The accession talks show that the EU holds a different attitude than that of those countries that consider it possible to export democracy through the instruments of war. The international community must support the Turkish reform and the process towards full membership, for the sake of peace and stability. A negative outcome of the accession negotiations would mean a serious blow from the western world to the whole Islamic community and would mean that 50 years of solidarity and cooperation were in vain.

As a close of this section, a quotation from the Report of the Independent Commission on Turkey: [British Council 2004]

> Admission of Turkey to the European Union would provide undeniable proof that Europe is not a closed "Christian Club". It would confirm the Union's nature as an inclusive and tolerant society, drawing strength from its diversity and bound together by common values of liberty, democracy, the rule of law and respect for human rights. In the great cultural debate of the twenty-first century, all too often fuelled by ignorance and prejudice and misused by criminal phenomena such as international terrorism, a multiethnic, multicultural, and multifaith Europe could send a powerful message to the rest of the world that the "Clash of Civilizations" is not the ineluctable destiny of mankind. Presenting an alternative model to the exclusive, sectarian and closed society propagated by radical Islamists, Europe could play an inestimable role in future relations between the "West" and the Islamic world. The Union would gain wide respect and credibility, enhancing its "soft power" in many parts of the globe. Turkish membership would further give evidence of the compatibility of Islam and democracy.

11.9 Recommended Reading

Recommended reading for this chapter are listed in Appendix C and are available on the web site companion to this book, at the URL http://stajano.deis.unibo.it/RQC.htm

Chapter 12
Budget of the European Union

12.1 Financial Perspectives

The Community budget is drawn up on an annual basis following a seven-year plan – the financial perspectives – that define the objectives and the financial envelop to achieve them. The budget follows a principle of balance establishing that expenditures must not exceed revenues. On a yearly basis, a budget draft for the next year is put forward by the Commission to Parliament and Council, which have authority over the budget and may approve, amend, or reject it. The Court of Auditors checks the legitimacy and validity of the Union's revenue and expenditure and verifies that the Union's budget is properly drawn up.

The Community budget has been growing from € 90 billion in 2000 to € 101 billion in 2004 and € 126.5 billion in 2007. To make a comparison, the federal budget of the U.S. in 2002 was $ 1.9 trillion. The Community budget is in 2006, 2.2% of the overall EU27 public spending[1] of member states. National budgets in the member states have been rising at a rate double of that of the EU in the past seven years. Figure 12.1 and Table 12.1 compare national budgets with the EU budget in the year 2006. National budgets take up an average of around 47% of national GDP, while the EU budget is only at 1.05% of the EU GDP; in absolute terms, the EU budget is comparable to the government expenditure of Austria or Poland.

The Union set a ceiling for her budget, which is 1.24% of the overall GDP of the member states. This ceiling amounted to € 105 billion in 2000 and to € 143 billion in 2006. Some countries are pushing for a reduction of current budget values. At the beginning of 2004, the *Financial Times* [Parker 2004-1] revealed that the Commission would suggest an increase in the budget in the period between 2007 and 2013, in order to increase investments in research by 300% and in transports by 400% and, in this way, fulfil the Lisbon 2000 objectives on competitiveness and employment.

---

[1] Public spending is what the central government, local authorities, and state-owned enterprises spend on goods, services, benefits, financial aid, capital formation, and the payment of interests on public debt.

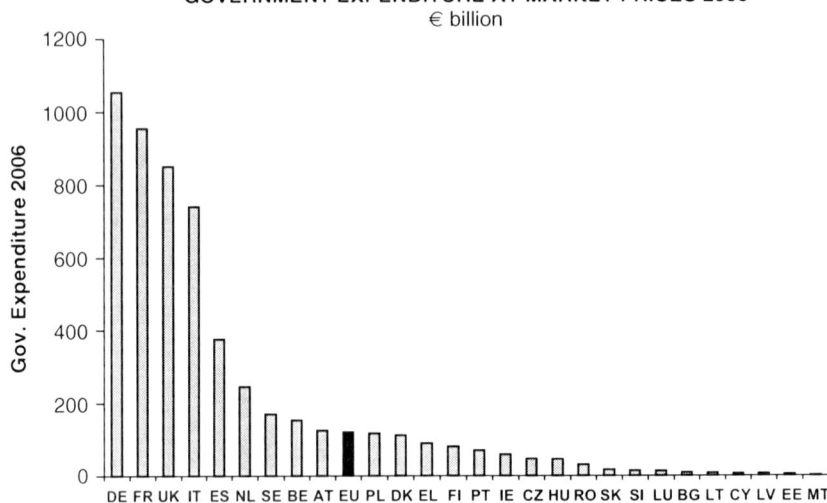

GOVERNMENT EXPENDITURE AT MARKET PRICES 2006
€ billion

Fig. 12.1 EU budget (*black*) compared to member states expenditure (*grey*) in the year 2006 (billion euro). (Eurostat 2007 and EC 2007)

Table 12.1 Comparison of EU budget with member states GDP and member states public expenditure for EU27 in the year 2006

Financial Item	Unit	Value	Year
GDP ppp EU27	€ trillion	11.0	2006
EU Budget	€ billion	112.0	2006
EU27 gov.ts expenditure	€ trillion	5.5	2006
EU27 gov.ts expenditure	% GDP	46.9	2006
EU Budget / EU27 GDP	%	1.02	2006
EU Budget / gov.ts exp.	%	2.06	2006

Source: Eurostat 2007

In February 2004, the Commission laid out a political project for the Union to tackle the key challenges facing Europe and her citizens until 2013. Its objective was to launch a forward-looking debate on the European Union's goals, and the tools required to make these goals happen. The target date of 1 January 2007 was proposed as the start of a new cycle of financial perspectives adapting the community budget to the changing conditions of the Union after enlargement [EC 2004-12].

The Commission did not propose any change in the ceiling on resources to be determined at EU level, which remains at 1.24% of gross national income, but only an increase in the resources used within that ceiling. The current phase of enlargement will add 5% to the Union's GDP – and to its revenues – but with 30% extra population. For example, enlargement means 4 million additional farmers, an increase of 50%. It follows that expenditure would increase more than revenue if the Community policies were not changed to cope with the challenges the Union will face in the years to come. To this end, agreements have been taken to modify the Common Agricultural Policy and the Cohesion Policy, as mentioned earlier in Section 11.6.2, "The Internal Market in the Enlarged Union".

The European Commission proposed in 2004 the Financial Perspectives for 2007–2013 presented in Table 12.2, using three criteria to identify policies where the action should be at Community level [EC 2004-12]:

- Effectiveness: cases where EU action is the only way to get results to create missing links, avoid fragmentation, and realize the potential of a border-free Europe
- Efficiency: cases where the EU offers better value for money, because externalities can be addressed, resources or expertise can be pooled, an action can be better coordinated
- Synergy: cases where EU action is necessary to complement, stimulate, and leverage action to reduce disparities, raise standards, and create synergies.

The financial perspective have been drawn with a view to supporting European competitiveness, education and training, sustainable growth, environment protection, and to taking on new tasks at EU level, in particular in the area of freedom, security, justice, and interventions on crises. The EC proposal was extensively discussed following the codecision procedure and was adopted in the form summarized in Table 12.3 [EC 2006-8].

Changes in the expenditure structure compared to the 2000-2006 period largely reflect the original Commission proposal, with a reorientation of expenditure in favor, in particular, of policies aimed at growth and employment. The changes by heading are as follows:

- 69% increase for growth and employment (sub-heading 1a), including:

  1. 139% increase for transport and energy (TENs)
  2. 81% increase for environment-friendly transport (Marco Polo II)
  3. 75% increase for research (7th Research Framework Program)
  4. 60% increase for the Competitiveness and Innovation Program (CIP)
  5. 52% increase for knowledge/training (Life Long Learning and Erasmus Mundus programs)

- 21% increase for cohesion for growth and employment (sub-heading 1b), including:

  1. 11% increase for structural funds

Table 12.2 Proposed commitment appropriations for the Financial Perspectives 2007–2013

Commitments	2007	2013	2007–2013
Sustainable growth	59.7	76.8	477.7
Competitiveness	12.1	25.8	132.8
Cohesion	47.6	51.0	345.0
Natural resources	57.2	57.8	404.7
Agriculture	43.5	42.3	301.1
Citizenship, freedom, security, and justice	1.6	3.6	18.4
EU as a global partner	11.4	15.7	95.5
Administration	3.7	4.5	28.7
Total	133.6	158.5	1025.1

Source: European Commission, [EC 2004-13]

Table 12.3 Finally adopted commitment appropriations for the Financial Perspectives 2007–2013

Commitments	2007	2013	2007–2013
1. Sustainable Growth	51.27	58.30	382.14
1a Competitiveness	8.40	12.96	74.10
1b Cohesion	42.86	45.34	308.04
2. Natural Resources	54.99	51.16	371.34
of which: agriculture	*43.12*	*40.65*	*293.11*
3. Citizenship, freedom,			
security, and justice	1.20	1.99	10.77
3a Freedom, Security, and Justice	.60	1.39	6.63
3b Citizenship	.60	.60	4.14
4. EU as a global player	6.20	8.03	49.46
5. Administration	6.64	7.61	49.80
6. Compensations	.42		.80
Total commitment	120.70	127.09	864.31
as a percentage of GDP	1.10%	1.01%	1.048%

Source: European Commission, [EC 2006-8].

    2. 74% increase for the Cohesion Fund

- 8% decrease for the preservation & management of natural resources (heading 2)
- 78% increase for citizenship, freedom, security and justice (heading 3)
- 8% increase for the EU as a global player (heading 4)

12.2 Resources

The Community budget financial resources are derived from

- agricultural levies
- customs duties on trade with third countries
- a 0.31% share of the value added tax (VAT) collected in the member states
- a contribution of the 0.73% of member states' GNP[2]
- resources unspent in previous years
- fines collected by member states or undertakings not complying to EU regulations
- taxes collected from EU civil servants

Over the years, the structure of budget revenue and expenditure has been radically modified along with the evolution of the Treaty on EU, with the objective of ensuring the necessary financial resources required by the Union to accomplish her new tasks. At the beginning, the EEC budget was funded by agriculture levies and customs duties on trade with third countries. Starting in 1979, a new budget revenue source was introduced, namely, a tax proportional to the VAT collected in the member states. Starting in 1988, one more resource was added, calculated on the basis of

---

[2] GNP (gross national product) is defined in the Glossary.

Where the money for the EU budget comes from

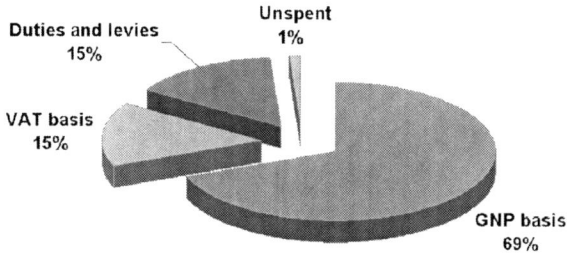

Fig. 12.2 Financial resources in the Community budget 2007. (EC 2007)

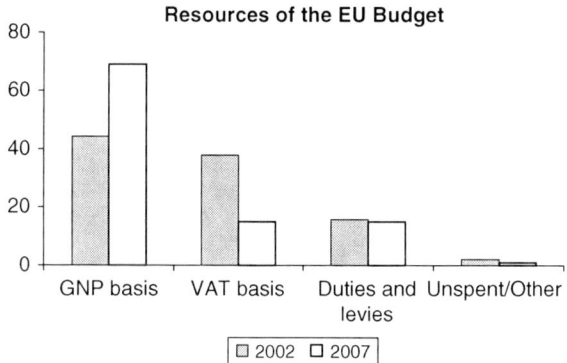

Fig. 12.3 Resources of the Community budget for the years 2002 and 2007, in percentages. (EC 2004)

Table 12.4 Member states contribution to the EU budget, € billion

Member states contribution to the EU Budget 2007			
Belgium	4.47	Luxembourg	0.26
Bulgaria	0.32	Hungary	0.92
Czech Republic	1.21	Malta	0.06
Denmark	2.33	Netherlands	6.35
Germany	22.35	Austria	2.38
Estonia	0.14	Poland	2.70
Greece	2.10	Portugal	1.55
Spain	10.82	Romania	1.07
France	18.18	Slovenia	0.32
Ireland	1.67	Slovakia	0.46
Italy	14.59	Finland	1.65
Cyprus	0.18	Sweden	3.01
Latvia	0.18	United Kingdom	14.23
Lithuania	0.27		

Source: Official Journal of the EU [OJ-EU 2007].

Member states contribution to the EU Budget

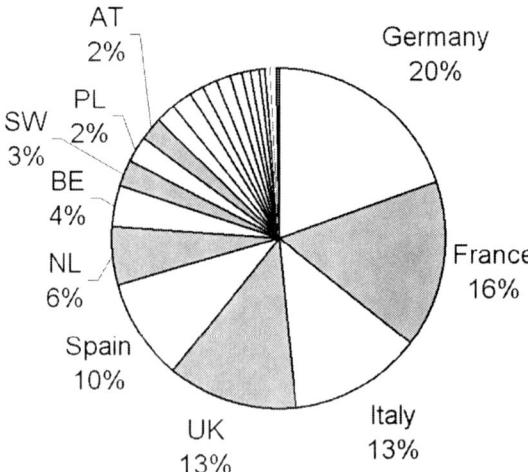

Fig. 12.4 Percentages of contributions from member states to the EU budget in the year 2007. The five largest economies contribute 70% of the budget; the 11 largest economies contribute to 90% of the EU budget; the 12 smallest economies contribute to 5% of the EU budget. The countries with a contribution smaller than that of Austria are, in decreasing order, Denmark, Greece, Ireland, Finland, Portugal, Czech Republic, Romania, Hungary, Slovakia, Bulgaria, Slovenia, Lithuania, Luxembourg, Latvia, Cyprus, Estonia, and Malta. (EC 2007)

the countries' GNPs. The burden of customs duties and levies has been progressively reduced, whereas VAT-based and GNP-based taxes have become more relevant as a budget resource (see in Fig. 12.2 the budget resources in 2007). The resources of the Community budget for the years 2002 and 2007 are presented (in percentages) in Fig. 12.3. The contributions of member states are shown in Table 12.4 and in Fig. 12.4.

12.3 Expenditure

The 2007 Community budget allocations are presented in Fig. 12.5. They are compared with the allocations in 2002 in Table 12.5. This accurate comparison was not easy because the budget lines are given different denominations over the years, perhaps to pretend that the picture is deeply changed and that there is more than just only a proportional increase of each line, which on the contrary appears to be the case. However, a big change has occurred within the definition of the policies: both agricultural policy and cohesion policy have been redirected. The agricultural expenditure, as a percentage of the total budget has been decreasing (see Fig. 12.6). The agricultural policy is moving away from price subsidies with a view

Commitment allocations in the EU Budget 2007

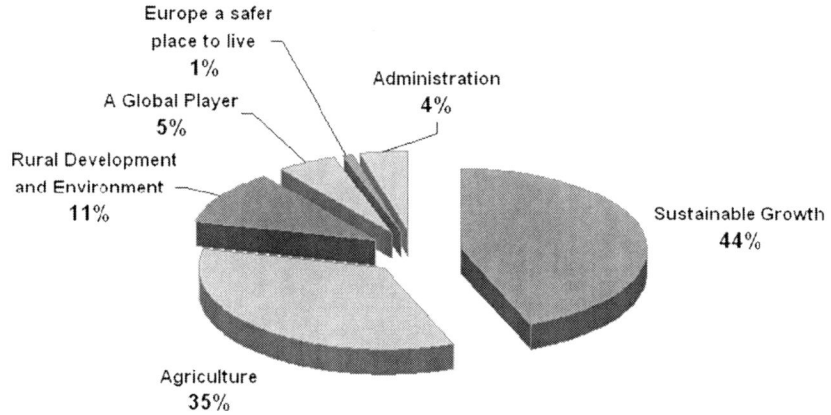

Fig. 12.5 EU Budget: percentages of allocations in 2007. The total expenditure is € 126.5 billion. The 44% of sustainable growth includes cohesion, namely the regional policy (35.5%). Rural development and environment together with agriculture contributes to a budget line "Natural resources". (EC 2007)

to supporting rural development, the environment and the competitiveness of market prices for quality products that protect the local specificities and vocations; the cohesion policy is intertwined with research and innovation and is oriented towards convergence to competitiveness and common high living standards. For details on these policies, see the recommended reading at the end of Chapter 10, "From the Treaty of Rome to the Reform Treaty of Lisbon".

Agricultural and regional policies currently account for the largest share of the Community budget (see Fig. 12.5), although Fig. 12.7 shows a tendency towards a percentage reduction of resources allocated to the agricultural policy and an increasing percentage of budget shares allocated to new policies.

Table 12.5 Community Budget allocations (percentages). The comparison between 2002 and 2007 is complicated by the change of terminology in the budget lines. For example,, external action and administration do not cover exactly the same budget items in the two years

Community Policy	2002		2007	
Agricultural policy	45		46	
Regional policy	35		36	
External action	8		5	
Internal policies	7		7	
including: Research		4		5
Administration	5		4	

Source: EC 2004, 2007.

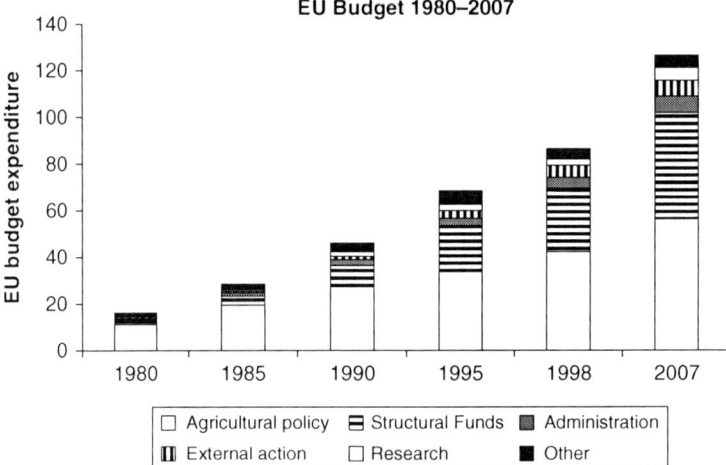

Fig. 12.6 Development of the EU budget in the period 1980–2007. The internal structure of allocations in the EU Budget shows a relative decrease of the allocations to the agricultural policy. Research accounts for about 5% of the total budget in 2007. Structural funds cover the cohesion and regional policy. (EC 2007)

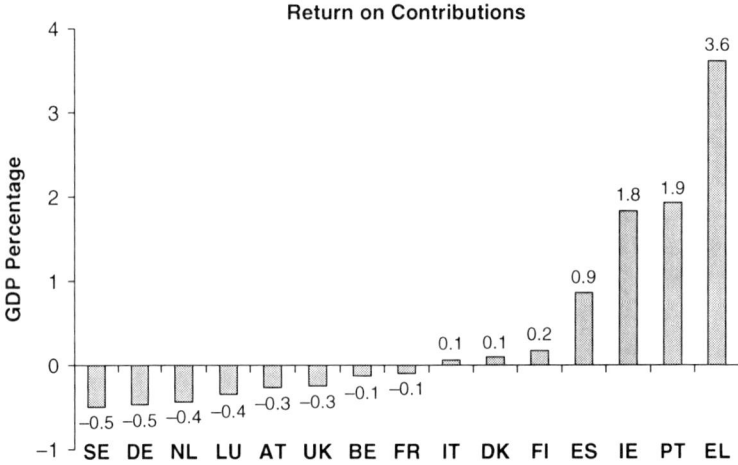

Fig. 12.7 Return on contributions recorded in the Budget in the year 2000 (GDP %). (EC 2002)

12.4 Return on Contributions

Financing of the Community budget is shared by the member states according to the criteria set in the Union's Treaty. In 2002, the five countries with the highest GDP contributed 76% of the Community budget, while the five countries with the lowest GDP only contributed a 6% of the Community budget. In 2007, the five largest economies contribute 70% of the EU budget; the 11 largest economies contribute

Table 12.6 Cumulative net balances of the main contributors and beneficiaries of the EU budget during the period 1986 to 2005 (€ billion)

Balance	DE	UK	FR	ES	EL	IE
Total net balance	−175.0	−40.0	−30.0	+104.0	+56.0	+41.0
Average annual net balance	−8.76	−2.0	−1.5	+5.2	2.8	+2.1

Source: Robert Schuman Foundation. March 2007 [Bréhon 2007].

to 90% of the EU budget; the 12 smallest economies contribute to 5% of the EU budget.

The actions promoted by the European Union are accompanied by a transfer of financial resources to the member states. These actions are not zero-sum, as they are specifically carried out to implement Community policies. In any case, there is a principle providing for a fair return on contributions. The enlargement of the Union goes hand in hand with a reallocation of resources, in order to favour the process of integration and convergence.

In Chapter 11, "Enlargement of the European Union", we presented the effect of present and future enlargements on financial contributions to EU15 countries under the provisions of CAP and Cohesion Policy. Enlargement lowers the average pc GDP of the EU and some regions that were rated "poor" by posting a pc GDP lower than three-fourths of the EU average no longer qualify for a Community financial intervention. This creates vibrations in several EU15 member states, as in some of the richer accession countries when discussing about further enlargements.

Some policies, such as regional and research policies are less directly dependent than others on the principle of fair return on contributions. Figure 12.7 clearly shows that regional policy in 2000 favoured less advantaged countries. The total and the average balance over the years 1986 to 2005 is presented in Table 12.6. The Italian government failed to seize the opportunities offered by the Community through the initiatives implemented under the regional and research policies. Conversely, other countries, such as Ireland, Greece, and Spain, managed to seize these opportunities and derived many benefits (not only financial) for their growth.

Given the importance of this topic, we shall look at *fair return* for research programs in more detail in Chapter 5, "EU Research: Objectives and Results", where we will be dealing with the financing of the European Union's Framework Program for research and technological development.

12.5 Recommended Reading

Recommended reading for this chapter are listed in Appendix C and are available on the web site companion to this book, at the URL http://stajano.deis.unibo.it/RQC.htm

Chapter 13
Internal Market and Competition

13.1 Internal Market

The unification of the member states' markets is one of the instruments for the EU economic integration. Article 2 of the EEC founding Treaty (1957) set the objective

> to promote an even and balanced development of economic activities within the Community, a continuous and balanced growth, increased stability, an ever-more rapid improvement of standards of living and enhanced relations among participating countries.

This goal was to be pursued through the implementation of two complementary measures. This first was the opening of borders, which would lead to free movement of persons, goods, capitals, and services. The second encouraged policies for sustainable growth, supported by financial tools and solidarity among the member states. The 1957 objective was in part fulfilled by the opening of a common market on 1 January 1993. As we shall argue, however, the internal market is a process in the making and the European Union continues to be vigilant against national or private interests that have created in the past years obstacles to its full implementation.

The internal market was instrumental for Europe to become a largely self-sufficient economic space. EU demand is met by 92 % by European supply, while the remaining 8 % is covered by imports from third countries. Moreover, Europe has recently managed to stabilize the level of prices and costs of production and her balance of trade both with the new industrialized countries and with North America is either balanced or favourable. The European Union, after the establishment of the Economic and Monetary Union (1998), has attracted an increasing amount of foreign investments, which jumped from € 20 billion yearly in the early 1990s to over € 140 billion at the beginning of the present decade. Her own foreign direct investments have reached over twice that volume: € 316 billion in the year 2000 [Eurostat 2002]. The change over to the euro in 2002 has attracted additional FDI in the Euro Area, where they increased in 2005, by 32 % over the value in the previous year up to $ 315 billion, while in the rest of the World they declined by 8 % [World Bank 2007-2], due to the economic slow down following 9/11.

The progress towards the creation of the internal market is based on four principles: free movements of people, goods, services, and capitals. Overall, the internal market has resulted in real benefits; for instance, in the 10 years since the completion

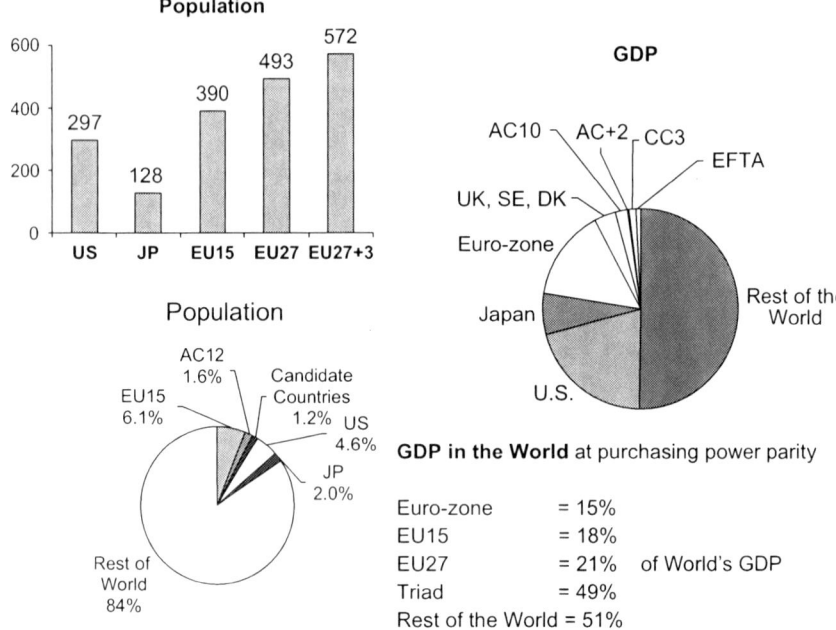

Fig. 13.1 EU in the world: Population (2004) and GDP (2006). EFTA is Iceland, Norway, the Swiss Confederation, and Liechtenstein. AC10 are the 10 countries that accessed the EU in 2004 and AC+2 the two that accessed in 2007. CC3 are the three countries for which accession negotiations are ongoing in 2007. The population of EU27 exceeds the sum of that of the U.S. and of Japan. These three regions (the Triad) produce together half of the world's GDP and have one-sixth of the world's population. The exchange rate of the dollar to the euro may introduce some differences from one year to the next in EU–U.S. comparisons. (International Monetary Fund [IMF 2007], Eurostat 2007, and Economist [Economist 2007-1])

of the first Single Market program in 1993, at least 2.5 million extra jobs have been created as a result of the removal of barriers. The increase in wealth attributable to the Internal Market in those 10 years is nearly € 900 billion; on average about € 6,000 per family in the EU. Intra-Community trade and competition have increased as companies find new markets abroad; prices have converged (in many cased downwards) and the range and quality of products available to consumers have increased. With its 493 million consumers, today the EU27 internal market is the world's largest market (Figs. 13.1 and 13.2). The size of the European foreign trade is comparable in value to that of the U.S., but the European trade balance is favourable, while the U.S. trade deficit reached $ 665 billion in 2004 and $ 759 billion in 2006 [EC 2007-8, Bivens 2004, US Census 2007].

13.1.1 Free Movement of People

The principle of *free movement of people* dates back to the creation of the European Economic Community. This principle was initially introduced to open labour

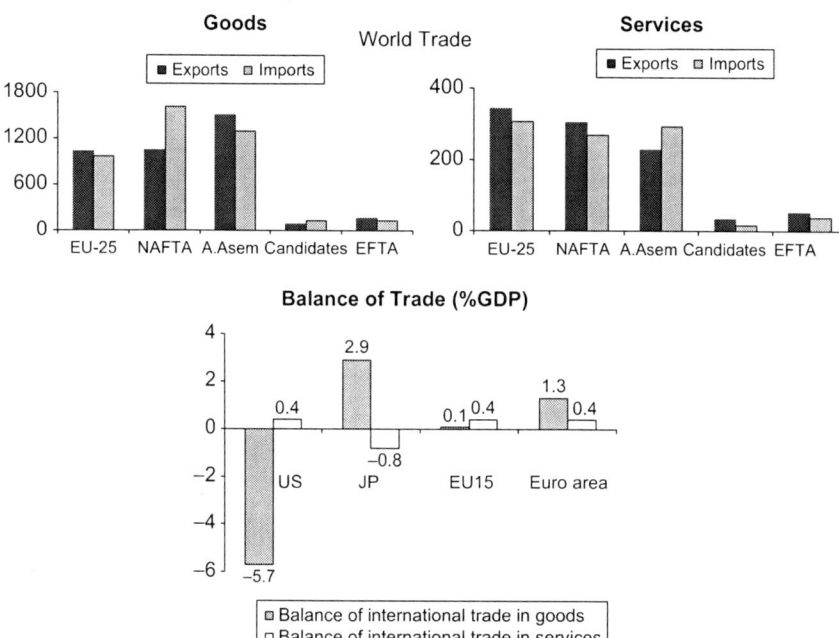

Fig. 13.2 EU and the world trade. Export and Import of good and services in € billion for 2004. The size of EU trade is comparable to that of North America, but the trade balance is favourable only in the EU, and notably in the Euro Area. The exchange rate of the dollar to the euro may introduce major differences from one year to the next in EU–U.S. comparisons. NAFTA is the North Atlantic Free Trade Association; A. Asem stands for Japan, China, Korea, and Southeast Asia; Euro Area are the 12 countries that adopted the euro in 2002 (see Chapter 14, "Economic and Monetary Union,"). (EUROSTAT 2007, WORLD BANK 2007, and European Commission 2007)

markets to migrant workers and, over the years, this right was extended to cover all categories of citizens. Today, with the lifting of most internal border controls, European citizens can move as freely around Europe as they could within their country and if they so choose, they can decide to study, work, or retire in another EU country. The right to cross-border mobility has become a reality that must be enjoyed in conditions of security so as to prevent criminals from taking advantage of a European space without frontiers. On the other hand, the remaining legal and practical obstacles must be removed in order to simplify benefiting from the freedom of movement and right to reside and work in another member state. The European policy on the mutual recognition of professional qualifications is an example of action undertaken by the Commission to remove the remaining legal and practical obstacles that deter people from benefiting from the freedom of movement.

13.1.2 Free Movement of Goods

The *free movement of goods* was achieved with the elimination of custom duties when trading goods within the EEC. A second step was made in 1982 with the customs agreement, an agreement on common external tariffs to trade with third countries. Member states may restrict the free movement of goods only in exceptional cases, for example, when there is a risk resulting from issues such as public health, environment, or consumer protection. The risks vary by product sector. For example, pharmaceuticals and construction products obviously present higher risks than office equipment or pasta. In order to minimize risks and ensure legal certainty across member states, EU legislation harmonizing technical regulations has been introduced in particular in the higher-risk product sectors. However, there remained *non-tariff barriers* – such as differences between the member states' safety and packaging requirements or between national administrative procedures. These differences in practice prevented manufacturers during many years from marketing exactly the same goods all over Europe.

13.1.3 Freedom to Provide Services

Services are crucial to the European Internal Market: they are everywhere, accounting for 60–70% of economic activity in EU25, and a similar (and rising) proportion of overall employment. Commissioner Frits Bolkestein proposed a directive (adopted by the Commission on 13 January 2004 [EC 2004-8]) on the liberalization of services in the internal market. This legal framework would eliminate obstacles to the establishment of service providers and to the *free supply of services* between the member states. The directive, which has been widely debated in the member states, was transmitted in 2004 to Council and Parliament, where there were animated discussions during 2005. An internal market for services is a prerequisite for the achievement of the Lisbon objectives and the enhancement of EU competitiveness. However, the directive has received some criticism because the proposed legal framework for the liberalization of services is perceived as an incentive to delocalize economic activities towards countries with cheaper labour costs. The *Polish plumber* became an epitome of the fears of the western European service providers. One of the concerns, expressed in particular in France and Italy, is the fear of social dumping. Unions fear that service providers established in EU15 might delocalize proposing work contracts to their present staff from those countries of AC10 with lower labour cost and weaker workers protection. They fear also that providers originally established in AC10 might create unsustainable competitive pressure when offering services in the countries of EU15 where there are higher standards. The National Secretariat of the Italian trade union FIOM-CGIL declared on 27 October 2004 [Fiom 2004]:

> The directive undermines the principles of solidarity and equality, the extension of social and labor rights, [. . .] and asserts the principle of the most savage competition in the field of services. [. . .] The worst lies in article 16, concerning the principle of the countries of origin. According to this new principle, a supplier of services is subject exclusively to the laws of the country where the headquarters of the companies are located and not to the ones of the country where the service is being provided. An enterprise can engage workers and

then transfer them to another state, maintaining the laws, contracts, security, and control rules of the country of origin.

On 19 March 2005, the European trade unions organized a parade against EU Commissioner Bolkestein, who was responsible for the proposed service directive. Sixty thousand people marched in the streets of Brussels, and cried out at the top of their voices "Bolkestein–Frankenstein" to defend the security of their service jobs and the Union's social model. However, other voices have spoken in favour of the directive [Badré et al. 2005]:

> Far from being a factor of weakening the Union's social model, an internal market enlarged to services is the opportunity for economic and social development and for enrichment. Tomorrow's Europe must be more united, and continue to be highly demanding also in the area of services.

The Services Directive 2006-123 was finally adopted by the European Parliament and the Council in December 2006 [OJ-EU 2007] and has to be transposed by the member states by the end of 2009. This directive is aimed at eliminating obstacles to trade in services, thus allowing the development of cross-border operations. It is intended to improve the competitiveness not just of service enterprises, but also of European industry as a whole. It will remove discriminatory barriers, cut red tape, modernize, and simplify the legal and administrative framework – also by use of information technology. This guarantees to EU companies the freedom to establish themselves in other member states and the freedom to provide services on the territory of another EU member state other than the one in which they are established.

13.1.4 Free Movement of Capitals

Free movement of capital is also an essential condition for the cross-border activities of firms and in particular of financial-services companies and for the free migration of workers. Not so long ago, Europeans were in principle obliged to manage and invest their money predominantly in their home country and migrant workers could not sell their home and transfer the capital to the hosting country. Now, further to the liberalization of capital movements and payments which has accompanied the consolidation of the single market, EU citizens can conduct most operations abroad, as diverse as opening bank accounts, buying shares in non-domestic companies, or purchasing real estate. However, the rules concerning some of these rights remain governed by national provisions which vary from one member state to another. Free movement of capital is an essential condition for the proper functioning of the Internal Market. It enables a better allocation of resources within the EU, facilitates trade across borders, favours workers mobility, and makes it easier for businesses to raise the money they need to start and to grow. Indeed, the effectiveness of EU initiatives in the financial services sector would be compromised if capital movements within the EU were subject to restrictions.

In 2005, the Commission completed the legislative phase of an action plan aimed at developing a true European-wide market in financial services and is now implementing a new strategy to deepen financial integration and deliver further benefits to industry and consumers alike. A more developed single market in financial services will provide consumers with a wider choice of financial products – such as loans, insurances, saving plans, and pensions – which they will be able to buy from anywhere in Europe. It will also make it easier and cheaper for companies to borrow money, bringing down the cost of capital, goods and services for everybody, and make member state administrations co-operate much more systematically. It will also strengthen the rights of users of services.

13.1.5 The Completion of the Internal Market

The Economic and Monetary Union and the single currency are great EU achievements that complete the construction of the internal market. They are addressed in next Chapter 14, "Economic and Monetary Union,". Other EU policies contribute to the internal market policy: consumer protection, standardization, transborder research, and not least, the competition policy, which is addressed in the next section.

European legislation does not suffice to implement the internal market: once the directives are adopted, they need to be transposed in the 27 national legislations and enforced. We saw in Chapter 10, "From the Treaty of Rome to the Reform Lisbon Treaty," that the delay in transposition and the control on application of Community legislation are problem areas addressed and monitored by the Commission. In July 2007, Charlie McCreevy, Commissioner responsible for the Internal Market, wrote in the Internal Market Scoreboard [EC 2007-6]:

> The Internal Market is a joint effort between the EU and member states, whose job it is to translate agreed Internal Market rules into national law. The Internal Market Scoreboard records twice a year whether member states have done this on time.
>
> This edition of the Scoreboard shows that the gap between the number of Internal Market laws adopted at EU level and those in force in the member states – known as the 'transposition deficit' – has risen to 1.6%. At first glance, this is disappointing news. It means that member states are relaxing their efforts again, having posted their best-ever result of 1.2% only six months ago.
>
> Nevertheless, I see reasons to be optimistic about the future. Member states will have fewer directives to implement than over the last six months, while nine member states have already reached the new target deficit of 1% decided recently by the European Council. On the whole there are signs that member states will be back on track in six months' time, when we publish the next Scoreboard.
>
> However, in far too many cases member states continue to apply Internal Market rules incorrectly, meaning in practice that citizens and businesses are being denied the very benefits to which their governments have themselves agreed. Infringement proceedings are costly and time-consuming, so wherever possible we try to avoid making use of them. For example, the EU's 'Solvit' network, set up in partnership with member states, offers a quick and effective means of redress to citizens and businesses affected by the misapplication of Internal Market rules.

The average transposition deficit for EU25 declined from 3.6% in December 2004 to 1.2% in December 2006, and then went up to 1.6% in July 2007, a value that goes to 1.8% for EU27 for the bad performance of the two new members. The average transposition delay is 8 months, with different performance in the various countries ranging from 3 to 12 months. In order for the internal market to function properly each directive must be enforced in every member state. The *fragmentation factor* records the percentage of the overall outstanding directives that have not been transposed in at least one member state. It went down to 8% in July 2007 from 27% in 2004. However, the number of infringements is increasing and averages 53 cases per country, with a maximum of 153 for Italy and a minimum of 17 for Slovenia. The average resolution time for infringements is 26 months for EU15 and 9 months for AC10.

13.2 Competition Policy

To compete means striving together to provide consumers with products that they want. It is a basic mechanism of the market economy that encourages innovation and pushes down prices. In order to be effective, competition needs suppliers who are independent of each other, each subject to the competitive pressure exerted by the others. EU competition policy ensures fair rivalry in the markets of all member states. It guarantees the unity of the internal market, prevents the creation of monopolies, and protects consumers.

The goals of the competition policy are closely linked to the realization of the internal market:

- promoting economic efficiency by creating an environment favourable to innovation and technical progress
- protecting consumers' interests by creating the best conditions for them to buy goods and services
- prohibiting businesses and national authorities from unfair competition

The EU had set different rules to regulate competition, which are presented in the following sections [EC 2007-7].

13.2.1 State Aid

The objective of state aid control is to ensure that government interventions do not distort competition and trade. The control is exercised on any type of financial government assistance, such as a grant, tax break, or trade barrier conceived to encourage the production or purchase of a good or service. The competition policy prohibits state aid when it is an advantage conferred by national public authorities to firms on a selective basis. On the contrary, subsidies granted to individuals or general measures open to all enterprises do not constitute state aid. We will see in

Section 13.6, "Research, the Internal Market, and Competitiveness" under which conditions funding industrial research is compatible with the state aid regulations of the competition policy.

The member state administrations are often tempted to propose state aid to defend national champions in the international arena or to support enterprises that might not win the international competition, for the purpose of protecting job security. Such measures, often inspired by a populist or patronage approach to economic policy, may offer short term relief to structural deficiencies, but they are flashes in the pan that generate worse problems in the medium term. In Italy, Alitalia and Fiat are typical examples of state aid that plugs a hole but buys problems for the future.

The EC Treaty pronounces the general prohibition of state aid, but there are provisions for exemptions. The European Commission is carrying out effective state aid control in all sectors, including as an example, transport (aid to companies in the road, rail, inland waterway, sea and air transport sectors), fisheries (the production, processing and marketing of fisheries and aquaculture products), and agriculture (the production, processing and marketing of agricultural products.

13.2.2 Antitrust

Antitrust exercises control over monopolies and over practices that restrict competition, such as price discrimination or exclusive dealing. The Treaty on EU defines two prohibition rules in the area of antitrust: agreements between two or more firms which restrict competition and abuse of dominant position with a view to increasing prices or eliminating competitors. The Commission is empowered to apply these prohibition rules and enjoys a number of investigative powers to that end. It may also impose fines on undertakings that violate EU antitrust rules. The fines collected when infringements are discovered contribute to the EU budget.

13.2.3 Mergers

A merger is the fusion into a new firm of two or more previously independent companies. The merger is called an acquisition if one firm has been bought by the other and is called a take-over if there has been resistance by the acquired undertaking. Mergers are referred to as *vertical* if they result of the union of a supplier and one of its customers; they are called *horizontal* if the two companies are in the same business. Both vertical and horizontal mergers may expand markets, bring benefits to the economy, and lead to economies of scale, more efficiency, and to reduction of production or distribution costs. Companies merge and join forces to face globalization and the increased competition within the European single market. Such reorganizations are welcome to the extent that they do not impede competition and hence are capable of increasing the competitiveness of European industry, improving the conditions of growth and raising the standard of living in the EU. However, some mergers may reduce competition in the market by creating a dominant position. This is likely to harm consumers through higher prices, reduced choice, or less innovation. The European Commission studies thoroughly proposed mergers to prevent harmful effects on competition. Mergers going beyond the national borders of any member state are examined at European level, allowing companies trading in different states to obtain clearance for their mergers in one go.

The European Commission examines mergers if the annual turnover of the combined businesses exceeds specified thresholds in terms of global and European sales no matter where in the world the merging companies have their registered office, headquarters, activities, or production facilities. In fact, even mergers between companies based outside the European Union may affect markets in the EU if the companies do business in the EU.

Mergers are prohibited if they would significantly impede effective competition in the EU. If they do not, they are approved unconditionally. If they do, and no commitments aimed at removing the impediment are proposed by the merging firms, they must be prohibited in the Community interests, with a view to protecting businesses and consumers from higher prices or a more limited choice of goods or services. Proposed mergers may be prohibited, for example, if the merging parties are major competitors or if the merger would otherwise significantly weaken effective competition in the market, in particular by creating or strengthening a dominant player. If the European Commission finds that a proposed merger could distort competition, the parties may commit to taking action to try to correct this likely effect and obtain conditional clearance for the merger to go ahead.

13.2.4 Cartels

A cartel is an association of producers to regulate prices by restricting output concluded between competitors who in coordination charge higher prices, restrict supply by limiting their sales or their production capacities, and/or divide up their markets or consumers. Cartels are illegal and harmful to consumers and business because they shield their participants from competition. Customers companies and consumers end up paying higher prices for lower quality and narrower choice. This also adversely affects the competitiveness of the economy as a whole.

The Treaty on European Union prohibits agreements and concerted practices between firms that distort competition with fines of up to 10 % of their worldwide turnover. Such fines reduce the tax burden on citizens because they finance the European Union's budget.

Cartel decisions by the Commission may be appealed against before the Court of First Instance and then before the Court of Justice in Luxembourg. These two courts are empowered to annul decisions in whole or in part and to reduce or increase fines, where this is deemed appropriate.

13.2.5 Liberalization

In the EU member states, transport, energy, postal services, banking services, and telecommunications had previously been provided by national organizations with exclusive rights to provide a given service. Such services became cheaper and more efficient since they have been open to competition. The European Commission has been instrumental in liberalizing these markets by opening them up. By opening up these markets to international competition, the EC empowered consumers to choose from a number of alternative service providers and products. Consumers benefit from lower prices and new services which are usually more efficient and consumer-friendly than before.

The European Commission issues the regulation required to ensure that in the liberalized market where new undertakings have replaced state-controlled national organizations, public services continue to be provided and that the consumer is not adversely affected. In air transport and telecommunications – the two markets which were opened up to competition first – average prices have dropped substantially. This is not yet case for electricity, gas, rail transport, and postal services, where prices have remained unchanged or have even increased.

13.2.6 International

With increasing globalization, more and more companies, mergers, and cartels are international. As a result, the activities of companies based outside the EU may affect competition within the EU. This has made international cooperation on competition policy essential. The EU has established bilateral agreements on competition, in particular, with its main trading partners. It has also been at the forefront of multilateral cooperation efforts, for example, being among the first to propose the inclusion of competition policy as a subject for discussion in the World Trade Organization, and playing a key role in the International Competition Network and at the OECD Competition Committee.

13.3 Transformations Introduced by the Internal Market

Trade in the European Community before the full accomplishment of the internal market was characterized by customs barriers, bureaucracy, and protectionism. The main obstacles to the free movement objectives set by the Treaty of Rome were the following:

- Customs inspections caused delays, higher costs, and inefficiency.
- Products had to comply with each member state's differing laws and standards. The enlargement of the market did not in fact produce economies of scale because products had to be custom-made to suit each market's requirements. For example, every car model had to be manufactured to meet the safety and signalling norms of each member state.
- Barriers created by state administrations protected basic services such as transport, telecommunications, energy, mail, banks, and insurance, which

consequently were not subject to competition. Such barriers were an obstacle to the reduction of tariffs and charges and to improvement in the services provided.

- Restrictions on change of residence, limits on the use of social security benefits in a member state other than the state of origin, restrictions on the export of savings, and nonrecognition of educational and professional qualifications hampered the free movement of citizens.
- Each country had its own currency and exchange rates fluctuated. Managing different currencies was quite expensive for transnational companies, which were subject to exchange risk. Trade and international tenders favoured national operators, who were not penalized by exchange risks and bank commissions on international payments. Price comparison was complicated by exchange rates and their fluctuation, which led to a loss in price transparency and to price differentials, each of which was a detriment to consumers.

13.4 Policies for the Creation of the Internal Market

The internal market creates all the necessary conditions for the free movement of labour, harmonizes standards, and prepares the ground for fiscal harmonization. The creation of the internal market must not neglect the respect and promotion of regional and local identities and the protection of local traditions, languages, and typical agricultural products and food. The realization of the internal market goes hand in hand with measures to improve the quality of life through the implementation of health controls and by ensuring law and order. These measures are necessary to make sure that the free movement of persons and goods does not lead to a rise in crime and in human and animal diseases.

The creation of the internal market is based on two pillars, namely, the principles of *subsidiarity* and *mutual recognition* [Bianchi 1999]. Subsidiarity is a principle suggesting that actions should be taken at the EU level only if the same goals cannot be achieved at a lower level, nearer to citizens. This principle also guides the attribution of powers to the local administrations, as opposed to the states' central governments. Subsidiarity is a tool to achieve harmonization, decentralization, and democratic participation. It makes the European administration much more streamlined than the national and local administrations; it makes possible a Community budget that is only 2 % of the overall public spending of member states and ensures that the cost of European institutions is lower than 0.1 % of the overall public spending of member states. The Treaty on EU identifies areas of exclusive competence of the EU, areas where the member states share their competence with the EU, and areas where EU action is limited to harmonization and coordination.

Mutual recognition is the extension of the application to the entire EU of rules originated in one member state. The adoption of mutual recognition results from a resolution of the Court of Justice dating back to 1979 on the *Cassis de Dijon* case [CoJ 1979]. This resolution provides for the enforceability throughout the Union of the rules promulgated by one state (e.g. the rules on consumer protection or security)

on issues decided by the European Council. Mutual recognition has proved to be an effective tool to dismantle some nontariff trade barriers artificially created to protect national products or enterprises. It follows that

- consumer protection starts at the source
- market segmentation is based on quality rather than origin
- competition is based on quality rather than price
- the dynamics of adaptation among the states leads to an upgrading of the standards and stimulates positive competition
- the most innovative states and enterprises are at the forefront of creativity and competition

Subsidiarity and mutual recognition create synergies and progress and give voice to all government levels, namely, the EU, the states, the regions, and the local authorities.

Because of some opposing political and economic forces that protected the dominant positions of the *national champions,* the realization of a borderless internal market was a slow and gradual process. The delay was caused primarily because some member states failed to implement Community directives on the creation of the internal market. At the end of the 1980s, the Commission decided to set a tight deadline, namely, the full accomplishment of the internal market by no later than 1992.

In 1988, Paolo Cecchini, a retired EC civil servant, published an influential study on the cost of *non-Europe* [Cecchini 1988]. He calculated that the elimination of nontariff barriers would stimulate domestic demand and lead to a decrease in consumer prices, an increase in volumes of domestic demand, and a € 200 billion in savings. Twenty years after the publication of Cecchini's book, the European Commission estimates that in 2006 the wealth creation due to the Internal Market was € 240 billion [EC 2007-8]. The virtuous circle described by Cecchini is characterized by reduced production costs, and increased competition and innovation, which gives a boost to the market and benefits final consumers.

In addition to the principles of subsidiarity and mutual recognition, the implementation of the internal market was also achieved through a more efficient decisional process within the EU and through policies for sustainable growth. The adoption of qualified majority voting in the Council avoided the risks that a single state could protect its national interests by the veto right. Sustainable development policies, in particular the research and innovation policy, created synergies with the internal market policy and accelerated its implementation, as we showed in the second part of this book.

13.5 Impact of the Internal Market

Trade in the European Community, in the 15 years since 1993, has not yet been characterized by the full accomplishment of the internal market, primarily because some national governments are not very keen on shared sovereignty and on losing their autonomy in areas such as economic and fiscal policy or in areas where the prosperity of national champions might be at risk. (In Section 13.8, "The Way Ahead" we will look closely at a specific example concerning the European car market.) Nevertheless, many improvements have been made that have proved beneficial to consumers. Among these are the following with respect to the list of obstacles presented in Section 13.3, "'Transformations Introduced by the Internal Market": at present

- Goods can normally be exported without any customs restrictions, resulting in increased efficiency, savings, and a reduction in delivery times.
- The harmonization of standards makes the enlargement of the market effective, facilitates the arrival of new players, and generates economies of scale.
- Basic services such as transport, telecommunications, energy, mail, banks, and insurance have been liberalized and are now subject to competition; consumers benefit from an improvement in the services provided and a reduction in tariffs and service charges.
- Free movement of persons is easier, as savings can be exported and educational and professional qualifications are recognized throughout the whole Union.
- The introduction of the single currency has eliminated exchange risk and commissions; the management of payments has been simplified to the advantage of transnational companies; the equal participation of all players in international trade and tenders is ensured; it is now easier to compare prices thanks to the single currency's stability; and transparency benefits consumers.

Additionally,

- competition has been enhanced and extended to the larger internal market;
- a wider range of products and services is available at lower prices, particularly in the liberalized sectors;
- there is increased labour mobility among member states; and
- because of contracts, competition, takeovers, and mergers, the process of company restructuring has been accelerated.

13.6 Research, the Internal Market, and Competitiveness

In the second part of this book, we examined the ways in which Community-financed research has played, and is still playing, an important role in the realization of the internal market. Here, we resume the main considerations:

- international cooperation on research kindles a process of adjustment towards higher standards;
- corporate business extends beyond national borders through cooperation on research, which results in a better knowledge of new markets, new types of demand, new commercial practices, and new legal frameworks;

- cooperation influences technology and leads to positive competition. There is a transfer of know-how and expertise from the most innovative and creative businesses;
- the pace of innovation, de facto standards, and competition rules are set by the most innovative countries and businesses;
- transnational R&D alliances accelerate the pace of corporate restructuring and stimulate mobility of the active population.

13.7 Competition Policy and R&D Funding

Public funding of R&D programs is compatible with the competition policy and does not constitute a state aid to the extent that the R&D programs are open to any undertaking without restrictions or discriminations, and that programs are aimed at standardization or are precompetitive in nature, namely, addressing strategic long term developments well ahead of design and development of products or processes. The competition policy is compatible with the funding of industrial collaboration in the development of standards and in basic technological research that may later lead the individual industrial groups to independently integrate new technological developments into their product lines. As a consequence, collaboration between industrial undertakings finalized to the feasibility study and demonstration of new technological breakthroughs are compatible with the competition policy, while the industrialization of prototypes, or any action connected to marketing and advertising are not compatible.

13.8 The Way Ahead

There is more to be accomplished. In 2002, the EU Commissioner for Competition, Mario Monti, recommended the creation of a single internal market in the car sector [Tarquini 2002]. In an interview in *La Repubblica* on 16 February 2002 [Polito 2002], Commissioner Monti explained that differences in car prices in the member states would be reduced by improving competition and liberalizing the car market. In 2002, the price gap between the same car models in two different countries was as high as 60 % of the retail price. Until then, each car manufacturer has adjusted the price of its models to fit different final markets in order to maximize market shares and profits. As an example, in 2002, the price of the Alfa Romeo 147 ranged from € 15,097 in Greece to € 19,060 in the UK (+26 %). Commissioner Monti proposed a regulation on the distribution and assistance of motor vehicles (CE 1400/2002) that allows the purchase of a car in a country other than the country of residence. This action was expected to harmonize and to lower car prices; however in 2007, it is still not possible to buy and insure a car in a different country from that of residence and the price for the same model, Alfa Romeo 147 ranges from € 14,562 in Bulgaria to

€ 19,283 in the UK (+33 %), although for some other models the spread of prices has narrowed.

13.9 Recommended Reading

Recommended reading for this chapter are listed in Appendix C and are available on the web site companion to this book, at the URL http://stajano.deis.unibo.it/RQC.htm

Chapter 14
Economic and Monetary Union

Economic and Monetary Union (EMU) is the natural completion of the internal market and it is a major political step forward in the process of European integration. The unification of currencies, which for many centuries had been symbols and instruments of sovereignty for many European countries, is an unprecedented achievement in history after the Roman Empire and a unique undertaking at the world level.

A few months before being induced to resign as Italian foreign minister, because of his disapproval of the attitude of some government ministers in the Berlusconi cabinet towards Economic and Monetary Union, Renato Ruggiero made the following statement on 31 October 2001 to the Italian Parliament in the run-up to the euro's arrival:

> The single currency is not a technical tool but rather a fact with a strong political impact. On the one hand, the euro will have a clear psychological effect on citizens as they will feel their European identity and sense of belonging much more strongly. On the other hand, the euro will require a tighter control over the economy. I don't think we need to set up new institutions. What we need is a much faster process of political coordination in order to ensure stability and growth everywhere in Europe and to avoid leaving the central banker alone, which is something completely different from ensuring his independence.

The euro benefits individuals and business. Individuals make savings by not having to change money when travelling within the Euro-zone. Consumers can compare prices more readily and this promotes price competition. Businesses reap these same benefits and more. Other countries will more readily accept invoices in euro than most of its predecessor currencies, thus reducing their foreign exchange risk.

14.1 History of Monetary Integration

In 1970, the Werner report [Council 1970] proposed the construction of a three-stage economic and monetary union that would take place within 10 years. However, the member states' will to create such a union was delayed by the first oil crisis.

The European exchange system, known as the *European monetary snake*, was established in 1972. In 1974, the Council adopted a decision concerning the creation of a tight macroeconomic convergence within the Community, as well as a directive concerning stability, growth, and full employment. However, growing economic instability gradually undermined the basis on which the system had been founded and the British Pound Sterling, the French Franc, and the Italian Lira were forced out of the monetary snake.

An area of monetary stability was eventually achieved in Europe after the adoption (1979) of the European Monetary System (EMS). The EMS favoured growth and investment and laid the ground for the Economic and Monetary Union.

The EMS was based on three main elements:

1. The European Currency Unit (*Ecu*, later on the *Euro*): Conceived as the core of the system, it is a weighted *basket* including a specified amount of all the member states' currencies.
2. *Exchange mechanisms:* Each currency was pegged to the Ecu. Central rates were used to set a grid of bilateral central rates. Until August 1993, fluctuation margins of 2.25 % (exceptionally, up to 6 %) were permitted for bilateral rates. Later on, such margins were widened to $\pm$ 15 %, following the profound transformations that had occurred in the exchange markets.
3. *Intervention mechanism:* As soon as a bilateral exchange rate reached the $\pm$ 15 % threshold, central banks had to intervene and take all possible measures to make sure that rates did not exceed this threshold.

The EMS could not achieve its full potential because some member states either did not join the exchange mechanism or participated with wider fluctuation margins. Owing to the inadequate convergence of budgetary policies and competitive devaluations, the unity of the internal market was seriously threatened.

In June 1989, based on the report presented by the Commission president Jacques Delors, the Madrid European Council established the general principles for the achievement of an EMU, namely, the objectives of a single currency and parallelism between monetary and economic policies.

The treaty signed in Maastricht in 1992 ensured an irreversible progress towards a single currency by fixing the criteria that member states have to meet to qualify for the adoption of the single currency. Finally, the Amsterdam Treaty (1997) enshrined a resolution that bound all member states to comply with a Stability and Growth Pact, which committed them to budget discipline. Such discipline was to be ensured by multilateral surveillance, sanctions for the defaulting parties, and by setting a limit to each state's budget deficit and public debt.

14.2 Criteria for Qualification

All EU member states are part of the Economic and Monetary Union, whose purpose is to integrate the economies of EU countries more effectively; all EMU members are eligible to adopt the euro, provided that they comply with the criteria

Fig. 14.1 Euro: Currency of the future. (European Commission, 2001)

outlined below. Also EU members that have not adopted the euro must comply with the provisions of the Stability and Growth Pact. In 2007, 13 countries have taken European integration a major step further by adopting the same currency, the euro (see Fig. 14.1). This group is called the Euro-zone or the Euro Area and is constituted by Austria, Belgium, Finland, France, Germany, Greece, Ireland, Italy, Luxembourg, Netherlands, Portugal, Slovenia, and Spain. Two countries, Cyprus and Malta introduced the euro on 1 January 2008. Denmark and the United Kingdom have opted to remain outside the euro for the time being, while no target date for joining is defined for Bulgaria, the Czech Republic, Estonia, Hungary, Latvia, Lithuania, Poland, Romania, Slovakia, and Sweden. Of this group, only Sweden qualified for adoption so far. The criteria for qualification for the adoption of the euro as set out in the Union's treaty are listed below. They were applied in 1998 in order to define which states would qualify for membership and are also applicable to those states that may ask to adopt the euro in the future (e.g. the United Kingdom or enlargement countries).

1. *Price stability*: In the member states seeking the adoption of the euro, the rate of inflation must not exceed by more than $1^1/_2$ percentage points that of the three best-performing member states.
2. *Budget towards balance:* The government budget deficit must not exceed 3 % of GDP.
3. *Public debt* must be under control: The ratio of government debt to GDP must not exceed 60 % at the end of the preceding financial year. If this is not the case, the ratio must have sufficiently diminished and must be approaching the reference value at a satisfactory pace.

4. National currency *exchange rate stability*: The Treaty stipulates "the observance of the normal fluctuation margins provided for by the exchange-rate mechanism of the European Monetary System, for at least two years, without devaluing against the currency of any other member state." The member state must have participated in the exchange-rate mechanism of the European monetary system without any break during the two years preceding the examination of the situation and without severe tensions.
5. Bank *interests in line* with the European average: The nominal long-term interest rate must not exceed by more than two percentage points that of the three best-performing member states.

14.3 Introduction Stages of the Euro

March 1998: The first group of member states of the European Union that qualified for membership in the economic and monetary union was defined. Fourteen states qualified: Austria, Belgium, Denmark, Finland, France, Germany, Ireland, Italy, Luxembourg, The Netherlands, Portugal, Spain, Sweden, and the United Kingdom. Greece qualified in January 2001. Two countries (Denmark and the United Kingdom) decided to opt out of the Euro-zone. Sweden did not fix so far a date for joining. Therefore, the *founding members* are 11, plus Greece, which joined in 2001.

Since 1 January 1999:

- exchange rates have been irrevocably fixed,
- currency markets have been transformed,
- the legislation on the EMU has entered into force,
- the European Central Bank has come into operation,
- bonds have been issued in Euro, and
- current accounts and bank transfers can be in Euro.

Since 1 January 2002:

- euro banknotes and coins have been put into circulation,
- national banknotes and coin (Marcs, Florins, etc.) have been replaced,
- all bank accounts have been in Euro, and
- the euro has been used for wages, pensions, and retailing.

Since 28 February 2002:

- the transition to the euro has been completed,
- national banknotes and coins have been out of circulation,
- cash dispensers have been in Euro, and
- only euro banknotes and coins have been in circulation.

During the transition process, 60 billion coins (about 300,000 tons) were minted and approximately the same amount of national coins was scrapped.

14.4   EMU and Budget Balance

For the member states, the 1990s were marked by a virtuous process triggered by the need to qualify for EMU membership: they moved towards readjustment of the budget towards balance; reduction of inflation; reduction and convergence of interest rates; and economic stability (see Figs. 14.2 and 14.3). Some structural measures were undertaken and a policy of financial discipline was implemented, as imposed by the Maastricht Treaty. The reduction in inflation brought some positive effects, such as a lower cost of money and, hence, lower interest rates for repayment of public debt (i.e. treasury bonds), for borrowing by productive enterprises and for loans to households to buy real estate.

In 2004, six years after the start of the EMU, six member states still did not comply with the criterion on the allowable level of public debt (see Fig. 14.4). Among

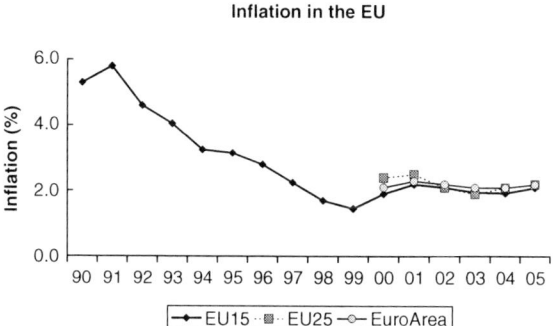

Fig. 14.2 Inflation is cut in the EU from 5–6 % to below 2 % during the 1990s. The good performance of the UK keeps the EU15 line below the Euro Area line. The inflation in AC10 is higher than in EU15 but the relative weight of the EU15 economies keeps the average inflation low. (Eurostat, 2000, 2007)

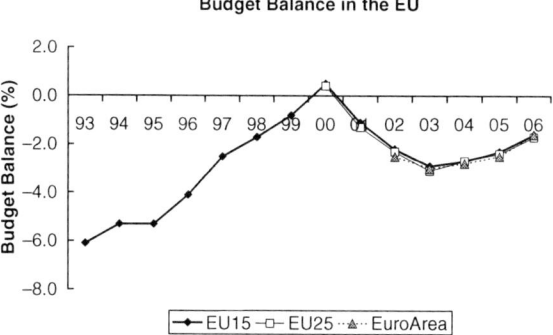

Fig. 14.3 Readjustment of budgets towards balance and move towards a culture of fiscal discipline in the EU member states during the 1990s. The trend towards an increase in deficit in the first years of the new century is reversed from 2003. (EUROSTAT, 2000, 2007)

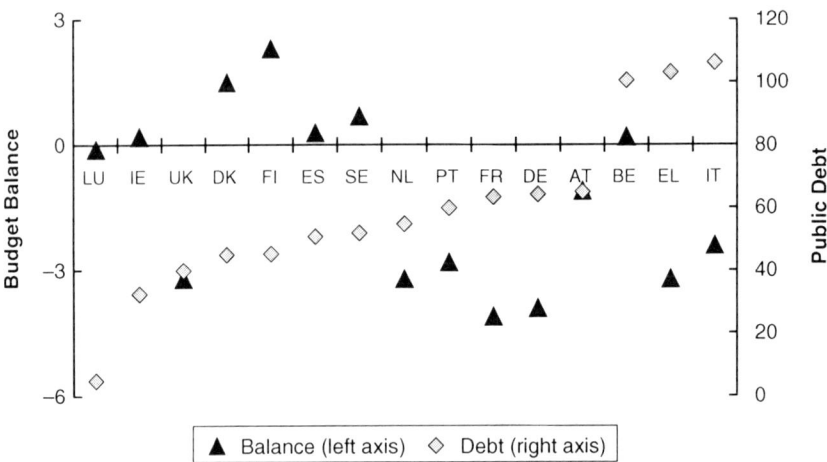

Fig. 14.4 Budget balance (*left axis*) and public debt (*right axis*) as a percentage of GDP for the EU15 member states in the year 2003. Member states are in order of increasing public debt. (Data from [ISTAT 2004], edited by the author)

these countries, Italy, Greece, and Belgium had extremely high levels of debt and out of these three countries, only Belgium had achieved the budget balance.

Inflation improved between 2001 and 2005 (see Fig. 14.5 for EU15 and Fig. 14.6 for AC12). There was both a reduction of the average (from 2.6 % to 2.2 % for EU15 and from 9.7 to 3.7 for AC12) and of the standard deviation (from 1.1 to 0.9 for EU15 and from 11.9 to 2.2 for AC12). Long-term interest rates had a positive

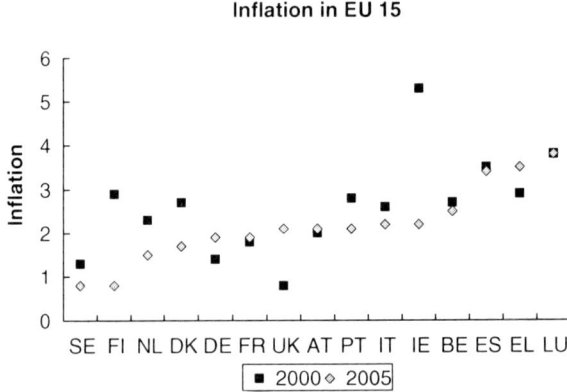

Fig. 14.5 Inflation for the EU15 member states in the years 2000 (*squares*) and 2005 (*diamonds*). A trend can be seen towards reduction of the average (from 2.6 to 2.2) and of their dispersion (from 1.1 to 0.9). (Eurostat 2007, edited by the author)

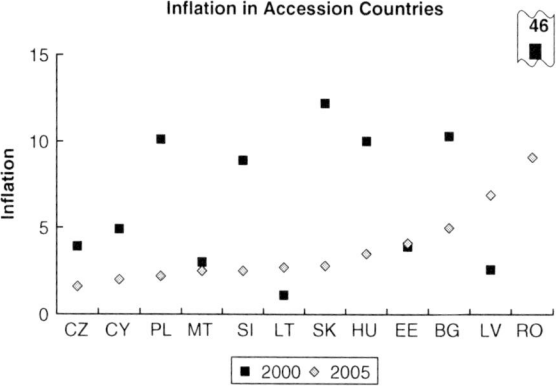

Fig. 14.6 Inflation for the accession countries in the years 2000 (*squares*) and 2005 (*diamonds*). A remarkable reduction can be seen both for the average (from 9.7 to 3.7) and the dispersion (from 11.9 to 2.2). Romania had the very high value of 46 % in 2000 but curbed inflation to one digit in 2005. (Eurostat 2007, edited by the author)

move (see Fig. 14.7) towards lower values from the year 2000 to the year 2005 (average from 5.51 % to 3.5 %) and lower dispersion (standard deviation from 0.2 to 0.09). In the accession countries, interest rates are higher and less homogeneous (see Fig. 14.8).

The positive effects of the discipline imposed by the Treaty on EU are particularly conspicuous in the case of Italy, which we will look at in some detail. Because of the economic policy pursued by the Italian governments in the 1970s and 1980s, until 1992 the Italian economy was characterized by high inflation rates (see Fig. 14.9) and was strongly affected by international crises and by its tremendously

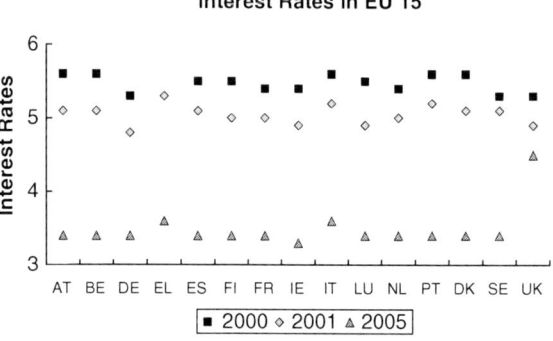

Fig. 14.7 Long-term interest rate in the EU member states in the years 2000 (*squares*), in the year 2001 (*diamonds*), and in the year 2005 (*triangles*). A trend can be seen towards reduction of the values and of their dispersion. The average values for interest rates in EU15 are in the three years: 5.5, 5.0, and 3.5. The standard deviations in the Euro Area are: 0.20, 0.14, and 0.09. A comparable interest rate across member states is a relevant contribution to a uniform market. (Data from Eurostat 2007, edited by the author)

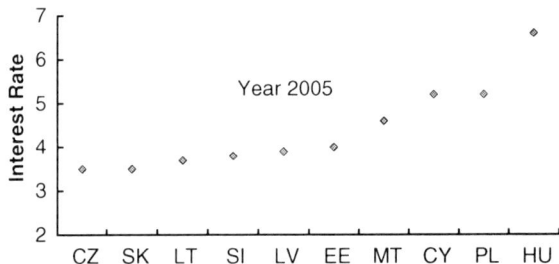

Fig. 14.8 Long-term interest rate in the ten 2004 accession states in the year 2005. The average value for interest rates in the accession states is 4.4 % and the standard deviation is 1.0. The average and the standard deviation in EU15 in the same year were 3.5 % and 0.09. (Data from Eurostat 2007, edited by the author)

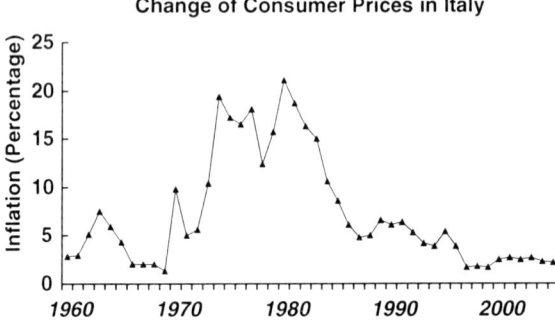

Fig. 14.9 Percentage annual change of consumer prices in Italy during the period 1960–2005. (Edited by the author from data published by *La Repubblica*, 2001 and by IsTAT 2005–2007)

burdensome public debt. Exports' competitiveness was supported through frequent devaluations of the Lira.

The first Giuliano Amato government (1992) started a policy of austerity and budget balance (see Fig. 14.10), which was carried on by Carlo Azeglio Ciampi (1994), suspended by the first Silvio Berlusconi government (May–December 1994), and then resumed by the governments led by Lamberto Dini (1995–1996), Romano Prodi (1996–1998), Massimo D'Alema (1998–2000), and by the second Amato government (2000–2001). The primary budget surplus rose to 6 % in 1997, when Ciampi was finance minister and net borrowing approached balance in 2000. It stayed above 4 % throughout the 1996–2001 parliamentary term.

However, these achievements were squandered by the second Berlusconi government (2001–2006): starting from 2001, the process towards budget balance was delayed (initially projected for achievement by 2003 and then, in 2001, put off until 2005, and in 2005 put off *sine die*). The primary budget surplus decreased year after year since 2001 and was down to 2.9 % in 2003 and to 0.1 in 2006; net borrowing

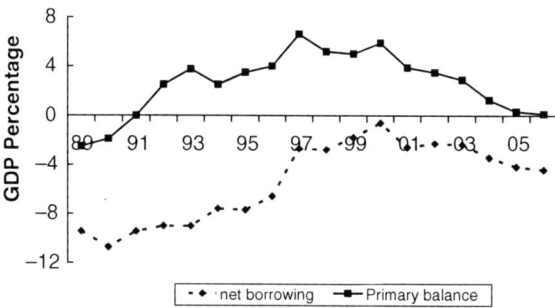

Fig. 14.10 Unbalance in the Italian budget in the period 1989–2006. The upper *continuous line* refers to the primary balance (percentage of GDP); the *dotted line* below refers to net borrowing (percentage of GDP) (Edited by the author from data published by Banca d'Italia and IsTAT 1998, 2005, and 2007)

was −4.2 in 2005 and −4.4 in 2006. Public debt reached € 1½ trillion in 2005, the highest among the EU25 both in absolute value and as percentage of GDP.

The present government chaired by Romano Prodi is in office from 2006 to 2011. The economy minister is Tommaso Padoa Schioppa, a former ECB board member. He imposed a strict budget discipline as is fighting tax evasion energetically to comply with the treaty criteria for budget deficit. The Economist anticipates that the deficit will be "comfortably below 3 % by year end 2007" [Economist 2007–19].

14.5   European Central Bank

Officially set up on 1 June 1998, the European Central Bank (ECB) belongs to the European System of Central Banks (ESCB) together with the EU27 member states' national central banks (NCBs). *Eurosystem*, the subset of ESCB including the NCBs of the Euro-zone countries, is the monetary authority of the Euro-zone. In the ECB statute, price stability is defined as the main objective to be pursued by the Central Bank. A second objective, dependent on the first one, is to provide development aid by reducing the cost of money. The declared core values of ECB are competence, effectiveness and efficiency, integrity, team spirit, transparency and accountability, and working for Europe. Some of the main tasks of the European Central Bank include the definition and implementation of the monetary policy in the Euro-zone and the management of the member states' currency reserves.

The NCBs of the member states that have not adopted the euro can implement their own national monetary policies, but they do not take part in the decision-making process and in the implementation of the monetary policy related to the Euro-zone. The EU member states that have not adopted the euro must, however, comply with a number of regulations set by the EMU, including the obligation to present yearly the Growth and Stability Pact and to follow the Broad Economic Policy Guidelines issued by the European Commission.

The role of the ECB has been criticized by some scholars that point to lack of democratic participation in the process of monetary control [Lintner 2000]; however the author thinks that these criticisms are unjustified because central banks should respect dependence from democratically controlled powers concerning the goals to achieve (which the ECB does) but attain them via instrumental independence [Frazer 1994].

14.6 Euro Versus Dollar

Currency exchange rates tend to fluctuate for many reasons, such as political changes and economic slumps. During the first semester of 1999, two main events led to a strong depreciation of the euro against the dollar, which the author does not forget because he was working in the U.S. with a salary in euro. First of all, the U.S. economy grew more than expected and more strongly than the economy in the Euro-zone. This situation encouraged investors to buy dollars and bonds in dollars, in view of a second event, namely, an increase in the U.S. interest rates and a reduction in the European interest rates. This is exactly what happened. The U.S. economy continued to grow faster than the economy in the Euro-zone for the greater part of the year 2000, once again to the advantage of the dollar. The exchange rate

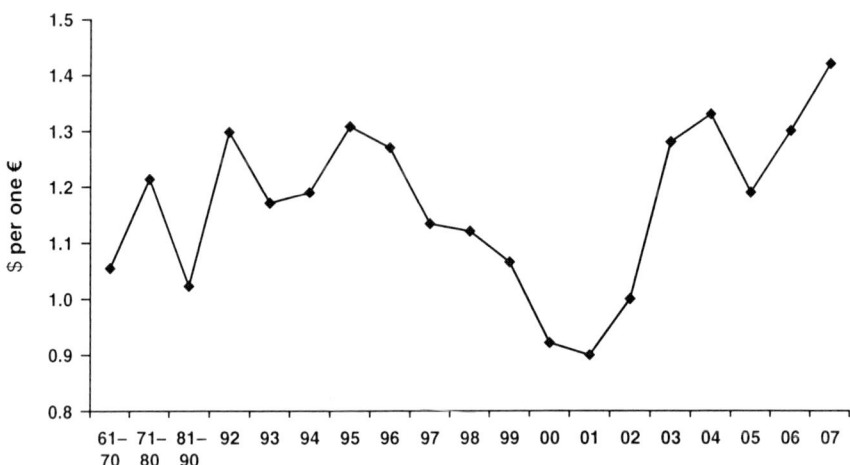

Fig. 14.11 U.S. Dollar per Ecu or Euro from 1960 to 2007. Before the introduction of the euro in 1998, the European Currency Unit or Ecu was a weighted basket including a specified amount of all member states' currencies. The time axis is not linear, the first three decades are compressed. (*La Repubblica*, 2002, 2007)

Fig. 14.12 Euro: A strong currency. (European Commission, 2001)

between the dollar and the euro became more stabilized in 2001, as the slowdown of the U.S. economy was becoming more marked.

The exchange rate relationship reversed in the following year. In 2002, the cost of money in the U.S. was lower than in Europe, where the resulting higher interest rates attracted foreign investments. This is one of the factors that have led to a rise of the euro against the dollar. Another factor resulting in the appreciation of the euro against the dollar is the smooth completion of the EMU transition phase and the introduction of the new currency. From 2003, with the objective of giving a boost to exports and reducing the huge trade deficit, the U.S. did not try to prevent the weakening of the dollar and kept interest rates lower than in Europe. As a result, the euro rose against the dollar (see Fig. 14.12) and continued to rise until the end of 2007, after an adjustment in 2005. Probably, this appreciation will not stop in the short term. Figure 14.11 gives a detailed graphical representation of the dollar-euro exchange rate over a long period of time, showing that appreciation/depreciation cycles were affecting the dollar-euro exchange rate well before the start of the EMU.

14.7 Prospects for the European Economy With a Strong Euro

A company's accounts will benefit from a weak currency in its own country if a high percentage of the turnover of that company is produced in foreign markets and if imports of raw materials and semifinished goods do not account for a high percentage of its budget. Vice versa, a company may benefit from a strong currency if it is heavily dependent on import of energy and raw materials and if its clients are in an economic area with the same reference currency. A strong currency attracts foreign investments, reduces the inflation rate, contributes to a favourable balance of trade, stimulates the stock exchange market, and results in higher profits for shareholders and investors, a lower cost of money for real investment, and a lower cost for imports of raw materials and energy.

In August 2001, General Motors' chief executive, John Devine, complained about the strong dollar [Burt et al. 2001], saying that it "destroys the competitiveness of American manufacturers." However, the opinion that a weak dollar would benefit American companies was not necessarily shared by everyone in the U.S. Many European entrepreneurs and political leaders made similar assertions in and after 2002.

At the end of 2003, the American monetary policy scenario had changed. Having seen the euro's appreciation in 2002, the monetary policy adopted by the Federal Reserve [Economist 2003–3] and by the new treasury secretary, John Show, aimed at depreciating the dollar on the international currency markets in order to stimulate economic recovery and cut trade deficit in view of the 2004 presidential elections ($ 478 billion deficit in 2002, of which $ 100 billion was with China).

On the basis of this new situation, the euro not only rose against the dollar but also became a reference currency in trade with third countries (see Fig. 14.13), just as the dollar had been from the end of World War II. Japan reacted to the rise in the exchange rate between the dollar and the yen by relevantly increasing their dollar reserves ($ 190 billion in 2003) and China pegged the (artificial) exchange rate of 8.28 renminbi per dollar by accumulating reserves up to $ 350 billion. The other major Asian economies (Hong Kong, Taiwan, and South Korea) also increased their dollar reserves in order to prevent their national currencies from rising too much (see Fig. 14.14). The ECB did not take any measures to halt the euro's appreciation.

Fig. 14.13 Euro: A worldwide currency. (European Commission, 2001)

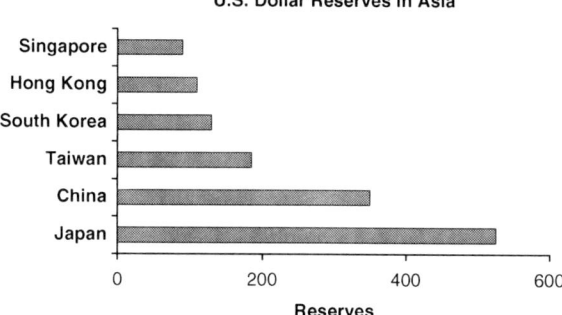

Fig. 14.14 U.S. Dollar reserves in Asia (2003). From 2001 reserves increased by 36 % in Japan, by 65 % in China, and by 49 % in Taiwan. (*The Economist* vol. 368, n. 8342 (2003))

The economies of member states can derive important benefits from the euro's appreciation. First of all, European citizens will find travelling in the U.S. and in those parts of the world where the dollar is widely accepted to be much less expensive now than in 2001. However, there are more profound effects on the economy, such as a reduction in inflation, stimulation of stock exchange markets, higher profits to shareholders and investors, reduction of the cost of money for investments, cuts in import costs of energy and raw materials, a lower cost of oil and microprocessors, and the shift in foreign investments. The 2007 Eurostat EU Foreign Direct Investments Yearbook says: ≪Foreign direct investment (FDI) is a category of international investment that indicates an intention to acquire a lasting interest in an enterprise operating in another economy. It covers all financial transactions between the investing enterprise and its subsidiaries abroad and it differs from portfolio investments, because the direct investor acquires at least 10 % of capital. Foreign direct investment acquires increasing importance as an indicator of the international economic climate and plays a key role in the globalization process as an important element of international relations and their development. Supplementing trade, FDI creates more direct and deeper links between economies. It is a source of extra capital, encourages efficient production, stimulates technology transfer, and fosters the exchange of managerial know-how. It is thus believed to improve the productivity of business and to make economies more competitive [EC 2007–9]≫. FDIs are beneficial to the hosting country if the investments are rooted into the national economy and answer to a demand of the market, including the local market. In this way, the danger of subsequent delocalization is reduced and additional economic activities are generated locally based on endogenous resources, skills, and vocation.

The Union as a whole registered a fivefold increase in foreign investments from 1990 to 2000. Figure 14.15 shows a leap forward in foreign investments in Europe since the irrevocable decision to build EMU was made. FDI in Europe continued in the following years with the 2004 slow down visible in Fig. 14.15. In that year, the total world's FDIs increased, with a boom of investments in emerging economies, to the detriment of Europe that recovered in 2005. The EU had 19 % of world FDI

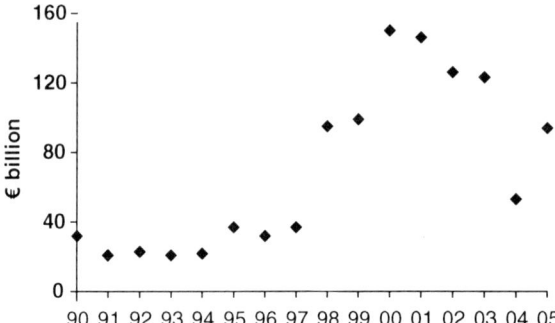

Fig. 14.15 Foreign investments (billion euro) in the EU member states from outside the EU from 1990 to 2005. A big jump is visible at the time of the irrevocable decision about EMU (1998). (Eurostat 2002)

inflows in 2005. World FDI inward flows increased in 2005 by 9 %; but EU FDI inward flows increased by 77 %, reaching € 94 billion for extra-EU flows in 2005. Intra-EU FDI inflows are about four times that value. The main investor country was the USA with 18 % of the inward flows [EC 2007–9].

The positive effects of a strong euro may be offset by a squeeze on export margins in trade with the U.S. However, Francesco Trapani, chief executive of the Italian company Bulgari, leader in the production of luxury articles and jewellery, made the following statement to *La Repubblica* on 9 January 2004 [Lonardi 2004]:

> There are both advantages and disadvantages for Bulgari. The disadvantages are linked to the fact that we must increase prices in the U.S. in order to maintain good profit margins but in this way, we become less competitive in the American market. The advantages are derived from the lower cost of raw materials, as gold, diamonds, and gems are purchased in dollars. Basically, for us advantages and disadvantages balance out.

Regarding the distribution of FDIs in European national economies, some countries attract more investments than others: Figure 14.16 shows that Italy is lagging behind all the large countries in Europe in terms of foreign investments both outside and inside Europe. A survey by Ernst & Young, published by *La Repubblica* [Pagni 2003, Bogo 2003], reveals that in 2002 foreign investments in Italy dropped by 44 % as compared with the previous year.

As far as Italy is concerned, from 1990 to 2000 foreign investments decreased from € 3.02 billion to € 2.98 billion [Eurostat 2002]. On 12 March 2004, the agency Adnkronos [Adnkronos 2004] wrote: "Italy lags behind the EU in the attraction of foreign investments."

A survey by Siemens, conducted by the consulting house Ambrosetti, reported these figures: the average of Italy's foreign direct investments in percentage of GDP (during the period 1996–2001) is only 0.5 % of the total, against 12.8 % of Ireland, 9.7 % of Sweden, 8.4 % of The Netherlands, 5.2 % of the United Kingdom, 2.9 % of Germany, 2.6 % of Spain, and 2.3 % of France [Adnkronos 2004]. Figure 14.17

Foreign Direct Investments 2004

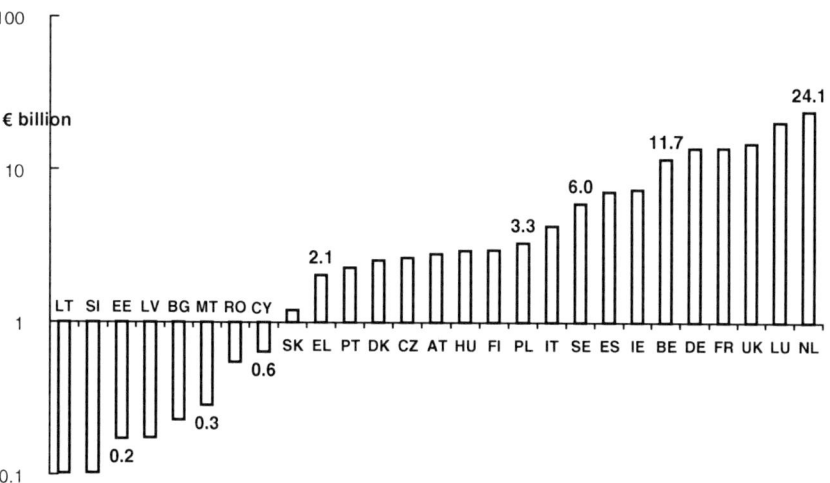

Fig. 14.16 Foreign investments (billion euro) in the EU member states from outside the EU in the year 2004 (Eurostat 2007)

presents the foreign investments stocks in the EU in the year 2004. The Netherlands, Luxembourg, Belgium, and Ireland are high up in the ranking, next to the larger EU27 countries. At the bottom of the ranking the weaker economies show a marked difference between the stock of FDI in their country and their own investments abroad.

A stronger euro means lower costs of energy and raw materials imported into Europe and contributes to the accomplishment of the ECB's key objective of keeping inflation below 2 %. To meet this objective, in a strong euro phase, the ECB could reduce discount rates and stimulate economic growth by enhancing internal demand and compensating a foreseeable decrease in exports. The strategy of getting into debt and spending more than what is internally produced, as the Americans tend to do, has been considered a possibility for the European economy to support recovery and growth, at least in the short term. However, constraints on budget deficit and on price stability do not leave much freedom of movement for this kind of policy and in addition, the American crisis triggered by sub-prime loans in 2006–2007 has proved that this strategy has major drawbacks.

A leap ahead in the European economy depends on the full accomplishment of the internal market, when citizens will finally be able to choose from the goods offered in the whole Euro-zone and buy the best products at the lowest price, as is already happening in the U.S. The internal market and monetary policy have not yet achieved their full potential, because each country continues to defend its own interests and is not actually ready to share sovereignty over economic policy. As we will explain later on, a strong blow was struck to the unification process and to the full accomplishment of the economic and monetary policy by the watering-down of

the Stability and Growth Pact in December 2003 by the European Council and the
changes to the pact adopted in 2005.

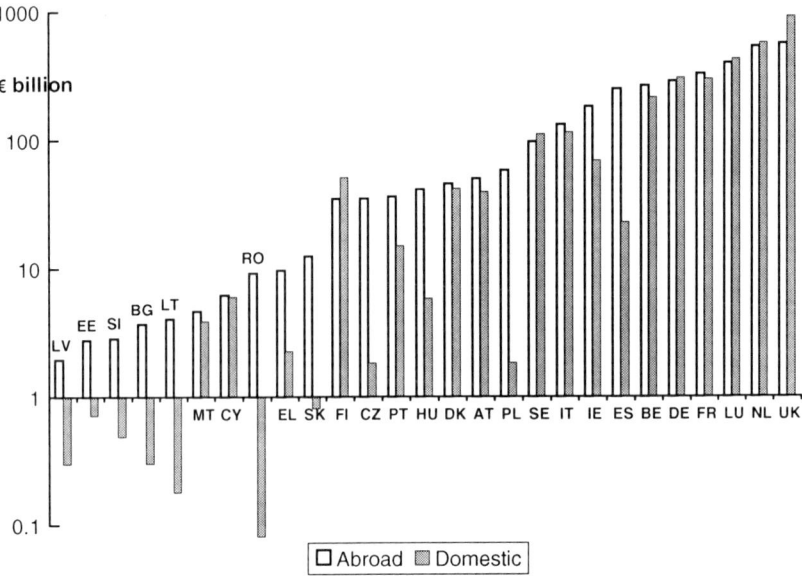

Fig. 14.17 Foreign investments stocks (billion euro) in the EU member states (domestic) and
abroad, in the year 2004. The largest economies have a balanced stock, while for weaker economies
foreign domestic investments exceed their own investments abroad. (Eurostat 2007)

14.8 Monetary Stability

The euro creates a vast area of monetary stability within the EMU to the benefit of
those countries that, in past decades, had experienced long periods of volatility and
currency misalignment. Despite the initial fluctuations against the U.S. dollar (see
Fig. 14.11), the euro contributes to the stability of international monetary relations
thanks to the strict budget policies that national states adopting the euro are forced
to comply with. The euro has the right credentials to become one of the main world
currencies, alongside the dollar and the yen (see Fig. 14.18).

14.8.1 Price Stability

The lowest possible inflation rate, or price stability, is the first goal of the Community's economic policy. Inflation discourages investments, erodes the value of
savings and pensions, and is a heavy burden for the worse-off population groups.

Fig. 14.18 Euro: A currency
with safe roots. (European
Commission, 2001)

There is price stability when the annual increase in prices is not above 2 % in the medium term. Fig. 14.2 shows that the 2004 enlargement has not interrupted the trend towards containment of inflation throughout Europe.

14.9 Economic and Monetary Union and Member States Sovereignty

The only aspect of national sovereignty – which is considered to be the country's power to act independently – that is directly affected by the economic and monetary union is monetary policy. This is entrusted to the Board of Directors of the European Central Bank, made up of the governors of the National Central Banks.

Euro-zone countries lose their power to manage their national currency exchange rates in an independent way. In practice, however, the countries that had joined the EMS exchange mechanism since its setup in 1979 had already given up such independence, piggybacking on the German Central Bank. This new situation is more advantageous for those countries with a weaker economy, because, instead of being forced to accept the decisions made by the German Central Bank, they can now take an active part in the ECB's decision-making process. The countries that have adopted the euro believe that less control over their inflation, interest, and exchange rates is a price worth paying to have the same currency as major trading partners, greater credibility in international financial markets and, consequently, greater flows of investment.

The EMU leads to an integration of the economies of participating countries and requires closer monitoring and coordination of national economies. Since 1 January 1999, the Ecofin Council has issued general guidelines relating to economic policy and the euro's exchange rate. Every year, in June, the European Council defines the annual general guidelines for economic policies, on the basis of the Commission's recommendations and following examination by the Ecofin Council. These general guidelines are meant to be the reference points for the Commission to monitor the economic situation in the member states. EU member states (not only those that have adopted the euro) continue to be in charge of their own economic policies and have to comply with the ceilings on budget deficit and public debt imposed by the Union's treaty and by the Stability and Growth Pact.

14.10 Stability and Growth Pact

The Stability and Growth Pact requires that every year EU member states submit to Community institutions their national forecast about public finances and other economic parameters for at least a three-year period. Through the Stability and Growth Pact, member states commit themselves to pursuing the right budget policy to achieve budget surplus or at least budget balance in the medium term. Member states can no longer modify their national currency exchange rates in their domestic monetary policy. The main economic policy tool is taxation. That is why it is so difficult for the Union to harmonize national fiscal policies, which are strongly protected by national sovereignty. This also helps explain why, as we saw in Chapter 10, "From the Treaty of Rome to the Reform Treaty of Lisbon", decisions on fiscal policy matters are made by the European Council by unanimity.

The Commission assesses the national stability and growth programs and puts forward a series of recommendations to the Council. If national policies diverge from the established guidelines, the Council can issue specific recommendations concerning the necessary corrective measures to be taken by a state. This procedure is applied to all EU member states. The fact that the Council can issue specific recommendations for a country before approving its stability program is a substantial step of progress in the coordination of the economic policy. If a member state does not comply with the Council recommendations, provisions are made in the Lisbon Treaty for the issuing of sanctions against that country (Article III 184).

In 2002 the Stability and Growth Pact was at the centre of a heated debate concerning its effectiveness and flexibility (see Fig. 14.19). One proposal to change the Pact was to make it more or less flexible according to the country, depending on the extent of its public debt.

In 2003, for the third consecutive year, France and Germany submitted growth programs with a budget deficit above the limit of 3 % of GDP. The Commission pointed out that, in this way, France and Germany were breaching the EMU agreements and suggested that some sanctions should be imposed on them in the Ecofin Council of 25 November 2003. The Council, under the presidency of the Italian finance minister, Giulio Tremonti, decided not to sanction France and Germany and avoided applying the measures envisaged by the Stability and Growth Pact in cases where the budget deficit exceeds the limit of 3 % of GDP during three years in a row. This decision actually implied a weakening or watering-down of the Pact. The *Financial Times*' headline on the front page on 8 January 2004 reads: "The ministers have broken the law on Stability Pact." As a matter of fact, some constitu-

Fig. 14.19 Stability and
Growth Pact: Hypotheses of
reform. (©Financial Times,
25 November 2002
reproduced with kind
permission)

tionalists argued that the Council did not have the authority to decide not to impose
the sanctions envisaged by the treaty, since, for the Council to be allowed to do so,
the member states should have previously approved derogation to the treaty through
a rather complex procedure. The European Commission appealed to the European
Court of Justice [Graham 2004] against France and Germany for having breached
the Stability and Growth Pact. The Court ruling supported the Commission's appeal,
and the *Financial Times* wrote on 14 July 2004 [Benoit 2004]:

> The European Union's highest court yesterday ruled that finance ministers acted illegally in
> suspending the threat of sanctions against France and Germany over their repeated breach
> of the EU's budget deficit rules.

The Italian presidency claimed that its interpretation of the pact was a flexible
and intelligent one. However, the detractors of Tremonti, and the author with them,
believe that the Italian minister actually wanted to smooth the way for Italy, since
Tremonti could anticipate that his economic policy would lead Italy to fail to comply
with the 3 % ceiling in 2004 and/or 2005, as it actually happened. Over the previous
three years, 2001–2003, the Italian government had managed to keep the budget
deficit under control, but only through one-off measures, such as amnesties, pardons
and securitizations and though accounting practices that have not been endorsed by
Eurostat or even by the national Court of Auditors. Italy did not apply measures to
curb inflation or to raise the primary budget balance, and, in this way, it postponed
the goal of cutting net borrowing. Not only Italy's budget deficit persisted but it
exceeded the 3 % limit from 2004 to 2006. On 13 April 2005, when the 2004 Italian
budget figures were still under scrutiny, the *Financial Times* online [Parker et al.

2005-1] headlined its coverage "EU in stern warning to Italy over budget deficit," and wrote:

> Mr. Silvio Berlusconi, Italy's prime minister, was warned by the European Commission of its intention to start formal action against Italy before the end of June 2005 over the country's rising budget deficit. The challenge could be even more uncomfortable for Mr. Berlusconi if the Commission concludes as expected that Italy has under-reported the size of its deficit in the past. Joaquin Almunia, the EU monetary affairs commissioner, regards Italy's economic situation as the most precarious in the 12-country Euro-zone.

The debate on the pact went on for over two years, with many EU leaders, including the German chancellor, Mr. Gerhard Schröder; the French president, Mr. Jacques Chirac; and the Italian prime minister, Mr. Silvio Berlusconi, wanting a much looser pact that would give them more freedom to run national economic policies without interference from Brussels [Parker 2005]. Luxembourg's EU presidency, on 20 March 2005, at an Ecofin meeting prior to the Spring 2005 European Council, proposed a more flexible Stability and Growth Pact, "weakening the enforcement powers against countries that breach the deficit limit but cracking down on debt and profligacy in upswings" [Parker 2005].

The overhaul of the pact keeps the 3 % GDP/deficit and 60 % GDP/debt ceilings but softens procedures against a defaulting country, introducing fuzzy concepts such as *pertinent factors*, *temporary* deficit, and *near to* the value of 3 % that make it possible to delay or dodge sanctioning procedures. The pact, which in 2002 – because of its rigidity – had been called "stupid" by Romano Prodi [Parker 2002], the then president of the European Commission, might become insidious and open up to creative budgetary discipline.

Editorial comments in the *Financial Times* of 22 March 2005 expressed British euroskeptic views by saying [FT 2005-1, FT 2005-2]:

> It is tempting to cheer the dismemberment of Europe's stability pact. [...] After all, fiscal sinners would continue to be penalized by markets.
>   European finance ministers finally agreed to relax the enforcement procedures of the pact, which underpins fiscal policy in the European Union. In fact, they weakened it so fundamentally as to render it practically worthless. [...] The immediate consequence will be an increase in creative accounting practices, as member states strive to take advantage of "due consideration" in case they face an excess deficit procedure.

On a similar tune, *The Economist* wrote [Economist 2005]:

> The agreement was, in essence, to rip up the existing fiscal rules for euro members and start again improving the operation of the Stability and Growth Pact. [...] But now the rules have been so loosened that they have been rendered almost entirely meaningless. [...] The trouble is that, largely at the behest of France and Germany, which have violated the 3 % ceiling for three years in a row, a raft of possible let-outs and exceptions have now been written into the pact. Governments can now avert the threat of sanctions by pointing to any recession, however shallow, or even just to a persistent period of slow growth.

The European Central Bank reacted with alarm to the change in the rules, expressing fears that confidence in public finances might be undermined. The Bundesbank, the German National Bank, said the pact's rules had become more complicated and less transparent, and it expressed fears about the backdrop against which future monetary policy decisions would be taken [Atkins 2005].

14.11 EMU and Enlargement

The value of the single European currency against other reference currencies does not change with the arrival of new member states in the Euro Area. When Greece adopted the euro on 1 January 2001, or when Slovenia did so on 1 January 2007, the external value of the single currency was not affected. The value of the euro is determined by the strength of Euro-zone's economy and by the action of the ECB to guarantee price stability.

Although the Euro-zone will certainly enlarge to include more countries, it seems quite unlikely that most of the countries from central, eastern, and southeastern Europe that accessed in 2004 or 2007 will qualify for the adoption of the euro in the short term. These countries must continue their process of structural, economic, and administrative reform and must build a sound market economy in order to comply with the EMU's objectives of budget discipline and price stability. Indeed, this is the only way in which they will permanently manage to satisfy convergence criteria.

Given that the adoption of the euro is only granted to those countries whose economic conditions are in line with those of the countries that are already in the Euro-zone, the euro adoption by enlargement countries (once they have qualified) should not produce significant effects on the external value of the euro. This conclusion is borne out by the fact that, as we shall see in Chapter 2, "Overview of Member States", the 10 new member states that joined the EU in 2004 and 2007 together account respectively for only 5.7 % and 1.1 % of the GDP produced by EU15.

14.12 Advantages of the Euro

The common currency and the EMU, of which the introduction of the euro is a very important component but not the only one, are generating advantages to consumers, undertakings, and the economy of member states. Some politicians have found it convenient to blame the euro for the bad performance of their economic policy, but had the EU not implemented the EMU, the economic slowdown at the turn of the century, and the international crises in various parts of the world would have had severe effects on EU national economies and on monetary stability.

14.12.1 Advantages for Consumers

Consumers gain many advantages from the EMU. Some of these are as follows:

- it is easier and less expensive to travel abroad (see Fig. 14.20), because there is no more need to change currencies and pay exchange commissions;

Fig. 14.20 Euro: For tourism. (European Commission, 2001)

- it is easier to compare prices and prices tend to converge;
- it is easier and less expensive to transfer capital from one country to another;
- prices have been reduced; thanks to transparency and competition;
- the foreign currency exchange risk has been eliminated for participating countries;
- consumer prices may be kept under control, which leads to a decrease in inflation, interest rates, and credit charges;
- if consumer prices are contained, then purchase power is maintained; and
- more sustainable economic growth is possible, which increases job security.

14.12.1.1 Consumer Protection

On two occasions, we have said that the containment of consumer prices is a positive effect produced by the introduction of the euro. Let us explain this point more clearly. The curb on inflation has actually been one of the main benefits to national economies gripped by runaway inflation (see Fig. 14.2 regarding the EU and Fig. 14.9 regarding Italy). In planning and implementing the transition to the euro, the Community had mainly focused attention on the issue of consumer protection. The changeover of the national currency units into euro was accompanied by all necessary legal precautions. The only conversion rate permitted is the one that was fixed on 1 January 1999 for each participating national currency. The agreed-upon conversion rate is the one determined by the market during the convergence period. It is unlawful to apply any other conversion rates. Consumers are also protected by a series of rounding rules after conversion.

Another aspect of consumer protection is the agreement reached at the Community level between consumers and retailers regarding a logo, to be displayed by retailers as a guarantee that they are actually complying with the rules on conversion and rounding.

However, Italian consumers have been complaining about the jump in retail prices since the summer of 2002, after the introduction of the single currency. At the end of 2003, two years after the introduction of the euro banknotes, consumers from some countries were still unhappy about the perceived steep increase in prices and kept blaming the transition to the euro for this. As a matter of fact, the increase in prices has moderated in those countries that have kept retailers under tighter control, thereby preventing speculation. In Spain and Italy, prices have increased twice as much as in Germany and in The Netherlands (see Table 14.1). Spanish and Italian consumers' conclusions from these findings about their respective governments' capacity to manage the transition to the euro and protect them from speculation may have contributed to the political shift in the elections that brought respectively in 2003 and 2006 to a change in the ruling party in the two countries.

The European Commission published a leaflet "Did the euro cause prices to rise?" that answers to frequently asked questions and gives an explanation of the perception that prices increased due to the switch over to euro [EC 2006-9]:

> A consumer price index is an economic indicator designed to measure changes in the prices that households pay for goods and services. There is a Harmonized Index of Consumer Prices (HICP) for the Euro Area.
>
> Monitoring the HICP basket before and after the introduction of euro banknotes and coins shows that a few items displayed unusual price increases. For example, some

Table 14.1 Increase in consumer prices from November 2002 to November 2003. Data for Greece are not available

	Food	Clothing	Alcohol Tobacco	Transport	Entertainment	Hotels restaurants	All goods
EU15	2.0	0.2	6.0	2.1	−0.5	3.0	2.9
Euro-zone	2.2	1.0	7.6	1.8	0.1	3.0	3.0
IE	3.3	−2.6	10.0	1.3	3.5	5.2	0.3
DK	1.4	1.9	−7.1	2.5	1.7	2.3	1.4
DE	1.3	−1.4	5.1	1.9	−0.9	0.4	1.5
SE	2.0	−2.1	1.4	1.6	1.5	2.6	1.5
NL	2.0	−2.0	2.0	2.9	0.4	2.6	1.6
LU	2.0	1.3	4.8	0.7	2.4	2.6	2.0
BE	1.8	0.9	4.7	1.5	0.7	3.4	2.1
FI	1.2	−0.5	1.2	−2.9	0.7	2.6	2.3
PT	2.3	1.4	3.5	1.8	0.3	3.5	2.5
UK	1.3	−4.3	1.6	3.7	−2.3	3.0	2.6
AT	1.3	0.5	2.3	0.6	0.1	2.7	2.9
FR	2.5	1.0	17.3	1.7	−0.7	2.9	3.0
IT	2.8	2.6	7.9	2.1	1.7	3.7	4.1
ES	2.9	2.3	2.6	1.6	0.3	4.1	4.3

Source: Eurostat, reported by *La Repubblica* on 16 January 2004.

restaurants, cafés and hairdressing salons showed significant price rises, as did car and house repair, and sports and recreation. These price rises were concentrated in local, neighborhood services that have little competition. They can be attributed to the euro as it seems that some small retailers took the opportunity to raise their prices at the same time that euro banknotes and coins were introduced. In contrast, many other prices remained steady or, in the case of a lot of manufactured products like computers, photographic and stereo equipment, continued to fall, partly because the technology became cheaper.

These unusual price increases affected a number of small items we buy frequently. Everyday, small, cash purchases determine our perceptions of price inflation more than less frequent purchases, such as a new computer, insurance or rental costs. However, their influence on the cost of living is less—they only contribute to a small part of the HICP basket. But our perception of price rises during the euro changeover corresponds more closely to the price inflation of these more frequently bought items than to that of the whole basket. This shows that while people have good reasons for the impression of widespread price increases during the switch to the euro, in reality these were limited and had only a small effect on the total cost of living.

14.12.2 Price Convergence

The euro can lead to price convergence for the same product in the different member states, thanks to the possibility of easy price comparison in the unified internal market, as they are expressed with one currency. A 1999 Commission report revealed that the average price dispersion (the gap between the highest and the lowest price for the same product) was around 16 %, as compared to 11 % in the United States.

The euro will strengthen competition among Euro-zone countries thanks to the greater transparency achieved when all prices are expressed in euro. This transparency will certainly have a positive effect on price convergence, as marked price differentials will no longer be acceptable.

In any case, just as some price variations are reported to exist today among various regions of the same country, it could well be the case that price differences among member states will still be there in the future. However, such differences, which could be due, for example, to different levels of indirect taxation, will be more difficult to sustain and, in general, will only reflect the application of different transaction costs.

14.12.3 Advantages for Enterprises

14.12.3.1 Advantages for Small and Medium-Sized Enterprises

The Economic and Monetary Union (EMU) offers wider choice and more trade opportunities to enterprises (Fig. 14.21), including small and medium-sized enterprises (SMEs), and makes many of them stronger and more competitive thanks to the following advantages:

- Elimination of the exchange risk in the majority of the main markets in the Union. In the past, smaller companies did not have the same access as larger

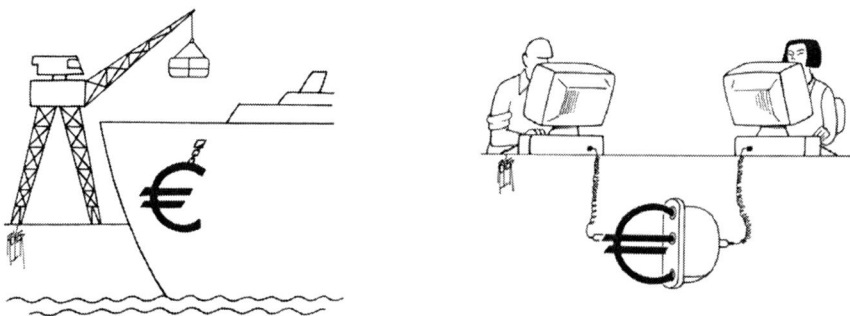

Fig. 14.21 Euro: A currency for trade and e-trade. (European Commission, 2001)

companies to the financial tools that could safeguard them from currency fluctuation.

- More competition in the banking sector and lower interest rates, which should allow SMEs to fund their own business activities at more advantageous costs.
- The possibility of doing business throughout the Euro-zone as in a single internal market. This means that SMEs will be able to sell their products and services in a much broader area (thereby creating economies of scale) and will have many more potential customers. In other words, exporting will become a much simpler internal market activity. In this new scenario, SMEs will be encouraged to introduce technological and commercial innovations, which will make them more competitive in an enlarged market and which would benefit the whole European economy.

14.12.3.2 Effects of the European and Monetary Union for Large Enterprises

Large enterprises doing business in Europe are experiencing a cost reduction, particularly concerning costs related to the exchange risk, which will almost disappear, but with reference to costs linked to cash management and foreign exchange transactions. However, the euro's effects are going to be much broader, as the new currency will affect employment policies and pricing and marketing policies and will play a part in the location of production and distribution centres. In some sectors, competition will certainly be fiercer and will lead to more price transparency. As a result, consumers will be able to buy some goods at a lower cost in certain markets.

14.12.4 Advantages for the Countries' Economies

The advantages of the EMU for participating countries are derived from the application of budget discipline, the curb on inflation, the completion of the internal market, and the benefits brought by harmonization and consolidation of the countries'

Fig. 14.22 Euro: A currency
protecting from international
crises. (European
Commission, 2001)

economic policies. Other important benefits are currently being experienced by
interrelated sectors:

- elimination of currency speculation based on foreign currency exchange among
  member states;
- boost to investment in the Euro-zone;
- less dependence on international crises (see Fig. 14.22); and
- less production and circulation of counterfeit banknotes and coins.

14.13 Recommended Reading

Recommended reading for this chapter are listed in Appendix C and are available on
the web site companion to this book, at the URL http://stajano.deis.unibo.it/RQC.htm

Appendix A
Steps in the Construction of the European Union

The main steps in the construction of the European Union (EU) are schematized in Table A.1, where only events that are significant from the point of view of the study of competitiveness and research policy are mentioned. The construction of the Union is masterfully summarized, covering all the policies, in [Fontaine 2003]. This text is available on the recommended reading web site that accompanies this book.

Table A.1 Steps in the construction of the European Union

Year	Event
1945	End of the Second World War
1947	Marshall Plan
1949	NATO: North Atlantic Treaty Organization
1950	ECSC: European Coal and Steel Community
1953	CERN: European Center for Nuclear Research
1957	EEC, Treaty of Rome: Belgium, France, Germany, Italy, Luxembourg, The Netherlands Euratom
	The EEC and Euratom treaties provide for R&D financing in three areas: nuclear, coal and steel, and agriculture.
1961	Building of the Berlin Wall
1962	The EU starts her "common agricultural policy" giving the countries joint control over food production. Farmers are paid the same price for their produce. The EU grows enough food for her needs and farmers earn well. The unwanted side-effect is overproduction with mountains of surplus produce. Since the 1990s, priorities have been to cut surpluses and raise food quality.
1968	In May, rioting by students and workers in France shakes the very foundations of the state. Milder student protests occur in other EU countries. They reflect frustration at remote and unresponsive governments as well as protests again the Vietnam War and the nuclear arms race.
	In June, the six EEC countries remove customs duties on goods imported from one another, allowing free cross-border trade for the first time.
1972	First oil crisis
	The Club of Rome defines the "Limits to growth" [Meadows 1972].
	The European Council extends the range of action of the Community to include industrial policy.

Table A.1 (continued)

Year	Event
	National policies support the national champions and define nontariff trade barriers to the creation of an internal market.
	In April, to maintain monetary stability, EU members decide to allow their currencies to fluctuate against each other only within narrow limits. This exchange rate mechanism (ERM) is a first step towards the introduction of the euro, 30 years later.
1973	Enlargement to Denmark, Ireland, and the United Kingdom of Great Britain
1974	To show their solidarity, EU leaders set up the European Regional Development Fund. Its purpose is to transfer money from rich to poor regions to improve roads and communications, attract investment, and create jobs. This type of activity later comes to account for one-third of all EU spending.
1979	Second oil crisis
	The European parliament is elected for the first time by direct universal suffrage.
	EMS monetary system
	The Commissioner Etienne Davignon starts negotiations with 12 large European companies that operate in the information and communications technology sector (ICT).
	The Court of Justice shapes the principle of mutual recognition.
1980	In summer, shipyard workers in the Polish city of Gdansk, led by Lech Walesa, strike for more rights. Other strikes follow across the country.
1981	Enlargement to Greece
1983	Pilot phase of the R&D program Esprit
1984	First Framework Program for research and technological development: Esprit I
	Objectives: development of the technological base to support industrial growth; across-the-border collaborations and synergies; European standards
1985	Jacques Delors is appointed European Commission president.
1986	Enlargement to Spain and Portugal
1987	Amendment to the Treaty of Rome called the Single European Act, establishing, by 1992, freedom of movement of persons, goods, services, and capital. R&D is part of the Treaty: "The Community has the objective of strengthening its scientific and technological bases, of encouraging it to become more competitive on an international level and of promoting all the research activities deemed necessary." First step towards majority decisions in the European Council.
1988	Second Framework Program (FP2) for research and technological development: in-depth consultations; Esprit II; Brite; Race
	Objectives of FP2 also include cohesion; small companies; from technology- driven research to market oriented research.
	Structural Funds reform
1989	The collapse of communism across central and eastern Europe, which began in Poland and Hungary, is symbolized by the fall of the Berlin Wall in November. Faced by a mass exodus of its citizens to West, the East German government throws open the gates.
1990	Enlargement to the eastern regions of Unified Germany
	Third Framework Program (FP3) for research and development
	Objectives: competitiveness, training, cohesion, environment, quality of life
1991	In the Balkans, Yugoslavia begins to break apart. Fighting erupts first in Croatia, then in Bosnia, and Herzegovina where Serbs, Croats and Muslims fight in a bloody civil war.
	In December, the USSR is dissolved into 15 independent states.

Table A.1 (continued)

Year	Event
1992	Signing of the Maastricht Treaty, defining the institutional structure, based on three pillars:
	1. Internal market, common agricultural policies, cohesion, trade, social affairs; economic and monetary union; environment, research, education; cooperation, and development
	2. Foreign policy, defence, and security
	3. Justice, home affairs; common policies concerning asylum and immigration; subsidiarity
	The power of the European Parliament increases.
	As regards industrial policy and research and technology policy, the Treaty formalizes on an institutional level what the Framework Programs had proposed and accomplished on a practical level
	End of the Cold War
	Transeuropean networks for transport, energy, and information
1993	Ratification of the Maastricht Treaty
	Completion of the internal market
	Cohesion Funds are added to the Structural Funds.
1994	Fourth Framework Program (FP4) for research and technological development: new objectives are added to the existing ones: international cooperation, mobility of researchers
1995	Enlargement to Austria, Sweden, and Finland
	Jacques Santer is appointed European Commission president.
	In March, the Schengen Agreement takes effect in seven countries – Belgium, Germany, Spain, France, Luxembourg, The Netherlands, and Portugal. Travellers of any nationality can travel between all these countries without any passport control at the frontiers.
1997	Signing of the Treaty of Amsterdam
	The main features of the Treaty are: Rights of workers and citizens; Elimination of barriers to free movement; Strengthening Security; Representative for the Common Foreign and Security Policy (CFSP); More efficient institutional structure
	Every year, the member states' governments must submit, according to the Stability and Growth Pact, a three-year forecast about their public finances.
	The Treaty enforces the Schengen Convention.
1998	Fourteen member states qualify for the economic and monetary union (EMU); 11 states choose to enter from the outset
	Fifth Framework Program (FP5) for research and technological development. Objectives are divided into four thematic programs:
	1. natural resources and ecosystem;
	2. information society, competitiveness, and sustainable growth;
	3. research and innovation in an international context; and
	4. small companies and human resources
1999	The European Parliament forces the Commission to resign following accusations of irregularities.
	The European Council appoints Romano Prodi as European Commission president
	Ratification of the Treaty of Amsterdam
	Start of the EMU and launch of the Euro (€)
	NATO enlargement to Hungary, Poland, and the Czech Republic

Table A.1 (continued)

Year	Event
2000	Romano Prodi is the European Commission president.
	In March, the European Council, committed the EU to become by 2010 "the most dynamic and competitive knowledge-based economy in the world, capable of sustainable economic growth with more and better jobs, greater social cohesion and respect for the environment." This commitment is known as the *Lisbon Strategy*.
	The European Council and Commission formally proclaim the Charter of Fundamental Rights of the European Union.
2001	Signing of the Treaty of Nice
	Greece qualifies for the EMU and becomes part of the Euro Zone.
2002	The euro currency replaces national currencies. Euro notes and coins arrive. Printing, minting, and distributing them in 12 countries is a major logistical operation. More than 80 billion coins are involved. Notes are the same for all countries. Coins have one common face, giving the value, while the other carries a national emblem.
	The European Convention on institutional reforms is set up.
	Sixth Framework Program (FP6) for research and technological development:
	seven areas of research: genomics, society of information, nanotechnologies, aerospace, food safety, sustainable development, social sciences.
	The negotiations for the enlargement of the Union to Cyprus, Malta, and ten states of central, eastern, and south-eastern Europe (Bulgaria, Czech Republic, Estonia, Hungary, Latvia, Lithuania, Poland, Romania, Slovakia, and Slovenia) are concluded. Their accession to the EU will follow in 2004 and 2007.
2003	On 31 March, as part of her foreign and security policy, the EU takes on peace-keeping operations in the Balkans, firstly in the Former Yugoslav Republic of Macedonia, and then in Bosnia and Herzegovina.
	The proceedings of the European Convention are concluded and the "Draft Treaty establishing a Constitution for Europe" is presented to the European Council on 20 June.
	The war of the U.S. and United Kingdom on Iraq creates a rift in Europe between the front that proposes to strengthen the action of the United Nations (France, Germany, Belgium) and the interventionist front (UK, Spain, Italy, Poland, and other states in line with the policy of the United States).
	The Accession Treaty is signed in Greece between the Union and the Czech Republic, Estonia, Cyprus, Latvia, Lithuania, Hungary, Malta, Poland, Slovenia, and Slovakia.
	A referendum on the entry of Malta, Slovenia, Hungary, Lithuania, Slovakia, Poland, the Czech Republic, Estonia, and Latvia into the Union is held in the respective countries; the majority of the citizens vote in favour of accession.
	EU–U.S. summit meeting is held in Washington; the EU and the U.S. decide to cooperate in the fight against terrorism and arms proliferation; they decide to share information concerning suspect bank accounts and to set up joint investigation teams.
	Italy takes over the Presidency of the Council of the Union in the second half of the year 2003.
	In September, Sweden holds a referendum on the Euro; the majority rejects joining the single European currency.
	In October the Inter-Governmental Conference (IGC) starts for the reform of the Treaty following the proceedings of the European Convention; various member states ask for changes to be made to the Draft of the European Constitution proposed by the Convention in July; the Italian presidency envisages that the IGC will reach an agreement before the next elections of the European Parliament to be held in June 2004.

Table A.1 (continued)

Year	Event
	Jean-Claude Trichet, former governor of the Banque de France, succeeds Willem F. Duisenberg as president of the European Central Bank.
	The Ecofin Council decides not to sanction France and Germany for the deficit in the state budget, which exceeds the limit fixed by the Stability and Growth Pact.
	In December, at the summit of the European Council no agreement is reached concerning the conclusion of the IGC, which will continue its proceedings in the year 2004.
2004	Ireland takes over the Presidency of the Council of the European Union.
	The European Commission turns to the Court of Justice against the European Council for not sanctioning France and Germany for exceeding the deficit limit of the state budget fixed by the Stability and Growth Pact.
	NATO enlargement to Bulgaria, Estonia, Latvia, Lithuania, Romania, Slovakia, and Slovenia
	On 1 May 2004, ten states access the Union: the Czech Republic, Cyprus, Estonia, Hungary, Latvia, Lithuania, Malta, Poland, Slovakia, and Slovenia.
	The sixth term of the European Parliament (EP) is elected by the citizens of the 25 member states; Joseph Burrell Fondles is elected EP president.
	The European Council of June 2004 adopts the Constitutional Treaty resulting from the agreement reached at the Inter-Governmental Conference.
	On 1 July, The Netherlands takes over the Presidency of the Council of the European Union.
	In September, the Kyoto Protocol, an international treaty to limit global warming and cut emissions of greenhouse gases, comes into force. The EU has consistently taken the lead in efforts to reduce the impact of climate change. The United States is not a party to the protocol.
	On 29 October, the heads of state and government and the EU foreign ministers sign in Rome the Treaty establishing a Constitution for Europe.
	A new European Commission with 25 members is appointed on 17 November for the 2004–2009 term; the Commission president is José Manuel Durão Barroso.
	On 17 December 2004, the European Council fixes the date of 3 October 2005 as the starting date for accession negotiations for Turkey.
	Lithuania and Hungary ratify the Constitutional Treaty during the second semester.
2005	Luxembourg takes over the Presidency of the Council of the European Union.
	Slovenia and Italy ratify the Constitutional Treaty.
	In Spain in a referendum on the Constitutional Treaty 76 % vote "yes"; a parliamentary ratification follows in May. During the first semester also, Bulgaria, Austria, Germany, Slovakia, Romania, Belgium and Cyprus ratified the Constitutional Treaty.
	The European Council in March agrees on a more flexible interpretation of the Stability and Growth Pact.
	On 29 May, the people of France and on 1 June the people of The Netherlands choose to say "no" to the ratification of the Constitutional Treaty. In the light of these results, the European Council (17 June), considered that "we do not feel that the date initially planned for a report on ratification of the Treaty, 1 November 2006, is still tenable, since those countries which have not yet ratified the Treaty will be unable to furnish a clear reply before mid 2007."
	The Lisbon strategy is revamped and the program "i–2010 – A European Information Society for growth and employment" is launched.
	The UK takes over the Presidency of the Council of the European Union during the second semester.

Table A.1 (continued)

Year	Event
	On 6 July, Malta ratifies the Constitutional Treaty with a parliamentary procedure. On 10 July, Luxembourg voters vote "yes" to ratification.
	On 3 October, European Union accession negotiations open with Turkey and Croatia.
2006	Austria takes over the Presidency of the Council of the European Union during the first semester.
	The European Parliament adopts, by a large majority, a first-reading report on legislation opening up the EU single market for services. The Services Directive, also known as the Bolkestein Directive, is a major issue for the European Union.
	On 9 May, Estonia ratifies the Constitutional Treaty.
	Finland takes over the Presidency of the Council of the European Union during the second semester. On 11 July, the Council adopted a decision allowing Slovenia to join the euro area as from 1 January 2007.
	The Croatian Prime Minister, Vivo Sanader, affirms on 1 August at a meeting in Salzburg on the Stability Pact for South East Europe, that Croatia wants to join the EU in 2008.
	On 26 September, the Commission issues a monitoring report on the state of preparedness for EU membership of Bulgaria and Romania.
	On 8 November, the Commission issues the yearly Turkey Progress Report on accession negotiations.
	The European Parliament adopts a second-reading report on legislation opening up the EU single market for services.
	On 29 November, the Commission presents its recommendation on the continuation of Turkey's accession negotiations: "today we confirm that these negotiations continue, although with a slower pace. We will be able to return to normal speed as soon as Turkey has fulfilled her obligations related to the Ankara Protocol."
	In December, the Finnish Parliament ratifies the Constitutional Treaty; this brings to 18 the number of states which have ratified it: 16 member states and two candidate countries that will access in 2007. Seven member states have suspended the ratification process. They are: the Czech Republic, Denmark, Ireland, Poland, Portugal, Sweden, and the United Kingdom.
	The seventh Framework Program, 2007–2013 is announced as a program to build the European Research Area (ERA) of knowledge for growth. It has been structured with four specific programs: Cooperation, Ideas, People, and Capacities. It is adopted with a budget of 52 billion; the first call of FP7 Cooperation and Ideas Programs were opened in December.
2007	Germany takes over the Presidency of the Council of the European Union during the first semester.
	Two more countries from eastern Europe, Bulgaria, and Romania, join the EU on 1 January, brining the number of member states to 27 countries. The European Parliament has 18 and 35 MPs from the new member states. The College of the European Commission is enlarged with two Commissioners from each of the two members. Croatia, the Former Yugoslav Republic of Macedonia, and Turkey are candidates for future membership.
	The population within the Union is now 492.8 million inhabitants. The EU now has 23 official languages, following the addition of Bulgarian, Romanian, and Irish.
	Slovenia successfully adopts the euro.
	The European Parliament elects Hans-Girt Pottering as its new President.
	On 25 March, Angela Merkel, President of the European Council declared in front of the Council at the official ceremony to celebrate the 50th anniversary of the signing

Table A.1 (continued)

Year	Event
	of the Treaties of Rome: "we are united in our aim of placing the European Union on a renewed common basis before the European Parliament elections in 2009."
	On 22 June, the European Council agrees to convene an Inter-governmental Conference (IGC) and invites the incoming Portuguese Presidency to draw up a draft Reform Treaty to be finalized by the IGC before the end of 2007.
	Portugal takes over the Presidency of the Council of the European Union during the second semester.
	On 10 July, Ecofin Council decides that Cyprus and Malta can adopt the euro as from 1 January 2008.
	On 23 July, the Portuguese Minister of Foreign Affairs, Luis Amador opens the IGC that will decide on the adoption, on common agreement, of the amendments to be made to the Treaties.
	On 19 October, the European Council agrees on the text of a Reform Treaty, ending a two-year crisis of confidence in Europe's future.
	On 6 November, the Commission issues the yearly Turkey Progress Report on accession negotiations.
	On 13 December, the 27 EU countries signed the Treaty of Lisbon, which amends the previous Treaties. It is designed to make the EU more democratic, efficient and transparent, and thereby able to tackle global challenges such as climate change, security and sustainable development. Before the Treaty can come into force, which is hoped to be before the next European Parliament elections in June 2009, it has to be ratified by each of the 27 member states.
	On 21 December, The Czech Republic, Estonia, Hungary, Latvia, Lithuania, Malta, Poland, Slovakia, Slovenia join the Schengen Treaty, lifting of internal border control. The EU members already belonging to the Schengen area are: Austria, Belgium, Denmark, France, Finland, Germany, Greece, Italy, Luxembourg, The Netherlands, Portugal, Spain, and Sweden.

Appendix B
University Courses on Research, Quality, and Competitiveness

From 1999 to 2007, the author held several courses on *Research and Technology Policy in the European Union* at the Georgia Institute of Technology in Atlanta, Georgia (1999); at the Faculty of Political Sciences at the University of Bologna, Italy (2000–2002), and the Faculty of Engineering at the University of Bologna (2002–2007), in first-level and second-level degree courses and in a Master's course; and at the Faculty of Economics at the University of Ferrara (2001–2005) in a first-level degree course. These courses provided the opportunity to think about and draft this book. Updated information on the author's courses is available at the URL http://stajano.deis.unibo.it

Since 2002, the above lectures have been integrated by asynchronous online e-learning activities inspired by the works by Palloff [Palloff et al. 1999, Palloff et al. 2001] within the framework of an experiment on educational technology conducted at the University of Bologna and described at the URL http://www.elearning.unibo.it. The author used the *Claroline* open-source platform of the University of Louvain-la-Neuve, Belgium, hosted by Dokeos (http://www.dokeos.com./hosting.php), a service company in Belgium run by the architect of Claroline, Dr. Thomas De Praetere. The user interface is particularly easy to use and the platform is continuously improved based on the sharing of some of the developments commissioned by users.

The author's approach to e-learning is meant to create a learning community in which students play an active role and are empowered to the creation of content, knowledge, know-how, and skills. The learning community is based on the extension of face-to-face lectures with an asynchronous online interaction.

The use of the e-learning platform in the author's courses has been designed to:

- ensure the utmost involvement of the students
- create the conditions for the creation of content and knowledge by the students to complement the role played by the professor
- create an online learning community, giving a voice to less extroverted and more reflective students than those who participate more in classroom discussions
- continue the classroom discussions through asynchronous online communication
- ensure a strong interaction with and between students

- let students introduce themselves and describe the reasons why they have enrolled and their expectations about the course
- structure asynchronous discussions in several forums on topics suggested both by the professor and by the students

The combination of face-to face, highly interactive classroom activity, and the asynchronous activities on the e-learning platform has resulted in intense involvement of all the students enrolled and in the achievement of the goal of letting students contribute to the development of content, knowledge, know-how, and skills.

The e-learning platform used offers many services and supports any educational approach. Among the services used, the most effective have been the forums, the groups, the agenda, the announcements, and the document repository. The high-quality educational service provided is very demanding in terms of the lecturer's time. In the academic year 2004–2005, the course lasted for 50 classroom hours over eight weeks, and in addition to time spent lecturing and preparing lectures, no less than two hours per day six days a week were required – with an attendance of 25 students – to monitor the online operations and guide and contribute to the discussions. The author was supported by a tutor, a researcher in political science, who worked part time during the course and the two months preceding it. The author confirms the figures suggested in the literature by saying that one tutor is required for every 25 students in order to guarantee a lively, fully controlled interaction on the platform, and to support students in the development of research papers.

Some examples of forums topics introduced in a 2004 course are provided in Table B.1.

A 50-hour classroom course (six credits) is described below. It is designed for the second year of the second-level degree in Electronics Engineering, Computer Engineering, and Telecommunications Engineering. The course (academic year

Table B.1 Topics of forums on the e-learning platform. The forum on Comments and Emerging Issues enables to ask for off-line explanations, thus giving a voice to less extroverted students than those who participate more in classroom discussions and expanding on topics that emerge through study and reading outside the class

Discussion forums
1   Comments and emerging issues
2   Why did I enroll in this course?
3   What if France or the UK does not ratify the Constitutional Treaty?
4   The profession of engineer in the information society
5   Fiat in the world market
6   Decline of Italy
7   War in Iraq and European integration
8   Renewable energy and sustainable development
9   Illegal immigration
10   Should the ban on nuclear power stations in Italy be lifted?
11   Environmental friendliness of renewable energies
12   One-off measures to balance the Italian budget

2004–2005) is part of a set of courses offered in English at the Faculty of Engineering at the University of Bologna, Italy. Any professor interested in producing teaching material based on this book will find in the in the recommended reading web site all the figures and tables of this book. These can be used for drawing up presentations in PowerPoint. Also, the PowerPoint presentations used by the author in the course described below are included as pdf files. Lecturers adopting the book may get additional and updated images in colour and PowerPoint ppt files, by contacting the author at the e-mail indicated in the preface.

The course, called *Research and Technology Policy in the European Union,* is documented at the URL http://stajano.deis.unibo.it/ptboEN2005.htm. A short description follows:

Objectives

Course students who will study the recommended books and take part in both the classroom discussions and the e-learning activities will be able to

- describe the institutional structure of the European Union its ongoing transformations and the fundamentals of its sustainable development policies
- describe the transformation of the European internal market and the competitiveness policy
- describe the sectors of excellence and competitiveness of the European industry both as a whole and in the single EU member states
- describe EU development policies and particularly the role of the R&D Framework Program in the promotion of cohesion, competitiveness, and industrial development in the EU
- describe the role of the e-Europe program in the promotion of the transformation of companies and public administrations within the context of the information society
- access the bibliographic sources and information on EU R&D available on the Internet

Content

The first part of the course focuses on the EU in general and describes the history of the European integration and the institutional transformation of the European Economic Community into the European Union outlining the relative policies and programs. The progressive enlargement from 6 to 27 members and the evolution from a customs agreement to the internal market are illustrated, together with the constitutional changes brought about by the Treaties from 1957 to 7. The principles on which the Community is funded, its policies and budget, and the roles of the institutions and member states are also described and discussed. Particular attention is paid to policies that are closely related to sustainable growth: internal market, competitiveness policy, Economic and Monetary Union, enlargement, and constitutional reform.

The second part of the course focuses on the comparative analysis between EU15, AC12, and the candidate countries, with particular attention to economic aspects and industry competitiveness, explaining the need for the European Union to compete on quality to face the competition in the Triad and from the emerging Asian economies.

The third part is the real core of the course. It focuses on industrial competitiveness, technology, research, and innovation policies within the framework of the new economy. The nature and peculiarities of the information society are studied and a number of success stories are analysed to show that small and large enterprises can find a way to success by coping with superabundant information and structuring it into corporate knowledge, which creates the new competitive advantage in the globalized and interconnected economy of the new century.

Course Structure

The course is a 50-hour course. It requires constant attendance, the reading of some books, the use of information sources available on the Internet, and the drawing up and discussion of a short research paper. Topics chosen for research papers in 2004 and 2005 are presented in Table B.2. Students are encouraged to read *The Economist* every week and the *Financial Times* every day. The university offers online access to these papers. After the first ten lectures, each student agrees with the professor the topic of a short research paper, which can be drawn up and presented either individually or in group (groups from two to four students). Such short papers can deal with one of the course topics or with hot issues in the economic, technological,

Table B.2 Topics of research papers are chosen by students on the basis of their interest in one subject addressed in the course. Students face both the challenge of writing a paper and presenting it in a fixed time, and of defending their arguments as they stand the test of discussion with other students

Research papers	
1	Accession countries: Economic and social challenges
2	Alternative energy
3	Brain drain
4	Cohesion and structural funds after the 2004 enlargement
5	Comparing Finland and Italy
6	Digital entertainment
7	Effects of undeclared work
8	Euro effects on prices
9	Ferrari: a model
10	IBM – Rises and falls
11	Information society and labour market
12	Ireland, the Celtic tiger
13	Italy and the rising of China's economy
14	Protection of typical products
15	Ryanair
16	The Alitalia case
17	The Fiat crisis
18	Turkey in the European Union

and political world if they are linked to the course content. The presentation and discussion of these short dissertations is an integral part of the course. Each student is both the author/co-author of a research paper and the *discussant* of another paper. The size of the working groups is adjusted to the number of students, which, over the years, has ranged from 15 to 90.

The e-learning platform gives the possibility of creating several types of exercises, i.e. multiple choice, true/false, or fill-in-the-gap exercises. This course includes exercises on EU geography, economy, and current history and on competitiveness, as well as a self-assessment test to measure the level of proficiency in English and assesses the oral, written, and comprehension skills of students who are thinking of taking part in the course, to see if they are fluent enough to feel at ease on the course.

Thanks to the e-learning platform, it is possible to trace students' activities and identify anyone who is for some reason in risk of losing contact with the class or moving at too slow a pace. Early intervention may rescue cases that would otherwise be lost. During the course, we checked that each student connected two or three times during each of the course days and mostly visited first the forums and then the online documentation sources and the agenda.

In conclusion, the integration of the classroom face-to-face course with asynchronous online activities supported by a good e-learning platform led to great involvement and cohesion between the work groups and completed the transfer of knowledge from the professor to the students through the creation of original content, knowledge, and skills by the students themselves.

The asynchronous tutoring has required daily availability on the part of the professor and of the tutor, who each took about twice the number of classroom hours to manage communication on the platform. The workload increases with the number of students, and a professor should be able to rely on the help of a tutor for every 25 students.

In 2004, students' comments highlighted the problem of the *digital divide* risk, which has been solved in 2005: the University can apply e-learning on a large scale without creating any discrimination only if it guarantees students an appropriate availability of workstations with Internet connection.

Appendix C
Contents of the Recommended Reading Web Site

The recommended reading web site at the URL http://stajano.deis.unibo.it/RQC.htm contains the following:

1. Recommended reading
2. Figures
3. Tables
4. Slides
5. Updates
6. Book reviews
7. A page for *ERRATA*.

C.1 Recommended Reading

Recommended reading is organized following the chapters of this book. The recommended reading prefixed with EC, EP, or EU is covered by copyright of the European Commission. They can be reproduced with acknowledgment of the source. "The Commission accepts no responsibility or liability whatsoever with regard to the material contained in the document." The papers drawn from the Italian daily *La Repubblica* are covered by copyright of *La Repubblica* and reproduced with kind permission. Further reproduction is not authorized. Other files are in the public domain and can be reproduced if the source is quoted.

A list of the recommended reading by chapter follows in Section C.8 below.

C.2 Figures

Figures in the book are presented for use by teachers for the production of course materials. Figures are displayed as HTM albums by chapter. Figures can be reproduced if the source is quoted.

C.3 Tables

Tables in the book are presented for use by teachers for the production of course materials. Tables can be reproduced if the source is quoted.

C.4 Slides

In the Slides folder there are pdf files of slides used in the 50-hour course described in Appendix A and taught by the author in the year 2005. These files can be reproduced if the source is quoted. Lecturers adopting the book may get additional and updated images and PowerPoint files, by contacting the author at the e-mail indicated in the preface.

C.5 Link to Updates

This link connects the reader with updates on developments relative to topics covered in the book, such as the progress of accession negotiations with the candidate countries.

C.6 Book Reviews

This section links to web pages provided by Springer and to web pages provided by the author containing book reviews and other book-related information.

C.7 Errata

This web page will contain correction to errors, if any, identified after the publication of the book. The author welcomes readers' e-mails mentioning any errors. The author's e-mail address is indicated in the Preface.

C.8 Recommended Reading by Chapter

C.8.1 Origins of the European Union

1.1. *History and Policies of the European Union*
      Pascal Fontaine, "Europe in 12 lessons" (2003)
      EC, The EU at a glance – The History of the European Union (2004)
      Pascal Fontaine, A new idea for Europe – The Schuman declaration – 1950–2000 (2000)
      50 Ways forward – Europe best successes (2007)

1.2.   *European Technology Policy*
Attilio Stajano, Technology Policy in the European Union, and European
Integration (1999)
See also: Recommended Reading of EC brochures at Chapter 9

C.8.2 Overview of Member States

2.1.   *European Commission*
EC, A Community of Fifteen. Key Figures (2000)
EC, EU Facts and Figures (2004)
EC, EU Facts and Figures (2005)
EC, EU Facts and Figures (2007)
2.2.   *Eurostat*
General government consolidated gross debt (1994–2005) as a percentage
of GDP
Gross domestic product at market prices (1996–2007)
Gross domestic product PC PPS per inhabitant (1997–2008)
Harmonized Indices Consumer Prices 2006
2.3.   *World Bank*
Economic overview and detailed economic data for accession and candidate
countries (2003–2007)
AC12 and CC3 data
2.4.   *International Monetary Fund*
GDP, constant prices, annual percent change for EU27 and CC3
GDP current prices and ppp per capita for EU27 and CC3
2.5.   Istat,
Italy in Figures (2004)
Italy in Figures (2005)
Italy in Figures (2007)
2.6.   *Summary data on EU27 CC3, EFTA, G8, BRIC, and other countries*
The World Factbook: Web pages of EU27, candidate countries, EFTA, G8,
emerging economies (BRIC) and other countries
World Bank: Economic overview and detailed economic data for accession
and candidate countries (2003–2007)
EC, Fact Sheets on accession and candidate countries (2005)
See also: Recommended reading of Chapter 11 "Enlargement of the European
Union."

C.8.3 Competitiveness of European Economy

3.1.   *European Commission, Competitiveness Reports*
Competitiveness Report 2007
Competitiveness Report 2006
Competitiveness Report 2004

Competitiveness Report 2003
Competitiveness Report 2002
Competitiveness Report 2001
Global Europe A Brochure: contribution to EU growth (2007)
Business demography (2007)
Monitoring Economic Activity (2007)
EU Productivity and Competitiveness (2004)

3.2. *Cohesion*
A more cohesive society (2007)

3.3. *SMEs*
European Charter for small enterprises (2006)
EIB, Helping SMEs to get finance (2006)
SMEs and entrepreneurship in the EU (2006)
SMEs and Co-operation (2003)

3.4. *Innovation*
Manual for Innovation Policy (2006)
Innovating regions in Europe (2005)
Innovation in services (2005)

3.5. *Enterprise*
Indicators 2004

3.6. *Ageing*
Age management in companies in Europe (2006)

3.7. *Information society*
Information society and enterprise (2005)
EU Productivity and Competitiveness: an Industry Perspective (2004)

3.8. *Strengthening competitiveness*
Enterprise Networking (2005)

3.9. *Internal Market*
OHIM Annual Report 2005
Community TradeMark (2006)
EU International Trade in Services in 2004

3.10. *Tourism*
Tourism and the Internet (2006)
Sustainable Development of Tourism (2006)
Sustainable Development of Tourism – Indicators (2006)
Tourism Trends (2006)
Identifying Declining Tourist Destinations (2004)
European Tourism Industry (2004)

3.11. *World Economic Forum*
Growth Competitiveness Annual Report and Ranking 2002
Growth Competitiveness Index Executive Summary 2002
Growth Competitiveness Annual Report and Ranking 2003–2004
Report 2003–2004 Growth Competitive Index
Growth Competitiveness 2003–04 Report Chapter One with GCI
components

Growth Competitiveness Annual Report and Ranking 2004–2005
Growth Competitiveness Annual Report 2004 Executive Summary
Growth Competitive Index Ranking 2004 and 2003
Extract from the Preface of the Competitiveness Report 2004–2005
Interviews: India, China, Employment, Corruption, Stability
(2003)
India: India's new breed of multinationals
China: China's challenge for the year ahead
Employment: Employment critical to the global balance
Corruption: Tackling corruption is good for business
Stability: Stability is crucial

3.12.   *Environment*
Towards Environmentally Sound Future (2004)

3.13.   *Kyoto Protocol*
EC, Kyoto protocol on climatic changes Italian (2003)
UN, Kyoto protocol on climatic changes English
EC, Kyoto protocol ratification list (2004)
UN, Climate Convention (1992)
EC, Environment Overview (2005)
EC, Environment Brochure (2002)

3.14.   *Broad Economic Policy Guidelines*
EC, Broad economic policy guidelines 1999
EC, Broad economic policy guidelines 2000
EC, Broad economic policy guidelines 2001
EC, Broad economic policy guidelines 2002
EC, Broad economic policy guidelines 2003–2005
EC, Broad economic policy guidelines 2005–2008

3.15.   *Indicators and Scoreboards*
EC, A Pocketbook of Enterprise Policy Indicators (2001)
EC, A Pocketbook of Enterprise Policy Indicators (2003)
EC, Benchmarking Enterprise Policy. Results from the 2001 Scoreboard
EC, Benchmarking Enterprise Policy Results from the 2002 Scoreboard
EC, Benchmarking Enterprise Policy Results from the 2003 Scoreboard
EC, Indicators for national research policies (2001)

3.16.   *BRICs*
Goldman Sachs: "Global Economic paper 99. Dreaming with BRICs, the Path to 2050" (2003)

3.17.   *Ranking of Top Universities*
Shanghai Jiao Tong University, "Top 100 European Universities" (2003)
Shanghai Jiao Tong University, "Top 101 World Universities" (2003)
The Times Higher Education World University Ranking (2004)

3.18.   *SMEs*
EC, Observatory of European SMEs (2002)
ECB, Helping SMEs to access Finance (2006)
EC, European Charter for SMEs (2006)

3.19. *Trade*
       Trade in high-tech products 2006
       Trade in high-tech products 2008
3.20. *Tax evasion*
       Gianluca Luzi "Berlusconi difende chi evade le tasse" *La Repubblica* 18
       February 2004, reproduced with kind permission
3.21. *Italy's decline*
       Roberto Petrini "L'Italia non cammina più" *La Repubblica* 14 February
       2004, reproduced with kind permission
       Carlo A. Ciampi "Discorso pronunciato alla consegna del premio Leonardo"
       (2003)
       See also: Recommended Reading of EC brochures at Chapter 9

C.8.4 Competitiveness and Quality

4.1. *Quality Competition*
      Karl Aiginger (editor), European Commission, Enterprise Papers, "Europe's
      position in quality competition" (2001)
4.2. *The Car industry*
      Confapi, "La crisi della Fiat Auto" (2002)
      Volkswagen and BMW increase their market share (2006)

C.8.5 EU Research: Objectives and Results

5.1. *Framework Programs FP6 and FP7*
      Cordis, the EC R&D Info Service: the seventh FP (2007)
      EC, Research and Innovation Overview (2005)
      EC, Innovation in FP6 (2005)
      EC, Guidelines for Proposal Evaluation and Selection Procedures FP6 (2003)
      EC, Building the ERA of knowledge for growth (FP7)
5.2. *European Research Council*
      Act establishing the ERC (2007)
      ERC Flyer (2007)
5.3. *European Institute of Innovation and Technology*
      Regulation establishing the EIT (2008)
      Member States' agreement on the European Institute of Technology (2007)
      EIT Flyer (2007)
5.4. *RDT-Info / research*eu the magazine of the European research area*
      Special Issue Eurobarometer November 2005
      no.51 December 2006
      Special Issue Eiroforum February 2007
      Special Issue FP7 June 2007

no.52 June 2007
no.53 September 2007
Special Issue "The Earth as a Work of Art" September 2007

5.5. *Reports of the DG Research (2003-2004)*
Slowdown of Transition towards the Knowledge-Based Economy Investment in Research and Development. Are we still on track?
International R&D Investment flows: the share of the EU15 in decline
From "European Paradox" to declining competitiveness?
Mapping Excellence in Science and Technology across Europe Life Sciences
Mapping Excellence in Science and Technology across Europe Nanoscience and Nanotechnology
Mapping Excellence in Economics

5.6. *Climate change*
Research on Climate in FP5 and FP6 (2006)

5.7. *Energy*
Energy research in FP7 (2007)
Energy futures: the role of R&D (2006)

5.8. *Health Care*
Stroke Research in the EU (2007)

5.9. *Information Society*
Towards a knowledge-based Europe (2002)
Information society technologies 2005–2006 Strategies for leadership (2006)

5.10. *SMEs*
Research-Results for SMEs (2005)

5.11. *Transport*
EU Research on Road Transport (2007)
See also: concerning results, Recommended Reading of Chapters 6, 7, and 8; and Recommended Reading of EC brochures at Chapter 9

C.8.6 *Framework Programs*

6.1. *FP5*
EC, FP5 (1997)
EC, Intermediate evaluation of the Fifth Framework Program (2000)
EC, Improving Quality of Life (2000)
EC, Information Society Technologies (2000)
EC, Environment, Energy (2000)
EC, Global Competition (2000)
EC, Human Capital (2000)
EC, Innovation and SMEs (2000)
EC, International Research (2000)

6.2. *FP6*
EC, FP6 in Brief (2002)
EC, Adoption of the Sixth Framework Program (2002)
EC, Specific Programs of the Sixth FP (2002)
EC, FP6 Networks of Excellence (2003)
EC, Instruments for PQ6 Integrated Projects (2003)
EC, EURATOM: The energy challenge (2003)

6.3. *Announcement of FP7*
Council approves FP7

6.4. *Description of FP7*

Cooperation	Predefined themes, refined FP6 instruments
Ideas	Frontier research, competition, individual grants
Capacities	Human potential, mobility
People	Infrastructure, SMEs, science and society
Euratom	Nuclear
Jrc	Non-nuclear
Jrc Program	Text of the Council decision

6.5. *Budget of FP7*
Community funding of Specific Programs in FP7

6.6. *Upcoming results from completed and ongoing projects*

6.7. *Reading on Nanotechnologies*
Christine Peterson, "Molecular Manufacturing" (2003)
Ralf Merkle, "Introduction to Nanotechnology" (2004)

6.8. *Leaflets on JRC (2004):*
Food and health
   Consumer protection and food
   Testing for BSE infection
   Biotechnology and GMOs
   Medical imaging
   Nanobiotechnology solutions for medical devices and biosensors
   Crop yield forecasts and production estimates
   Cancer therapy
   Validation of alternative methods
Life science and environment
   Indoor pollution monitoring
   Life sciences
   Oil discharge monitoring
   Mining waste management
   Air pollution and climate change
   Water quality
   Systems for alternative fuels
   Global land cover monitoring
   Clean technologies
   Vehicle pollution

    Nuclear
      Natural hazards
      Nuclear power
      Transuranus code
      Thermophysics and nuclear materials science
      Basic actinide research
      Nuclear safeguards
      Measuring environmental radioactivity
    Other subjects
      Laser transmutation studies
      European media monitoring
      High precision maps
      Enlargement futures
      Privacy in the information society
      Land parcel identification systems
      Technology transfer
      JRC overview / JRC quick tour
    See also: Recommended Reading of Chapters 5, 6, and 7

C.8.7 *Entrepreneurship, Innovation, and Competitiveness*

7.1.   *Competitiveness and Innovation*
      Innovation Policy Studies (2002)
      Competitiveness and Innovation Framework Program (2005)
7.2.   *Venture Capital*
      Venture Capital Investments (2007)
7.3.   *Innovation in Energy Savings*
      Seven stories of innovative projects in Europe (2006)
7.4.   *Scoreboards and Statistics*
      European Innovation Scoreboards 2005 by Sector
      European Innovation Scoreboard 2005 – Innovation and Economic
      performance
      Innovation in Services (2006)
      Value added of high-tech manufactures (2007)
      Innovation in European Banks (2007)
7.5.   *Patents and IPRs*
      EC, The Competitiveness (Internal Market, Industry and Research) Council
      reaches agreement on Community Patent on 3 March 2003 (1999)
      Attilio Stajano, "Making Academia aware of Intellectual Property Rights
      (IPR) – comparing US and EU experiences" (1999)
7.6.   *Reviews*
      Innovation Strengths and Weaknesses by Country (2005)
      Methodology Report (2006)

Evaluating and Comparing the innovation performance of the U.S. and the EU (2005)

EC, Final Report on Skills and Mobility (December 2001) (1999)

Unice, Benchmarking report The reNEWed Economy (2001)

C.8.8 *Competitiveness in the Knowledge-Based Society*

8.1.   *e-Commerce*
       EC, Electronic Commerce—An Introduction (1998)
8.2.   *e-Europe*
       EC, e-Europe 2005: An Information Society for all (2002)
       EC, e-Europe 2002 Benchmarking Report [EC 2002-4] (2002)
8.3.   *Information Society*
       EC, Information Society Overview (2005)
8.4.   *Lisbon Strategy*
       EC, A new start 2005
       EC, Lisbon Strategy for Growth and Employment (2004)
       EC, i2010 – A European Information Society for growth and employment (2005)
       UNICE, Lisbon Strategy Time is running out (2003)
       Eurostat, Statistics on the Information Society in Europe (2003)
8.5.   Strategy of Selected Enterprises in the Knowledge-based Society
       GE
       Siemens
       Cemex
       7-Eleven
       Ryanair
       Michelin
8.6.   *Speeches and Opinions*
       Viviane Reding, "La stratégie de Lisbonne et le rôle moteur du secteur des TIC" (2004)
       Janez Potočnik, "Competitiveness and Economic Growth: R&D policies and the Lisbon Agenda" (2004)
       Joaquín Almunia, "Remettre l'Europe sur le chemin de la croissance" (2004)
       José Manuel Barroso, "The Lisbon Commission" (2004)
       EC, Remarks of the President of the European Council, Dutch Prime Minister Jan Peter Balkenende, about the Kok report (2004)
       EC, Tripartite Social Summit – joint press release Presidency / European Commission (2004)
8.7.   *Other recommended reading from The Economist (2001)*[1]
       The Economist, vol. 359 n. 8228, "Older, wiser, webbier" 30 June 2001

---

[1] These articles are not reproduced on the recommended reading web site because they are covered by copyright.

The Economist, vol. 359 n. 8224, "Electronic glue" 2 June 2001
The Economist, vol. 359 n. 8222, "While Welch waited" 19 May 2001
The Economist, vol. 359 n. 8226, "The Cemex way" 14 June 2001

C.8.9 The Position of the EU in the Quest for Competitiveness and Beyond

9.1. *The World Economic Forum*
The World Economic Forum, 2005–2006 Report Executive Summary (2005)
9.2. *European Commission Brochures*

- *General on EU*

  - 50 Ways forward - Europe best successes (2007)
  - A community of cultures – The European Union and the arts (2001)
  - A new idea for Europe – The Schuman declaration-1950–2000 (2000)
  - Europe's agenda 2000 (1999)
  - How the European Union works (2007)
  - Panorama of the European Union (2007)
  - Serving the people of Europe – What the EC does for you (2005)
  - Many tongues, one family – Languages in the EU (2004)
  - Traveling in Europe (2007)

- *Internal Market*

  - It's a better life How the EU's single market benefits you (2002)
  - Better off in Europe – How the EU's single market benefits you (2005)
  - Living, learning and working anywhere in Europe (2003)
  - Customs policy (2003)

- *Research and Knowledge-based Society*

  - Europe: an area for research (2000)
  - Looking beyond tomorrow: scientific research in the EU (2004)
  - Towards a knowledge-based Europe (2003)

- *Enlargement*

  - The EU biggest enlargement (2003)
  - The EU: still enlarging (2003)

- *Environment*

  - A quality environment – How the EU is contributing (2005)
  - Combating climate change – The EU leads the way (2007)

- *Transport*

  - Europe at a crossroads – The need for sustainable transport (2003)

- *Budget*

  - Going for growth (2003)
  - Investing in our future – The budget of the European Union (2006)

- *Agricultural Policy*

  - From farm to fork-Safe food for Europe's consumers (2004)
  - Healthy food for Europe's consumers (2004)

- *Foreign Policy*

  - The EU in the world – The foreign policy of the EU (2007)

- *Foreign Trade*

  - The EU and World Trade (2003)

- *Social Policy*

  - European employment and social policy: a policy for people (2000)

- *Justice and Home Affairs*

  - Freedom, security and justice for all – Justice and home affairs in the European Union (2004)

- *Tax Policy*

  - Tax policy in the European Union (2000)

C.8.10 *From the Treaty of Rome to the Reform Treaty of Lisbon*

10.1. *Selected European Union Policies*
     Common Agricultural Policy (2001)
     Fact sheet: Simplification of the Common Agricultural Policy (2006)
     Regional Policy (2001)
     Enterprise (2001)
10.2. *The Constitutional Treaty*
     EU, A Constitution for Europe. A Brochure (2004)
     EU, A Constitution for Europe. Presentation to citizens (2004)
     EU, Constitutional Treaty Full Text (2004)
     EU, Constitutional Treaty Short Summary (2004)
     European Council, Charter of the Fundamental Rights of the EU (2000)
     Lucio Levi, Carta dei diritti e Costituzione europea (2003)
10.3. *The Schengen Convention (1989)*
     Schengen (Agreement and Convention)
10.4. *The European Commission 2004–2009*
     EC, The College of Commissioners (2004)
     EC, The Initially Designated Commissioners (2004)

10.5.   *Fifty Year Anniversary of the Treaty of Rome*
Declaration on the occasion of the 50th anniversary of the signature of the
Treaties of Rome
Speech by Ms Angela Merkel, President of the European Council
Hans-Gert Pöttering on the 50th anniversary of the signature of the Treaties
of Rome
Barroso on the 50th anniversary of the signature of the Treaties of Rome
10.6.   *Press cuttings on the revamping of the Constitutional Treaty*
Constitutional Treaty – key elements – 20 April 2007
Europe set for constitutional revamp? – 26 March 2007
UK and The Netherlands back slimmed-down EU Treaty – 17 April 2007
Poland and Czech Republic get tough – 19 April 2007
Merkel wants EU leaders to stick to 2009 treaty deadline – 20 April 2007
Barroso presses for new EU Treaty by 2009 – 20 April 2007
Sarkozy: "France is back in Europe" – 7 May 2007
10.7.   *IGC 2007*
Draft Preamble as of 24 July 2007
Presidency Conclusions – Brussels, 21/22 June 2007 and draft IGC mandate
Opening the IGC
IGC Final Act
10.8.   *Treaty of Lisbon*
Treaty of Lisbon amending the Treaty on EU and the Treaty establishing the
European Community
10.9.   *Press cutting on the Lisbon Treaty*
Javier Solana welcomes the Lisbon Treaty
See also: Recommended Reading of EC brochures at Chapter 9

C.8.11 Enlargement of the European Union

11.1.   *Images of AC12 and CC3 countries*
EC, Images of accession and candidate countries
11.2.   *Web pages and overviews on accession and candidate countries*
CIA, The World Factbook
EUROPA, Web pages of accession and candidate countries
EC, Progress: Strategic papers about ongoing negotiations
WorldBank, Economic overview and detailed economic data for accession
and candidate countries.
EC, InfoEnlarg: General information on Enlargement, Leaflet, Documents,
Images, and FAQs on Enlargement
11.3.   *The EU view on Enlargement*
More unity and more diversity. The European Union's biggest enlargement
(2003)

Europe's Next Frontiers Olli Rehn, Enlargement Commissioner 20 October 2006

Questions and answers on the Union's enlargement strategy and its integration capacity Brussels, 8 November 2006

Frequently asked questions on Instrument for Pre-Accession Assistance (IPA) Brussels, 8 November 2006

Commission proposes renewed consensus on Enlargement Olli Rehn, Enlargement Commissioner 8 November 2006

Enlargement Strategy and Main Challenges 2006–2007, 8 November 2006

Enlargement Strategy and Main Challenges 2007–2008 Brussels, 6. November 2007

11.4.   *Eurobarometer*
Special Issue November 2005 ISSN 1024-0802

11.5.   *Romania and Bulgaria*
Monitoring report on the state of preparedness for EU membership of Bulgaria and Romania 26 September 2006

Bulgaria's progress on accompanying measures following Accession Brussels, 27 June 2007

Romania's progress on accompanying measures following Accession Brussels, 27 June 2007

11.6.   *Croatia and the Former Yugoslav Republic of Macedonia*
Key findings of the 2006 progress reports on the candidate countries

Croatia 2006 Progress Report 8 November 2006

Council Decision on the principles, priorities and conditions contained in the Accession Partnership with Croatia Brussels, 6 November 2007

Croatia 2007 Progress Report, Brussels, 6 November 2007

The former Yugoslav Republic of Macedonia 2006 Progress Report

The former Yugoslav Republic of Macedonia 2007 Progress Report

Council Decision on the principles, priorities and conditions contained in the Accession Partnership with the former Yugoslav Republic of Macedonia Brussels, 6 November 2007

11.7.   *Turkey*
Barroso on the opening of accession negotiations with Turkey 3 October 2005 Turkey Economic profile (2005)

Commission presents its recommendation on the continuation of Turkey's accession negotiations 29 November 2006

Turkey 2006 Progress Report on accession negotiations 8 November 2006

Proposal for a Council Decision of 6 November 2007 "on the principles, priorities and conditions contained in the Accession Partnership with Turkey"

Turkey 2007 Progress Report on accession negotiations 6 November 2007

A. Stajano, "Turkey as a member of the EU: an open ended negotiation for over forty years," Proceedings of the International Conference on the 50th Anniversary of the Treaty of Rome and EU-Asia Relations, Tamkang University, Taipei, Taiwan 21 November 2007.

See also: Recommended Reading of EC brochures at Chapter 9

C.8.12 Budget of the European Union

12.1.   *EC, Budget*
        Financing the EU (1998)
        Introduction (2001)
        Allocation of EU Expenditure (2004)
        EU Budget Meeting today's (2006) needs. Building a foundation for the
        future
        The budget of the European Union: Investing in our common future (2007)
        EU Budget 2005
        EU Budget 2006
        EU Budget 2007
        Proposed EU Budget 2008
12.2.   *Financial perspectives*
        A Financial Perspective for Europe's future
        Financial perspectives 2007–13 (full text)
        See also: Recommended Reading of EC brochures at Chapter 9

C.8.13 Internal Market and Competition

13.1.   *Industrial Policies*
        Patrizio Bianchi, "Industrial Policies and Economic Integration," Routledge
        1998,[2] Chapters 5 and 6
13.2.   *Internal Market*
        EC, Internal Market Overview
        EC, Bolkenstein Directive on Services (2004)
        The Single Internal Markey for Services
        General Principles
        Common Position adopted by the Council with a view to the adoption of a
        Directive of the European Parliament and of the Council on services in the
        internal market (17 July 2006)
13.3.   *Competition*
        EC, Competition Overview
        State Aid
        Study on the enforcement of State Aid law at national level (2006)
13.4.   *Press Cuttings on the Internal Market*
        Antonio Polito, "La competizione farà bene a imprese e consumatori," *La
        Repubblica*, 16 February 2002 (reproduced with kind permission)
        Andrea Tarquini, "Concessionarie libere, guerra ai monopoli," *La Repub-
        blica,* 16 February 2002 (reproduced with kind permission)
        See also: Recommended Reading of EC brochures at Chapter 9

---

[2] This reading is not available on the recommended reading web site because it is protected by
copyright.

C.8.14 *Economic and Monetary Union*

14.1.   *European Central Bank*
        How the euro became our money. A short history of the euro banknotes
        and coins
        The Eurosystem – The European System of Central Banks
        European Central Bank Report 2004
        European Central Bank Report 2005
        European Central Bank Report 2006
        European Central Bank Convergence Report 2007
14.2.   *European Commission Brochures*
        A short Guide to the Euro (2006)
        Monitoring Consumer Prices - Did the euro cause prices to rise? (2005)
        EC, Stability and Growth Pact
14.3.   *European Commission Economic Outlook*
        European Commission Convergence Report 2006
        EC, Economic Outlook for the Euro Area in 2005–06 (2004)
        EC, Economic Forecast 2007
        EC, Economic Forecast 2006
        EC, Economic Forecast 2005
        EU, Economy 2007 Review
        EU, Economy 2004 Review
        EU, Economy 2003 Review
        EU, Economy 2002 Review
        EU, Economy 2001 Review
        EU, Economy 2000 Review
        EU, Economy 1999 Review
14.4.   *European Commission Broad economic policy guidelines*
        EC, Broad economic policy guidelines 1999
        EC, Broad economic policy guidelines 2000
        EC, Broad economic policy guidelines 2001
        EC, Broad economic policy guidelines 2002
        EC, Broad economic policy guidelines 2003–2005
        EC, Broad economic policy guidelines 2005–2008
14.5.   *Press Cuttings*
        Luca Pagni, "Prodi: Addio investitori esteri l'Italia rischia l'isolamento,"
        La Repubblica, 29 June 2003, reproduced with kind permission
        Federico Rampini, "La sindrome cinese del governo Berlusconi,"*La Repub-
        blica*, 22 August 2003, reproduced with kind permission
        Federico Rampini, "La moneta cinese nuova sfida per l'Occidente," *La
        Repubblica*, 27 August 2003, reproduced with kind permission
        Fabio Bogo, "Industria: stranieri in fuga l'Italia è ultima in Europa," *La
        Repubblica,* 7 June 2003, reproduced with kind permission
        Elena Polidori, "L'Euro vola e sorpassa il Dollaro," *La Repubblica*, 16 July
        2002, reproduced with kind permission.

References

[Abramson et al. 1997] H. Norman Abramson et al., "Technology Transfer Systems in the U.S. and Germany." Fraunhofer Institute for Systems and Innovation Research and National Academy of Engineering, National Academy Press ISBN 0-309- 05530-X, Washington, D.C. (1997)

[Adnkronos 2004] Adnkronos, "Ue: Italia fanalino di coda per attrazione investimenti esteri" URL: http://www.adnkronos.com/IGNDispacci/20040312/ADN20040312171913.htm (2004)

[Adnkronos 2007] Adnkronos, "Turkey: EU Keeps Monetary, Economic Policy Out Of Membership Talks" (Monday 25 June 2007- 16:01)

[Agentur 2003] Salzburg Agentur Institut der deutschen Wirtschaft "Cost of Labor" URL: http://www.salzburgagentur.at/C1256D7B005829AE/o/5387956681101644C1256DCE004F 727B/$file/CostLabor.V.1.final_eng.pdf (2003)

[Ahern 2004] Bertie Ahern, Estratti della dichiarazione del Taoiseach (primo ministro irlandese) davanti al Dáil Éireann (parlamento irlandese), "La sesta presidenza irlandese dell'Unione europea," URL: http://europa.eu.int/futurum/index_it.htm (20 January 2004)

[Aiginger (ed.) 2001] Karl Aiginger (ed.), European Commission, Enterprise Papers, "Europe's position in quality competition," Office for Official Publications of the EC, Luxembourg (2001)

[Albert 2004] Dénes Albert, "A new era begins," Hongrie en direct, Numéro spécial élargissement de Commission Européenne en direct (15 March 2004)

[Arikan 2003] Harun Arikan, "Turkey and the EU: An awkward candidate for EU membership?", Ashgate, London (2003)

[Arlen 2002] Stéphane Arlen, "Giscard d'Estaing/Turquie/Vatican: les dessous de l'affaire," Fairelejour.org, (9 November 2002)

[Atkins 2005] Ralph Atkins, "ECB fears threat to confidence in public finances," *The Financial Times* (22 March 2005)

[Badré et al. 2005] Denis Badré, Marie-Thérèse Hermange, Robert Bret, and Serge Lagauche, "What should we think of the Bolkenstein directive?" Republique Française Assemblée Nationale Rapport 2098 http://www.assemblee-nationale.fr/12/pdf/rap-info/i2098.pdf (15 February 2005)

[Balls 2003] Andrew Balls, "Why G7 must soon make way for the *Brics*," quoting Goldman Sachs 2003 "Global Economic paper 99. Dreaming with BRICs, the Path to 2050," *The Financial Times* (7 October 2003)

[Balmary 1999] Marie Balmary, "Abel, ou la traversée de l'Eden" Grasset (1999)

[Barber 2005] Tony Barber, "Berlusconi seeks new cabinet. Italian premier submits government resignation," *The Financial Times* (21 April 2005)

[Barnier 2004] Michel Barnier, "Dichiarazione in merito alla costituzione europea," Parlamento europeo, Bruxelles, published on the URL: http://europa.eu.int/futurum/index_it.htm (28 January 2004)

[BdI 2003] Banca d'Italia, "Considerazioni finali del Governatore alla Assemblea Generale Ordinaria," (31 May 2003)

[Benoit 2004] Bertrand Benoit and George Parker, "EU deficit sanctions decision *illegal*," *The Financial Times* (14 July 2004)

[Bianchi 1999] Patrizio Bianchi, "Industrial Policies and Economic Integration," Routledge 1998. Also in Italian: "Le politiche industriali dell'Unione europea," Il Mulino 2nd edition (1999)

[Bivens 2004] L. Josh Bivens "Debt and the dollar," EPI Issue Brief Issue Brief #203 Economic Policy Institute (December 14, 2004)

[Bobinski 2004] Krzysztof Bobinski, "On the eve of accession," Pologne en direct Numéro spécial élargissement de Commission Européenne en direct (26 April 2004)

[Bocca 2004] Giorgio Bocca, "L'invasione dei campi," *La Repubblica* (24 February 2004)

[Bogo 2003] Fabio Bogo, "Industria: stranieri in fuga l'Italia è ultima in Europa," *La Repubblica* (7 June 2003)

[Boland 2004-1] Vincent Boland, "Lasting growth is now within reach," *The Financial Times* (27 April 2004)

[Boland 2004-2] Vincent Boland, "After 41 years, a gulf still divides Turkey and Europe," *The Financial Times* (18 December 2004)

[Boland 2005] Vincent Boland, "Strong government plays role in winning over sceptics in Turkey's business sector," *The Financial Times* (21 January 2005)

[Boland et al. 2006] Vincent Boland and Daniel Dombey, "Turkey in fresh move to end Cyprus impasse", *The Financial Times* (24 January 2006)

[Bolkestein 2006] Frits Bolkestein, "The Limits of Europe" Gibson Square Books (2006)

[Bomberg et al. (eds.) 2003] Elisabeth Bomberg and Alexander Stubb (eds.), "The European Union: How does it work?" Oxford University Press (2003)

[Bonanni 2004] Interview with Mario Monti by Andrea Bonanni, *La Repubblica* (24 October 2004)

[Boyer 2004] Peter J. Boyer, "Paul Wolfowitz defends his war," The New Yorker (1 November 2004)

[Branscomb 1998] Lewis M Branscomb, et al. (eds.) "Investing in Innovation. Creating a research and innovation policy thaw works," ISBN 0 262 02446-2 The M.I.T. Press Cambridge, MA (1998)

[Brazauskas 2004] Algirdas Brazauskas, "Not just a consumer, also a contributor," Lituanie en direct Numéro spécial élargissement de Commission Européenne en direct (24 May 2004)

[Bréhon 2007] Nicolas-Jean Bréhon, "Policy Paper" Robert Schuman Foundation. March 2007 quoted in "The European budget and the "fair return" principle: What is it about?" La Note de Veille 21 May 2007 no. 59 (2007)

[British Council 2004] British Council and the Open Society Institute, Turkey in Europe: More than a promise? Report of the Independent Commission on Turkey (September 2004)

[Broughton et al. 2002] Sally Broughton and Eran Fraenkel "Macedinia: Extreme challenges for the model of multiculturalism," in [van Tongeren et al. (eds.) 2002]

[Browne 2004] Anthony Browne, "EU in crisis as Parliament rejects new Commission," *Times online* (28 October 2004)

[Brown-Humes 2004] Christopher Brown-Humes, "Is production at Nokia in Finland at risk?" *The Financial Times* (15 May, 2004)

[Bryant 2007] Steve Bryant "Turkey Needs Budget Rule to Cap Spending, Yilmaz Says" http://www.bloomberg.com Updated: New York (14 July 2007 at 04:36)

[Burgess 2003] Kate Burgess, "China effect convulses commodity markets," *The Financial Times* (15 November 2003)

[Burrell et al., eds. 2005-1] Alison Burrell and Arie Oskam (eds.), "Turkey in the European Union: Implications for agriculture, food and structural policy", CABI Wallingford, UK and Cambridge, MA (2005)

[Burrell 2005-2] Alison Burrell, "Turkey's Foreign Trade" in [Burrell et al., eds. 2005-1]

[Burrel 2005-3] Alison Burrell, "Opportunities, Threads, and Challenges", in [Burrell et al., eds. 2005-1]

[Burt et al. 2001] Tim Burt and Nikki Tait, "U.S. carmakers confront a second invasion," *The Financial Times* (24 August 2001)

[Caglioti 2003] Luciano Caglioti, "Il Fenomeno Italia. L'autolesionismo come missione," Gangemi Editore (2003)

[Caracostas et al. 1998] Paraskevas Caracostas, and Muldur Uğur "Society, the endless frontier," Office for Official Publications of the EC, ISBN 92 828-1186-7 (1998)

[Çarkoğlu et al. eds. 2003] Ali Çarkoğlu and Barry Rubin (eds.), "Turkey and the European union: Domestic politics, economic integration and international dynamics", F. Cass, London (2003)

[Çarkoğlu 2003] Ali Çarkoğlu "Who wants Full Membership?", in [Çarkoğlu et al. eds. 2003]

[Castle 2007] Stephen Castle and Graham Bowley "Treaty on Running European Union Is Signed," The New York Times (14 December 2007)

[Çayhan 2003] Esra Çayhan, "Towards a European Security and Defense Policy: With or without Turkey?" in [Çarkoglu et al. eds. 2003]

[CoJ 1979] The Court of Justice, Case 120/78 Judgment "Cassis de Dijon," (20 February 1979)

[Cecchini 1988] Paolo Cecchini, "La sfida del 1992," Sperling & Kupfler (1988)

[Cemex 2007] Cemex, Annual Report 2006 (2007)

[Christensen 2002] Clayton M. Christensen, "The Innovator's dilemma," HarperBusiness (2002)

[Christensen 2003] Clayton M. Christensen, "The Innovator's Solution," Harvard Business School Press (2003)

[Christou 2007] Jean Christou, "The end of Cyprus' leverage on Turkey?" Cyprus Mail (13 May 2007)

[Chong-ko Peter Tzou 2008] Chong-ko Peter Tzou, Proceedings of the "International Conference on the 50th Anniversary of the Treaty of Rome and EU-Asia Relations," Tamkang University, Taipei, Taiwan 21 November 2007. Printed in August 2008.

[Ciampi 2003-1] Carlo Azeglio Ciampi, "Intervento del Presidente della Repubblica Carlo Azeglio Ciampi alla cerimonia di consegna del *Premio Leonardo* e dei *Premi Leonardo Qualità Italia*," Palazzo del Quirinale, 4 December 2003, URL: http://www.quirinale.it/Discorsi/Discorso.asp?id=23723 (2003)

[Ciampi 2003-2] Carlo Azeglio Ciampi, "Messaggo di fine anno del Presidente della Repubblica Carlo Azeglio Ciampi agli italiani," 31 December 2003, URL: http://www.quirinale.it/Discorsi/Discorso.asp?id=23858 (2003)

[Ciampi 2004] Carlo Azeglio Ciampi, "Intervento del Presidente della Repubblica Carlo Azeglio Ciampi in occasione della consegna al Presidente Patrick Cox delle insegne di Cavaliere di Gran Croce dell'Ordine al Merito della Repubblica Italiana," (24 March 2004), URL: http://www.quirinale.it/Discorsi/Discorso.asp?id=24486 (2004)

[CNN 2004] CNN Anomymous "China's TCL posts big profit jump," URL: http://cnn.com (2 March 2004)

[COGR 1999] U.S. Council on Governmental Relations, "Bayh-Dole Act. A guide to the law and implementing regulations," URL: http://www.ucop.edu/ott/bayh.html (1999)

[Confapi 2002] CONFAPI, "La crisi della Fiat Auto," URL: http://confapi.org/fiat_confapi.htm (2002)

[Coppola 2007] Alessandra Coppola, "Turchia: un plebiscito per Erdoğan," IL Corriere della Sera (23 July 2007)

[Council 1970] 116th Council Session of 8–9 June 1970, T 347/70 (11 June 1970)

[Council 1994] High Level Group on the Information Society Europe, "Recommendation to the European Council" Brussels, Presented by the EC on the occasion of the Corfu Summit of Heads of State (26 May 1994)

[Council 2000-1] "Convention Implementing the Schengen Agreement," Official Journal L 239, 22/09/2000 P. 0019–0062 (2000)

[Council 2000-2] Consiglio europeo, "Carta dei diritti fondamentali dell'Unione europea," (2000)

[Council 2002] Consiglio europeo, "Convenzione europea Dichiarazione di Laeken sul Futuro dell'Unione europea" (2002)

[Council 2003] Competitiveness Council, "Community Patent" Brussels (3 March 2003)

[Cram et al. 1999] Laura Cram, Desmond Dinan and Neil Nugent (eds.), "Developments in the European Union," Macmillan Press Ltd. (1999)

[Davies 2004] Norman Davies, "Contrary to popular belief, this could be an opportunity," *The Financial Times* (1 May 2004)

[Davies 2007] Paul J. Davies, "Successful debt sale underlines Fiat's robust turnround," *The Financial Times* (6 June 2007)

[Dempsey et al. 2004] Judy Dempsey and George Parker, "European Union commissioner argues case for refusing Turkey membership," *The Financial Times* (8 March 2004)

[Devuyst 2002] Youri Devuyst, "The European Union at the Crossroads: An Introduction to the EU's Institutional Evolution," Brussel, Presses Interuniversitaires Européennes (2002)

[Dombey 2004-1] Daniel Dombey, "Turkey talks hit stumbling block over EU conditions," *The Financial Times* (17 December 2004)

[Dombey et al. 2004-2] Daniel Dombey, George Parker and Vincent Boland "Turkey accepts EU's offer of accession talks," *The Financial Times* (17 December 2004)

[Dosi et al. 2005] Giovanni Dosi, Patrick Llerena, andMauro Sylos Labini, "Evaluating and Comparing the innovation performance of the United States and the European Union," report prepared for the TrendChart Policy Workshop 29 June 2005 (2005)

[Dyer 2005] Geoff Dyer, "DaimlerChrysler considers plans to build cars in China," *The Financial Times* (22 April 2005)

[Dzurinda 2004] Mikuláš Dzurinda, "Message from the Prime Minister," in Slovaquie en direct, Numéro spécial élargissement de Commission Européenne en direct (19 April 2004)

[EC 1964] EEC-Turkey Association Agreement (1963), Official Journal No. 217 (29 December 1964)

[EC 1993] European Commission, "Growth Competitiveness, Employment," White Paper ISBN 92-826-7000-7 OOP-EC, Luxembourg (1993)

[EC 1995] European Commission "Accession Criteria as set by the European Council in Copenhagen (1993) and Madrid (1995)," http://ec.europa.eu/enlargement/enlargement_ process/accession_process/criteria/index_en.htm (1995)

[EC 1996-1] European Commission, "Applying Information Technology: 101 Success Stories from Esprit," ISBN 92-827-6975-5 OOP-EC, Luxembourg (1996)

[EC 1996-2] European Commission, "First action plan for innovation in Europe," ISBN 92 827 9110-6 OOP-EC CD-99-96-681-EN-C, Luxembourg (1996)

[EC 1996-3] European Commission, "Information Technology Solutions for Business. Case studies from Esprit," ISBN 92 827 8389-8 OOP-EC CO-95-96-916-EN-C, Luxembourg (1996)

[EC 1997-1] European Commission, "Esprit Success Stories for the Information Society" ISBN 92-828-1633-8 OOP-EC CO - 06-97-771-EN-C, Luxembourg (1997)

[EC 1997-2] European Commission, "The present state of the patent system in the European Union," Office for Official Publications of the EC. ISBN 92 826-9555-7 CD-NA-17014-EN-C, Luxembourg (1997)

[EC 1997-3] Commissione europea, direzione generale dell'industria "European competitiveness report" (1997)

[EC 1998-1] European Commission, "Building The Info Society. Perspectives from Esprit," ISBN 92 828 3344-5 OOP-EC CO 1698 926 EN C, Luxembourg (1998)

[EC 1998-2] European Commission, "The Fifth Framework Programme for R&D" (1998)

[EC 1998-3] Commissione europea, direzione generale dell'industria "European competitiveness report" (1998)

[EC 1999-1] European Commission, "Information Society Techologies Challenges and Opportunities," 1999 ISBN 92-828-7979-8 OOP-EC CD-24-99-033-EN-C, Luxembourg (1999)

[EC 1999-2] Commissione europea, direzione generale dell'impresa "European competitiveness report" (1999)

[EC 1999-3] Commissione europea, direzione generale della ricerca "Second European Report on S&T Indicators. Key Figures" (1999)

[EC 1999-4] Commissione europea, direzione generale della società dell'informazione "Accelerating e-commerce in Europe," 2nd edition ISBN 92-828-5985-1 (1999)

[EC 2000-1] Commissione europea, "Quinto Programma quadro Programma Specifico Società dell'informazione" (2000)

[EC 2000-2] Commissione europea, "Quinto Programma quadro Programma Specifico Qualità della vita" (2000)

[EC 2000-3] Commissione europea, "Quinto Programma quadro Programma Specifico Sviluppo sostenibile" (2000)

[EC 2000-4] Commissione europea, "Quinto Programma quadro Programma Specifico Ambiente ed energia" (2000)

[EC 2000-5] Commissione europea, "Quinto Programma quadro Programma Specifico Ricerca internazionale" (2000)

[EC 2000-6] Commissione europea, "Quinto Programma quadro Programma Specifico Innovazione e PMI" (2000)

[EC 2000-7] Commissione europea, "Quinto Programma quadro Programma Specifico Capitale umano" (2000)

[EC 2000-8] Commissione europea, direzione generale dell'impresa "European competitiveness report" (2000)

[EC 2000-9] Commissione europea, "Esame intermedio del quinto Programma quadro" (2000)

[EC 2000-10] European Council, "Council Decision of 20 December 2000 on a Program for Enterprises and Entrepreneurship, in Particular for SMEs (2001–2005)." Official Journal of the EC L 333/84 (29 December 2000)

[EC 2001-1] Commissione europea, direzione generale dell'impresa "European competitiveness report" (2001)

[EC 2001-2] Commissione europea, direzione generale dell'impresa "A pocketbook of enterprise policy indicators. How member states rank in the 2001 Enterprise Scoreboard" (2001)

[EC 2001-3] Commissione europea, direzione generale dell'impresa, "Benchmarking Enterprise Policy. Results from the 2001 Scoreboard" (2001)

[EC 2001-4] Commissione europea, direzione generale della ricerca, "Key Figures 2001: Indicators for benchmarking national research policies" (2001)

[EC 2001-5] Commissione europea, direzione generale dell'impresa, "Competitiveness, innovation and enterprise performance" (2001)

[EC 2001-6] Commissione europea, "Sintesi del Trattato di Nizza" (2001)

[EC 2001-7] Commissione europea, "Indirizzi di massima per le politiche economiche" (2001)

[EC 2001-8] Commissione europea, direzione generale della Società dell'informazione, "e-Europe: una società dell'informazione per tutti" (2001)

[EC 2001-9] European Commission "Culture 2000 Programme: General Objectives" Official Journal of the European Communities, C21, 24 January 2001

[EC 2002-1] European Commission, "IST 2002. Partnership for the Future," ISBN 92-894-4139-9 OOP-EC KK-45-02-272-EN-C, Luxembourg (2002)

[EC 2002-2] Commissione europea, direzione generale dell'impresa, "European competitiveness report" (2002)

[EC 2002-3] Commissione europea, direzione generale dell'impresa "Observatory of European SMEs Highlights from 2002 Survey" (2002)

[EC 2002-4] Commissione europea, "Sesto Programma quadro" (2002)

[EC 2002-5] European Commission, "e-Europe 2002 Benchmarking Report" (2002)

[EC 2002-6] European Commission, "eEurope 2005: An information society for all" (2002)

[EC 2002-7] Commissione europea, "Indirizzi di massima per le politiche economiche" (2002)

[EC 2002-8] Commissione europea, "Il trattato di Amsterdam," URL: http://europa.eu.int/scadplus/printversion/it/lvb/a09000.htm (2002)

[EC 2002-9] European Council, "Presidency Conclusions Barcelona European Council" (15 and 16 March 2002)

[EC 2002-10] European Commission, "Declaration of the European Ministers of Vocational Education and Training, and the European Commission, convened in Copenhagen on 29 and 30 November 2002, on enhanced European cooperation in vocational education and training" (2002)

[EC 2003-1] European Commission, "IST 2003 The Opportunities ahead," ISBN 92-894-5752-X
OOP-EC KK-53-03-330-EN-C, Luxembourg (2003)

[EC 2003-2] European Commission, "Guidelines for Proposal Evaluation and Selection Procedures FP6," COM C 2003-883 (2003)

[EC 2003-3] European Commission DG Trade, "U.S.," URL: http://europa.eu.int/comm/trade/
index_en.htm (2003)

[EC 2003-4] European Commission, Press Release after the 12–13 December 2003 European Council, reported by Act4Europe in Bulletin 20. URL: http://www.ngo-eu.cz/ doku-
menty/ustava.doc (15 December 2003)

[EC 2003-5] European Commission, "European competitiveness report" (2003)

[EC 2003-6] European Commission, "Benchmarking enterprise policy: Results from the 2003
Scoreboard" (2003)

[EC 2003-7] Commissione europea, Direzione generale della ricerca. "Cosa fa l'Europa?" Una
serie di opuscoli divulgativi. URL: http://europa.eu.int/comm/research/leaflets/index_it.html
(2003)

[EC 2003-8] Commissione europea, "Decisione del Consiglio 2002/358, Protocollo di Kyoto sui
cambiamenti climatici" (2003)

[EC 2003-9] Commissione europea, "Indirizzi di massima per le politiche economiche" (2003)

[EC 2004-1] European Commission, "External Trade Statistics," URL: http://europa.eu.int/comm/
trade/issues/bilateral/data.htm (2004)

[EC 2004-2] European Commission, "Strategic Paper on progress in the enlargement process,"
COM(2004) 657 final (6 October 2004)

[EC 2004-3] European Commission, "2004 Regular Report on Bulgaria's progress towards accession," SEC(2004) 1199 (6 October 2004)

[EC 2004-4] European Commission, "2004 Regular Report on Romania's progress towards accession," SEC(2004) 1200 (6 October 2004)

[EC 2004-5] European Commission, "2004 Regular Report on Turkey's progress towards accession," SEC(2004) 1201 (6 October 2004)

[EC 2004-6] European Commission, "Issues arising from Turkey's membership perspective,"
SEC(2004) 1202 (6 October 2004)

[EC 2004-7] European Commission, "European competitiveness report 2004" (2004)

[EC 2004-8] European Commission, "Directive on Services in the internal market," COM (2004)
2final/3 (2004)

[EC 2004-9] European Commission, "Communication COM (2004) 101 final/2, "Building our
common future, policy challenges and budgetary means of the enlarged union 2007–2013" (26
February 2004)

[EC 2004-10] European Commission "Commission recommendation on the 2004 update of the
Broad Guidelines of the Economic Policies of the Member States and the Community (for the
2003–2005 period)" COM (2004) 238 (2004)

[EC 2004-11] European Council, "The policy of the EU with regard to the Turkish Cypriot community" General Affairs Council on 26 April (2004)

[EC 2004-12] European Commission, "Financial perspectives 2007–2013—Added value and
delivery instruments," COM(2004) 487 final (2004)

[EC 2004-13] Communication from the Commission to Council and Parliament "Building our
common Future" Brussels, 26.2.2004 COM(2004) 101 final/2

[EC 2004-14] European Council, "Presidency Conclusions 16–17 December 2004 Policy challenges and Budgetary means of the Enlarged Union 2007–2013," COM (2004) 101 final 26
February (2004)

[EC 2004-15] European Commission, "Facing the challenge The Lisbon strategy for growth and
employment" Report from the High Level Group chaired by Wim Kok (November 2004)

[EC 2004-16] European Commission, "A Constitution for Europe," Brochure ISBN: 92-894-6114-
4 (2004)

[EC 2004-17] European Commission, "A Constitution for Europe-Presentation to Citizens," Brochure ISBN: ISBN: 92-894-0751-4 (2004)

[EC 2004-18] European Commission, "Treaty establishing a Constitution for Europe," IGC 87/1/04 (2004)

[EC 2004-19] European Commission, "Cohesion policy: The 2007 watershed," FactSheet KN-61-04-274-EN-J (2004)

[EC 2004-20] European Commission, "The Common Agricultural Policy Explained," Brochure ISBN 92-894-8204-4 (2004)

[EC 2005-1] European Commission, "Building the ERA of knowledge for growth," COM (2005) 118 final (2005)

[EC 2005-2] European Commission Portal to Gate to Growth Initiative, URL: http://www.gate2-growth.com (2005)

[EC 2005-3] European Commission, Innovation and SMEs, an online brochure at the URL: http://europa.eu.int/comm/research/sme/leaflets/en/intro01.html (2005)

[EC 2005-4] European Commission, "Turkey 2005 Progress Report (COM 2005- 561 final)" Brussels (9 November 2005)

[EC 2005-5] European Commission, "The Lisbon Special European Council (March 2000): Towards a Europe of Innovation and Knowledge," 2005 update http://europa.eu/scadplus/leg/en/cha/c10241.htm (2005)

[EC 2005-6] European Commission, "Council Recommendation 2005/601/EC of 12 July 2005 on the broad economic policy guidelines of the Member States and the Community (2005-2008)" Official Journal L 205 (6.8.2005)

[EC 2005-7] European Commission, "Life sciences and biotechnology. A strategy for Europe," COM 2005 286 final (2005)

[EC 2005-8] European Commission, "PVC Recycling, a profitable opportunity for the industry," http://ec.europa.eu/ (2005)

[EC 2005-9] European Commission, "Competitiveness and Innovation Framework Programme (2007–2013)" {SEC (2005) 433} COM (2005) 121 final (2005)

[EC 2005-10] European Commission, "i2010 – A European Information Society for growth and employment" COM (2005) 229 final(2005)

[EC 2005-11] European Commission, "Common Actions for Growth and Employment: The Community Lisbon Programme" COM (2005) 330 final (2005)

[EC 2005-12] European Commission, "eContentplus (2005–2008)" Decision No 456/2005/EC of the European Parliament and of the Council of 9 March 2005 Official Journal L 79 of (24.03.2005)

[EC 2006-1] European Commission, "EU Enlargement. Turkey 2006 Progress Report" (2006)

[EC 2006-2] European Commission, "Eurobarometer 66 Public Opinion in the EU" (Autumn 2006)

[EC 2006-3] European Commission, "Monitoring report on the state of preparedness for EU membership of Bulgaria and Romania," (26 September 2006)

[EC 2006-4] European Commission, "The former Yugoslav Republic of Macedonia 2006 Progress Report," SEC (2006) 1387 (8 November 2006)

[EC 2006-5] European Commission, "MEMO/06/411 Key findings of the progress reports on the candidate countries: Croatia, the former Yugoslav Republic of Macedonia and Turkey," Brussels (8 November 2006)

[EC 2006-6] European Commission, "Myths and Facts about Enlargement," (4 August 2006)

[EC 2006-7] European Commission, "MEMO/06/410 Frequently asked questions on Instrument for Pre-Accession Assistance," Brussels (8 November 2006)

[EC 2006-8] European Commission, "Interinstitutional Agreement on Budgetary Discipline and Sound Financial Management 2007–2013 MEMO/06/204" Brussels (17 May 2006)

[EC 2006-9] European Commission, "Did the euro cause prices to rise?" KC-70-05-956-EN-D (2006)

[EC 2006-10] European Commission, "European competitiveness report" (2006)

[EC 2006-11] European Commission, "Information Society Technologies 2005–2006. Strategies for Leadership," (2006)

[EC 2006-12] European Commission, "Information Society and Transport," (2006)

[EC 2006-13] European Commission, "Making sense of content – Quality content for all," (2006)

[EC 2006-14] European Commission, "A new start for the Lisbon Strategy (2005)" (29.11.2006)

[EC 2006-15] European Commission, "Online learning: eLearning Programme (2004–06)" Decision No 1720/2006/EC of the European Parliament and of the Council of 15 November 2006 Official Journal L 327 of 24.11.2006 (2006)

[EC 2006-15] European Commission, "European Economy, Economic Forecasts 2006" (2006)

[EC 2006-16] European Commission, "Energy futures: The role of research and technological development" ISBN 92-79-01639-3 (2006)

[EC 2006-17] European Commission, "Energy research on climate change" ISBN 92-79-01713-6 (2006)

[EC 2007-1] European Commission: Web site of the Delegation of the EC to Turkey, 2007 http://www.avrupa.info.tr/DelegasyonPortal.html (2007)

[EC 2007-2] European Commission: Report to the European Council and the European Parliament "On Bulgaria's progress on accompanying measures following Accession" COM (2007) 377 final (27.6.2007)

[EC 2007-3] European Commission: Report to the European Council and the European Parliament "On Romania's progress on accompanying measures following Accession" COM (2007) 378 final (27.6.2007)

[EC 2007-4] European Commission: "Commission monitoring report on Bulgaria and Romania: Extracts from the press conference by EC Vice-President Franco Frattini," (27 June 2007)

[EC 2007-5] European Commission: "EU – the former Yugoslav Republic of Macedonia relations," http://ec.europa.eu/enlargement/ (2007)

[EC 2007-6] European Commission: "Internal Market Scoreboard n. 16" (July 2007)

[EC 2007-7] European Commission: "Competition" at the URL: http://ec.europa.eu/comm/ competition/index_en.html (2007)

[EC 2007-8] European Commission: "Internal Market" at the URL: http://ec.europa.eu/ internal_market/index_en.htm (2007)

[EC 2007-9] European Commission, Eurostat, "EU FDI Yearbook 2007" (2007)

[EC 2007-10] European Commission, The Directorate General for Enterprise and Industry, "EU industrial structure Competitiveness and Economic Reforms," (2007)

[EC 2007-11] European Commission, "Budget of the Seventh Framework Programme,"in Cordis at http://cordis.europa.eu/fp7/budget_en.html (2007)

[EC 2007-12] European Commission, "The Seventh Framework Programme, Taking the European Reserch to the Forefront," (2007)

[EC 2007-13] European Commission, "Seventh Framework Programme,"in Cordis at http://cordis.europa.eu/fp7/home_en.html (2007)

[EC 2007-14] European Commission, "Seventh Framework Programme –Concentrating Solar Power," (2007)

[EC 2007-15] European Commission, "Funding and Priorities for ITC Research," (2007)

[EC 2007-16] European Commission, "FET-OPEN Activity Report of the 6th Framework Program," (2007)

[EC 2007-17] European Commission, "European research in action – Recycling vehicles," (2007)

[EC 2007-18] European Commission, "European research in action – Aeronautics," (2007)

[EC 2007-19] European Commission, "The players in the fusion game," research*eu-the magazine of the European research area Special Issue June 2007 (2007)

[EC 2007-20] European Commission, "The former Yugoslav Republic of Macedonia 2007 Progress Report," SEC (2007) 1432 Brussels (6.11.2007)

[EC 2007-21] European Commission, "Enlargement Strategy and Main Challenges 2007–2008," COM (2007) 663 final Brussels (6.11.2007)

[EC 2007-22] European Commission Research Communication Unit, "European Research in Action" http://ec.europa.eu/research/leaflets/index_en.html (2007)

[EC 2007-23] European Commission, "Europe for lifelong learning," (2007)

[EC 2007-24] European Commission, "Progress towards the Lisbon objectives 2010 in education and training 2007 Indicators and Benchmarks" SEC 2007/1284 (2007)

[EC 2007-25] European Commission, "Europe's Information society Thematic Portal" http://ec.europa.eu/information_society/soccul/egov/index_en.htm (2007)

[EC 2007-26] European Commission, "A Handbook for Citizen-centric eGovernment" (December 2007)

[EC 2007-27] European Commission, "Economic Forecast Autumn 2007" ISSN 0379-0991 (2007)

[EC 2007-28] European Commission, "Science and Technology Indicators for the European Research Area," Cordis, (2007)

[EC 2007-29] European Commission, "Energy research in road transport" ISBN 92-79-02945-2 (2007)

[EC 2007-30] European Commission, "Energy research in the seventh Framework Program" ISBN 978-92-79-03891-4 (2007)

[EC 2007-31] European Commission, "Stroke research in the EU" ISBN 978-92-79-05049-7 (2007)

[EC 2007-32] European Commission, "Treaty of Lisbon Amending the Treaty on European Union and the Treaty Establishing the European Community," IGC 14/07 (3 December 2007)

[ECB 2004] European Central Bank, "Annual Report 2004" (2004)

[Economist 1999-1] Anonymous, "Finance and economics: Black hole," *The Economist*, vol. 352, n. 8134 (28 August 1999)

[Economist 1999-2] Anonymous, "The rad to riches," *The Economis*, vol. 353, n. 8151 (23 December 1999)

[Economist 2001-1] Anonymous, "Older, wiser, webbier," *The Economist*, vol. 359, n. 8228 (30 June 2001)

[Economist 2001-2] Anonymous, "Electronic glue," *The Economist*, vol. 359, n. 8224 (2 June 2001)

[Economist 2001-3] Anonymous, "While Welch waited," *The Economist*, vol. 359, n. 8222 (19 May 2001)

[Economist 2001-4] Anonymous, "The Cemex way," *The Economist*, vol. 359, n. 8226 (14 June 2001)

[Economist 2001-5] Anonymous, "Over the counter, e-commerce," *The Economist*, vol. 359, n. 8223 (26 May 2001)

[Economist 2001-6] Anonymous, "The Plot Thickens," *The Economist*, vol. 359, n. 8224 (2 June 2001)

[Economist 2001-7] Anonymous, "Islands of Quality," *The Economist*, vol. 359, n. 8224 (2 June 2001)

[Economist 2001-8] Economic and financial indicators, "Trade" *The Economist*, vol. 361, n. 8250 (1 December 2001)

[Economist 2002-1] Technology Quarterly "Thanksgiving for Innovation," *The Economist*, vol. 364, n. 8291 (19 September 2002)

[Economist 2002-2] Anonymous, "Demography," *The Economist*, vol. 364, n. 8287 (17 August 2002)

[Economist 2002-3] Anonymous, "Italy, Silvio Berlusconi, and the law. He's sitting pretty," *The Economist*, vol. 364, n. 8282, (13 July 2002)

[Economist 2002-4] Anonymous "Too big for Europe?" *The Economist*, vol. 365, n. 8299 (14 November 2002)

[Economist 2003-1] Anonymous, "Working Hours," *The Economist*, vol. 368, n. 8338 (9 August 2003)

[Economist 2003-2] Anonymous, "American productivity growth Paradox lost," *The Economist*, vol. 368, n. 8341 (30 August 2003)

[Economist 2003-3] Anonymous, "John Snow comes adrift," *The Economist*, vol. 367, n. 8325 (10 May 2003)

[Economist 2003-4] Anonymous, "Why Slovenia is not the Balkans," *The Economist*, vol. 369, n. 8351 (20 November 2003)

[Economist 2003-5] Anonymous, "Lithuania Baltic tiger," *The Economist*, vol. 368, n. 8333 (17 July 2003)

[Economist 2003-6] *The Economist* – "Il Mondo in cifre 2003," International-Profile Books (2003)

[Economist 2004-1] Anonymous, "A survey of India," in *The Economist*, vol. 370, n. 8363 (31 January 2004)

[Economist 2004-2] Anonymous, "Italy's Mafia.A capo's annual report," in *The Economis*, vol. 370, n. 8365 (14 February 2004)

[Economist 2004-3] Anonymous, "Ataturk's long shadow," *The Economist*, vol. 373, n. 8403 (30 November 2004)

[Economist 2004-4] Anonymous, "How terrorism trumped federalism," *The Economist*, vol. 373, n. 8395 (2 October 2004)

[Economist 2004-5] Anonymous, "What the IBM/Lenovo deal says about Chinese firms' overseas ambitions," *The Economist*, vol. 373, n. 8403 (9 December 2004)

[Economist 2004-6] Anonymous, "Migration in the European Union," *The Economist*, vol. 370, n. 8358 (15 January 2004)

[Economist 2005-1] Anonymous, "European Union leaders tear up fiscal rules even as they fight for their new constitution," *The Economist*, vol. 374, n. 8419 (26 March 2005)

[Economist 2005-2] Anonymous, "General Motors and Fiat Valentine's Day divorce," *The Economist*, vol. 374, n. 8414 (17 February 2005)

[Economist 2005-3] Anonymous, "Eastern Europe's economies. Hot and bothered," *The Economist*, vol. 377, n. 8448 (13 October 2005)

[Economist 2005-4] Anonymous, "Meet the neighbours," *The Economist*, vol. 375, n. 8432 (23 June 2005)

[Economist 2006-1] Anonymous, "A Country to run from. But only for a while," *The Economist*, vol. 379, n. 8477 (13 May 2006)

[Economist 2006-2] Anonymous, "Pocket president. Slovenia eyes a rare moment of glory," *The Economist*, vol. 381, n. 8504 (16 November 2006)

[Economist 2006-3] Anonymous, "Romania, Bulgaria and the European Union We're off on a European odyssey," *The Economist*, vol. 380, n. 8497 (28 September 2006)

[Economist 2006-4] Anonymous, "Hungarian dances," *The Economist*, vol. 380, n. 8496 (21 September 2006)

[Economist 2006-5] Anonymous, "Survey: Climate Change – The economics of living with climate change," *The Economist*, vol. 380, n. 8496 (21 September 2006)

[Economist 2006-6] Anonymous, "Lithuania's government Politics exam," *The Economist*, vol. 379, n. 8481 (8 June 2006)

[Economist 2006-7] Anonymous, "Eastern Europe's stars. The dynamic duo," *The Economist*, vol. 381, n. 8508 (13 December 2006)

[Economist 2007-1] "Pocket World in Figures 2007" (2007)

[Economist 2007-2] Leader: "The European Union summit. Trick or treaty?" *The Economist*, vol. 383, n. 8535 (30 June 2007)

[Economist 2007-3] Charlemagne: "Nobody was happy with the summit deal on a new European Union treaty?" *The Economist*, vol. 383, n. 8535 (30 June 2007)

[Economist 2007-4] Economist Intelligence Unit Source: Country Views-Wire "Factsheet", 3 July 2007

[Economist 2007-5] Leader, "Estonia and Russia. The right to be wrong", *The Economist*, vol. 383, n. 8527 (3 May 2007)

[Economist 2007-6] Economist Intelligence Unit Source: Country Views, "Poland- Economic Data", (11 July 2007)

[Economist 2007-7] Anonymous, "Poland's unsteady government" *The Economist*, vol. 384, n. 8537 (14 July 2007)

[Economist 2007-8] Anonymous, "A swamp of paranoid nostalgia" *The Economist*, vol. 384, n. 8536 (7 July 2007)

[Economist 2007-9] Economist Intelligence Unit Source: Country Views, "Slovakia-Forecast" (11 July 2007)

[Economist 2007-10] Economist Intelligence Unit Source: Country Views, "Slovenia- Economic Data" (6 August 2007)

[Economist 2007-11] Economist Intelligence Unit Source: Country Views, "Bulagaria's Politics" (5 April 2007)

[Economist 2007-12] Economist Intelligence Unit Source: Country Views, "Romania-Factsheet" (3 July 2007)

[Economist 2007-13] Economist Intelligence Unit Source: Country Views, "Romania-Forecast" (3 July 2007)

[Economist 2007-14] Anonymous, "Romanian politics-A captain faces the storm," *The Economist*, vol. 383, n. 8526 (26 April 2007)

[Economist 2007-15] Economist Intelligence Unit Source: Country Views, "Romania gets ready," (31 July 2007)

[Economist 2007-16] Economist Intelligence Unit Source: Country Views, "Czech Republic Forecast," (22 April July 2007)

[Economist 2007-17] Anonymous, "Romanian politics-A captain faces the storm," *The Economist*, vol. 384, n. 8541 (12 August 2007)

[Economist 2007-18] Anonymous, "A battle for the future," *The Economist*, vol. 384, n. 8538 (21 July 2007)

[Economist 2007-19] Economist Intelligence Unit Source: Country Forecast- Italy (21 August 2007)

[Economist 2007-20] Economist Intelligence Unit Source: Country Forecast- Japan (21 August 2007)

[Economist 2007-21] Anonymous, "Special Report on Brazil" *The Economist*, vol. 383, n. 8524 (12 April 2007)

[Economist 2007-22] Anonymous, "The EU and Russia, Enter, pursued by a bear" *The Economist*, vol. 384, n. 8546 (15 September 2007)

[Economist 2007-23] Economist Intelligence Unit Source: Country Forecast- Russia (8 May 2007)

[Economist 2007-24] Economist Intelligence Unit Source: Country Forecast- South Korea (27 April 2007)

[Economist 2007-25] Anonymous, "Fit at 50? Report on the European Union" *The Economist*, vol. 382, n. 8520 (17 March 2007)

[Economist 2007-26] Anonymous, "Something new under the Sun" *The Economist*, vol. 382, n. 8515 (8 February 2007)

[Economist 2007-27] Anonymous, "Who are the champions?" *The Economist*, vol. 385, n. 8550 (11 October 2007)

[Economist 2007-28] Anonymous, "Generation Game" *The Economist*, vol. 385, n. 8550 (11 October 2007)

[Eder 2003] Mine Edre, "Implementing the Economic Criteria of EU Membership: How Difficult it is for Turkey", in [Çarkoglu et al. eds. 2003]

[Eder 2004] Mine Eder, "Populism as a barrier to integration in the EU. Rethinking the Copenhagen Criteria" in [Uğur et al., eds. 2004]

[Etienvre 2005] E. Etienvre, "Le Groupe Michelin :strategie de croissance," Centrale des Cas et de Medias Pedagogiques (2005)

[EITO 2003] European Information Technology Observatory "EITO Report" (2003)

[EITO 2004] European Information Technology Observatory, "EITO Report" (2004)

[EP 2000] European Parliament Archives http://www.europarl.eu.int/charter/default_en.htm (2000)

[Euractiv 2007] Anonymous, "Cyprus inches closer to reunification as symbolic wall is razed", http://www.euractiv.com/ article-162350 (9 March 2007)

[Eurostat 2001] Eurostat, "Yearbook" (2001)

[Eurostat 2002] Eurostat, "Yearbook" (2002)

[Eurostat 2003] Eurostat, Statistics on the information society in Europe. Data from 1966–2003

[Eurostat 2006] Eurostat, Statistics in Focus 14/2006 "Trade in High-tech Products" (2006)

[Eurostat 2007-1] Eurostat, Statistics on Harmonised unemployment rate (2007)

[Eurostat 2007-2] Eurostat, "Yearbook 2006–2007" (2007)

[Eurostat 2008] Eurostat, Statistics in Focus 7/2008 "Trade in High-tech Products" (2008)

[Factbook 2007] http://www.cia.gov/cia/publications/factbook/ (2007)

[Fagerberg (ed.) 2004] Jan Fagerberg (ed.), "The Oxford Handbook of Innovation," Oxford University Pres (2004)

[Fantová 2004] Veronika Fantová, "Doing business," in République tchèque en direct Numéro spécial élargissement de Commission européenne en direct 29 March 2004

[Faris 2007] Stephan Faris, "The Turnaround," Fortune vol. 155, n. 8 (14 May 2007)

[Fischer 2006] Manfred M. Fischer, "Innovation, Networks, and Knowledge Spillovers," (Springer 2006)

[Fiom 2004] The National Secretary of the trade union FIOM-CGIL, "FIOM calls for stopping the EU's new Directives on Working Time and Bolkenstein Directiv," URL: http://www.fiom.cgil.it/uf_sind/osservatorio/c_271004.htm (27 October 2004)

[Flam 2004] Harry Flam, "Turkey and the EU: Politics and Economics of Accession" CESifo Economic Studies 50, 171–210 (2004)

[Fontane 1998] Pascal Fontaine, "10 lezioni sull'Europa," Un opuscolo della Commissione europea. ISBN 92-828-3328-3 (1998)

[Fontaine 2003] Pascal Fontaine, "Europe in 12 lessons," A booklet of the European Commission. ISBN 92-894-6785-1 (2003)

[Forrester 2004] Forrester Research, "Convergence and the Digital Wold," in EITO Observatory (2004)

[Frazer 1994] B.W. Frazer, "Central Bank Independence: What Does It Mean?" Reserve Bank of Australia Bulletin (December 1994)

[FT 2004-1] Anonymous, "Ferrari chief to do Confindustria circuit," The Financial Times (17 March 2004)

[FT 2004-2] Anonymous Leader, "Fate of Fiat: The car company highlights wider plight of Italian economy," The Financial Times (16 December 2004)

[FT 2005-1] Lex Column, "Euro-zone II," The Financial Times (22 March 2005)

[FT 2005-2] Anonymous, "Death of a pact," The Financial Times (22 March 2005)

[GE 2007] General Electric, Annual Report 2006 (2007)

[Gerino 2005] Claudio Gerino, "La Cina tecnologica alla scalata del mondo sbarca in forze nel Vecchio Continente," La Repubblica, Affari e Finanza 21 March 2005

[Getler 2006] Getler, Michael, "Documenting and Debating a 'Genocide' ", The Ombudsman Column, PBS, April 21, 2006 http://www.pbs.org/ombudsman/2006/04/documenting_and_- debating_a_genocide.html (2006)

[Giscard d'Estaing 2004] Valery Giscard d'Estaing, "A better European bridge to Turkey," The Financial Times (25 November 2004)

[Goldman Sachs 2003] Goldman Sachs, "Global Economic paper 99. Dreaming with BRICs, the Path to 2050" (2003)

[Graham 2004] Robert Graham, "France's budget deficit 4.1 % in 2003," The Financial Times (2 March 2004)

[Griffiths 2007] John Griffiths, "Heading in the right direction Fiat has high hopes for two of its new models – and with good reason," The Financial Times (14 July 2007)

[Groom 2004] Brian Groom, "Message of Hope," The Financial Times (14 May 2004)

[Guzzetti 1995] Luca Guzzetti, "A brief history of EU Research Policy," Office for Official Publications of the EC. ISBN 92 827 5353-0 (1995)

[Higher 2004] *The Times*, Higher education supplement, "World University Rankings" (November 2004)

[Hoekman et al. (eds.) 2005] Bernard Hoekman and Subidey Togan, (eds.) "Turkey: Economic reform and accession to the European Union", Washington, DC: World Bank; Centre for Economic Policy Research (2005)

[Holm 2001] Erik Holm, "The European Anarchy: Europe's hard road into high politics," Copenhagen Business School Press (2001)

[Hui 2003] Wang Hui, quoted by: *People's Daily*, "Foreign Quality Standards Largest Obstacle to China's Export," (10 June 2003)

[Hope 2007] Kerin Hope, "Cyprus barricade demolished", *The Financial Times* (10 March 2007)

[IlSole24Ore 2003] Anonymous, "I numeri," *Il Sole 24 Ore* (29 January 2003)

[IMF 2006] International Monetary Fund "World Economic and Financial Surveys" (April 2006)

[IMF 2007] International Monetary Fund GDP current prices based on ppp for years 2000 to 2006 (2007)

[Istat 1999] Istat, "L'Italia in cifre" (1999)

[Istat 2001] Istat, "L'Italia in cifre" (2001)

[Istat 2002] Istat, "L'Italia in cifre" (2002)

[Istat 2004] Istat, "Italy in figures" (2004)

[JMM 2002] Jupiter Media Matrix, Inc "Market Size and Trends for e-commerce," URL: http://www.jmm.com/xp/jmm/press/industryProjections.xml (2002)

[Ker-Lindsay 2005] James Ker-Lindsay, "EU Accession and UN Peacemaking in Cyprus", Palgrave-Macmillan (2005)

[Kroeber 2004] Arthur Kroeber, "China's high-tech success story is pure fiction," *The Financial Times* (12 November 2004)

[Kynge 2003] James Kynge et al., "China's sprint for modernisation leaves India panting in its trail," *The Financial Times* (22 September 2003)

[Lacovou 2004] George Lacovou, "A win-win situation for all" in Chypre en direct, Numéro spécial élargissement de Commission européenne ed direct, 10 May 2004

[LaGro et al. (eds.) 2007] Esra Lagro and Knud Erik Jørgensen, (eds.) "Turkey and the European Union: Prospects for a difficult encounter", Palgrave Macmillan (2007)

[Lejour et al. 2004] A.M. Lejour et al. "Assessing the economic impact of Turkey accession to the EU" CPB # 56 CPB-NL Bureau for Economic Policy Analisys, The Hague March 2004, cited in [Burrell et al., eds. 2005–1]

[Lembke 2002] Johan Lembke, "Competition for Technological Leadership. EU Policy for High Technology", Edward Elgar, Cheltenham, UK (2002)

[Levi 2003] Lucio Levi, "Carta dei diritti e Costituzione europea" URL: http://www.cittadinanzaeuropea.net/cartadeidiritti/commenti.asp# (2003)

[Lintner 2000] Valerio Lintner, "Controlling EMU" in C Hoskins & M Mewman Democracy and the European Union, Manchester UP (2000)

[Lonardi 2004] Giorgio Lonardi, "Il declino dell'Europa frena la ripresa del lusso" *La Repubblica* (9 January 2004)

[Luce 2003] Edward Luce et al., "Visas and the west's *hidden agenda*," *The Financial Times* (9 April 2003)

[Luzi 2004] Gianluca Luzi, "Berlusconi sfida sinistra e alleati. Mi candido, niente lista unica. Il premier difende chi evade le tasse" *La Repubblica* (18 February 2004)

[MacLellan 2007] Stephanie MacLellan, "OECD praises Slovak economy", The Slovak Spectator, vol. 13, n. 15 (16 April 2007)

[Maddison 2001] Angus Maddison, "The World Economy/A millennium perspective," OECD Code: 412001011P1 (2001)

[Mahlich et al., eds. 2007] Jörg Mahlich and Pascha Werner, "Innovation and Technology in Korea," (Springer 2007)

[Majone 1996] Giandomenico Majone "Regulating Europe," Routeledge, 29 west 35th Street. New York, NY 10001 (1996)

[Malkani et al. 2003] Gautam Malkani and Susanna Voyle, "Dotcom investors confidently rub hands," *The Financial Times* (10 May 2003)

[Malta Info 2004] Malta Government, Department of Information Press Release, "Address by the Hon, Lawrence Gonzi, Prime Minister during the Conference organized by the Malta Federation of Industry on: "Repositioning our SMEs in theGlobal Market"," (30 March 2004)

[Malta-EUIC 2005] The Malta–EU Information Centre, "The EU has to say on marine pollution and maritime safety," http://www.mic.org.mt/ (2005)

[Martin 2002] Peter Martin, "Decline of the Tourin empire," *The Financial Times* (14 May 2002)

[McGregor 2003] Richard McGregor, "China to be VW export base," *The Financial Times* (29 November 2003)

[McLaren et al. 2003] Laren M McLaren and Melten Müftüler-Baç" Turkish Parliamentarians' Perspective on Turkey's Relation with the EU", in [Çarkoğlu et al. eds. 2003]

[Meadows 1972] D H Meadows, "Limits to Growth: A Report for the Club of Rome's Project on the Predicament of Mankind," ISBN: 0856440086 Earth Island (1972)

[Merkle 2004] Ralf Merkle, "A Brief Introduction to Nanotechnology" URL: http://www.zyvex.com/nano/ (2004)

[Michaels 2007-1] Adrian Michaels, "Fiat awarded an investment grade rating," *The Financial Times* (19 June 2007)

[Michaels et al. 2007-2] Adrian Michaels and John Reed, "Fiat shifts gear in Chinese alliance," *The Financial Times* (8 August 2007)

[Michelin 2007-1] Michelin, "2006 Annual Report," (2007)

[Michelin 2007-2] Michelin, "Performance and Responsibility Report 2005–2006," (2007)

[Michellone 2007] Gian Carlo Michellone "Centro Ricerche Fiat Key Achievements," in http://www.crf.it/ (2007)

[Monti 2005] Mario Monti, "Toughen up EU reform agenda and make it count," *The Financial Times* (21 March 2005)

[Morgan-Stanley 2005] Morgan-Stanley, "Ryanair/Significant Upside Potential for Growth Scenarios", published on www.ryanair.com (2 March 2005)

[Moules 2004] Jonathan Moules, "Businesses that embrace expansion to stay alive," *The Financial Times* (30 April 2004)

[Muldur et al. 2006] Muldur, U., Corvers, F., Delanghe, H., Dratwa, J., Heimberger, D., Sloan, B., Vanslembrouck, S., "A New Deal for an Effective European Research Policy," (Springer 2006)

[Nike 2005] Nike: Innovate for a better world; Improve conditions in our contract factories. See URL http://www.nikeresponsibility.com/#home URL:

[OJ-EU 2007] Official Journal of the European Union "Directive 2006/123/EC of 12 December 2006 on services in the internal market," L 77 vol. 50 (16 March 2007)

[Özben 2006] Riza Tunç Özben, "Da 'Mamma! Li turchi!' a 'Mamma!? Gli italiani!?' " Università Yeditepe (2006)

[OECD 2001] OECD, "Enhancing SME Competitiveness" The OECD Bologna Ministerial Conference, OECD (2001)

[OECD 2004] OECD, "Economic Survey – Turkey", (2004)

[OMS 2007] OMS/WHO World Health Organization report quoted by [Michelin 2007-2]

[Oscam 2005] Arie Oscam, "Consequences for the EU27 of Enlargement to Turkey", in [Burrell et al., eds. 2005-1]

[Pagni 2003] Luca Pagni, "Prodi: Addio investitori esteri; l'Italia rischia l'isolamento" *La Repubblica* (29 June 2003)

[Palloff et al. 1999] Rena M. Palloff and Keith H. Pratt, "Building Learning Communities in Cyberspace: Effective Strategies for the Online Classroom" Jossey-Bass (1999)

[Palloff et al. 2001] Rena M Palloff and Keith H. Pratt, "Cyberspace Classroom Online Teaching: Realities of Online Teaching" Jossey Bass Wiley (2001)

[Paolini 1989] Edmondo Paolini "Altiero Spinelli, Diario Europeo" Il Mulino (1989)

[Paolini 1996] Edmondo Paolini "Altiero Spinelli, dalla lotta antifascista alla battaglia per la Federazione europea" Il Mulino (1996)

[Paolini 2005] Edmondo Paolini, private communication (2005)

[Parker 2002] George Parker, "Commission chief hints that pact is on last legs," *The Financial Times* (18 October 2002)

[Parker 2004-1] George Parker, "EU budget plan sets up clash with big states," *The Financial Times* (19 January 2004)

[Parker 2004-2] George Parker, "EU's *big bang* would never be repeated," *The Financial Times* (3 May 2004)

[Parker 2004-3] George Parker, "Four new EU states signal early Euro interest," *The Financial Times* (4 May 2004)

[Parker 2005] George Parker, "Germany's deficit divides EU finance ministers," *The Financial Times* (21 March 2005)

[Parker et al. 2005-1] George Parker and Tony Barber, "EU in stern warning to Italy over budget deficit," *The Financial Times* (13 April 2005)

[Parker et al. 2005-2] George Parker and Daniel Dombey, "Europe divided over whether to press ahead with treaty if France votes No," *The Financial Times* (22 April 2005)

[Parts 2004] Juhan Parts, "New possibilities for Estonia," in Estonie en direct, Numéro spécial élargissement de Commission Européenne en direct (3 May 2004)

[Peel 2004] Peel, Quentin, "Italy's transatlantic balancing act," *The Financial Times* (14 October 2004)

[People's Daily 2002] Anonymous, "China's TCL Buys Bankrupt German Schneider," (8 October 2002)

[Peterson et al. 1998] John Peterson and Margaret Sharp, "Technology Policy in the European Union," St. Martin's Press, Inc. (1998)

[Peterson 2003] Christine Peterson, "Molecular Manufacturing Societal Implications of Advanced Nanotechnology,". URL: http://www.house.gov/science/hearings/full03/apr09/peterson.htm (9 April 2003)

[Petrini 2004-1] Roberto Petrini, "Il declino dell'Italia" Laterza, 2° edizione (2004)

[Petrini 2004-2] Roberto Petrini, "Crescita zero per l'economia nel quarto trimestre 2003. L'Italia non cammina più" *La Repubblica* (14 February 2004)

[Polidori 2002] Elena Polidori, "L'Euro vola e sorpassa il Dollaro" *La Repubblica* (16 July 2002)

[Polito 2002] Antonio Polito, "La competizione farà bene a imprese e consumatori" *La Repubblica* (16 February 2002)

[Potočnik 2004] Janez Potočnik, "Slovenia: Small is beautiful" Slovénie en direct, Numéro spécial élargissement de Commission Européenne en direct (14 June 2004)

[Prodi 2004] Romano Prodi, "Europe: Adding value, changing quickly," speech at the London School of Economics, 19 January 2004. URL: http://europa.eu.int/futurum/index_it.htm (2004)

[Prodi 2004-2] Romano Prodi, "Torniamo uniti in un continente senza frontiere" *La Repubblica* Album 28 April 2004 page 3

[Radaelli 2000] Claudio M. Radaelli "Whither Europeanization? Concept stretching and substantive change" European Integration online Papers (EIoP) vol. 4 (2000) N° 8; http://eiop.or.at/eiop/texte/2000-008a.htm (2000)

[Rall 2002] Ted Rall, "My government went to Afghanistan and all I got was this stupid pipeline," URL: http://www.citypaper.net (2002)

[Ramonet 2004] Ignatio Ramonet, "Editorial : Débat sur l'éventuelle entrée dans une douzaine d'années de la Turquie," Le Monde Diplomatique (November 2004)

[Rampini 2003-1] Federico Rampini, "La sindrome cinese del governo Berlusconi" in *La Repubblica* (22 August 2003)

[Rampini 2003-2] Federico Rampini, "La moneta cinese nuova sfida per l'Occidente" *La Repubblica* (27 August 2003)

[Rampini 2005-1] Federico Rampini, "A chi fa paura il made in China" *La Repubblica* (12 March 2005)

[Rampini 2005-2] Federico Rampini, "Il secolo cinese" ISBN 88-04-54482-1 Mondadori (2005)

[Rampini 2007] Federico Rampini, "Cina primo fornitore dell'Europa. È caduta la leadership americana," *La Repubblica* (10 April 2007)

[RTD Info 2006-1] RTD Info – Magazine on European Research Published by the European Commission, "The changing face of the family," n. 49 (May 2006)

[RTD Info 2007-1] RTD Info – Magazine on European Research Published by the European Commission, "Inside the Seventh Framework Programme," Special Edition June 2007

[RTD Info 2007-2] RTD Info – Magazine on European Research Published by the European Commission, "Climate Change,"Issue number 52-June 2007

[Repubblica 2003] Anonimo, "Crescono i lavoratori in nero al Sud sono quasi un quarto" *La Repubblica* (28 November 2003)

[Rifkin 2004] Jeremy Rifkin, "The European Dream," Penguin (2004)

[Rossi 2004] Lucia Serena Rossi, "What if the Constitutional Treaty is not ratified?" The European Policy Centre (6 July 2004)

[Rubython 2002] Tom Rubython, "Agnelli, the final act" in Eurobusiness, vol. 4, n. 5 (2002)

[Ryanair 2007] Ryanair, Annual Report 2007 at http://www.ryanair.com

[Sbragia 2000] Alberta M. Sbragia, "Environmental Policy," in [Wallace et al. (eds.) 2000]

[Sbragia 2003] Alberta M. Sbragia, "Key Policies," in [Bomberg et al. (eds.) 2003]

[Shanghai Univ. 2004] Shanghai Jiao Tong University Institute of Higher Education, Ranking of Top World Univesities, URL: http://ed.sjtu.edu.cn/ranking.htm (2004)

[Siemens 2007] Siemens, Annnual report 2007 (2007)

[Spinelli 1941] Altiero Spinelli, "Il Manifesto di Ventotene" Il Mulino 1991 (1941)

[Stajano 1999-1] Attilio Stajano, "Technology Policy in the European Union, and European Integration." A paper in the proceedings of the EU-U.S. Science and Technology Policy Conference. Atlanta, GA 9 and 10 April 1999 GeorgiaTech EU Center/USG Atlanta, GA 30332-0610, USA (1999)

[Stajano 1999-2] Attilio Stajano, "Making Academia Aware of Intellectual Property Rights (IPR) – Comparing U.S. and EU Experiences." A paper in the proceedings of Pattinova '99, an EC conference held at the Sani Convention Center at Kassandra, Greece on 20–22 (October 1999)

[Stajano 2008] Attilio Stajano, "Turkey as a member of the European Union: An open-ended negotiation for over forty years," in [Chong-ko Peter Tzou 2008]

[Sylos Labini 2003] Paolo Sylos Labini, "Berlusconi e gli anticorpi. Diario di un cittadino indignato," Laterza (2003)

[Syrett et al. 2002] Michel Syrett and Jean Lammiman, "Successful Innovation: How to encourage and shape profitable ideas," *The Economist* Books, (2002)

[Tarquini 2002] Andrea Tarquini, "Concessionarie libere, guerra ai monopoli," *La Repubblica* (16 February 2002)

[TDN 2007-1] Turkish Daily News, 10 July, 2007, page 5

[TDN 2007-2] Turkish Daily News, 11 July, 2007, page 1

[TDN 2007-3] Turkish Daily News, 23 July, 2007

[Temel 2005] Turgul Temel "Expected consequences for Turkey of EU Entry in 2015"in [Burrell et al., eds. 2005-1]

[Tessa 2004] Mariangela Tessa, "Per Pasquale Pistorio si apre la sfida Indiana," *La Repubblica* Affari e Finanza, 1 March 2004

[Tornhill 2002] John Tornhill, "Challenges thrown up by China's continued economic growth," *The Financial Times* (13 May 2002)

[Uğur et al. (eds.) 2004] Mehmet Uğur and Nergis Canefe, eds., "Turkey and European Integration – Accession prospects and issues" Routledge (2004)

[Uğur 2004-1] Mehmet Uğur, "Economic mismanagement and Turkey's troubled relation with the EU", in [Uğur et al., eds. 2004]

[Uğur 2004-2] Mehmet Uğur and Nergis Canefe "Turkey and European Integration" in [Uğur et al. (eds.) 2004]

[UN 2003] United Nations, "Convention on Climate Change," URL: http://unfccc.int/resource/kpthermo_if.html (2003)

[Unice 2001] Union of Industrial and Employers' Confederation of Europe "The Renewed Economy. The Unice Bench-marking Report," URL: www.unice.org (2001)

[Unice 2002] Union of Industrial and Employers' Confederation of Europe, "The Lisbon Strategy Status 2003," URL: www.unice.org (2002)

[US Census 2007] U.S. Census Bureau, Foreign Trade Statistics. 2007

[van Ark et al. 2003] Bart van Ark, Robert Inklaar and Robert H. McGuckin, "ICT and Productivity in Europe and the United States Where Do the Differences Come From," CESifo Economic Studies, vol. 49, 3/2003, 295–318 (2003)

[van Tongeren et al. (eds.) 2002] Paul van Tongeren, et al. (eds.), "Searching for peace in Europe and Eurasia : An overview of conflict prevention and peacebuilding activities", Lynne Rienner, Bouler, CO (2002)

[Vikke-Freiberga 2004] Vaira Vikke-Freiberga, "A success story of change" Lettonie en direct, Numéro spécial élargissement de Commission Européenne en direct (24 May 2004)

[Verhofstadt 2006] European Parliament, Debate on the future of Europe with the participation of the Belgian Prime Minister Guy Verhofstadt, Speech at the EP in Brussels on 31 May 2006 (2006)

[Wallace et al. (eds.) 2000] Hellen Wallace and William Wallace, eds., "Policy-making in the European Union," Oxford University Press, 4th edition (2000)

[WEF 2003] World Economic Forum, "Growth Competitiveness Report," 2002–2003 (2003)

[WEF 2004] World Economic Forum, "Growth Competitiveness Report" 2003–2004 (2004)

[WEF 2005] World Economic Forum, "Growth Competitiveness Report" 2004–2005 (2005)

[WEF 2006] World Economic Forum, Annual Report 2005–2006 (2006)

[Wicksteed 1985] Bill Wicksteed (ed.), "The Cambridge Phenomenon/The growth of high technology industry in a University Town," Segal Quince Wicksteed (1985)

[Wicksteed 2000] Bill Wicksteed (ed.), "The Cambridge Phenomenon/Revisited," Segal Quince Wicksteed (2000)

[Wojtkun 2006] Matt Wojtkun "Americans, Europeans share Increased Fears of Islamic Terrorism", in http://www.transatlantictrends.org/trends/showdoc.cfm?id=104 (6 September 2006)

[Wolf 2003] Martin Wolf, "China is now shaking the world," The Financial Times, 12 November 2003

[Wolleh 2002] Oliver Wolleh, "Cyprus: A civil society caught up in the question of recognition" in [van Tongeren et al., eds. 2002]

[World Bank 2007-1] World Bank, "Macedonia-Economy. Developments since independence," (2007)

[World Bank 2007-2] World Bank, "FDI in the Euro Area," (2007)

[World Bank 2007-3] World Bank, Country Briefing, "Croatia" (2007)

[WorldFactbook 2005] World Factbook Country Profiles, https://www.cia.gov/library/publications/ (2005)

[WorldFactbook 2007] World Factbook Country Profiles, https://www.cia.gov/library/publications/the-world-factbook/print/mk.html (2007)

Glossary

Balance of payments: The balance of payments (BoP) is a record of a country's international transactions with the rest of the world. It is composed of the current account and the capital and financial account. The current account is itself subdivided into goods, services, income, and current transfers; it registers the value of exports (credits) and imports (debits). The difference between these two values is the 'balance'.

Consumer Prices: Harmonized Indices of Consumer Prices (HICPs) are used in the assessment of inflation convergence as required under Article 121 of the Treaty of Amsterdam. They form the basis for the Monetary Union Index of Consumer Prices (MUICP), the European Index of Consumer Prices (EICP), and the European Economic Area Index of Consumer Prices (EEAICP). HICPs are compiled on the basis of a legislated methodology, binding for all member states. The common classification for Harmonized Indices of Consumer Prices is the COICOP (Classification Of Individual COnsumption by Purpose). A version of this classification (COICOP/HICP) has been specially adapted for the HICPs. Sub-indices published by Eurostat are based on this classification.

Current account: The current account of the balance of payments is the sum of the balance of trade (exports minus imports of goods and services), net factor incomes (such as interest and dividends), and net transfer payments (such as foreign aid). A current account surplus increases a country's net foreign assets by the corresponding amount and a current account deficit does the reverse.

Dependency ratio: *See* Old-age-dependency ratio.

FDI: *See* Foreign direct investment

Foreign direct investment (FDI): It is a category of international investment that indicates an intention to acquire a lasting interest in an enterprise operating in another economy. It covers all financial transactions between the investing enterprise and its subsidiaries abroad and it differs from portfolio investments because the direct investor acquires at least 10 % of capital. Foreign direct investment acquires increasing importance as an indicator of the international economic climate and plays a key role in the globalization process as an important element of international relations and their development.

GDP: *See* **Gross domestic product**

General Government debt: The general government consolidated gross debt is usually presented as a percentage of GDP. The general government sector comprises the subsectors of central government, state government, local government, and social security funds; debt is valued at nominal (face) value, and foreign currency debt is converted into national currency using end-year market exchange rates (though special rules apply to contracts).

General government: The general government sector includes all institutional units whose output is intended for individual and collective consumption, and mainly financed by compulsory payments made by units belonging to other sectors, and/or all institutional units principally engaged in the redistribution of national income and wealth. The general government sector is subdivided into four subsectors: central government, state government, local government, and social security funds.

GNI: *See* Gross national income

GNP: *See* Gross national income

Government expenditure: According to the European Commission definition, total general government expenditure comprises the following categories: intermediate consumption; gross capital formation; compensation of employees; other taxes on production; subsidies payable; property income; current taxes on income, wealth, etc.; social benefits other than social transfers in kind; social transfers in kind related to expenditure on products supplied to households via market producers; other current transfers; adjustment for the change in net equity of households in pension fund reserves; capital transfers payable; and acquisitions less disposals of non-financial non-produced assets.

Government revenue: According to the European Commission definition, total general government revenue comprises the following categories: market output; output for own final use; payments for the other non-market output; taxes on production and imports; other subsidies on production receivable; property income; current taxes on income, wealth, etc.; social contributions; other current transfers; and capital transfers.

Gross domestic product (GDP): It is an indicator for a nation's economic situation. It reflects the total value of all goods and services produced less the value of goods and services used for intermediate consumption in their production plus taxes less subsidies on products. Gross domestic product may be expressed at market prices or in purchasing power standards. Gross domestic product at market prices is converted into the artificial currency unit PPS (purchasing power standard) through a special conversion rate called PPP (purchasing power parity). The GDP in PPS represents pure volume, after price-level differences between countries have been removed by the special conversion rate PPP. Gross domestic product may be expressed on a per head basis (GDP pc or pc GDP, where pc stands for *per capita*) allows for the comparison of economies different in size.

Gross national income (GNI) or Gross national product (GNP): Gross national income or gross national product equals GDP minus primary income payable by resident units to non-resident units, plus primary income receivable from the rest of the world. The GNI definition by the World Bank national accounts data and OECD National Accounts, Atlas method (formerly GNP), is: "the sum of value added by all resident producers plus any product taxes (less subsidies) not included in the valuation of output plus net receipts of primary income (compensation of employees and property income) from abroad. Data are in current U.S. dollars. GNI, calculated in national currency, is usually converted to U.S. dollars at official exchange rates for comparisons across economies, although an alternative rate is used when the official exchange rate is judged to diverge by an exceptionally large margin from the rate actually applied in international transactions. To smooth fluctuations in prices and exchange rates, a special Atlas method of conversion is used by the World Bank. This applies a conversion factor that averages the exchange rate for a given year and the two preceding years, adjusted for differences in rates of inflation between the country, and through 2000, the G-5 countries (France, Germany, Japan, the United Kingdom, and the United States). From 2001, these countries include the Euro Zone, Japan, the United Kingdom, and the United States."

Harmonized Index of Consumer Prices (HICP): See Inflation

HICP: See Inflation

Inflation: Harmonized Indices of Consumer Prices (HICPs) are used in the assessment of inflation convergence as required under Article 121 of the Treaty of Amsterdam. They form the basis for the Monetary Union Index of Consumer Prices (MUICP), the European Index of Consumer Prices (EICP) and the European Economic Area Index of Consumer Prices (EEAICP). HICPs are compiled on the basis of a legislated methodology, binding for all member states. The common classification for Harmonized Indices of Consumer Prices is the COICOP (Classification Of Individual COnsumption by Purpose). A version of this classification (COICOP/HICP) has been specially adapted for the HICPs. Sub-indices published by Eurostat are based on this classification.

Labour productivity: Various measures of labour productivity are available. For the structural indicators this measure is based on GDP in PPS either relative to the number of persons employed or to the number of hours worked; in both cases it is then expressed as an index.

Long-term unemployment: Long-term unemployed are persons who have been unemployed for one year or more. Unemployed persons are defined as persons aged 15–74 (in Spain, the United Kingdom, Iceland, Norway: 16–74) who were without work during the reference week, were currently available for work, and were either actively seeking work in the last four weeks or had already found a job to start within the next three months. The duration of unemployment is defined as the duration of a search for a job or as the length of the period since the last job was held (if this period is shorter than the duration of the search for a job). This definition follows the guidelines of the International Labor Organization (ILO).

Market capitalization: *See* Stock market capitalization

Net borrowing: Net borrowing (+)/net lending (−) of general government is the difference between the revenue and the expenditure of the general government sector. The general government sector comprises the following subsectors: central government, state government, local government, and social security funds. GDP used as a denominator is the gross domestic product at current market prices.

Net lending: Net borrowing (+)/net lending (−) of general government is the difference between the revenue and the expenditure of the general government sector. The general government sector comprises the following subsectors: central government, state government, local government, and social security funds. GDP used as a denominator is the gross domestic product at current market prices.

Old-age-dependency ratio: The ratio of the number of elderly persons of an age when they are generally economically inactive to the number of persons of working age.

Population: The inhabitants of a given area on 1 January of the year in question (or, in some cases, on 31 December of the previous year). The population is based on data from the most recent census adjusted by the components of population change produced since the last census, or based on population registers.

PPS: *See* Purchasing power standards

Purchasing power parities (PPPs): Monetary exchange rates should not be used to compare the volumes of income or expenditure because they usually reflect more elements than just price differences (e.g. volumes of financial transactions between currencies, expectations in the foreign exchange markets). In contrast, purchasing power parities (PPPs) are established to eliminate the differences between the price levels in different countries. Therefore, they truly reflect the differences in the purchasing power, for example, of households. Purchasing power parities are obtained by comparing the price levels for a basket of comparable goods and services that is selected to be representative of consumption patterns in the various countries.

Purchasing power parities convert every national monetary unit into a common artificial currency unit, the purchasing power standard (PPS).

Purchasing power standards (PPS): Expressing GDP in PPS (purchasing power standards) eliminates differences in price levels between countries, and calculations on a per head basis allows for the comparison of economies significantly different in absolute size.

Return on Investment (ROI): A financial ratio showing profit gained in an investment as a percentage of capital employed. ROI does not indicate how long an investment is held.

Small and medium-sized enterprises (SMEs): According to Commission Recommendation 2003/361/EC adopted on 6 May 2003, small and medium-sized enterprises are classified with regard to the number of employees, its annual turnover, and the firm's independence. For statistical purposes, small and medium-sized enterprises are generally defined as those enterprises employing fewer than 250 people: micro enterprises (less than 10 persons employed); small enterprises (10–49 persons employed); medium-sized enterprises (50–249 persons employed); while large enterprises are defined as those with 250 or more persons employed.

SME: *See* Small and medium-sized enterprises

Stock market capitalization: It is a measurement of corporate or economic size equal to the share price times the number of shares outstanding of a public company.

Trade integration of goods as a percentage of GDP (gross domestic product): Average of imports and exports of the item goods of the balance of payments divided by GDP. If the index increases over time it means that the country/zone is becoming more integrated within the international economy.

Unemployed person: Unemployed persons are persons aged 15–74 (in Spain, the United Kingdom, Iceland, Norway: 16–74) who were without work during the reference week, were currently available for work and were either actively seeking work in the last four weeks or had already found a job to start within the next three months. This definition follows the guidelines of the International Labor Organization.

Unemployment rate: Unemployed persons as a percentage of people in the labour force.

Acronyms and Abbreviations

.

Table 1 Standard acronyms for country names

Acronym	Name
	European Union Countries
AT	Austria
BE	Belgium
BG	Bulgaria
CY	Cyprus
CZ	Czech Republic
DE	Germany
DK	Denmark
EE	Estonia
EL	Greece
ES	Spain
FI	Finland
FR	France
HU	Hungary
IE	Ireland
IT	Italy
LT	Lithuania
LU	Luxembourg
LV	Latvia
MT	Malta
NL	The Netherlands
PL	Poland
PT	Portugal
RO	Romania
SE	Sweden
SI	Slovenia
SK	Slovakia
UK	United Kingdom

Table 1 (continued)

Acronym	Name
	EU Candidate Countries
HR	Croatia
MK	Former Yugoslav Republic of Macedonia
TR	Turkey
	EFTA Countries
CH	Switzerland
IS	Iceland
LI	Liechtenstein
NO	Norway
	Accession 2004 Countries
CY	Cyprus
CZ	Czech Republic
EE	Estonia
HU	Hungary
LV	Latvia
LT	Lithuania
MT	Malta
PL	Poland
SK	Slovakia
SI	Slovenia
	Accession 2007 Countries
BG	Bulgaria
RO	Romania
	G7 Countries
CA	Canada
DE	Germany
FR	France
IT	Italy
JP	Japan
UK	United Kingdom
U.S.	United States of America
	BRIC Countries
BR	Brazil
CN	China
IN	India
RU	Russia

Table 1 (continued)

Acronym	Name
	Other Countries
AU	Australia
HK	Hong Kong
ID	Indonesia
IL	Israel
KR	South Korea
MY	Malaysia
NZ	New Zealand
SG	Singapore
TH	Thailand
TW	Taiwan

Table 2 Acronyms and abbreviations

Acronym	Name
AA	Ankara Agreement
A.Asem	Japan, China, Korea, and Southeast Asia
AC10	Ten countries that accessed the EU in 2004
AC12	Twelve countries that accessed the EU in 2004 or in 2007
AC+2	Two countries accessing the EU in 2007: Bulgaria and Romania
AC+4	Four countries that in 2004 were candidate for accession in or after 2007: Bulgaria, Romania, Croatia and Turkey
ACTS	EU research program on Telecommunications
ADAS	Advanced Driver Assistance Systems
ADSL	Asymmetric Digital Subscriber Line
AIDS	Acquired Immuno-Deficiency Syndrome
AKP	Turk Justice and Development Party
ARPA	Advanced Research Projects Agency
ASEAN	Association of Southeast Asian Nations
AUTM	Association of University Technology Managers
B2B	Business-to-Business
B2C	Business-to-Consumer
Benelux	Belgium, The Netherlands, Luxembourg
BERD	Research and Technological Development in the Business Enterprise
BGL	Bulgarian national currency, Lev
Brite	EU research program on materials and industrial processes
CAP	Common Agricultural Policy
CC3	Three countries candidate for accession in 2007: Croatia, The Former Yougoslav Republic of Macedonia, and Turkey
CEEC	Central and Eastern European Countries
CEO	Chief Executive Officer
CERN	European laboratory for research in high energy physics
CFC	Chlorofluorocarbon
CGIL	Confederazione Generale Italiana del Lavoro, an Italian trade union
CFO	Chief Financial Officer
CFSP	Common Foreign and Security Policy

Table 2 (continued)

Acronym	Name
Cocobu	Committee on Budgetary Control
CoG	Court of Justice
Coreper	Committee of permanent representatives
COST	Cooperation in Science and Technology
CPI	Consumer Price Index
CU	Customs Union
CYP	Cypriot national currency, Pound
CZK	Czech national currency, Koruna
DAE	Dynamic Asian Economies
EC	European Commission
ECB	European Central Bank
Ecofin	Council of economic and finance ministers
ECSC	European Community for Steel and Coal
ECU	European Currency Unit
EDI	Electronic Data Interchange
EEA	European Economic Area
EEC	European Economic Community
EEK	Estonian national currency, Kroon
EFDA	European Fusion Development Association
EFTA	European Free Trade Association
EIB	European Investments Bank
EITO	European Information Technology Observatory
EMBL	European Molecular Biology Laboratory
EMBO	European Molecular Biology Organization
EMS	European Monetary System
EMU	Economic and Monetary Union
EP	European Parliament
ERA	European research area
ESA	European Space Agency
ESCB	European System of Central Banks
ESF	European Science Foundation
ESO	European astronomical observatory
Esprit	European Strategic Program for R&D in Information Technologies
ESSI	European Software Systems Initiative
ESRF	European Syncrotron Radiation Facility
EU	European Union
EU15	The 15 EU member states from 1995 to 2004
EU19	The 15 EU15 member states and the four EFTA States
EU25	The 25 EU member states from 2004 to 2007
EU29	The 25 EU member states and the four candidate countries
Euratom	European Program for Energy from Nuclear Fission
Eureka	Pan-European Market Oriented R&D Program
FDI	Foreign direct investment
FIOM	Federazione Impiegati ed Operai Metallurgici, an Italian Trade Union of Metal Workers
FP4	Framework Program for R&D, 1994–1998

Table 2 (continued)

Acronym	Name
FP5	Framework Program for R&D, 1998–2002
FP6	Framework Program for R&D, 2002–2006
FP7	Framework Program for R&D, 2007–2013
FUSE	First User Action
FT	The Financial Times
G6	G7 minus Canada
G7	The seven major industrial democracies: Canada, France, Germany, Italy, Japan, the United Kingdom, and the United States
G8	G7 plus the Russian Federation
GATT	General Agreement on Tariffs and Trade
GCI	Growth Competitiveness Index
GDP	Gross Domestic Product
GEIS	General Electric Information Services
GM	General Motors
GMES	Global Monitoring for Environment and Security
GMO	Genetically Modified Organism
GNP	Gross National Product
GNSS	Global Navigation Satellite System
HICP	Harmonized Index of Consumer Prices
HRK	Croatian national currency, Kuna
HUF	Hungarian national currency, Forint
ICT	Information and Communication Technology
ICTY	International Criminal Tribunal for the former Yugoslavia
IFO	Institut für Wirtschaftsforschung
IGC	Intergovernmental Conference
ILL	Langevin Institute for Physics, Chemistry
IMF	International Monetary Fund
IOC	International Olympic Committee
IPA	Instrument of Pre-Accession
IPR	Intellectual Property Rights
ISDN	Integrated Services Digital Network
JRC	Joint Research Centre
LLL	Lifelong Learning
LLLP	Lifelong Learning Program
LLN	Louvain-la-Neuve
LTL	Lithuanian national currency, Lita
LVL	Latvian national currency, Lat
MEI	Macroeconomic Environment Index
MTL	Maltese national currency, Lira
NAFTA	North Atlantic Free Trade Association
NATO	North Atlantic Treaty Organization
NCB	National Central Bank
NGO	Non-Governmental Organization
NSF	National Science Foundation
OECD	Organization for Economic Cooperation and Development
OEEC	Organization for European Economic Cooperation

Table 2 (continued)

Acronym	Name
OMI	Open microprocessors iniziative
PAPI	Public Access-Points to the Internet
pc GDP	per capita (per person) Gross Domestic Product
PC	Personal Computer
pc	per capita, per person
PIAP	Public Internet Access Points
PII	Public Institutions Index
PKK	Kurdistan Workers' Party
PLN	Polish national currency, Zloty
POS	Point of Sale
ppp	purchasing power parity
PPS	Position in Price Segments
R&D	Research and Development
RACE	R&D for Advanced Communication in Europe
RISC	Reduced Instruction Set Computer
ROL	Romanian national currency, Leu
RQE	Revealed Quality Elasticity
S&T	Science and Technology
SIDS	Sudden Infant Death Syndrome
SIT	Slovene national currency, Tolar
SKK	Slovak national currency, Koruna
SMEs	Small and Medium Enterprises
STM	ST-Microelectronics
STREP	Specific Targeted Research Project
TABD	Transatlantic Business Dialogue
TBC	Tuberculosis
Telematics	EU research program on ICT applications
TI	Technology Index
TRNC	Turkish Republic of Northern Cyprus
TRL	Turkish national currency, Lira
TV	Television
UN	United Nations
Unesco	United Nations Educational Scientific and Cultural Organization
Unficyp	United Nations Forces in Cyprus
URL	Uniform Resource Locator
USSR	Union of Soviet Socialist Republics
VAT	Value Added Tax
VGE	Valéry Giscard d'Estaing
VW	Volkswagen
WEF	World Economic Forum
WTO	World Trade Organization

Index

Printed in the United States
126361LV00001B/49/P